Understanding Present and Past Arctic Environments

Understanding Present and Past Arctic Environments

An Integrated Approach From Climate Change Perspectives

Edited by

NELOY KHARE
Ministry of Earth Sciences, Govt. of India,
New Delhi, India

ELSEVIER

Elsevier
Radarweg 29, PO Box 211, 1000 AE Amsterdam, Netherlands
The Boulevard, Langford Lane, Kidlington, Oxford OX5 1GB, United Kingdom
50 Hampshire Street, 5th Floor, Cambridge, MA 02139, United States

Notices
Knowledge and best practice in this field are constantly changing. As new research and experience broaden our understanding, changes in research methods, professional practices, or medical treatment may become necessary.

Practitioners and researchers must always rely on their own experience and knowledge in evaluating and using any information, methods, compounds, or experiments described herein. In using such information or methods they should be mindful of their own safety and the safety of others, including parties for whom they have a professional responsibility.

To the fullest extent of the law, neither the Publisher nor the authors, contributors, or editors, assume any liability for any injury and/or damage to persons or property as a matter of products liability, negligence or otherwise, or from any use or operation of any methods, products, instructions, or ideas contained in the material herein.

British Library Cataloguing-in-Publication Data
A catalogue record for this book is available from the British Library

Library of Congress Cataloging-in-Publication Data
A catalog record for this book is available from the Library of Congress

ISBN: 978-0-12-822869-2

For Information on all Elsevier publications
visit our website at https://www.elsevier.com/books-and-journals

Publisher: Candice Janco
Acquisitions Editor: Marisa LaFleur
Editorial Project Manager: Ruby Gammell
Production Project Manager: Kumar Anbazhagan
Cover Designer: Miles Hitchen

Typeset by MPS Limited, Chennai, India

Dedication

Dedicated to
Late Prof. Satyendra Nath Bhalla

(July 1, 1940–May 10, 2021)

Prof. Bhalla was born on July 1, 1940 at Basti, Uttar Pradesh, India. Having completed his early education at Colvin Taluqdars' College, Lucknow, he obtained his B.Sc. and M.Sc. (Geology) degrees from Aligarh Muslim University (AMU), Aligarh. He was awarded his Ph.D. on the foraminifera from the intertrappean beds of Rajahmundry in 1964, under the supervision of the famous Indian micropaleontologist, late Prof. S. B. Bhatia from Punjab University, Chandigarh.

Prof. Bhalla joined the Department of Geology at AMU as a lecturer. He established the Micropaleontology Laboratory in the department and pioneered research on Jurassic foraminifera of Kutch, within a few years elevating it to a leading center in India for research on Jurassic foraminifera and Recent foraminifera from the Indian coasts. The Department of Geology got fresh impetus in academics and research under his chairmanship from 1990 to 1994. He was the first one to be awarded a Doctorate in Science (D.Sc.) from the Geology Department of AMU. He had a distinguished research career of almost 60 years. Though his main focus was taxonomy and distribution of Recent and Jurassic foraminifera, he also worked on other microfossil groups, namely, shark teeth and ostracods. He also used microfossils in stratigraphic studies, geochronology, paleoenvironmental and paleobiogeographic studies, pollution studies, and petroleum exploration. He pioneered work on the east coast of Gondwana and predicted the possibility of finding oil and gas in the onshore and offshore Andhra Pradesh, way back in 1970. This was confirmed by the Oil and Natural Gas Commission, Government of India, during their drilling operations in 1982. He was appointed as the Chairman of the Research Advisory Council of Birbal Sahni Institute of Palaeobotany, Lucknow, by the Government of India in 2011.

Prof. Bhalla published 105 research papers and supervised 6 doctoral and several M.Phil. students. He was invited by the Elsevier Science Publishers, Holland, in 1983 to contribute to the book, *Phanerozoic Geology of the World: II. The Mesozoic.* He traveled worldwide upon invitations from British Council, London (1971 and 1975), Royal Society London (1994), and regularly from German Academic Exchange Service Bonn (1972, 1975, 1995, and 1997) to Germany for collaborations in advanced research in foraminifera and marine environment. In 1985, he visited the Museum of Natural History in Geneva, Switzerland, for advanced research on Jurassic foraminifera. In 1987, the Japan Society for Promotion of Science invited him to deliver lectures at different universities and institutes in Japan to further research on Recent benthic foraminifera of the Pacific Ocean. Besides, he visited and had research collaborations with the United Kingdom, Germany, France, Holland, Switzerland, the United States, Canada, Japan, the United Arab Emirates, Kuwait, Oman, and Nepal.

Prof. Bhalla was a meticulous researcher, an innovative thinker, and an excellent teacher. He was modesty and dignity personified and a valued, trusted, friend, philosopher, and guide to his associates. Unfortunately, Prof. Bhalla passed away in the early hours of May 10, 2021, leaving a vacuum in the Micropaleontology fraternity forever.

Contents

List of contributors

Y.N. Ahammed
Department of Physics, Yogi Vemana University, Kadapa, India

K. Amrutha
Department of Geology, Central University of Punjab, Bathinda, India

B.C. Arya
Radio and Atmospheric Sciences Division, National Physical Laboratory, New Delhi, India

Kato K. Ayemi
Geological Survey of India, Faridabad, India

Ruchika Bajpai
Department of Agronomy, Assam Agricultural University, Jorhat, India

Felix Bast
Department of Botany, Central University of Punjab, Ghudda-VPO, Bathinda, India

Himanshu Chaurasia
Department of Physics, Govt. College Rehti, Sehore, India

Shabnam Choudhary
Ministry of Earth Sciences, Govt. of India, New Delhi, India; National Centre for Polar and Ocean Research (NCPOR), Goa, India

Suryanshu Choudhary
Department of Physics, Rabindranath Tagore University, Bhopal, India

Moumita Das
Mahila Mahavidyalaya, Banaras Hindu University, Varanasi, India

Sweta Baidya Das
National Institute of Disaster Management, New Delhi, India

Supriya Dey
Department of Earth Sciences, Indian Institute of Science Education and Research Kolkata, Mohanpur, India

Amit Dharwadkar
Geological Survey of India, Faridabad, India

Som Dutt
Wadia Institute of Himalayan Geology, Dehradun, India

Deepak Y. Gajbhiye
Geological Survey of India, Faridabad, India

Sasi Kiran Gera
National Centre for Seismology, Ministry of Earth Sciences, New Delhi, India

M. Ghatak
DGCO, Geological Survey of India, New Delhi, India

Sachin D. Ghude
Indian Institute of Tropical Meteorology, Pashan, Pune, India

Anil K. Gupta
Department of Geology and Geophysics, Indian Institute of Technology Kharagpur, Kharagpur, India

Kriti Gupta
Department of Botany, DAV College, Bathinda, India

Poorti Gusain
National Centre for Seismology, Ministry of Earth Sciences, New Delhi, India

A.K. Gwal
Department of Physics, Rabindranath Tagore University, Bhopal, India

S.L. Jain
Radio and Atmospheric Sciences Division, National Physical Laboratory, New Delhi, India

Shridhar D. Jawak
Svalbard Integrated Arctic Earth Observing System (SIOS), SIOS Knowledge Centre, Svalbard Science Centre, Longyearbyen, Norway

Dhanasree Jayaram
Department of Geopolitics and International Relations and Co-coordinator, Centre for Climate Studies, Manipal Academy of Higher Education, Manipal, India

Ratan Kar
Birbal Sahni Institute of Palaeosciences, Lucknow, India

Tushar Kaushik
Biodiversity and Palaeobiology, Agharkar Research Institute, Pune, India

Neloy Khare
Ministry of Earth Sciences, Govt. of India, New Delhi, India

Rajni Khare
Delhi Public School, Greater Noida, India

K.P. Krishnan
National Centre for Polar and Ocean Research, Ministry of Earth Sciences, Vasco da Gama, India

Arun Kumar
Radio and Atmospheric Sciences Division, National Physical Laboratory, New Delhi, India

Avinash Kumar
National Centre for Polar and Ocean Research, Ministry of Earth Sciences, Government of India, Vasco-da-Gama, India

Naresh Kumar
Laboratory of Cyanobacterial Systematics and Stress Biology, Department of Botany, Institute of Science, Banaras Hindu University, Varanasi, India

Pankaj Kumar
ÁMS & Geochronology Group, Inter-University Accelerator Centre, New Delhi, India

Pradeep Kumar
Geological Survey of India, Faridabad, India

Vikas Kumar
National Centre for Seismology, Ministry of Earth Sciences, New Delhi, India

Alvarinho J. Luis
Polar Remote Sensing Section, National Centre for Polar and Ocean Research, Ministry of Earth Sciences, Vasco da Gama, India

Kirti K. Mahanta
Odisha Space Applications Centre (ORSAC), Department of Science & Technology, Government of Odisha, Bhubaneswar, India

Ravikant Mahto
National Centre for Seismology, Ministry of Earth Sciences, New Delhi, India

Kirtiranjan Mallick
Department of Geology, Utkal University, Bhubaneswar, India

Abhijit Mazumder
Birbal Sahni Institute of Palaeosciences, Lucknow, India

Amit K. Mishra
Birbal Sahni Institute of Palaeosciences, Lucknow, India

Kriti Mishra
Birbal Sahni Institute of Palaeosciences, Lucknow, India

O.P. Mishra
National Centre for Seismology, Ministry of Earth Sciences, New Delhi, India

Rahul Mohan
National Centre for Polar and Ocean Research, Ministry of Earth Sciences, Government of India, Vasco-da-Gama, India

Sahil Narwal
Department of Botany, Central University of Punjab, Ghudda-VPO, Bathinda, India

G.N. Nayak
Marine Sciences, School of Earth, Ocean and Atmospheric Sciences, Goa University, India

Sagarika Pal
Laboratory of Cyanobacterial Systematics and Stress Biology, Department of Botany, Institute of Science, Banaras Hindu University, Varanasi, India

S.K. Patil
K.S. Krishnan Geomagnetic Research Laboratory, Indian Institute of Geomagnetism, Prayagraj, India

Jitendra Kumar Pattanaik
Department of Geology, Central University of Punjab, Bathinda, India

S.M. Pednekar
National Centre for Polar and Ocean Research, Ministry of Earth Sciences, Headland Sada, Vasco-da-Gama, India

Alagappan Ramanathan
School of Environmental Sciences, Jawaharlal Nehru University, New Delhi, India

Rashmi Ramesh
School of Conflict and Security Studies, National Institute of Advanced Studies, Bengaluru, India

Archita Rana
Department of Earth Sciences, Indian Institute of Science Education and Research Kolkata, Mohanpur, India

P.S. Ranhotra
Birbal Sahni Institute of Palaeosciences, Lucknow, India

Alka Rani
Department of Botany, Central University of Punjab, Ghudda-VPO, Bathinda, India

Syed Mohammad Saalim
National Centre for Polar and Ocean Research (NCPOR), Goa, India

Khem Chand Saini
Department of Botany, Central University of Punjab, Ghudda-VPO, Bathinda, India

Sayantan Sarkar
Department of Earth Sciences, Indian Institute of Science Education and Research Kolkata, Mohanpur, India; School of Engineering, Indian Institute of Technology Mandi, Kamand, India

Shashank Shekhar
National Centre for Seismology, Ministry of Earth Sciences, New Delhi, India

Ankush Shrivastava
Department of Geology, Mohanlal Sukhadia University, Udaipur, India

Devesh Kumar Shukla
Environmental Science and Bio-medical Metrology Division, National Physical Laboratory, New Delhi, India

Ashutosh K. Singh
Department of Geology, Centre of Advanced Studies, University of Delhi, Delhi, India

Jagvir Singh
Ministry of Earth Sciences, Govt. of India, New Delhi, India

Jaswant Singh
Department of Environmental Sciences, Dr. R.M.L. Avadh University, Ayodhya, India

Kajal Singh
Birbal Sahni Institute of Palaeosciences, Lucknow, India

Prashant Singh
Laboratory of Cyanobacterial Systematics and Stress Biology, Department of Botany, Institute of Science, Banaras Hindu University, Varanasi, India

Priya Singh
National Centre for Seismology, Ministry of Earth Sciences, New Delhi, India

Raj K. Singh
School of Earth, Ocean and Climate Sciences, Indian Institute of Technology, Bhubaneswar, India

Vikram Pratap Singh
Department of Geology, Indira Gandhi National Tribal University, Amarkantak, India

Devesh K. Sinha
Department of Geology, Centre of Advanced Studies, University of Delhi, Delhi, India

Mohd Soheb
School of Environmental Sciences, Jawaharlal Nehru University, New Delhi, India

Prakash Kumar Srivastava
Geological Survey of India, Faridabad, India

Rohit Srivastava
National Centre for Polar and Ocean Research, Ministry of Earth Sciences, Government of India, Vasco-da-Gama, India

Anurag Tiwari
National Centre for Seismology, Ministry of Earth Sciences, New Delhi, India

Anand V.
Department of Geopolitics and International Relations, Manipal Academy of Higher Education, Manipal, India

Vandana
National Centre for Seismology, Ministry of Earth Sciences, New Delhi, India

S. Venkatachalam
National Centre for Polar and Ocean Research, Ministry of Earth Sciences, Vasco da Gama, India

Abhishek Verma
Geological Survey of India, Faridabad, India

Juhi Yadav
National Centre for Polar and Ocean Research, Ministry of Earth Sciences, Government of India, Vasco-da-Gama, India

About the editor

Dr. Neloy Khare, presently Adviser/Scientist G to the Government of India at MoES, has a very distinctive acumen not only of administration but also of quality science and research in his areas of expertise covering a large spectrum of geographically distinct locations such as the Antarctic, Arctic, Southern Ocean, Bay of Bengal, Arabian Sea, and the Indian Ocean. Dr. Khare has almost 30 years of experience in the field of paleoclimate research using paleobiology (paleontology)/teaching/science management/administration/coordination for scientific programs (including Indian Polar Programme), etc. Having completed his doctorate (PhD) on tropical marine region and Doctor of Science (DSc) on Southern High latitude marine regions toward environmental/climatic implications using various proxies, including foraminifera (microfossil), has made significant contributions in the field of paleoclimatology of Southern high latitude regions (the Antarctic and Southern Ocean) using Micropaleontology as a tool. These studies coupled with his paleoclimatic reconstructions from tropical regions helped understand causal linkages and teleconnections between the processes taking place in southern high latitudes with that of climate variability occurring in tropical regions. Dr. Khare has been conferred Honorary Professor and Adjunct Professor by many Indian Universities. He has a very impressive list of publications to his credit (121 research articles in national and international scientific journals; 3 special issues of national scientific journals as Guest Editor; edited special issue of *Polar Science* as its Managing Editor. Authored/edited many books, 130 abstracts have been contributed to various seminars, 23 popular science articles, 5 technical reports). The Government of India and many professional bodies have bestowed him with many prestigious awards for his humble scientific contributions to past climate changes/Oceanography/Polar Science and Southern Oceanography, the most coveted award being Rajiv Gandhi National Award, 2013 conferred by Honorable President of India. Others include ISCA Young Scientist Award, Boyscast Fellowship, CIES French Fellowship, Krishnan Gold Medal, Best Scientist Award, Eminent Scientist Award, ISCA Platinum Jubilee Lecture, and IGU Fellowship. Dr. Khare has made tremendous efforts to popularize ocean and polar sciences across the country by way of delivering many invited lectures and radio talks, and publishing popular science articles. Many books authored/edited on thematic topics and published by the reputed International Publishers are testimony to his commitment to popularize science among the masses.

Dr. Khare has sailed in the Arctic Ocean as a part of "Science PUB" in 2008 during the International Polar Year campaign for scientific exploration and became the first Indian to sail in the Arctic Ocean.

Foreword

The Arctic is an astounding region with many superlatives. It has a unique landscape, nature, and giant expanses of ice and snow. The incredible and bizarre icebergs floating and drifting across the Arctic create a most beautiful sight. Nevertheless, both global and local processes generate many harmful environmental impacts on the Arctic, leading to severe climate changes. It is well established that the Arctic region is warming up at a much faster pace compared to the tropics.

The unprecedented reduction in Arctic summer sea ice's areal extent has emerged as a precursor of the greater danger of rising sea levels and the release of greenhouse gases due to permafrost's thawing. Due to Arctic Sea ice's continuous melting, new avenues and opportunities are emerging for future contestation among the stakeholders. It will also pose an in-built conflict of interest. Therefore geopolitical, geoeconomic, and geoecological issues are of paramount importance. It also necessitates that ensuing economic opportunities should be explored with sustainable conservation of the Arctic's ecosystems, otherwise exposed to high risk.

It is most appropriate to spread awareness among the masses towards climate change issues of the Arctic region. The present book *Understanding Arctic Environment: An Integrated Approach From Climate Change Perspectives* is aptly brought out with 20 dedicated chapters, where each chapter has its specific importance.

The book provides an overview of the ongoing climate changes over the Arctic. It describes its impacts and assessment by Choudhary et al. On the other hand, Krishnan and Venkatachalam highlight India's scientific endeavors in the Arctic with particular reference to the climate change–related studies, who also project India's future perspectives in Arctic Climate Science. The importance of various lakes and fjords—for assessing the ongoing climate change—has been explicitly dealt with by Choudhary et al. About past climate assessment over the Arctic region, Kar et al. have ably covered the changing climate and environment since the Last Glacial Maximum. While Singh et al. have explained the critical role, that ocean plays in modulating the temperature and climate over the Arctic region.

Interestingly, Gupta et al. have attempted a teleconnection during the Holocene between tropical Indian monsoon and Arctic climate.

Verma et al. have reconstructed the past climatic conditions over the Arctic region. It is based on sediment's diamictite, granulometry, and geochemistry. On the other hand, Amrutha et al. reconstructed the Arctic Quaternary climate using ^{10}Be as a proxy. One of the most severe impacts of the ongoing climate changes over the Arctic region is the induced pressure on biodiversity and arctic life. To understand the implications of the climate change for the Arctic living entity and the Arctic life's responses toward changing climates, Singh and Khare studied ultraviolet radiation protecting pigments in plants growing in the Arctic region. Cyanobacterial population in the Arctic region has been studied by Singh et al. who have ably covered diversity, adaptation, and taxonomic problems of the *Cyanobacteria* in the Polar regions. These *Cyanobacteria* have a long evolutionary history and strong adaptive tendencies to survive and thrive in the harsh Arctic conditions. On the contrary, Rani et al. have provided an in-depth and detailed assessment of how the climate apocalypse will adversely impact Arctic biodiversity and how best such an alarming situation could be averted or minimized.

To tackle climate variability and its impacts on the Arctic region, Soheb and Ramanathan have attempted to understand the effects of climate variability on Arctic ice mass balance in the Arctic's glaciers the VestreBroggerbreen and the Feiringbreen. Similarly, Luis et al. report snowmelt onset variability across Svalbard inferred from scatterometer data (2000–17). On the other hand, Kumar et al. have assessed the pattern and shift in the Arctic Sea ice over the last four decades with plausible forcing of ocean–atmosphere interaction.

The influence of global climate variability on the Arctic atmosphere has been covered by Gwal et al. who investigated the Arctic ionosphere's space weather phenomenon. Simultaneously, Rana et al. put forth their detailed brown carbon aerosol assessment in the Norwegian Arctic region. Simultaneously, Arya et al. provide the essential carbon monoxide measurements over the Arctic and compared it with the southern high-latitude polar atmosphere. Additionally, the spatial and temporal variability of physical parameters around the Ny-Ålesund region of the Arctic has been reported by Pednekar.

We are witnessing drastic environmental, economic, and societal changes that not only have regional implications but are also likely to have profound global consequences. Issues dealing with climate change's political and economic impacts in the Arctic region must be flagged and addressed appropriately. It becomes essential for healthy participation, equity, and transparency envisaged in future commercial activities over the Arctic region. Dhanasree Jayaram et al. attempt to detail vital aspects of environment, economic and societal changes which carry significant impact in this book. An assessment of the glacial mass change-induced earthquakes in the Himalayan region has been made by Mishra et al., and the results have bearing to understand Arctic glaciers dynamics.

This book presents various diversified scientific findings and recent data related to the ongoing climate change in and around the Arctic region. It commensurates recent insights to explain multiple causal linkages of such climate change events and their influence not only on the Arctic ecosystem but even to address its far-reaching impacts that distant regions can perceive.

This book provides a panorama of climate issues from the past through the present to the future. It will be a useful reference book for polar scientists and serve its purpose to help generate interest among upcoming researchers to get an insight into the recent happening over the Arctic region.

Neloy Khare is complimented for bringing out a very timely book dealing with Arctic climate issues and their impact. All the authors deserve appreciation for focusing on key topics appropriately.

May 2021

Harsh K. Gupta

President, Geological Society of India, Hyderabad, India

Preface

Sustaining life in the harsh Polar Regions is not only complicated but also full of challenges. Extreme chilling cold winds with ambient long-lasting extreme lower temperatures make these regions inhospitable. Surprisingly, both the Arctic and the Antarctic cherish rich biodiversity and wildlife comfortably on land and ocean, having attained adequate adaptation, evolution, and protection mechanisms to face the tyranny of extreme climatic conditions over Polar Regions. However, various Arctic species will face changes due to global warming, which radically shifts their habitats in a few decades.

Although the Antarctic landscape is seemingly barren, millions of people live in the Arctic with permanent inhabitants. About 4 million people live in the Arctic, out of which about one-tenth are Indigenous. Geographically, the Arctic is a deep ocean of drifting pack ice situated at the North, which is encircled by continents and archipelagos, making it the planet's most extensive and least fragmented inhabited region. The Arctic, a semienclosed ocean and almost surrounded by land, does not allow ice to float around; consequently, ice floes bump into each other and pile up into thick ridges.

In the Arctic region, three main environmental issues are apparent: changes in biodiversity, climate change, and the use of toxic chemicals. It is a consensus that global warming causes Arctic ice to melt. Increasing air and sea temperatures whip across the landscape and prompt the ice to melt faster, causing a threat to its ecosystem. Undoubtedly, the Arctic environment and climate's visible changes due to warmer temperatures, thawing permafrost, and melting sea ice and glaciers pose severe impacts on Arctic land, water, plants and animals, and the indigenous people who are enduring these effects.

The Arctic changes will also have global consequences and the world is already feeling the effects as the Arctic temperature increase almost double than that of the worldwide average. Such an alarming situation further worsens in the absence of timely mitigation practices globally. If acceptable practices remained sidelined, warming the Arctic will continue to affect the world by rising sea levels; increasing severe weather events; changes in climate and precipitation patterns; and loss of fish stocks, marine mammals, and birds. The multifaceted positive and negative immediate impacts of ongoing Arctic climate changes are intensive. Simultaneously, the arctic ecosystem, habitat, and wilderness are adversely affected by thawing permafrost and sea ice melting. Yet, it offers new economic opportunities to the Arctic as oil and gas, shipping, and tourism, in the North. Such economic potential of the Arctic has warranted a new scramble for territorial rights among the Arctic nations.

The Arctic is the fastest warming region globally, and its climate change is not a confined, local issue; rather it embraces the entire world. Due to its global repercussions, we need to understand not only the current environmental and climatic setup of the Arctic but also require a deep insight into the past climatic changes over the Arctic region in the geological period.

Understanding past climate history is quite useful; examining different historical perspectives is critical and arguing scientifically different perspectives is vital to understanding climate patterns and scenarios over the Arctic in the time ahead. India is in Arctic Science since 2007 for scientific pursuits and has been putting efforts to understand Arctic climate through an integrated approach.

With such intentions, *Understanding Present and Past Arctic Environments: An Integrated Approach From Climate Change Perspectives* has been put forth to help understand shifting climates over the Arctic region.

The book begins with an overview of the ongoing climate changes over the Arctic by Choudhary et al. covering its impacts and assessment. Over a decade, India's scientific endeavors in the Arctic have been ably highlighted by Krishnan and Venkatachalam. The role of various lakes and fjords in climate change has been explicitly addressed by Choudhary et al. Further, Kar et al. have attempted to cover the past Arctic changing climate and environmental conditions since the Last Glacial Maximum. Singh et al. have evaluated the driving role of the ocean in modulating temperatures over the Arctic region. Similarly, Gupta et al. have explored a teleconnection between tropical Indian monsoon and Arctic climate during the Holocene. On the other hand, Verma et al. on the basis of sediment's diamictite, granulometry, and geochemistry reconstructed the Arctic Quaternary Climate. Amrutha et al. used ^{10}Be as a proxy to reconstruct the past climatic conditions over the Arctic region.

It is well established that the Arctic's biodiversity is going to be adversely impacted owing to prevailing climatic condition. Therefore it becomes imperative to assess the response of the Arctic biodiversity toward ongoing climate variability. Singh and Khare detailed ultraviolet radiation protecting pigments in plants growing in the Arctic region; on the contrary, Singh et al. focused on diversity, adaptation, and taxonomic issues of the Cyanobacteria in the Polar regions. They further put forth the critical stress factors that test their survival capabilities, including temperature, freeze—thaw cycles, desiccation, and intense light and ultraviolet radiations. An in-depth and detailed assessment of the impact of the climate on biodiversity has been made by Rani et al.

Undoubtedly, the rising temperature over Arctic region has severe consequences on the Arctic sea ice and glacier with bearing on the global climate scenario and, therefore, these aspects ought to be studied in detail. Accordingly, Soheb and Ramanathan have attempted to understand the impact of climate variability on the snow—ice mass balance over the Arctic region. Similarly, the snowmelt onset variability across Svalbard inferred from scattero-meter (2000—17) has been put forth by Luis et al. On the other hand, the shift in the Arctic sea ice over the last four decades has been dealt by Kumar et al. The atmospheric and space research over the Arctic region have emerged as the front ranking area of investigation because of their importance to understand the causal link for the on-going climate changes. In this direction, Gwal et al. investigated the space weather phenomenon of the Arctic ionosphere. Rana et al. put forth their detailed brown carbon aerosol assessment in the Norwegian Arctic region. Simultaneously, Arya et al. provided the essential carbon monoxide measurements over the Arctic region.

The spatio-temporal variability of physical parameters over the Arctic region has been focused by Pednekar. The melting Arctic sea ice due to global warming is opening various new opportunities to global players having stake in the Arctic Region. Jayaram et al. have attempted to detail vital aspects of environmental, economic, and societal changes which carry significant impact. On the other hand, Mishra et al. have detailed assessment on Glacial mass change-induced earthquakes in the Himalayan region of South Asia, to help understand the dynamics of Arctic glaciers due to tele-connectivity between these two regions.

This book highlights the updated scientific findings and recent insights pertaining to climate change. It also provides possible causal linkages of such climate change events.

This book will serve its purpose and help generate interest among upcoming researchers to know the recent changes over the Arctic region.

Date: May 2021 **Neloy Khare**
Place: New Delhi

Acknowledgments

Kind supports and help received from all contributing authors are gratefully acknowledged as without their valuable inputs on various facets of the Arctic climate, the book would not have been possible. The book has been greatly benefitted by the critical review of the learned experts who have offered their valuable suggestions and comments on different chapters.

Dr. M.N. Rajeevan, Secretary of the Ministry of Earth Sciences, Government of India, is gratefully acknowledged for his support and encouragement. Prof. Harsh K. Gupta, Former Secretary to Government of India, is sincerely acknowledged for kindly agreeing to write the foreword of the book.

Akshat Khare and Ashmit Khare have unconditionally supported enormously during various stages of this book. Dr. Shabnam Chaudhury and Shri. Hari Dass Sharma from the Ministry of Earth Sciences, New Delhi (India) have helped immensely in formatting the text and figures of this book and bring the book to its present form. The publisher (Elsevier) has also done a commendable job and is sincerely acknowledged.

Neloy Khare

Date: May 2021
Place: New Delhi

CHAPTER 1

Climate change over the Arctic: impacts and assessment

Shabnam Choudhary[1,2], Syed Mohammad Saalim[2] and Neloy Khare[1]
[1]Ministry of Earth Sciences, Govt. of India, New Delhi, India
[2]National Centre for Polar and Ocean Research (NCPOR), Goa, India

1.1 Introduction

The Polar regions are more susceptible to any change in the Earth's climate than the rest of the planet. Because of ongoing global warming, the poles are warming faster than lower latitudes. The impacts of climate change are being observed earlier in the Arctic, and with more immediate and severe consequences, than in the rest of the world. The Arctic is experiencing rapid climate change in response to changes in greenhouse gases, aerosols, and other climate drivers.

The Arctic is warming at a rate almost twice the global average (IPCC, 2014) and reductions in Arctic sea ice and permafrost and weather changes are increasingly visible. The effects of global warming or climate change in the Arctic include rising atmospheric temperatures, ice and melting of the Greenland ice sheet with a related cold-temperature anomaly (Bitz, 2006; Richard, 2007; Screen and Simmonds, 2010a,b). Related impacts include changes in ocean circulation, increased input of freshwater, and ocean acidification. Indirect effects through potential climate teleconnections to mid-latitudes may result in a greater frequency of extreme weather events (flooding, fires, and drought), (Cohen et al., 2014) ecological and biological changes, biological migrations and extinctions (Grebmeier, 2012), natural resource stresses as well as human health, displacement, and security issues. Potential methane releases from the region, primarily through the thawing of permafrost and methane clathrates, may occur (Schuur et al., 2015). Presently, the Arctic is warming twice as fast as the rest of the world (http://www.sciencemag.org/news/2013/08/scienceshot-arctic-warming-twice-fast-restworld). The pronounced warming signal, the amplified response of the Arctic to global warming, is often

Understanding Present and Past Arctic Environments
DOI: https://doi.org/10.1016/B978-0-12-822869-2.00011-6

seen as a leading indicator of global warming. Greenland's ice sheet's melting is linked to polar amplification (http://news.uga.edu/releases/article/study-2015-melting-greenland-ice-faster-arctic-warming-0616; Tedesco et al., 2016). The warming in the Arctic is primarily caused by human-caused greenhouse gas emissions such as carbon dioxide, methane, and nitrous oxide.

This phenomenon's primary cause is *ice-albedo feedback* (Fig. 1.1) whereby melting ice uncovers darker land or ocean beneath, which then absorbs more sunlight, causing more heating (Richard, 2007; Bitz, 2006; Screen and Simmonds, 2010a,b). As bright and reflective ice melts, it gives way to a darker ocean; this amplifies the warming trend because the ocean surface absorbs more heat from the Sun than the surface of snow and ice. In more technical terms, losing sea ice reduces the Earth's albedo: the lower the albedo, the more a surface absorbs heat from sunlight rather than reflecting it back to space. Sea ice plays a fundamental role in a polar

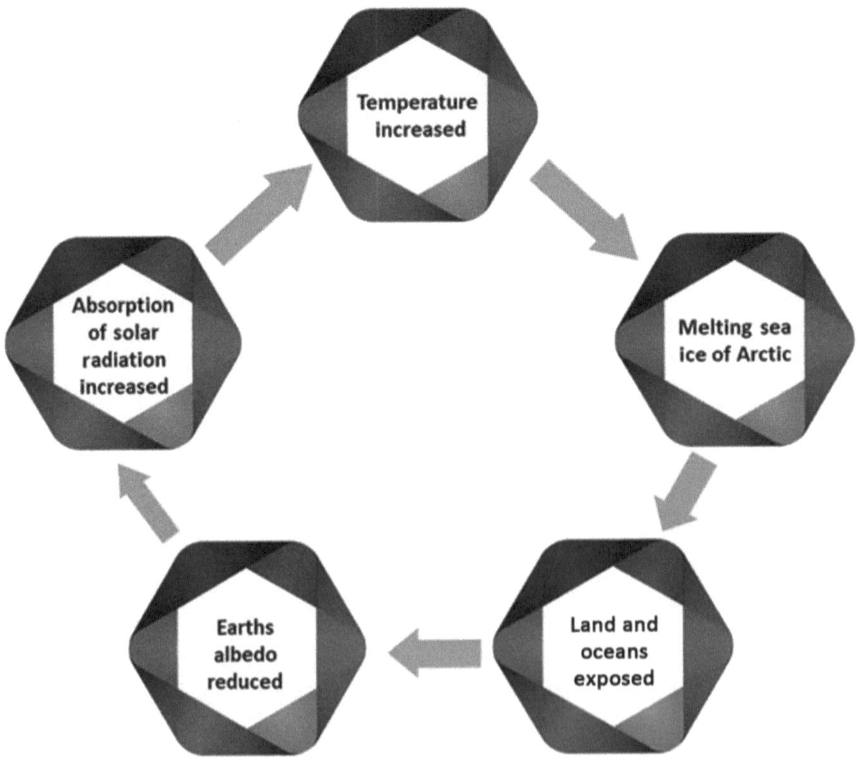

Figure 1.1 Schematic showing ice-albedo feedback.

environment. The loss of the Arctic sea ice may represent a tipping point in global warming.

1.2 Sea ice and its role in the polar environment

Sea ice is frozen seawater that floats on the surface of the ocean (Fig. 1.2). Its formation takes place in the winter season in both hemispheres, that is, the Arctic and the Antarctic. As ice crystals form at the ocean surface, they expel salt and increase the underlying waters' salinity. This cold, salty water is dense and can sink to the ocean floor, where it flows back toward the equator. The sea ice layer also restricts wind and wave action near coastlines, reducing coastal erosion and protecting ice shelves. It retreats in the summer but does not disappear completely (source: https://earthobservatory.nasa.gov/features/SeaIce/14/12/2020). The most crucial clarification regarding sea ice is that it does not influence sea level. Because it is already floating in the ocean, sea ice has already displaced water equivalent to its weight. Melting sea ice will not raise the ocean level anymore.

Arctic sea ice keeps the Polar regions cool and helps moderate global climate. Sea ice also creates an insulating cap across the ocean surface,

Figure 1.2 Sea ice in the Arctic.

which reduces evaporation and heat loss to the atmosphere. As a result, the weather over ice-covered areas tends to be colder and drier than it would be without ice. This floating ice profoundly influences the polar environment, affecting ocean circulation, weather, and regional climate (Serreze and Barry, 2011). Sea ice has a high albedo. Therefore the influence of sea ice on the Earth is not confined to the regional level, but acts globally. Once sea ice begins to melt, it exposes more dark water, and the water absorbs more sunlight. The sun-warmed water then melts more ice. Over several years, this positive-feedback cycle (the ice-albedo feedback) can influence the global climate.

Furthermore, a retreating sea ice cover insulates the cold Arctic atmosphere less from relatively warm ocean waters advected to the high northern latitudes by the North Atlantic current. The reduced insulation increases the vertical heat transfer to the atmosphere (*ice-insulation feedback*) (Fig. 1.3). Due to this sensitivity, the Arctic may be regarded as an "early warning system" in the context of the progressing climate change

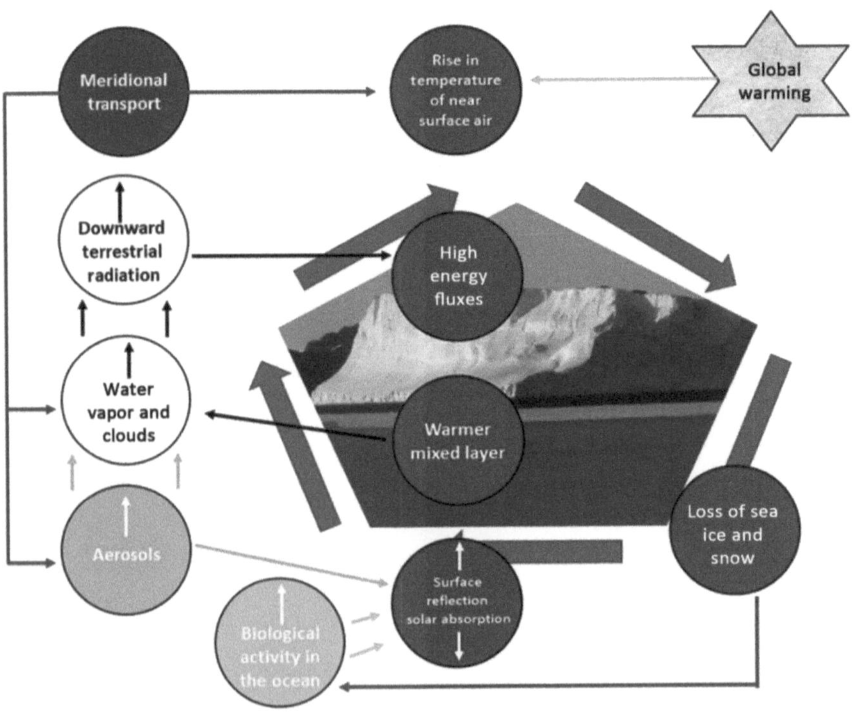

Figure 1.3 Feedback processes that amplify an initial near-surface air temperature rise caused by global warming.

(Sherwood et al., 2011). The Arctic reacts very sensitively to climatic and oceanographic changes as positive-feedback mechanisms strengthen climate trends in high latitudes, particularly during the current climate warming.

1.3 Role of black carbon aerosol in the warming of the Arctic

Over the past few decades, a lot of attention has been paid to greenhouse gases, such as carbon dioxide and methane, and their role in warming the Earth's climate. However, several aerosols also significantly impact regional and global climate; one of these is black carbon (BC). It is essentially the soot emitted from natural [the incomplete combustion of fossil fuels like petroleum, coal, biofuels, and biomass (wood, animal dung, etc.)] and human-induced (industry, transport, and household) sources, and is the most potent climate-warming aerosol (Johnson et al., 2011). The aerosol enters the atmosphere from automobiles (those with diesel engines being particularly high emitters), coal-fired power plants, wood stoves, kerosene lamps, and forest fires. If produced continuously, and in large quantities, these can have an enormous impact on the climate, especially in the cryosphere. It is found that BC is second only to carbon dioxide as far as major contributors to climate change. Soot produced by fossil fuels and biofuels combined may contribute to about 16% of gross global warming (Jones et al., 2007). After being emitted, BC particles can remain suspended in the atmosphere anywhere from 1 to 4 weeks before they are "washed out" and return to the Earth via settling or precipitation. While suspended in the atmosphere, the presence of BC particles has a net warming effect. Unlike greenhouse gases, which trap infrared radiation already in the Earth's atmosphere, BC particles, being black, absorb all wavelengths of sunlight and then reemit this energy to the surrounding environment as infrared radiation. BC particles continue to impact once they reach the ground in the cryosphere, particularly in the Arctic, which is polluted by BC aerosol transported through the atmosphere from its source, the highly industrialized areas of the Northern Hemisphere. Pure snow is white and thus effective at reflecting solar radiation from the Earth's surface. If soot particles, even in small concentrations as small as 10 parts per billion (ppb), enter the snow, this reduces the snow's albedo, allowing it to absorb more solar radiation and to melt more readily (source: https://DiscoveryofGlobalWarming//Carbon Dioxide Green house Effect// accessed May 24, 2013).

It is believed that BC has been responsible for significant warming in the Arctic. As much as 30% of the warming in the Arctic since preindustrial times can be attributed to anthropogenic BC, according to the Earth System Modelling Facility at the University of California, an expert in modeling atmospheric aerosol transport to the Arctic (Junker and Liousse, 2008). During Arctic winters, when there is little sunlight, the presence of dark soot particles in the snow does not make much of a difference. Only when solar radiation begins to increase rapidly in the spring and the melting season begins does BC have its most significant snow cover impact. The presence of BC on the snow makes it absorb more solar radiation; otherwise, it would lead to more intense melting earlier in the spring. As snow and ice melt, this exposes darker, less reflective surfaces such as land and open water and creates a positive-feedback situation, leading to accelerated melting, commonly referred to as snow-albedo feedback (Lee et al., 2011). According to the Intergovernmental Panel on Climate Change, Fifth Assessment Report (IPCC, 2014), black carbon aerosols are considered to be the third largest anthropogenic forcing agent after CO_2 and CH_4, with a present-day radiative forcing of $\sim 0.40\ W\ m^{-2}$ relative to the preindustrial period, which is $\sim 25\%$ of the present-day CO_2 radiative forcing (Koch et al., 2007). The warming of the Arctic region has recently gained considerable attention due to its projected impacts on the global climate system. The effect of anthropogenic BC aerosol on snow is of enduring interest due to its role in aerosol radiative forcing and further consequences for the Arctic and global climate change.

1.4 Causes and effects of rapid warming in the Arctic

The higher rise in atmospheric temperatures in the Arctic than the rest of the world has made the Arctic one of the most interesting regions on Earth for meteorologists and climate scientists. There climate changes currently take place at an unprecedented pace and intensity, and the reported dramatic changes have not been completely projected. Arctic amplification, or ice-albedo feedback, is a process identified by Arrhenius. In simpler terms, initial warming will melt some of the Arctic's highly reflective (high-albedo) snow and ice cover, exposing darker underlying surfaces that readily absorb solar energy, leading to further warming and retreat of snow and ice cover (Fig. 1.3) (Arrhenius, 1897). This feedback can work in reverse order where initial cooling leads to amplification and is presently understood to have a series of causes, operating on different temporal and spatial scales. Prominent among these are expansion or retreat of the Arctic sea ice cover altering vertical heat fluxes between the Arctic Ocean and the overlying atmosphere (Serreze et al., 2009; Screen and Simmonds, 2010a,b) changes in atmospheric and oceanic heat flux convergence (Hurrell, 1996; Graversen et al., 2008; Chylek et al., 2009; Yang et al., 2010), and changes in cloud cover and water vapor content that affect the downward long-wave radiation flux (Francis and Hunter, 2006) arising from processes either within the Arctic or in response to alterations in atmospheric energy flux convergence (Abbot et al., 2009; Graversen and Wang, 2009). Other studies point to the impacts of soot on snow (Hansen and Nazarenko, 2004) and of heat-absorbing BC aerosols in the atmosphere (Shindell and Faluvegi, 2009). Different processes can work simultaneously. A change in atmospheric heat flux convergence that leads to warming may result in reduced sea ice extent that furthers the warming. Arctic amplification has consequences for nearby land ice, too. As land ice melts, it adds freshwater to the oceans causing sea levels to rise, and Greenland's surface melt increases. Warmer, sea ice-free oceans could also increase melting from Greenland's glaciers.

The Arctic region's increasing temperature is thawing frozen ground in the Arctic—known as "permafrost." CO_2 and methane released from the carbon-rich permafrost could cause additional warming by adding to greenhouse gases already present in the atmosphere. This is considered as another positive feedback (Lee et al., 2011). Frozen methane also exists in methane hydrates on the seafloor. However, there is a lot of uncertainty about how much carbon these frozen deposits hold and how quickly they are thawing. So far, only a small amount of methane in the atmosphere can be traced to thawing permafrost (Goddard et al., 2016). At present, warming is mainly driven by CO_2 and other greenhouse gas emissions. However, it is expected that in coming years sea ice losses will cause land and ocean to warm in the Arctic and the release of methane from thawing of permafrost regions could contribute more to warming. Another impact of Arctic and Greenland ice melt could be that the freshwater runoff into the ocean disrupts part of an effective circulation system known as the Atlantic Meridional Overturning Circulation (AMOC). The AMOC carries warm surface water northward from the tropics, giving western Europe its mild climate. Climate models have predicted that the AMOC could slow down by an average of 25% by 2100 (Buckley and Marshall,

2016). This is mainly due to changes in temperature and precipitation patterns, and freshwater from melting ice could become increasingly important with climate change. Some studies suggest that the AMOC has already weakened by 15% since the middle of the last century. From the modeling, it has been suggested that a slowdown in the AMOC could cool the northern hemisphere this century, but the effect is likely to be exceeded by greenhouse gas warming (Macias Fauria and Johnson, 2008). The AMOC is just one part of a giant conveyor belt that currently redistributes warmth worldwide, so losing ice could affect regional climates worldwide. Changes in the strength of the AMOC can also affect sea ice melt by transporting more or less warm surface water into the region (Rossby, 1996). It is a complicated picture, and so more research is needed to understand the full extent of the impacts.

Furthermore, melting sea ice can also influence winter weather. The change in pressure gradient can also alter atmospheric circulations, including the jet stream—a stream of fast-flowing air high up in the atmosphere. Research suggests that this makes the jet stream's path meander more, allowing warm air to penetrate north. Cold air can plunge south into the mid-latitudes (source: https://e360.yale.edu/features/as_arctic_ocean_ice_disappears_global_climate_impacts_intensify_wadhams). Larger waves also mean the systems moving from west to east tend to travel more slowly, increasing the likelihood of persistent periods of very cold winter weather or heat extremes in the summer.

1.5 Global effects of warming in the Arctic

1.5.1 Effects on inhabitants, wildlife, and the ecosystem

Arctic communities are experiencing some of the most dramatic effects of receding sea ice, but the thawing contributes to complex global patterns that affect the whole world's population. The impact of disappearing ice extends far beyond the Arctic Circle and its inhabitants (Seimon et al., 2007). Melting sea ice is likely to have global consequences by unlocking new shipping routes and exposing more fossil fuel reserves. Although Arctic sea ice covers only a small fraction of the Earth's surface, there may be severe climate-related impacts (Beaugrand et al., 2002). The loss of sea ice will also have consequences for people and ecosystems in the Arctic region. Such ongoing changes have an adverse impact on the food chain, from phytoplankton to marine mammals such as seals, walrus, whales, and polar bears (Fig. 1.4). Polar bears are a poignant symbol of climate change.

Figure 1.4 Polar bear, seals, and reindeer are under threat due to changing climate.

Locally, there is a high probability that ice-dependent species like polar bears and reindeer will experience a population decline as their food options disappear. Narwhals, polar bears, and walruses are at increasing risk with shrinking sea ice cover. By 2100, polar bears could face starvation and reproductive failure even in far northern Canada. Walruses depend on sea ice as a platform for feeding and resting, and the warming Arctic is disrupting their normal patterns (Parmesan, 1997). Over the past decade, earlier melting of sea ice in the summer has forced abnormally large numbers of Pacific walruses ashore on Russia and Alaska's coasts. These "haul outs" of up to 35,000 individuals can be deadly, particularly for walrus calves that are crushed in stampedes provoked by disturbances. As sea ice melts, walrus populations will be exposed to more industrial activity, like shipping and oil and gas exploration (Seimon et al., 2007). In addition to direct impacts like icebreaking ships and disturbance to walruses on shore, there is an increased risk of oil spills, which are virtually impossible to clean up on ice. There has been a shift observed in vegetation pattern due to warming in the Arctic region. Due to change in climate, new species of plants are growing in the Arctic, with more woody plants. High precipitation and warmer winters cover plants with layers of ice instead of snow, resulting in unavailability of food for reindeer and caribou (Lucht et al., 2002). These animals need to adapt their movements to the availability of food. They are migratory, and their habitat crosses territorial and national borders. Southern species pose a risk to existing Arctic species. On the tundra, rising temperatures have resulted in a new competitor—the Arctic fox's much larger cousin, the red fox. Not only does the newcomer colonize their dens, but it can also kill the smaller Arctic foxes. In the ocean, both scientists and Inuit have observed that orcas are increasing in numbers, and after a certain period, they may be abundant in the Arctic (Martins et al., 1998). Orcas prey on narwhals and bowhead whales. Birds are being affected, including the breeding population of ivory gull in the Canadian Arctic falling 70% since the 1980s. As a result of climate change, fish stocks in the Barents Sea are moving north at up to 160 km per decade. Fish are sensitive to water temperature changes, posing a risk for commercial and subsistence fisheries that may see fish resources move away from where they can be harvested. The moving fish also change the ecosystems into which they move. Plants and animals frozen in the ground begin to decay as permafrost thaws (Korontzi et al., 2006). When they decompose, they release carbon dioxide and methane back to the atmosphere to contribute to further warming. The changing vegetation of the Arctic also affects the brightness of the surface, which then influences warming. Humans in the Arctic have become more vulnerable to coastal erosion and are forced to relocate their communities away from coastlines. Thawing permafrost has led to more wildfires and affected infrastructure, while rising ocean temperatures and acidification have already altered marine fisheries. Arctic communities are experiencing some of the most dramatic effects of receding sea ice, but the thawing contributes to complex global patterns that affect the whole world's population. Melting Arctic sea ice will also contribute to sea-level rise, both as it melts into the oceans and as the warming water expands. It is believed that ice sheets moderate weather by reflecting heat away from the Earth, and their disappearance will contribute to more extreme weather in the coming decades (Kupiainen and Klimont, 2007). As a result, communities and governments are beginning to plan for these changes, and exploring what can be done to adapt to these new conditions.

The increasingly rapid rate of recent climate change poses new challenges to the resilience of Arctic life. In addition to the impacts of climate change, many other stresses brought about by human activities are simultaneously affecting life in the Arctic, including air and water contamination, overfishing, increasing levels of ultraviolet radiation due to ozone depletion, habitat alteration and pollution due to resource extraction, and increasing pressure on land and resources related to the growing human population in the region. The sum of these factors threatens to overwhelm the adaptive capacity of some Arctic people and ecosystems.

1.6 How to slow down the warming in the Arctic

The Arctic needs near-term temperature control. It is warming at a higher rate than the global mean (IPCC, 2014). It is experiencing climate impacts, such as thawing permafrost and rapidly melting sea ice, that have adverse effects on the region and the global climate system. Reductions in methane emission are essential to decrease the rate of near-term warming and are an essential component of an overall greenhouse gas-mitigation strategy.

The Arctic Council is an ideal forum to address methane emissions. Arctic nations have a common and heightened stake in methane mitigation. Methane reductions would slow down warming globally; also, they would disproportionately benefit the Arctic, by avoiding the warming two to three times to the global average (Lee et al., 2009). Arctic Council member and observer nations also account for a significant portion—42%—of global anthropogenic methane emissions. Reductions in methane emissions from Arctic Council participants alone could make a substantial difference to global methane levels.

The Arctic Council is aware that methane is a critical concern for its region. Since 2009, the forum has convened a task force focused on BC and methane. The current incarnation of the task force is developing strategies to reduce methane and BC emissions. The recommendations have been reported at the Arctic Council Ministerial meeting in Iqaluit, Canada, in 2015 (Law and Stohl, 2007). The Center for American Progress has advocated that the United States make climate change the overarching theme of its Arctic agenda. Making methane reductions a vital component of this plan, the United States could bring about a meaningful decrease in the rate of near-term Arctic and global warming. Reductions in methane emissions are a critical means of reducing near-term warming and are necessary for an overall climate strategy (Junker and Liousse, 2008).

The increasingly rapid rate of recent climate change poses new challenges to the resilience of Arctic life. In addition to the impacts of climate change, many other stresses brought about by human activities are simultaneously affecting life in the Arctic, including air and water contamination, overfishing, increasing levels of ultraviolet radiation due to ozone depletion, habitat alteration and pollution due to resource extraction, and increasing pressure on land and resources related to the growing human population in the region. The sum of these factors threatens to overwhelm the adaptive capacity of some Arctic people and ecosystems (ACIA, 2014). The science suggests that responding to this challenge will require two sets of actions: mitigation, to slow the speed and amount of future climate change by reducing greenhouse gas emissions; and the other, called adaptation, to attempt to limit adverse impacts by becoming more resilient to the climate changes that will occur while society pursues the first set of actions. The scope of this assessment did not include an evaluation of either of these sets of activities. These are being addressed by efforts under the auspices of the United Nations Framework Convention on Climate Change and other bodies (ACIA, 2014).

References

Abbot, D.S., Walker, C., Tziperman, E., 2009. Can a convective cloud feedback help eliminate winter sea ice at high CO_2 concentration? J. Clim. 22, 5719–5731.

ACIA, 2014. Impacts of a Warming Arctic: Arctic Climate Impact Assessment, A Report of the ACIA (Arctic Climate Impact Assessment) Produced in 2004 by a Large International Panel of Scientists.

Arrhenius, S., 1897. On the influence of carbonic acid in the air upon temperature of the earth. Astron. Soc. Pac. 9 (54), 14.

Beaugrand, G., Ried, P.C., Ibanez, F., Alistair Lindley, J.A., Edwards, M., 2002. Reorganization of North Atlantic copepod biodiversity and climate. Science 296, 1692–1694.

Bitz, C., 2006. Polar Amplification, RealClimate.org.

Buckley, M.W., Marshall, J., 2016. Observations, inferences, and mechanisms of the Atlantic Meridional Overturning Circulation: a review. Rev. Geophys. 54.1, 5–63.

Chylek, P., Folland, C.K., Lesins, G., Dubey, M.K., Wang, M.-Y., 2009. Arctic air temperature change amplification and the Atlantic multidecadal oscillation. Geophys. Res. Lett. 36, L14801.

Cohen, J., Screen, J.A., Furtado, J.C., Barlow, M., Whittleston, D., Coumou, D., Francis, J., Dethloff, K., Entekhabi, D., Overland, J., Jones, J., 2014. Recent Arctic amplification and extreme mid-latitude weather. Nature geoscience 7 (9), 627–637.

Francis, J.A., Hunter, E., 2006. New insight into the disappearing Arctic sea ice. EOS Trans. Am. Geophys. Union 87, 509–511.

Graversen, R.G., Wang, M., 2009. Polar amplification in a coupled climate model with locked albedo. Clim. Dyn. 33, 629–643.

Graversen, R.G., Mauritsen, T., Tjernstrom, M., Kallen, E., Svensson, G., 2008. Vertical structure of recent Arctic warming. Nature 451, 53–56.

Grebmeier, J.M., 2012. Shifting patterns of life in the Pacific Arctic and sub-Arctic seas. Annual review of marine science 4, 63–78.

Hansen, J., Nazarenko, L., 2004. Soot climate forcing via snow and ice albedo. Proc. Natl. Acad. Sci. 101, 423–428.
Hurrell, J.W., 1996. Decadal trends in the North Atlantic oscillation: regional temperatures and precipitation. Geophys. Res. Lett. 23, 665–668.
IPCC, 2014. (Pachauri, R.K., Meyer, L., Plattner, G.K. and Stocker, T.) Climate change 2014: synthesis report. IPCC, Geneva, Switzerland.
Johnson, M.R., Devillers, R.W., Thomson, K.A., 2011. Quantitative field measurement of soot emission from a large gas flare using Sky-LOSA. Environ. Sci. Technol. 45, 345–350.
Jones, A., Haywood, J.M., Boucher, O., 2007. Aerosol forcing, climate response and climate sensitivity in the Hadley Centre climate model. J. Geophys. Res. 112, D20211.
Junker, C., Liousse, C., 2008. A global emission inventory of carbonaceous aerosol from historic records of fossil fuel and biofuel consumption for the period 1860-1997. Atmos. Chem. Phys. 8, 1195–1207.
Koch, D., Bond, T.C., Streets, D., Unger, N., Van der Werf, G.R., 2007. Global impacts of aerosols from particular source regions and sectors. J. Geophys. Res. 112, D02205.
Korontzi, S., McCarty, J., Loboda, T., Kumar, S., Justice, C., 2006. Global distribution of agricultural fires in cropland from 3 years of Moderate Resolution Imaging Spectroradiometer (MODIS) data. Glob. Biogeochem. Cycles 20, GB2021.
Kupiainen, K., Klimont, Z., 2007. Primary emissions of fine carbonaceous particles in Europe. Atmos. Environ. 41, 2156–2170.
Law, K.S., Stohl, A., 2007. Arctic air pollution: Origins and impacts. science 315(5818), 1537–1540.
Lee, D.H., Behera, S.K., Kim, J.W., Park, H.S., 2009. Methane production potential of leachate generated from Korean food waste recycling facilities: a lab-scale study. Waste Management 29 (2), 876–882.
Lee, S., Gong, T., Johnson, N., Feldstein, S.B., Pollard, D., 2011. On the possible link between tropical convection and the Northern Hemisphere Arctic surface air temperature change between 1958 and 2001. Journal of Climate 24 (16), 4350–4367.
Lucht, W., Prentice, I.C., Myneni, R.B., 2002. Climatic control of the high-latitude vegetation greening trend and Pinatubo effect. Science 296, 1687–1689.
Macias Fauria, M., Johnson, E.A., 2008. Climate and wildfires in the North American boreal forest. Philos. Trans. R. Soc. B 363, 2315–2327.
Martins, J.V., Hobbs, P.V., Weiss, R.E., Artaxo, P., 1998. Sphericity and morphology of smoke particles from biomass burning in Brazil. J. Geophys. Res. 103, 32051–32057.
Parmesan, C., 1997. Ecological and evolutionary responses to recent climate change. Annu. Rev. Ecol. Syst. 37, 637–669.
Richard, B., 2007. Earth – melting in the heat? BBC News Retrieved January 3, 2008.
Rossby, T., 1996. The North Atlantic Current and surrounding waters: At the crossroads, 827 Rev. Geophys. 34 (4), 463–481, https://doi.org/10.1029/96RG02214
Schuur, E.A., McGuire, A.D., Schädel, C., Grosse, G., Harden, J.W., Hayes, D.J., Hugelius, G., Koven, C.D., Kuhry, P., Lawrence, D.M., Natali, S.M., 2015. Climate change and the permafrost carbon feedback. Nature 520 (7546), 171–179.
Screen, J.A., Simmonds, I., 2010a. The central role of diminishing sea ice in recent Arctic temperature amplification. Nature 464, 1334–1337.
Screen, J.A., Simmonds, I., 2010b. Increasing fall-winter energy loss from the Arctic Ocean and its role in Arctic temperature amplification. Geophys. Res. Lett. 37, L16707.
Seimon, T.A., et al., 2007. Upward range extension of Andean anurans and chytridiomycosis to extreme elevations in response to tropical deglaciation. Glob. Change Biol. 13, 288–299.

Serreze, M., Barry, R., 2011. Processes and impacts of Arctic Amplification. Glob. Planet. Change 77 (1–2), 85–96.
Serreze, M.C., Barrett, A.P., Stroeve, J.C., Kindig, D.M., Holland, M.M., 2009. The emergence of surface-based Arctic amplification. Cryosphere 3, 11–19.
Sherwood, S., et al., 2011. Robust tropospheric warming revealed by iteratively homogenized radiosonde data. J. Clim. 22 (20), 5336–5352.
Shindell, D., Faluvegi, G., 2009. Climate response to regional radiative forcing during the twentieth century. Nat. Geosci. 2, 294–300.
Tedesco, M., Mote, T., Fettweis, X., Hanna, E., Jeyaratnam, J., Booth, J.F., et al., 2016. Arctic cut-off high drives the poleward shift of a new Greenland melting record. Nat. Commun. 7, 11723.
Upstill-Goddard, R.C., Barnes, J., 2016. Methane emissions from UK estuaries: Re-evaluating the estuarine source of tropospheric methane from Europe. Marine Chemistry 180, 14–23.
Yang, X.-Y., Fyfe, J.C., Flato, G.M., 2010. The role of poleward energy transport in Arctic temperature evolution. Geophys. Res. Lett. 37, L1480.

CHAPTER 2

India's scientific endeavors in the Arctic with special reference to climate change: the past decade and future perspectives

K.P. Krishnan and S. Venkatachalam
National Centre for Polar and Ocean Research, Ministry of Earth Sciences, Vasco da Gama, India

2.1 Introduction

The Arctic region is known as one of the fastest-warming regions of the world, and current predictions indicate that the ongoing environmental changes are greater than any other part of the Earth. The Arctic sea ice is shrinking at the rate of approximately 8%—10% per decade since the late 1990s (Fadeev et al., 2018). Climate models project that the Arctic Ocean may experience ice-free summers in the next 100 years (Polyakov et al., 2017). Recent observations suggest increasing temperatures of the Atlantic water inflow in the Arctic shelf-waters (Walczowski et al., 2017; Jain et al., 2019). Similarly, Arctic terrestrial ecosystems are also undergoing rapid changes (e.g., thawing of permafrost and melting of glaciers). These unprecedented changes have already proven to have a profound impact on the ocean biogeochemical process, food web structure, biodiversity changes, and associated human livelihood. Monitoring and documenting these changes in the Arctic marine and terrestrial environments are critical to forecast and respond to changes. Over the last 10 years, researchers worldwide have developed model systems that can help in real-time observation and prediction of climate change impacts. This has facilitated substantial progress in the understanding of Arctic ecosystems. India has a key interest in the Arctic climatic research, due to the critical connection between the Arctic ice melt and the Indian Monsoons. However, the exact effect remains yet to be understood entirely. It is observed that rapid melting of Arctic ice and the consequent increase of freshwater in the region prevents heat from escaping to the atmosphere thereby providing

Understanding Present and Past Arctic Environments
DOI: https://doi.org/10.1016/B978-0-12-822869-2.00006-2

warmer waters to the Indian Ocean, which alters Indian monsoons (Prabhu et al., 2012; Nanda and Council, 2019). Given its significance and after decades of experience in Antarctic research, Indian scientific endeavors in the Arctic realm commenced from 2007. Ministry of Earth Sciences launched a national level Indian Arctic program and all the expedition-related activities are spearheaded by National Centre for Polar and Ocean Research (NCPOR), formerly known as National Centre for Antarctic and Ocean Research. Here in this review, we briefly summarize the contribution of Indian research activities in the Arctic realm.

2.2 Establishment of permanent field station at Ny-Ålesund

Ny-Ålesund is a former coal-mining colony located (at 78°55′N, 11°56′E) on the west coast of Spitsbergen in the Svalbard Archipelago. After the closure of mining activities, this town has been converted into an international science village by the Norwegian government. This small town is a home away from home for many researchers. A total of 12 countries have their stations to undertake research activities around this area. All the research infrastructure, including marine laboratory and dining facilities, are maintained by Kings Bay AS, a Norwegian company. India established its first field station named "Himadri" during the year of 2008, which can accommodate up to eight researchers at a given time to conduct field studies (Fig. 2.1A). Indian Arctic operations are usually performed for around 8 months in a year, during summer season (May–October) and winter/spring season (March–April) respectively. To date, about three hundred researchers from various institutes and universities across India have visited Ny-Ålesund to conduct field studies. The research activities can be broadly

(A) (B)

Figure 2.1 (A) Arctic field station Himadri and (B) Gruvebadet Atmospheric Laboratory located at Ny-Ålesund, Svalbard.

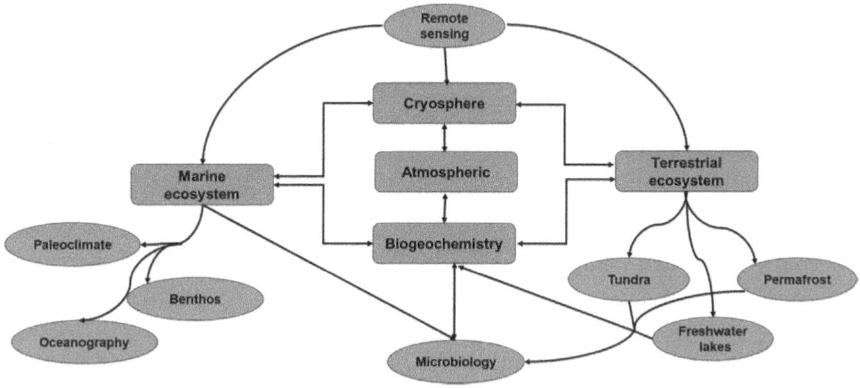

Figure 2.2 Schematic representation of India's scientific domains involved characterizing various ecosystems in the Arctic realm.

categorized into the following domains, (1) atmospheric studies, (2) cryospheric studies, and (3) biogeochemical studies (Fig. 2.2).

2.3 Atmospheric research actives around Ny-Ålesund

During the course of the years, India has established well-equipped infrastructure facilities for Atmospheric research at Gruvebadet laboratory in Ny-Ålesund (Fig. 2.1B). Research activities are mainly focused on studying atmospheric aerosols, precipitation, and cloud rainfall patterns around the Ny-Ålesund area. Atmospheric aerosols play a crucial role in the global climate system, especially in changing the surface-atmosphere system's radiation budget, inducing both direct and indirect radiative effects (Rajan and Krishnan, 2016). The transport of aerosols from low-latitude regions to the Arctic atmosphere and subsequent deposition in Arctic snow/glaciers and ice sheets have been a major scientific concern in recent years. Also, the Arctic is considered as a net sink for black carbon (BC), where deposition exceeds emission.

2.3.1 Monitoring of black carbon ratios

BC aerosols are among the most substantial contributors to global warming after carbon dioxide. It absorbs radiation and warms the atmosphere (Gogoi et al., 2016, 2018). Incomplete combustion of coal, fossil fuel, and wood emits large BC and organic carbon (OC) into the atmosphere. Deposition of BC on snow/ice results in decreased surface albedo and a reduction in snow cover by enhanced melting. However, the magnitude of this mechanism is very uncertain, and to improve the understanding of BC variability in Polar regions, NCPOR has initiated the Polar Aerosol Network Programme, which not only includes measurements from the Arctic but also from Antarctic, Himalaya, and Southern Ocean.

Measurements of atmospheric aerosols are specifically targeted to study Arctic summer aerosols source, their concentrations, and new aerosols particle generation processes in the Arctic region. This long-term observation envisions quantification of the aerosols physical and ocean optical properties and associated processes during the summer season and estimates the aerosol radiative forcing over the region. Results indicated that during springtime enhancement in the atmospheric BC concentrations, the Svalbard location of Arctic was observed with a climatological (2010–13) seasonal mean value of $\sim 50.3 \pm 19.5$ ng m^{-3} (Gogoi et al., 2016). Daily mean BC concentrations in the Arctic snow (from snow sampling studies at Ny-Ålesund) ranged from the 0.6 to 4.1 ppb during the 2012 spring season.

2.3.2 Monitoring of Arctic precipitation and clouds

Monitoring Arctic precipitation and clouds over Ny-Ålesund and its relation with the rapidly changing atmosphere is one of the thrust areas of research. Precipitation in the form of snow and rain is an integral part of the hydrological system as it modulates energy, hydrological cycle, and ecosystem. Hence this has an enormous socioeconomic impact as well. Though precipitation is by and large caused by regional processes, knowledge of precipitation, its underlying processes along with its assessment for spatiotemporal variations can provide important leads to global energy and water cycle studies. For this purpose, precipitation measurements using microrain radar, upper atmospheric measurements using microwave radiometer, and cloud measurements using ceilometer are being conducted at Gruvebadet laboratory since 2014. The main focus is to understand atmospheric circulation in the past four decades in the Northern Hemisphere to ascertain its changes and linkages with the precipitation and air temperature at Ny-Ålesund (Nuncio et al., 2020). Due to its geographical location, Ny-Ålesund acts as a unique natural laboratory to observe changes associated with heat and moisture transport from the mid-latitude Atlantic. A study by Nuncio et al. (2020) revealed a high-pressure anomaly influenced daily variability of temperature and precipitation over northwestern Europe as well as by the cyclones near Iceland. This study also correlates with the extreme events during the year of 2015–16 at Ny-Ålesund where higher temperature and precipitation in summer season was observed, which could be the result of teleconnections from tropics and local processes like the cyclones.

2.4 Cryosphere studies

The Arctic cryosphere is an essential component of the earth system, which has an active role in regulating global and regional climate systems. Substantial loss of sea ice and melting of ice sheets and glaciers are shown to have critical impacts in the marine and terrestrial carbon cycles in numerous interconnected ways (Parmentier et al., 2017). Svalbard archipelago in the high Arctic is home to several large bodies of mountain and tidewater glaciers. These glaciers respond significantly to the atmospheric processes and the changing climate in general. They also impact the local hydrological cycle and fjord ecosystems. Currently, components of the Arctic cryosphere are subjected to dramatic change due to global warming. The considerable ice mass stored in the Arctic can have a significant role in modulating future changes in the global environment, including contribution to sea-level rise, which leads our need to study the cryosphere dynamics processes. The objective of the Arctic glaciological program is to understand the dynamics and mass budget of Arctic glaciers in the context of climate change.

To understand the interaction between surface mass balance and ice flow dynamics, the bed slope, surface movement, ice temperature, and ice thickness are important factors. Indian researchers are involved in monitoring two glaciers, namely, Vestre Broggerbreen and Feiringbreen (Fig. 2.3) for the mass balance dynamics of glacier ice in the context of climate change (Patel et al., 2019). In the Vestre Broggerbreen, the minimum snow depth for the years 2012–17 varied from 16 to 76 cm while the maximum varied from 118 to 310 cm. For the Feiringbreen glacier, the average snow depth was 133.7 ± 56.6 cm, ranging from 22 to 235 cm with a 28.3 cm/100 m.

The study further revealed a significant reduction in snow depth across the glacier during the past 6 years, even though the snowpack's average mean density is almost similar throughout the studied period. Snow water equivalent across the glacier has reduced significantly due to the increase in liquid precipitation during winter governed by the southerly winds over the Ny-Ålesund region, which has a crucial impact over the net annual surface mass balance of the glaciers in this region.

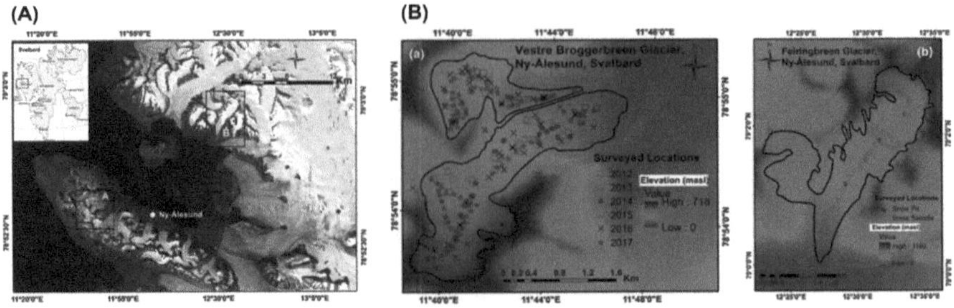

Figure 2.3 (A) Location of the selected glaciers in Svalbard and (B) surveyed locations of Vestre Broggerbreen and Feiringbreen glacier. *Adapted from Patel, L., Sharma, P., Thamban, M., 2019. Spatio-temporal variability of snow water equivalent over the Vestre Broggerbreen and Feiringbreen glaciers, Ny-Ålesund, Svalbard. J. Earth Syst. Sci. 128, 183.*

2.5 Biogeochemical studies

2.5.1 Marine ecosystem

Fjords are the ideal and natural systems to study glacier meltwater and oceanic influence interactions and their associated physical, biological, and geochemical processes. Kongsfjorden located at Ny-Ålesund is an established reference site for Arctic marine studies, considering the uniqueness of the fjord, its proximity to the Arctic and the Atlantic water masses. During the year of 2010, a flagship program was initiated to monitor Kongsfjorden. This program is multiinstitutional and multidisciplinary that addresses the possible response of this fjord to climate variabilities at different time scales, like seasonal, annual, and long-term changes in its hydrography, hydrochemistry, and microbiota. The objectives of the program are to address (1) the variability in the Arctic/Atlantic climate signal by understanding the interaction between freshwater from the glacial runoff and Atlantic water from west Spitsbergen current; (2) the effect of interaction between warm Atlantic water and the cold glacial melt freshwater on the biological productivity and phytoplankton species composition and diversity within the fjord; (3) winter convection and its role in the biogeochemical cycling; and (4) the production and export of OC in the fjord to quantify the CO_2 flux.

As part of the project mandate, 9–13 discrete sampling stations were identified within Kongsfjorden and sampled every year to collect the baseline data and its dynamics to identify its response to climate change (Fig. 2.4). Over the years, the study revealed that Kongsfjorden is experiencing variations in the hydrographic properties at different seasons.

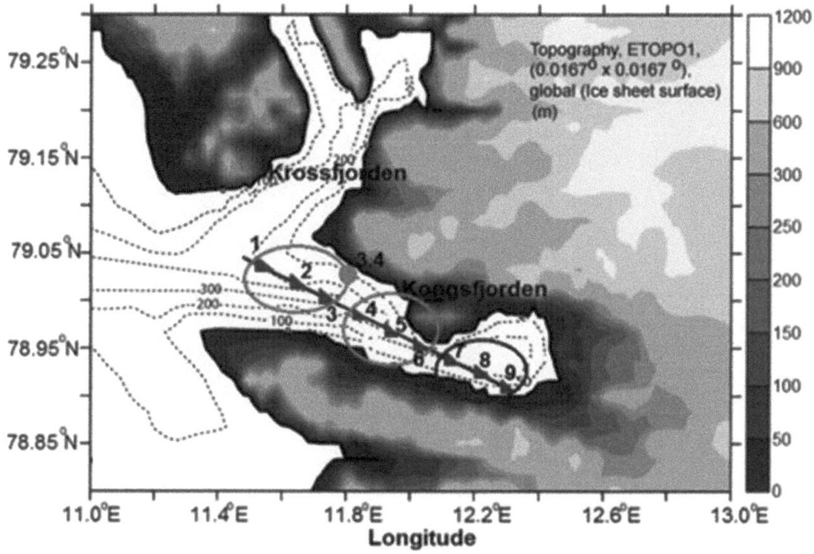

Figure 2.4 Sampling stations of Kongsfjorden for the long-term monitoring program. *Adapted from David T and Krishnan, 2017. Recent variability in the Atlantic water intrusion and water masses in Kongsfjorden, an Arctic fjord. Polar Sci. 11, 30-41.*

During summer, freshwater discharge in the form of precipitation, river runoff, glacier ablation, and ice melt contributes to the water column (Noufal et al., 2017). An increased volume of warm Atlantic water was found inside the fjords during winter seasons. The observed dynamics could have a significant impact on diversity and distribution of microbiota and their associated biogeochemical processes.

One of the significant accomplishments in the monitoring program was establishing a year-round ocean observation platform for measuring the water column's physical and chemical properties. During 2014, a team of scientists from the NCPOR and National Institute of Ocean Technology successfully deployed IndARC (Fig. 2.5), the country's first multisensor moored observatory in the Kongsfjorden (78°57′N 12°01′E). These sensors are programmed to collect real-time data on seawater temperature, salinity, current, and other vital biochemical parameters of the fjord (Venkatesan et al., 2016). The data acquired would be of critical importance to climate researchers and the international fraternity to understand the dynamics of Kongsfjorden. Further, deployment has also proven the capability of Indian researchers to design, develop, deploy, and maintain moored ocean observatories in the Polar regions.

Figure 2.5 Deployment of IndARC mooring ocean observatory in Kongsfjorden, Ny-Ålesund.

2.6 Microbial community composition of Kongsfjorden system

Microbes are the primary drivers of any ecosystem and a precursor to the food web cycle, which play a critical role in the ecosystem's functioning. As a part of Kongsfjorden flagship program, monitoring microbial community dynamics is undertaken in Kongsfjorden. The microbial community composition was investigated by using high-throughput 16S and 18S rRNA gene amplicon sequencing. Studies have revealed that microbial community compositions are significantly different spatially within the systems inner and outer fjord (Sinha et al., 2017a,b). Further, within the microbial communities, the composition of free-living (0.2–1.2 Pm) and particle ($>1.2\ \mu m$) bacterial communities was investigated along with mapping their spatial variation between the outer and inner part of Kongsfjorden. Results suggest the clear spatial variation among free-living bacterial communities and particle-associated communities within the Kongsfjorden system (Jain and Krishnan, 2017; Jain et al., 2019). Further studies have also provided significant insights into microbes role in degrading the complex organic substrates.

Studies have also investigated subsurface microbial communities and their bioprospecting potentials. Several bacterial strains belonging to members of *Colwellia aestuarii, Marinobacter maritimus, Photobacterium frigidiphilum, Pseudoalteromonas nigrifaciens, Pseudoalteromonas tetraodonis, Pseudoalteromonas translucida, Pseudomonas simiae, Psychrobacter aquimaris, Zobellia laminariae, Brevundimonas vesicularis, Pseudoalteromonas paragorgicola*, and *Pseudoalteromonas prydzens* were isolated by using different types of media from the fjord sediments and are being screened for cold-active lipases, amylases, and proteases (Srinivas et al., 2009; Prasad et al., 2014). Studies related to the effect of anthropogenic activities have also been undertaken (Nandan et al., 2013; Kabeer et al., 2019). Evidence of heavy metal pollution in the Arctic environment has been the topic of intense research for the past few decades. Several Arctic inland ecosystems have naturally high concentrations of heavy metals. The elevated levels of heavy metals in the food chain can pose a severe threat to the balance of the Arctic ecosystem (Mohan et al., 2017). Metal-tolerant bacteria are used as bioindicators of the polluted environment. Bacterial strains were tested for their ability to tolerate toxic metals such as mercury (Hg, 1–100 ppm), lead (Pb, 10–100 ppm), cadmium (Cd, 1–100 ppm), iron (Fe, 500–5000 ppm), manganese (Mn, 500–5000 ppm), nickel (Ni, 50–200 ppm), cobalt (Co, 50–200 ppm), and copper (Cu, 10–100 ppm) (Balan et al., 2018) and several isolates were recovered that could tolerate heavy metals at higher concentrations.

2.7 Benthic studies

Benthic foraminifera are established indicators of climate change and have been used extensively to infer immediate past climatic variations. Studies carried out in Kongsfjorden system reported that abundance of foraminifera increases from the innermost fjord to the mouth in Kongsfjorden (Saraswat et al., 2018). A study by Shetye et al. (2011) revealed dominant benthic foraminifera belonging to members of *Nonionellina labradorica, Elphidium excavatum, Cassidulina reniforme, Quinqueloculina stalkeri*, and *Islandiella islandica* in the fjord systems. *N. labradorica* was found to be the predominant species in the outer fjord, whereas *E. excavatum* and *C. reniforme* were dominant in the inner fjord. Further the study also revealed melting water, glacial calving, and sedimentation rates affect the community composition of benthic foraminifera.

2.7.1 Terrestrial ecosystem

The terrestrial and marine ecosystems in the Arctic region are closely interconnected. Any implications in one of the systems have been shown to have adverse effects on the functioning of other ecosystems (Murphy et al., 2016). Particularly in the northern hemisphere, glaciers of Svalbard Archipelago are melting at record braking rates due to the rapid warming of oceans, which triggers the retreat of glaciers and thawing permafrost (with the formation of thermokarst) in the terrestrial ecosystems. Accelerated thermokarsts forming due to climate change have shown to impact hydrology, ecology, and biogeochemistry (Olefeldt et al., 2016). This scenario is being hypothesized to make two significant changes in the community dynamics of the ecosystem viz. (1) retreating of glaciers provides a unique opportunity for the colonization of microbes to succession at higher trophic level (vegetation shifts in the glacier forelands) and (2) heavier flows of freshwater runoff, resulting from glacial retreat can boost the amount of sediment delivered into the fjords, which in turn alters the biogeochemical dynamics of the fjord ecosystems including the community composition of marine life forms. In both said cases, the community-level changes in the ecosystem are evident, along with the associated biogeochemical processes within the ecosystem. Several novel bacteria and yeast including *Winogradskyella psychrotolerans* (Begum et al., 2013), *Cyclobacterium qasimii* (Shivaji et al., 2012), *Arcticibacter svalbardensis* (Prasad et al., 2013), *Iodobacter arcticus* (Srinivas et al., 2013), *Subsaxibacter sediminis* (Sharma et al., 2018), *Rhodotorula svalbardensis* (Singh et al., 2014) have been isolated from different habitats of Arctic terrestrial environment. Culture-based studies from different habitats such as glaciers, cryoconite holes, and glacier forelands systems have been carried out in the past to isolate industrially critical microbial strains for biopharmaceutical and enzyme industries (Reddy et al., 2009; Srinivas et al., 2009; Pathan et al., 2010; Mohan Singh et al., 2011; Singh et al., 2013). At present, studies are being carried out to delineate critical components of various terrestrial ecosystems to unravel microbial community composition and their role in the functioning of systems by using recent advancements in genome sequencing techniques. A recent study by Thomas et al. (2019) investigated the bacterial community composition of Vestre Broggerbreen glacier, its proglacial channels, and the associated downstream fjord ecosystem to understand the transport of microbes from glacier to fjord system.

2.8 India's engagement in Arctic-related international scientific bodies

The Ministry of Earth Sciences, Government of India through its autonomous institute and NCPOR have entered into an MoU with the Norwegian Polar Institute for scientific and logistic cooperation Arctic. Also, India in the form of NCPOR is already a member of the Ny-Ålesund Science Managers Committee (NySMAC)—the apex body which oversees enhanced cooperation and coordination among research activities at Ny-Ålesund, since 2008. India has also been admitted to the International Arctic Science Committee (IASC) and is a part of the several scientific working groups. India is represented in the terrestrial, marine, cryosphere, atmosphere, and social and human working groups since 2012. Of late, NCPOR has also become an associate partner in the Svalbard Integrated Arctic Earth Observing System (SIOS)—the Norwegian initiative for integrated studies of the Svalbard archipelago, which forms a part of the roadmap of the European Strategy Forum on Research Infrastructures (ESFRI). NCPOR is also a council member of the University of Arctic. India's bid for observer status in the Arctic Council was accepted in May 2013, along with five other countries—China, Italy, Japan, South Korea, and Singapore at a meeting of the Arctic Council in Sweden.

2.9 Need for India to enhance its present engagement in Arctic affairs

Today, when the Arctic is growing both in environmental and geopolitical relevance, it would be most prudent to consider the importance of the region. While it may be unable to benefit from it in the same way as other Asian nations, certain efforts would definitely yield results in a sustainable way. It will be wise to consider the Arctic as "the common heritage of mankind." There is adequate room to work within this framework both in the Arctic Council and in the region. With measures focused on increased cooperation and collaboration with other nations, it will improve its current research facilities, economic activities, and relations. India's environmental focus will receive international recognition, thereby preventing climate change issues from being downplayed by other nations' political and strategic focus.

Moreover, tangible benefits are likely to mount up, such as in collaboration, even in scientific research's limited capacity. There could be multiple scenarios that arise in the region with global implications, but what is certain is that the Indian understanding of the Arctic will be enhanced with its presence in the region. Over the years, India has demonstrated its capability to be a world leader in the responsible use of nature and its resources. This could be well extended to the Arctic for its sustainable use and development. As an observer in the Arctic Council and as a nation that visions to engage in an absolute position in global governance, India could utilize the meaningful platform provided by developments in the Arctic region to display its competence in areas outside of its immediate neighborhood.

2.10 India's future plans for Arctic research

The NCPOR under the aegis of Ministry of Earth Sciences, Government of India is continuously expanding its Arctic operations to undertake research activities in different geographical locations with new scientific domains and subdisciplines through various forms of collaborative network programs. An agreement has been signed with SIOS to share and disseminate the observation data among its partners, to operate field expeditions cost-effectively. An MoU has been signed with Polar knowledge Canada to access its field station, Canadian High Arctic Research Station (CHARS) to initiate research activities in the Canadian Arctic (Cambridge Bay).

Acknowledgments

The authors wishes to thank the Secretary, Ministry of Earth Sciences, Government of India and Director, National Centre for Polar and Ocean Research for the support provided for implementation of Indian Arctic program.

References

Balan, B.M., Shini, S., Krishnan, K.P., Mohan, M., 2018. Mercury tolerance and biosorption in bacteria isolated from Ny-Ålesund, Svalbard, Arctic. J. Basic Microbiol. 58, 286–295.

Begum, Z., Srinivas, T., Manasa, P., Sailaja, B., Sunil, B., Prasad, S., et al., 2013. *Winogradskyella psychrotolerans* sp. nov., a marine bacterium of the family

Flavobacteriaceae isolated from Arctic sediment. Int. J. Syst. Evol. Microbiol. 63, 1646–1652.

David, T.D., Krishnan, K.P., 2017. Recent variability in the Atlantic water intrusion and water masses in Kongsfjorden, an Arctic fjord. Polar Sci. 11, 30–41.

Fadeev, E., Salter, I., Schourup-Kristensen, V., Nöthig, E.-M., Metfies, K., Engel, A., et al., 2018. Microbial communities in the east and west Fram strait during sea ice melting season. Front. Mar. Sci. 5, 429.

Gogoi, M.M., Babu, S.S., Moorthy, K.K., Thakur, R.C., Chaubey, J.P., Nair, V.S., 2016. Aerosol black carbon over Svalbard regions of Arctic. Polar Sci. 10, 60–70.

Gogoi, M.M., Babu, S.S., Pandey, S.K., Nair, V.S., Vaishya, A., Girach, I.A., et al., 2018. Scavenging ratio of black carbon in the Arctic and the Antarctic. Polar Sci. 16, 10–22.

Jain, A., Krishnan, K.P., 2017. Differences in free-living and particle-associated bacterial communities and their spatial variation in Kongsfjorden, Arctic. J. Basic Microbiol. 57, 827–838.

Jain, A., Krishnan, K.P., Singh, A., Thomas, F.A., Begum, N., Tiwari, M., et al., 2019. Biochemical composition of particles shape particle-attached bacterial community structure in a high Arctic fjord. Ecol. Indic. 102, 581–592.

Kabeer, F.A., Jabir, T., Krishnan, K., Abdulla, M.H., 2019. Metagenomic data of fungal community in Kongsfjorden, Arctic using Illumina next generation sequencing. Data Brief. 22, 195–198.

Mohan, M., Chandini, P., Krishnan, K., Gopikrishana, V., Sajinkumar, K., Kannan, V., 2017. Mercury Fractionation in the Sediments of Kongsfjorden, an Arctic Fjord, Svalbard. Int. J. Mar. Sci. 7.

Mohan Singh, S., Sahab Yadav, L., Kumar Singh, S., Singh, P., Nath Singh, P., Ravindra, R., 2011. Phosphate solubilizing ability of two Arctic *Aspergillus niger* strains. Polar Res. 30, 7283.

Murphy, E.J., Cavanagh, R.D., Drinkwater, K.F., Grant, S.M., Heymans, J., Hofmann, E. E., et al., 2016. Understanding the structure and functioning of polar pelagic ecosystems to predict the impacts of change. Proc. R. Soc. B: Biol. Sci. 283, 20161646.

Nanda, D., A. Council, 2019. India's Arctic Potential. Occasional Papers 186.

Nandan, S.B., Asha, C., Vijay, A., Jayachandran, P., 2013. Benthic faunal assemblage-diversity and time scale changes in the ice cold Kongsfjorden Arctic ecosystem. Clim. Change Himal. Inform. 183.

Noufal, K., Najeem, S., Latha, G., Venkatesan, R., 2017. Seasonal and long term evolution of oceanographic conditions based on year-around observation in Kongsfjorden, Arctic Ocean. Polar Sci. 11, 1–10.

Nuncio, M., Chatterjee, S., Satheesan, K., Chenoli, S.N., Subeesh, M.P., 2020. Temperature and precipitation during winter in NyÅlesund, Svalbard and possible tropical linkages. Tellus A: Dyn. Meteorol. Oceanogr. 72, 1–15.

Olefeldt, D., Goswami, S., Grosse, G., Hayes, D., Hugelius, G., Kuhry, P., et al., 2016. Circumpolar distribution and carbon storage of thermokarst landscapes. Nat. Commun. 7, 1–11.

Parmentier, F.-J.W., Christensen, T.R., Rysgaard, S., Bendtsen, J., Glud, R.N., Else, B., et al., 2017. A synthesis of the arctic terrestrial and marine carbon cycles under pressure from a dwindling cryosphere. AMBIO 46, 53–69.

Patel, L., Sharma, P., Thamban, M., 2019. Spatio-temporal variability of snow water equivalent over the Vestre Broggerbreen and Feiringbreen glaciers, Ny-Ålesund, Svalbard. J. Earth Syst. Sci. 128, 183.

Pathan, A.A.K., Bhadra, B., Begum, Z., Shivaji, S., 2010. Diversity of yeasts from puddles in the vicinity of midre lovenbreen glacier, arctic and bioprospecting for enzymes and fatty acids. Curr. Microbiol. 60, 307–314.

Polyakov, I.V., Pnyushkov, A.V., Alkire, M.B., Ashik, I.M., Baumann, T.M., Carmack, E.C., et al., 2017. Greater role for Atlantic inflows on sea-ice loss in the Eurasian Basin of the Arctic Ocean. Science 356, 285–291.

Prabhu, A., Mahajan, P., Khaladkar, R., 2012. Association of the Indian summer monsoon rainfall variability with the geophysical parameters over the Arctic region. Int. J. Climatol. 32, 2042–2050.

Prasad, S., Manasa, B.P., Buddhi, S., Pratibha, M.S., Begum, Z., Bandi, S., et al., 2013. *Arcticibacter svalbardensis* gen. nov., sp. nov., of the family Sphingobacteriaceae in the phylum Bacteroidetes, isolated from Arctic soil. Int. J. Syst. Evol. Microbiol. 63, 1627–1632.

Prasad, S., Manasa, P., Buddhi, S., Tirunagari, P., Begum, Z., Rajan, S., et al., 2014. Diversity and bioprospective potential (cold-active enzymes) of cultivable marine bacteria from the subarctic glacial fjord, Kongsfjorden. Curr. Microbiol. 68, 233–238.

Rajan, S., Krishnan, K., 2016. India's scientific endeavours in the Arctic. Asia and the Arctic. Springer, pp. 43–48.

Reddy, P.V.V., Rao, S.S.S.N., Pratibha, M.S., Sailaja, B., Kavya, B., Manorama, R.R., et al., 2009. Bacterial diversity and bioprospecting for cold-active enzymes from culturable bacteria associated with sediment from a melt water stream of Midtre Lovénbreen glacier, an Arctic glacier. Res. Microbiol. 160, 538–546.

Saraswat, R., Roy, C., Khare, N., Saalim, S.M., Kurtarkar, S.R., 2018. Assessing the environmental significance of benthic foraminiferal morpho-groups from the northern high latitudinal regions. Polar Sci. 18, 28–38.

Sharma, A., Jani, K., Feng, G.-D., Karodi, P., Vemuluri, V.R., Zhu, H.-H., et al., 2018. *Subsaxibacter sediminis* sp. nov., isolated from Arctic glacial sediment and emended description of the genus *Subsaxibacter*. Int. J. Syst. Evol. Microbiol. 68, 1678–1682.

Shetye, S., Mohan, R., Shukla, S.K., Maruthadu, S., Ravindra, R., 2011. Variability of *Nonionellina labradorica* Dawson in surface sediments from Kongsfjorden, West Spitsbergen. Acta Geol. Sin.-Eng. Ed. 85, 549–558.

Shivaji, S., Reddy, P.V.V., Rao, S.N., Begum, Z., Manasa, P., Srinivas, T., 2012. *Cyclobacterium qasimii* sp. nov., a psychrotolerant bacterium isolated from Arctic marine sediment. Int. J. Syst. Evol. Microbiol. 62, 2133–2139.

Singh, P., Tsuji, M., Singh, S.M., Roy, U., Hoshino, T., 2013. Taxonomic characterization, adaptation strategies and biotechnological potential of cryophilic yeasts from ice cores of Midre Lovénbreen glacier, Svalbard, Arctic. Cryobiology 66, 167–175.

Singh, P., Singh, S.M., Tsuji, M., Prasad, G.S., Hoshino, T., 2014. *Rhodotorula svalbardensis* sp. nov., a novel yeast species isolated from cryoconite holes of Ny-Ålesund, Arctic. Cryobiology 68, 122–128.

Sinha, R.K., Krishnan, K.P., Hatha, A.A.M., Rahiman, M., Thresyamma, D.D., Kerkar, S., 2017a. Diversity of retrievable heterotrophic bacteria in Kongsfjorden, an Arctic fjord. Braz. J. Microbiol. 48, 51–61.

Sinha, R.K., Krishnan, K.P., Kerkar, S., Thresyamma, D.D., 2017b. Influence of glacial melt and Atlantic water on bacterioplankton community of Kongsfjorden, an Arctic fjord. Ecol. Indic. 82, 143–151.

Srinivas, T., Rao, S.N., Reddy, P.V.V., Pratibha, M., Sailaja, B., Kavya, B., et al., 2009. Bacterial diversity and bioprospecting for cold-active lipases, amylases and proteases, from culturable bacteria of Kongsfjorden and Ny-Ålesund, Svalbard, Arctic. Curr. Microbiol. 59, 537–547.

Srinivas, T., Manasa, P., Begum, Z., Sunil, B., Sailaja, B., Singh, S., et al., 2013. *Iodobacter arcticus* sp. nov., a psychrotolerant bacterium isolated from meltwater stream sediment of an Arctic glacier. Int. J. Syst. Evol. Microbiol. 63, 2800–2805.

Thomas, F.A., Sinha, R.K., Krishnan, K., 2019. Bacterial community structure of a glacio-marine system in the Arctic (Ny-Ålesund, Svalbard). Sci. Total Environ. 135264.

Venkatesan, R., Krishnan, K., Arul Muthiah, M., Kesavakumar, B., Divya, D.T., Atmanand, M., et al., 2016. Indian moored observatory in the Arctic for long-term in situ data collection. Int. J. Ocean Clim. Syst. 7, 55–61.

Walczowski, W., Beszczynska-Möller, A., Wieczorek, P., Merchel, M., Grynczel, A., 2017. Oceanographic observations in the Nordic Sea and Fram Strait in 2016 under the IO PAN long-term monitoring program AREX. Oceanologia 59, 187–194.

CHAPTER 3

Lakes and fjords of Polar regions—potential indicators of climate change

Shabnam Choudhary[1,2], G.N. Nayak[3] and Neloy Khare[1]
[1]Ministry of Earth Sciences, Govt. of India, New Delhi, India
[2]National Centre for Polar and Ocean Research (NCPOR), Goa, India
[3]Marine Sciences, School of Earth, Ocean and Atmospheric Sciences, Goa University, India

3.1 Introduction

Polar regions are characterized by various glacial landforms out of which lakes and fjords have great potential to unravel sedimentary processes, depositional environment, and climate change. Despite the occurrence of numerous lakes and fjords along the coastlines of high-latitude regions (Hjelstuen et al., 2009) and their potential to provide an understanding of sedimentary and depositional processes, these landforms have received considerably less attention.

Lakes and fjords are local response systems, which act as high-resolution archives of local and regional change. They help in understanding the source of sediment and their depositional environment. The lakes integrate the processes of their surrounding catchments, and therefore they are susceptible to the changes in the landscape. High-latitude lakes (Arctic) and their catchments experience persistent low temperature, intense seasonality, and severe freeze–thaw cycles. These characteristics are likely to increase their sensitiveness to landscape and environmental change. A large variety of lakes exist in the Arctic region, and one such lake is depicted in Fig. 3.1. When these lakes were compared with freshwater lakes of other areas, it was found that these lakes are subjected to different physical, chemical, and biological processes. Lake type, water chemistry, and physical mixing characteristics of these lakes make them different from the lakes of other parts of the globe. The long period of ice cover and complete darkness coupled with several months of 24-hour light during the summer is one of the distinguishing features of these lakes.

Understanding Present and Past Arctic Environments
DOI: https://doi.org/10.1016/B978-0-12-822869-2.00007-4

Figure 3.1 Lake environment in Ny-Ålesund, Svalbard.

During this period, lakes are exposed to the atmosphere that influences the biogeochemistry of its ecosystem. Therefore the extent and thickness of ice cover in the catchment area greatly influence the biogeochemistry of Arctic lakes (Wharton et al., 1993; Fritsen and Priscu, 1999). Conversely, in lower latitudes, terrestrial inputs through river discharge, saline water influence, plenty of vegetation, and considerable anthropogenic impacts play an important role in regulating the ecosystem of the lakes (Padma and Periakali, 1999; Jeelani and Shah, 2006; Chakraborty et al., 2015; Nazneen and Raju, 2017). The sedimentation in high-latitude lakes is primarily a product of sediment input through glacial meltwater and biological productivity.

High-latitude lakes have many similar characteristics with those of temperate latitude and are subjected to similar landscape controls. These lakes are modern day analogues of paraglacial and periglacial lakes that were common during glacial periods in temperate latitudes (Phartiyal et al., 2011). The changes in the lakes and their catchments during the past are well archived in sedimentary records. Sediment column of a modern lake indicates that the sediment deposited at its bottom consists of relatively older sediments than the surface. Therefore a sediment column represents the deposition of sediments in a span of time. These lakes have gained importance in recent years as they are a source for paleoarchives, present pristine conditions, and are easily accessible in ice-free areas. In general, lake sediments are ideal repositories of eolian and fluvial materials. Lakes preserve catchment area soil, which is a product of weathering and is used to track past climatic and environmental changes (Peck et al., 2004). The study on lakes along with fjords gained importance in the recent years as the climate change has severely impacted the terrestrial, marine, and freshwater ecosystems in the Polar regions (IPCC, 2014).

A fjord is a U-shaped glacial valley formed by the glacier advance and retreat (Fig. 3.2) and is located in high-latitude regions from 40°N to 40° S. They act as a nexus between the terrestrial (source) and the oceanic domain (sink), providing the potential for continuous exchange between the fjord and the coastal waters on the adjacent shelf (Cottier et al., 2005). These landforms did not receive much attention of sedimentologists despite their great potential in unraveling the character of different sedimentary processes and mechanisms responsible for the abrupt and long-term changes in sedimentary processes.

One such fjord system studied in Arctic region is Krossfjord–Kongsfjord system situated on the west coast of the Svalbard Archipelago, equilibrates Atlantic, Arctic, brine, and freshwater input, which are the indicators of environmental changes (Nilsen et al., 2008). This fjord system is mainly affected by the Western Spitsbergen Current, which brings in warm and saline Atlantic water at one side. From the other side, it is affected by extensive glacial runoff from different tidewater glaciers. Rapid sedimentation in the fjord system has resulted in the sedimentary record of high resolution (Griffith and Anderson, 1989), which can be used to reconstruct past environmental changes. Factors such as seafloor topography, oceanographic conditions, sea ice availability, size of the drainage area, proximity to the source of sediment,

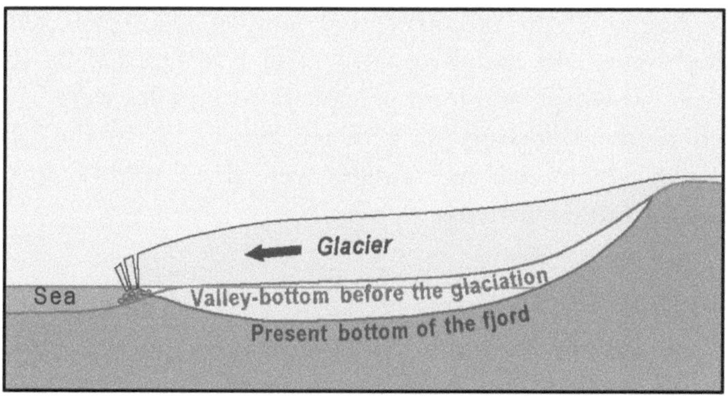

Figure 3.2 Formation of a fjord. Courtesy: https://commons.wikimedia.org.

and climate have controlled the sedimentation in fjords (Munoz and Wellner, 2016). In addition to these factors, glaciers have also contributed by eroding and transporting lot of terrigenous influx to the fjords. Therefore glacial fjord system has been regarded as a natural laboratory for the study of climate change in the high-latitude regions (Fendeng et al., 2018).

The sediment characteristics such as grain size, total organic carbon (TOC), total nitrogen (TN), total phosphorus (TP), and biogenic silica (BSi) were used earlier as proxies in order to understand the biogeochemistry, productivity, and environmental behavior of the lakes and surrounding ocean (Mortlock and Froelich, 1989; Muller and Schneider, 1993; Colman et al., 1995) and major and trace elements have been used to elucidate the paleo-processes that occurred within a lake and fjord system. The present study focuses on understanding the sedimentary processes, depositional environment, and climate change through the sediments of lakes and fjords. Proxies such as grain size are used to understand the source, transportation, and depositional processes. Organic element proxies such as organic carbon, nitrogen, phosphorus and their ratios are used to identify the source. Vertical profiles of major and trace elements provide information on weathering, depositional, and postdepositional processes and mobilization of metals.

3.2 Materials and methods

3.2.1 Study area and sample collection

The samples were collected from Ny-Ålesund, the International Arctic research base, located in the Spitsbergen Island of Svalbard archipelago (Fig. 3.3) as part of the Indian Arctic program (summer phase) during the month of August 2016. Sediment core samples were collected from a shallow freshwater lake and Krossfjord. The lake was fed by meltwater from glaciers, snowdrift, and local precipitation. Samples were labeled and transported to the laboratory in a frozen condition. In the laboratory, each core was subsampled, and samples were dried at 60°C in the oven and utilized for further analysis.

3.2.2 Sample analysis

For grain-size analysis portion of the sediment sample was washed with distilled water to check and remove the salinity. Then 10% sodium hexametaphosphate was added to dissociate clay particles and hydrogen

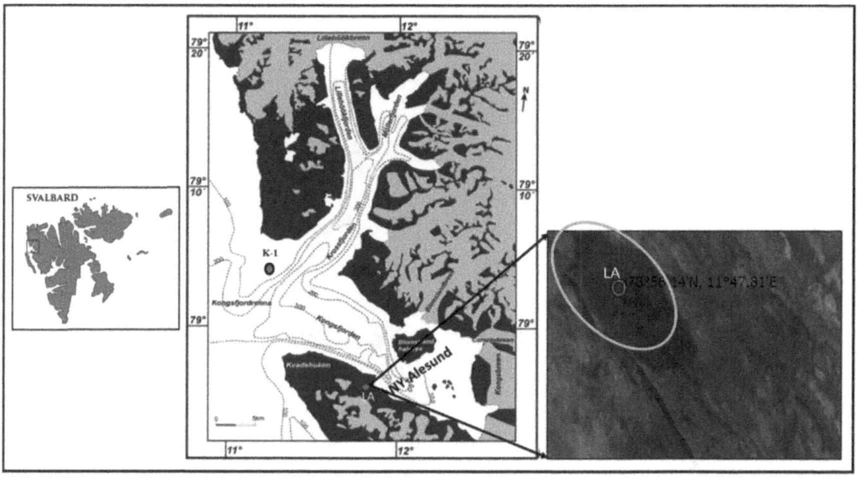

Figure 3.3 Map showing the study area. *Modified after Svendsen, H., Beszczynska-Moller, A., Hagen, J.O., Lefauconnier, B., Tverberg, V., Gerland, S., et al., 2002. The physical environment of Kongsfjorden–Krossfjorden, an Arctic fjord system in Svalbard. Pollut. Res. 21 (1), 133–166.*

peroxide to oxidize the organic matter. The sediment sample was analyzed by the pipette method (Folk, 1974), which is based on Stoke's settling velocity principle. Another portion of each subsample was powdered and homogenized in an Agate mortar and used to determine total carbon (TC) using the elemental analyzer. The analytical precision was $\pm 0.30\%$ (1σ standard deviation), which was attained using sulfanilamide as the standard and repeated after every 10 samples. Total inorganic carbon (TIC) was analyzed using UIC carbon coulometer. TIC was subtracted from TC to obtain TOC. Calcium carbonate was computed as TIC *8.33. The freeze-dried sediment sample was also analyzed for TN concentration in the Marine Stable Isotope Lab (MASTIL) at National Centre for Antarctic & Ocean Research, Goa, India using an Elemental Analyzer (EA)—Isoprime, Vario Isotope Cube. The precision for N% was $\pm 0.63\%$ (1σ standard deviation) obtained by repeatedly running sulfanilamide as the standard. BSi from the freeze-dried sample was extracted using 25 mL of 1% Na_2CO_3 in an 85°C water bath for 5 hours and measured by the wet alkaline extraction method, modified by Mortlock and Froelich (1989) and Muller and Schneider (1993) where the intensity of blue silico-molybdenum complex was measured at 810 nm using UV-1800 (Shimadzu) visible spectrophotometer. Duplicate measurements were conducted on each sample, and relative error was noted to be less than 3%. The sediment sample was digested using HF:HNO_3:$HClO_4$ mixture for TP analysis and brought to liquid phase as adopted by Yu et al. (2013) and further determined following the procedure given by Murphy and Riley (1962) where the intensity of phosphomolybdenum blue complex was measured at 880 nm using UV-1800 (Shimadzu) visible spectrophotometer. The accuracy of phosphorus analysis was determined using a digested sample of JLK-1, and relative error was noted to be less than 4%.

Further, 0.2-g ground sediment samples were digested in Teflon beakers using HF, HNO_3, and $HClO_4$ acid mixture (Jarvis and Jarvis, 1985) with a ratio of 7:3:1 for total metal analyses. The metals Fe, Mn, Al, Co, Zn, Ba, and Ti were analyzed using flame atomic absorption spectrophotometer and Cd, Pb, and Cr using graphite furnace atomic absorption spectrophotometer (Thermo Scientific-SOLAAR M6 AAS model).

3.2.3 Quality assurance and quality control

Together with the samples, certified reference standards JLK-1 from the Geological Survey of Japan were digested and run, to test the analytical accuracy of the method. The average recoveries were 94% for Ba and Ti; 95% for Mn and Co; 96% for Fe, Cu, Ni, and Al; and 99% for Zn and Pb. Merck's internal chemical standards were used to calibrate the instrument, and recalibration checks were performed at regular intervals.

3.3 Paleoenvironmental indicators

3.3.1 Sediment components (sand, silt, and clay)

Grain size is susceptible to changes in climate and remains unaffected by the biological activities. Therefore sediment grain size is a robust proxy for the reconstruction of climatic changes. Earlier studies have reported that gravel, sand, silt, and clay occurred and distributed in lacustrine sediments from the bank to the center, suggesting decreasing grain size with increasing water depth and decreasing hydrodynamic energy (Sly, 1978; Sun et al., 2001). Thus it is considered that a larger sediment grain size indicates a shorter distance of sampling location from the bank. While smaller sediment grain size suggests a farther distance of sampling location from the bank. The coarse-grained particles are transported from the catchment area into the lake basin by glacial meltwater as reported by

Tatur et al. (2004) and Yoon et al. (2006) indicating warmer conditions. Therefore a high percentage of coarser grain sediments deposition in the lake indicates warmer conditions in the region. Higher clay showed the influx of finer material to the lakes by ice meltwater suggesting lower melting in the region (Choudhary et al., 2018a,b). The statement "Coarser sediment grain size indicates warmer conditions and higher melting while finer sediments suggest relatively colder conditions and lower melting in the region," was also applied by Govil et al. (2011).

In core LA of the present study, collected from the lake, the upper portion of the core (Fig. 3.4A) consists of higher than the average values of sand suggesting the dominance of mechanical (glacial) weathering/frost weathering processes (Jiang et al., 2011) releasing coarse-grained material to the lake basin in the recent years. The sand grains exhibited angular to subangular and rounded to subrounded shape supporting their mechanical origin, as observed by Kar et al. (2018). Along with the glacial activity, eolian processes must have also played an important role in supplying finer sediments to the lake basin. Due to the warming conditions in the region, glaciers retreated and the area was with less ice allowing intense winds to

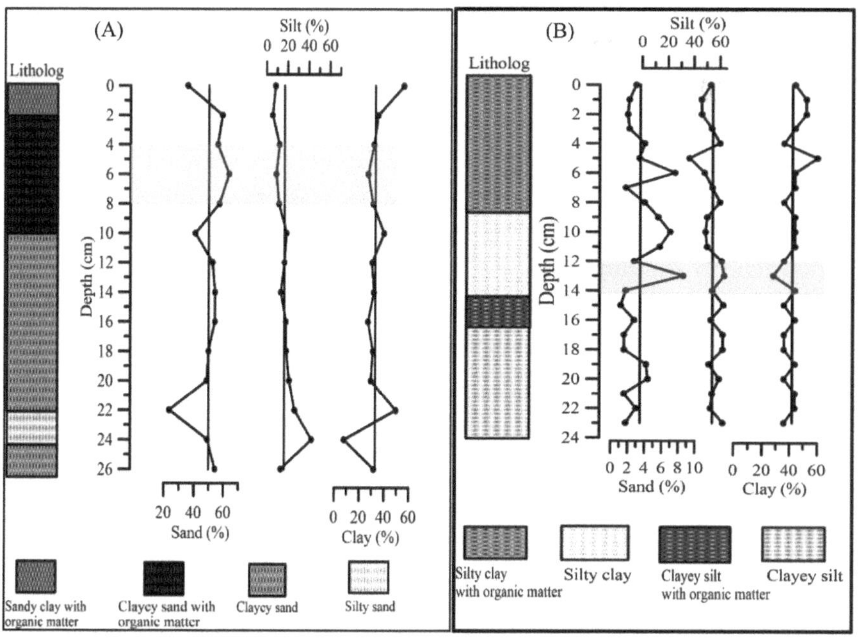

Figure 3.4 Variations of sediment components along with depth. (A) LA and (B) K-1.

rework the sediments (Kar et al., 2018). The lower portion of core LA consists of silt higher than the average value, which is attributed to the deposition of finer sediments by snowmelt water.

The rate of sedimentation from Knudsenheia Lake based on ^{210}Pb was determined to be 0.01–0.03 cm per year (Jiang et al., 2011) and is applied to the present data set using the formula given in Eq. (3.1), though the sedimentation rate from one lake to the other varies considerably. Core LA of length 26 cm represented 1040 years of sediment deposition, that is, from year 976 to 2016.

$$\text{Age} = \frac{\text{Depth}}{\text{Sedimentation rate}} \tag{3.1}$$

In core K-1 collected from the Krossfjord, the lower portion of the core (Fig. 3.4B) from 24 to 14 cm depth, sand showed lower than the average values with overall decreasing trend while silt showed higher than the average values indicating deposition of fine-grained sediment through suspension mode. In the middle section of the core from 14 to 6 cm depth, sand showed higher than the average value attributed to the warming conditions in the region leading to the melting of the icebergs. This released coarse-grained material trapped at the bottom of the icebergs (Dowdeswell and Forsberg, 1992) must have transported to the fjord as it is the main mechanism transporting coarse-grained material to the deeper waters (Dowdeswell et al., 1998). Further, in the upper portion from 6 cm to the surface sand showed lower than average values with the increasing trend toward the surface. In contrast, clay showed higher than the average values with decreasing toward the surface. The higher concentration of clay in the upper portion as compared to the lower sections may be due to the temperature-salinity stratification resulting in aggregation of finer particles and/or due to the hydrodynamic conditions responsible for transportation of fine-grained sediments from offshore as suggested by Zaborska et al. (2006).

Based on ^{210}Pb the rate of sedimentation from Kongsfjord was determined to be 0.1–0.5 cm per year (Zaborska et al., 2010; Kumar et al., 2018) and mean sedimentation rate (0.3 cm per year) is applied to the present data set, though the sedimentation rate from one fjord to the other varies considerably. Core K-1 of length 24 cm represented 80 years of sediment deposition, that is, from the year 1936 to 2016. Based on these chronology climate conditions with time for core LA and K-1 developed is presented in Table 3.1.

Table 3.1 Characteristic features, estimated dates, and climatic variations for core LA and K-1.

Sampling site	Water depth (m)	Sample interval (cm)	Characteristic features	Estimated dates (YBP)	Climatic variations
Core LA	4 m	0–2	High clay, TOC, Pb, Cd, and Zn	80-Recent	Recent warming
		2–10	High sand, TOC, Cu, Pb, and Cd	400	Little Ice Age (LIA)
		10–22	High sand, TOC, TP, Al, Ti, Fe, Mg, Cr, Cu, Ba, and Zn	880	
		22–24	High clay, Al, Ti, Fe, Co, Ni, Ba, and Zn	960	
		24–26	High silt	1040	
Core K-1	260.2 m	0–9	High clay and almost all the elements	30-Recent	Recent warming with climate fluctuations after the termination of LIA
		9–14	High sand, silt, TOC, TN, C/N and almost all the elements	46	
		14–16	High silt, TOC, TN with organic matter Almost all the elements low	53	
		16–24	Fluctuating silt, high TN Low concentrations of all the elements	80	

3.3.2 Organic components (TOC, TN, and TP)

Organic element proxies such as organic carbon, nitrogen, and phosphorus are indicators of the status of past productivity. C/N ratio is a widely used indicator of the source of the organic matter and helps to understand the biogeochemistry of the environment. Depth-wise distribution of the organic elements is presented in Fig. 3.5. In core LA, TOC showed lower than the average values in the lower depths range from 26 to 20 cm and higher than the average values in the upper portion from 8 to 6 cm and at 2 cm with an overall increasing trend toward the surface (Fig. 3.5A). TN

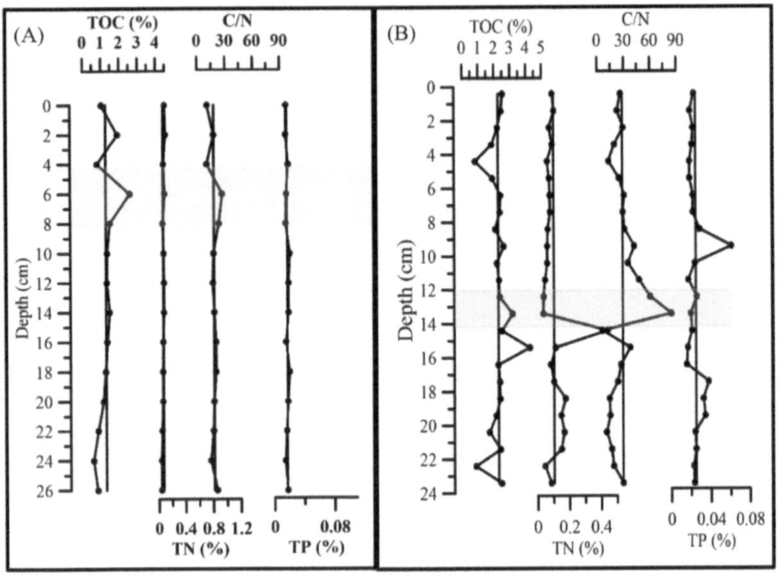

Figure 3.5 Depth-wise distribution of organic components of cores (A) LA and (B) K-1.

and TP showed uniform distribution throughout the core. C/N ratio fluctuated around the average line up to a depth of 6 cm, and further, it showed lower than the average values suggesting the source of organic matter to be autochthonous.

In core K-1, TOC fluctuated around the average line throughout the core. TOC showed negative peaks at depth 22 and 4 cm similar to that of silt and clay, respectively, and a positive peak at 15 cm similar to that of silt indicating that these sediment components regulated the distribution of organic matter at these depths. Further, TOC showed an increasing trend from 4 cm to the surface similar to that of clay toward the surface, suggesting increased primary productivity. Fluctuating trend with depth indicates changing rate of supply of organic matter through changing processes (Choudhary, 2019). Organic matter concentration in the sediments of fjord mainly depends on the location of sampling. Whether the sampling location lies in the proximity of high primary production areas or the glaciers are available in the vicinity, supplying land-derived material. The high concentration of land-derived material causes high turbidity resulting in the shallow photic zone. Regions receiving a high amount of freshwater discharge from glaciers and rivers fed by glaciers, generally, have low primary production due to the high turbidity (Gorlich et al., 1987; Zajączkowski and Włodarska-Kowalczuk, 2007). The marine organic matter gets diluted with suspended material discharged by glaciers. Therefore glaciers play an important role in regulating the distribution of organic matter in the fjord system. TN and TP showed higher concentration in the lower depth range from 23 to 17 cm and an increasing trend toward the surface in the upper portion. TN showed a positive peak at 14 cm similar to that of clay and TP at 9 cm similar to that of TOC. TN does not show similarity with TOC in its distribution throughout the core indicating their different source.

C/N ratio of core K-1 varied from 5.60 to 84.03 (average 28.52) indicating a mixed source of organic matter derived from marine as well as terrestrial. C/N ratio showed lower than average values in the lower depth range from 23 to 14 cm while at 13 cm C/N ratio was found to be high (Fig. 3.5B), similar to that of sand indicating ice-free conditions and increased glacial meltwater, which must have delivered terrestrial organic matter to the fjord. The high primary production may lead to nitrogen limitation in the surface water. Also, the nitrogen loss in sediments can be possible due to diagenesis, which must have resulted in a relatively large range of C/N ratio. Further, it decreased up to 4 cm and then increased toward the surface.

3.3.3 Major and trace elements

Vertical profiles of major and trace metals provide information on weathering, depositional, and postdepositional processes and mobilization of metals as sediments is considered as a suitable medium for regulating the concentrations of metals. They perform an important function of buffering higher metal concentration by adsorption and precipitation. The elements in the sediments may be available either as silicates in the rock fragments or Fe-Mn oxides and hydroxides. Oxides/hydroxides have tendency to take up different elements either by incorporating them into their crystal lattice or by adsorbing on the surface (Burdige, 1993). In addition, variation in geomorphological and physicochemical parameters such as topography, weathering, hydrodynamic conditions, salinity, and climate change may also control the distribution of metals.

In core LA (Fig. 3.6A), from 24 to 18 cm elements such as Al, Ti, Fe, Co, Ni, Ba, and Zn showed higher than the average value. Mn and Ca remained constant. Elements such as Al, Ti, Fe, Cr, Co, Ni, Ba, and Zn

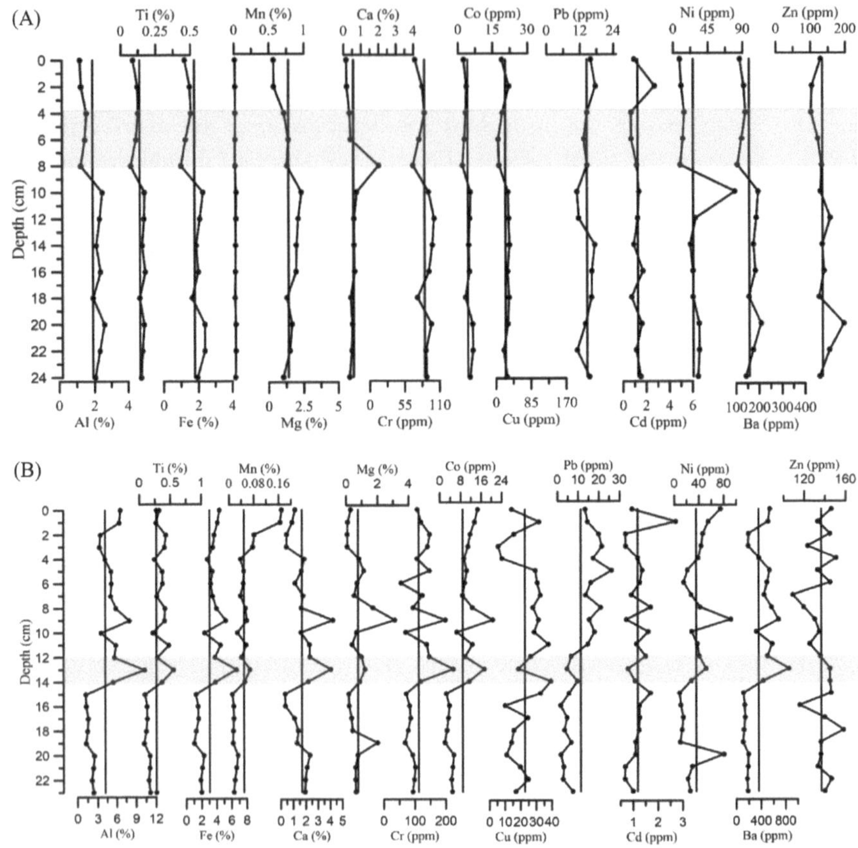

Figure 3.6 Distribution of elements Al, Ti, Fe, Mn, Mg, Ca, Cr, Co, Cu, Pb, Cd, Ni, Ba, and Zn in sediment with the depth of (A) core LA and (B) core K-1.

showed a decrease from 20 to 18 cm. From 18 to 8 cm, elements such as Al, Ti, Fe, Mg, Cr, Cu, Ba, and Zn showed higher than the average values. Mn, Co, and Cu showed an almost constant trend. In this zone, most of the elements showed an increase from 18 to 16 cm, and further remained constant up to 10 cm followed by a decrease up to 8 cm. Al, Ti, Fe, Cr, Co, Cu, Ba, and Ni showed an increasing trend from 8 to 4 cm followed by a decreasing trend. Ca and Mg showed a decreasing trend. Pb, Cd, and Zn showed decreasing trend up to 4 cm and then slightly increased. Overall, Al, Ti, and Fe showed a similar decreasing trend toward the surface similar to that of sand indicating their association with the coarser sediments.

In core K-1 (Fig. 3.6B) from 24 to 15 cm, all the elements are either constant or decrease. Further, from 15 to 8 cm almost all the elements except Mn and Zn showed higher concentration similar to that of sand, clay, and TOC suggesting these sediment components have regulated the distribution of these elements. From 8 cm to the surface, elements such as Ti, Cr, Cu, Pb, and Cd showed a decreasing trend while other elements viz. Al, Fe, Mn, Ca, Mg, Co, Ni, Zn, and Ba showed an increasing trend similar to that of silt, TOC, TN, TP, and BSi toward the surface due to increased influx of freshwater in recent years facilitating high concentration of elements along with the silt and organic elements.

3.4 Interpretation of paleoenvironmental changes in lakes and fjords

In lake LA, TOC and C/N ratios showed a peak between 8 and 4 cm coinciding with a higher concentration of sand and a lower concentration of almost all the metals except Ca, which corresponds to the Little Ice Age (LIA CE 1696−1856). The LIA was an extensive cooling event that took place in the Arctic region and has been supported by a large number of evidences from ice cores, marine sediments, and lake sediments (Birks et al., 2004; Isaksson et al., 2005). On the basis of δ18O records Isaksson et al. (2005) suggested that a significant cooling occurred in Svalbard from about CE 1500 to the end of the 1800s, and the most negative δ18O values (coldest temperatures) occurred between about CE 1760 and 1900 in Svalbard. According to Jiang et al. (2007), many records from marine sediments suggest a cooling of surface and bottom waters started between CE 1200 and 1350, followed by an intensified cooling after CE 1550 or 1600. High TOC and C/N ratio with high sand indicated cold climatic conditions suggesting relatively higher input of terrestrial organic matter resulted in a reduction in lake production. Katabatic winds can move sediments up to sand size and contribute to the deposition of coarse-grained particles to the lake (Doran et al., 2002). Sediment supply to the high-latitude lakes is mainly by glaciofluvial meltwater and intense wind (Katabatic winds) that helps in transporting coarse particles, which gets deposited on lake ice and percolates through cracks in the lake ice and get deposited on the lake floor (Spaulding et al., 1997).

Further, C/N ratio decreased and reached its lowest toward the surface suggesting the source of organic matter to be autochthonous, which agrees well with the termination of LIA as also observed by Jiang et al.

(2011) in Knudsenheia Lake. This is also supported by major and trace elements exhibiting low concentration at this time period. Further showed an increasing trend toward the surface, indicating increased melt-water influx due to warmer conditions in the region or diagenetic remobilization of metals in recent years.

In core K-1, sand, silt, TOC, TN, C/N, and almost all the elements showed a positive peak at 13 cm representing the time period CE 1973 indicating recent warming with climate fluctuations after the termination of LIA, which led to the melting in the region. Due to higher melting, high input of terrestrial material is observed in the fjord depositing coarser grain-size particles, terrestrial organic matter, and high concentration of metals from the weathering of rocks available in the catchment area.

The present study on sediment cores from Polar regions revealed that the sedimentation rate in the lakes was low while it was found to be high in fjords. Sedimentation rate in glacial fjords is classified as subpolar and found to be usually high, which can reach up to 10 cm per year (Zaborska et al., 2006). In these fjords, the rate of sedimentation is influenced by the rate of production of sediment from the glacier front, fjord walls and side streams, the topographic influence on glacier retreat causing sediment trapping and hydrological conditions as tides and currents (Zaborska et al., 2006). Core LA represents LIA and recent warming (1040 years), while core K-1 represents recent warming after the termination of LIA (80 years). Sediment column of a lake or fjord indicates that the sediment deposited at its bottom consists of relatively older sediments than the surface and a sediment column represents the deposition of sediments in a span of time. Therefore lakes and fjord system has been regarded as potential indicators of past climatic changes in the high-latitude regions.

3.5 Conclusions

A detailed study was carried out on sediment cores from the lake and mouth of the Kongsfjord from Svalbard, Arctic. On comparison of two cores from different environments, it is concluded that the sedimentation rate in the lakes is low as the lakes are a closed system and mostly affected by the snow available in the catchment area. Core LA of length 26 cm represented ~1040 years of sediment deposition. While the rate of sedimentation is found to be high in fjords as it is mainly affected by the sediments derived by the glacier front. Core K-1 of length 24 cm represented

80 years of sediment deposition, that is, from the year 1936 to 2016. In lake LA, TOC and C/N ratios showed a peak between 8 and 4 cm coinciding with higher sand concentration and a lower concentration of almost all the metals except Ca, which corresponds to the LIA (CE 1696–1856). In core K-1, sand, silt, TOC, TN, C/N, and almost all the elements showed a positive peak at 13 cm representing the time period CE 1973 indicating warming conditions after the termination of LIA. Thus lakes and fjords are potential indicators of climate change.

Acknowledgments

The authors place on record to thank the Director, National Centre for Polar and Ocean Research (NCPOR), Goa, for providing an opportunity to one of the authors (SC) to participate in Indian Arctic programme and Ministry of Earth Sciences (MoES) for providing the logistic support required for the collection of samples. One of the authors (SC) thanks the University Grant Commission (UGC) for awarding a national fellowship (F1-17.1/2015-16/MANF-2015-17-UTT-51739). Another author (GNN) thanks Council of Scientific and Industrial Research (CSIR) for awarding Emeritus Scientist position.

References

Birks, H.J.B., Jones, V.J., Rose, N.L., 2004. Recent environmental change and atmospheric contamination on Svalbard as recorded in lake sediment – an introduction. J. Paleolimnol. 31, 403–410.

Burdige, D.J., 1993. The biogeochemistry of manganese and iron reduction in marine sediments. Earth-Sci. Rev. 35, 249–284.

Chakraborty, P., Sarkar, A., Vudamala, K., Naik, R., Nath, B.N., 2015. Organic matter—a key factor in controlling mercury distribution in estuarine sediment. Mar. Chem. 173, 302–309.

Choudhary, S., 2019. Multiproxy Paleoclimate Reconstruction from the Sediments of High Latitude Regions (Unpublished Ph.D. thesis). Goa University, India.

Choudhary, S., Tiwari, A.K., Nayak, G.N., Bejugam, P., 2018a. Sedimentological and geochemical investigations to understand source of sediments and processes of recent past in Schirmacher Oasis, East Antarctica. Polar Sci. 15, 87–98.

Choudhary, S., Nayak, G.N., Tiwari, A.K., Khare, N., 2018b. Sediment composition and its effect on the productivity in Larsemann Hills, East Antarctica. Arab. J. Geosci. 11 (15), 416.

Colman, S.M., Peck, J.A., Karabanov, E.B., Carter, S.J., Bradbury, J.P., King, J.W., et al., 1995. Continental climate response to orbital forcing from biogenic silica records in Lake Baikal. Nature 378, 769–771.

Cottier, F., Tverberg, V., Inall, M.E., Svendsen, H., Nilsen, F., Griffiths, C., 2005. Water mass modification in an Arctic fjord through cross-shelf exchange: the seasonal hydrography of Kongsfjorden, Svalbard. J. Geophys. Res.: Ocean 110, 1–18.

Doran, P.T., McKay, C.P., Clow, G.D., Dana, G.L., Fountain, A.G., Nylen, T., et al., 2002. Valley floor climate observations from the McMurdo dry valleys, Antarctica, 1986–2000. J. Geophys. Res. 107, 4772.

Dowdeswell, J.A., Forsberg, C.F., 1992. The size and frequency of icebergs and bergy bits derived from tidewater glaciers in Kongsfjorden, Northwest Spitsbergen. Polar Res. 11, 81–91.
Dowdeswell, J.A., Elverhøi, A., Spielhagen, R., 1998. Glacimarine sedimentary processes on the polar north Atlantic margins. Quat. Sci. Rev. 17, 243–272.
Fendeng, S., Xeufa, S., Xin, S., Xisheng, F., Yenghua, W., Zhenbo, C., et al., 2018. Clay minerals in Arctic Kongsfjorden surface sediments and their implications on provenance and paleoenvironmental change. Acta Oceanol. Sin. 37 (5), 29–38.
Folk, R.L., 1974. Petrology of Sedimentary Rocks. Hemphill, Austin, TX, p. 182.
Fritsen, C.H., Priscu, J.C., 1999. Seasonal change in the optical properties of the permanent ice cover on Lake Bonney, Antarctica: consequences for lake productivity and phytoplankton dynamics. Limnol. Oceanogr. 44 (2), 47–454.
Gorlich, K., Weslawski, J.M., Zajaczkowski, M., 1987. Suspension settling effect on macrobenthos biomass distribution in the Hornsund Fjord, Spitsbergen. Polar Res. 5, 175–192.
Govil, P., Mazumder, A., Tiwari, A., Kumar, S., 2011. Holocene climate variability from lake sediment core in Larsemann Hills, Antarctica. Quat. Int. 235 (1), 128–136.
Griffith, T.W., Anderson, J.B., 1989. Climatic control of sedimentation in bays and fjords of the northern Antarctic Peninsula. Mar. Geol. 85, 181–204.
Hjelstuen, B.O., Haflidason, H., Sejrup, H.P., Lysa, A., 2009. Sedimentary processes and depositional environments in glaciated fjord systems—evidence from Nordfjord, Norway. Mar. Geol. 258 (1–4), 88–99.
IPCC, 2014. (Pachauri, R.K., Meyer, L., Plattner, G.K., Stocker, T.) Climate Change 2014: Synthesis Report. IPCC, Geneva.
Isaksson, E., Kohler, J., Pohjola Van, V., Moore, J., Igarashi, M., Karlof, L., et al., 2005. Two ice-core delta O-18 records from Svalbard illustrating climate and sea-ice variability over the last 400 years. Holocene 15, 501–509.
Jarvis, I., Jarvis, K.E., 1985. Rare-earth element geochemistry of standard sediments: a study using inductively coupled plasma spectrometry. Chem. Geol. 53 (3–4), 335–344.
Jeelani, G., Shah, A.Q., 2006. Geochemical characteristics of water and sediment from the Dal Lake, Kashmir Himalaya: constraints on weathering and their implications on provenance and paleoenvironmental change. Acta Oceanol. Sin. 37 (5), 29–38.
Jiang, H., Ren, J., Knudsen, K.L., Eiriksson, J., Ran, L.H., 2007. Summer sea-surface temperatures and climate events on the North Icelandic shelf through the last 3000 years. Chin. Sci. Bull. 52, 789–796.
Jiang, S., Liu, X., Sun, J., Yuan, L., Sun, L., Wang, Y., 2011. A multi-proxy sediment record of late Holocene and recent climate change from a lake near Ny-Ålesund, Svalbard. Boreas 40 (3), 468–480.
Kar, R., Mazumder, A., Mishra, K., Patil, S.K., Ravindra, R., Ranhotra, P.S., et al., 2018. Climatic history of Ny-Alesund region, Svalbard, over the last 19,000 year: Insights from quartz grain microtexture and magnetic susceptibility. Polar Sci. 18, 189–196.
Kumar, V., Tiwari, M., Rengarajan, R., 2018. Warming in the Arctic captured by productivity variability at an Arctic Fjord over the past two centuries. PLoS One 13 (8), 201456.
Mortlock, R.A., Froelich, P.N., 1989. A simple method for the rapid determination of biogenic opal in pelagic marine sediments. Deep Sea Res. Part A, Oceanogr. Res. Pap. 36 (9), 1415–1426.
Muller, P.J., Schneider, R., 1993. An automated leaching method for the determination of opal in sediments and particulate matter. Deep Sea Res. Part I: Oceanogr. Res. Pap. 40 (3), 425–444.

Munoz, Y.P., Wellner, J.S., 2016. Local controls on sediment accumulation and distribution in a fjord in the West Antarctic Peninsula: implications for paleoenvironmental interpretations. Polar Res. 35 (1), 25284.
Murphy, J., Riley, J.P., 1962. A modified single solution method for the determination of phosphate in natural waters. Anal. Chim. Acta 27, 31–36.
Nazneen, S., Raju, N.J., 2017. Distribution and sources of carbon, nitrogen, phosphorus and biogenic silica in the sediments of Chilika lagoon. J. Earth Syst. Sci. 126 (13).
Nilsen, F., Cottier, F., Skogseth, R., Mattsson, S., 2008. Fjord–shelf exchanges controlled by ice and brine production: the interannual variation of Atlantic Water in Isfjorden, Svalbard. Cont. Shelf Res. 28 (14), 1838–1853.
Padma, S., Periakali, P., 1999. Physico-chemical and geochemical studies in Pulicat lake, east coast of India. Indian J. Geo-Mar. Sci. 28, 434–437.
Peck, J.A., Green, R.R., Shanahan, T., King, J.W., Overpeck, J.T., Scholz, C.A., 2004. A magnetic mineral record of Late Quaternary tropical climate variability from Lake Bosumtwi, Ghana. Palaeogeogr. Palaeoclimatol. Palaeoecol. 215 (12), 37–57.
Phartiyal, B., Sharma, A., Bera, S.K., 2011. Glacial lakes and geomorphological evolution of Schirmacher Oasis, east Antarctica, during late Quaternary. Quat. Int. 235 (1), 128–136.
Sly, P.G., 1978. Sedimentary processes in lakes. In: Lerman, A. (Ed.), Lakes: Chemistry, Geology, Physics. Springer-Verlag, Berlin, p. 1978.
Spaulding, S.A., McKnight, D.M., Stoermer, E.F., Doran, P.T., 1997. Diatoms in sediments of perennially ice-covered Lake Hoare, and implications for interpreting lake history in the McMurdo dry valleys of Antarctica. J. Paleolimnol. 17 (4), 403–420.
Sun, Q.L., Zhou, J., Xiao, J.L., 2001. Grain-size characteristics of Lake Daihai sediments and its paleoenvironment significance. Mar. Geol. Quat. Geol. (Chin.) 21 (1), 93–95.
Svendsen, H., Beszczynska-Moller, A., Hagen, J.O., Lefauconnier, B., Tverberg, V., Gerland, S., et al., 2002. The physical environment of Kongsfjorden–Krossfjorden, an Arctic fjord system in Svalbard. Pollut. Res. 21 (1), 133–166.
Tatur, A., Valle, R.D., Barczuk, A., Martinez-Macchiavello, J., 2004. Records of Holocene environmental changes in terrestrial sedimentary deposits on King George Island, Antarctica; a critical review. Ocean Polar Res. 26 (3), 531–537.
Wharton, R.A., Lyons, W.B., Des Marais, D.J., 1993. Stable isotopic biogeochemistry of carbon and nitrogen in a perennially ice-covered Antarctic lake. Chem. Geol. 107 (1), 159–172.
Yoon, H., Khim, B., Lee, K., Park, Y., Yoo, K., 2006. Reconstruction of postglacial palaeoproductivity in Long Lake, King George Island, West Antarctica. Pol. Polar Res. 27 (3), 189–206.
Yu, Y., Song, J., Li, X., Yuan, H., Li, N., Duan, L., 2013. Environmental significance of biogenic elements in surface sediments of the Changjiang estuary and its adjacent areas. J. Environ. Sci. 25 (11), 2185–2195.
Zaborska, A., Pempkowiak, J., Papucci, C., 2006. Some sediment characteristics and sedimentation rates in an Arctic Fjord (Kongsfjorden, Svalbard). Annu. Environ. Prot. 8, 79–96.
Zaborska, A., Mietelski, J.W., Carroll, J., Papucci, C., Pempkowiak, J., 2010. Sources and distributions of 137Cs, 238Pu, 239,240 Pu radionuclides in the north-western Barents Sea. J. Environ. Radioact. 101 (4), 323–331.
Zajączkowski, M., Włodarska-Kowalczuk, M., 2007. Dynamic sedimentary environments of an Arctic glacier-fed river estuary (Adventfjorden, Svalbard). Flux, deposition, and sediment dynamics. Estuar. Coast. Shelf Sci. 74 (1–2), 285–296.

CHAPTER 4

Deciphering the changing climate and environment around Ny-Ålesund, Svalbard, since the Last Glacial Maximum: a multiproxy approach

Ratan Kar[1], Amit K. Mishra[1], Abhijit Mazumder[1], Kriti Mishra[1], S.K. Patil[2], P.S. Ranhotra[1], Ruchika Bajpai[3] and Kajal Singh[1]
[1]Birbal Sahni Institute of Palaeosciences, Lucknow, India
[2]K.S. Krishnan Geomagnetic Research Laboratory, Indian Institute of Geomagnetism, Prayagraj, India
[3]Department of Agronomy, Assam Agricultural University, Jorhat, India

4.1 Introduction

The rapid climate change of the present times is unprecedented, with average global temperatures estimated to rise by 2.5°C during 2000–2100 (IPCC, 2001). As compared to the global averages, winter Arctic temperature is projected to rise as high as 4°C–7°C for 2000–2100 (IPCC, 2001). Over the last few decades, the Arctic has been experiencing extreme environmental changes (Serreze et al., 2000; Comiso and Parkinson, 2004). Winter temperature in the Arctic region has increased by 2°C–4°C during the last 50–60 years (Hassol, 2004). The sharp increase in temperature has resulted in the rapid decline in sea-ice cover (Serreze et al., 2003; Rigor and Wallace, 2004; Meier et al., 2005; Comiso et al., 2008; Stroeve et al., 2008; Polyak et al., 2010; Stranne et al., 2014), with assumptions that the Arctic could even experience ice-free summers in the very near future (Stroeve et al., 2008). Importance of the Arctic region is also due to the fact that it has one of the strongest impacts on the Earth's climate system, especially in the Northern Hemisphere (ACIA, 2005a,b).

The higher latitudes are extremely sensitive to even subtle climatic oscillations; and Svalbard, a high Arctic region, provides good natural archives to study such changes. Though, there are good climatic records for the Arctic region from the marine deposits (Stroeve et al., 2008;

Understanding Present and Past Arctic Environments
DOI: https://doi.org/10.1016/B978-0-12-822869-2.00020-7

Polyak et al., 2010; Stranne et al., 2014), robust data from the terrestrial realm are still somewhat lacking (Schweger, 1982; Anderson, 1985; Guthrie, 2001). A comparison of past climate change data, from both the marine and terrestrial archives, can provide a better understanding of the recent climatic conditions, and could possibly help in simulations for future climate change predictions. This chapter attempts to assess the changes in the environment and climate, since the Last Glacial Maximum (LGM: c.19,000 yr BP), around Ny-Ålesund, Svalbard, and integrates multiproxy data from the terrestrial archives, involving palynology, quartz grain microtexture, and mineral magnetism.

Palynology, the study of organic-walled microfossils of plant origin, especially the pollen grains and spores, provides important information to reconstruct the past vegetation and concurrent climatic changes. As compared to other fossils, pollen grains are better preserved within the sediments in various depositional environments, due to their prolific abundance, minute size, and resistant nature. Since vegetation is climatically controlled and is sensitive to even minor fluctuations in their surroundings, therefore, pollen produced by them are highly indicative of the environment in which they are deposited. Paleopalynology, among the various climatic proxies, has proved to be one of the most potent tools for the reconstruction of vegetation-based past climates (Erdtman, 1952). Over the years, they have been successfully used for the reconstruction of past vegetation and contemporary climatic conditions (Faegri and Iversen, 1989; Birks, 1995; Kar and Quamar, 2019). However, the pollen-vegetation relation is not always linear, as some taxa produce far greater quantities of pollen, which are more widely dispersed and preserved than others. Therefore pollen data require careful interpretation as representation of respective pollen in the pollen spectrum is shaped by differences in pollen productivity, dispersal, and preservation (Birks and Birks, 2005; Bajpai and Kar, 2018). In the present investigations, palynological studies involve preparation of reference slides from the pollen collected from the flowers; aeropalynology of the airborne pollen; interpreting the pollen-vegetation relationship from the surface sediments; and palynological analysis of the subsurface sediments from a trench.

Scanning electron microscopy (SEM) was employed to study the grain morphology and surface microtexture of the quartz grains in the sediments (Krinsley and Donahue, 1968; Margolis and Kennett, 1971; Krinsley and Doornkamp, 1973; Margolis and Krinsley, 1974; Amstrong-Altrin and Natalhy-Pineda, 2014). Predominance of glacial activities is indicated by the microtextures having angular fragments, high reliefs, conchoidal fractures, subparallel linear fractures, steps, and crescentic gouges (Whalley and Krinsley, 1974; Mahaney et al., 1996; Hart, 2006; Mathur et al., 2009). Aqueous influence is discernible in the form of etching patterns in some grains that had long-term contact with water. Eolian activities are indicated by the absence of sharp edges and angularity, and a general roundness of the quartz grains. In the present study, quartz grain microtexture studies were undertaken to decipher the depositional setting (transport mechanism of the sediments) and environmental changes, around Ny-Ålesund, Svalbard, since the LGM.

Magnetic properties of the sedimentary sequences have been recognized as potential indicators to provide information related to paleoclimatic investigations. The foundation of environmental magnetism is based on the principle that the concentration of the magnetic minerals is dependent on climatic processes that influence weathering, erosion, and transportation of sediments in the depositional environment (Nolan et al., 1999; Liu et al., 2012). Climatic changes affect weathering and sedimentation processes in a way that it can give rise to distinct magnetic variations. Magnetic susceptibility is one of the common and simple methods employed in paleoclimatic research and primarily refers to the concentration of magnetic minerals in the sediments. The variations in the amplitudes of magnetic susceptibility, within a stratigraphic column, can be used as proxy indicators of the intensity of weathering and erosion, which are climatically controlled. In glaciated regions, environmental magnetic studies have been used in climatic reconstructions to evaluate the extent of ice (Walden and Ballantyne, 2002) and to mark the magnitude of glacial retreat (Gurney and White, 2005). Magnetic susceptibility data have been generated from the trench sediments for interpreting the past climatic changes, in the Ny-Ålesund area, over the last c.19,000 years.

4.2 Area of study

The archipelago of Svalbard is situated between 76 and 81°N latitudes and is regarded as a high Arctic region (Fig. 4.1A). It is characterized by a broad range of geological deposits, ranging from Precambrian to the Cenozoic, which very well reflect its past geological and tectonic history (Dallmann, 1999). The region has a predominant glacial environment, with an average topographic height of 2000 m amsl (Pelto et al., 1990). The varied rock formations with their differing weathering patterns, the

Figure 4.1 (A) Geographical position of Ny-Ålesund, Svalbard. (B) Schematic map of the Brogger Peninsula showing the location of sampling sites. (C) Panoramic view of the Ny-Ålesund settlement.

general topography, and the glaciers have broadly influenced the Quaternary sedimentation in the area.

The climate in this region is mainly cold and dry, with an annual average range of temperature between −14°C (February) and 5°C (July). The Svalbard archipelago is surrounded by the Barents Sea in the south and southeast, the Norwegian Sea in the southwest, the Greenland Sea in the west, and the Arctic Ocean in the north (Fig. 4.1A). This western part of the archipelago is washed both by the warm and cold currents that provide a unique climate in the region (Ahlmann, 1953; Lamb, 1977; Manley, 1995; Houghton et al., 2001; Humlum et al., 2005; Boitsov and Karsakov, 2012; Eriksen et al., 2015). However, the warm Norwegian Current (a branch of the North Atlantic Drift) has the major influence on the climate in the region and keeps it comparatively warmer, relative to its high latitudinal location. The climate sensitivity of this region, caused by characteristic air and water circulation patterns, coupled with rapid fluctuations in sea-ice volume, provides a distinctive environment to study climatic changes of the recent times on Earth (Benestad et al., 2002).

The area of the present study, Ny-Ålesund, is situated on the west coast of Spitsbergen Island and is considered the northernmost permanent settlement on Earth (78° 45′N to 79° 20′N). It is located on the peninsula of Brøggerhalvøya at Kongsfjorden (Fig. 4.1B). The settlement of Ny-Ålesund, with its research stations, is sandwiched between the fjord in the north and the glaciers in the south (Fig. 4.1C). The area is a long stretch of strandflat (~25 km), encompassing gravelly beaches, small freshwater bodies, terraces, and raised shore deposits. The glaciers, on the southern side, terminate on land and the meltwater streams crisscross the strandflat while flowing out into the Kongsfjorden. The landscape around Ny-Ålesund provides a spectacular view of snow-capped mountains and huge glaciers surrounding the fjord. Four glaciers terminate into the sea at the head of Kongsfjorden, occasionally calving into the sea. The coal deposits in the area were extensively mined from 1920 to 1962 and laid the foundation for the settlement of Ny-Ålesund. Some studies have been undertaken from the region, pertaining to past climatic and environmental changes during the late Pliestocene-Holocene (Rozema et al., 2006; Singh and Ravindra, 2011; Singh et al., 2011; Kar et al., 2018). In the chapter, a multiproxy approach has been presented to decipher the changing climate and environment around Ny-Ålesund over the last c.19,000 years.

4.3 Material and methods

4.3.1 Sampling

Soil and sediment samples were the primary repositories for the studies undertaken; polleniferous materials (flowers) were collected for making reference pollen slides and airborne pollen were also monitored.

4.3.1.1 Surface soils

Surface soil samples were collected in a linear transect covering the entire peninsula (Fig. 4.1B). The surface soil samples were collected by hand-pick method, by skimming the upper surface for about 1 cm with a spatula. About 20 g of sample each was collected, for palynological analysis, from all the geomorphological features, covering the complete length of the peninsula (Fig. 4.2; Table 4.1). At each sampling point, the surrounding vegetation was studied, the coordinates noted, and the soil samples were packed in zip-locked polythene bags. Palynological studies of the surface samples are required to understand the pollen-vegetation

Figure 4.2 The different geomorphic features where surface sampling was done. (A) Glacial outwash with glacial terrace in the background. (B) Close-up of the glacial terrace showing luxuriant ground vegetation. (C) Beach along the Kongsfjord. (D) Raised shoreline with the cliff of Stuphallet in the background.

relationship in the region, for developing the modern analogues, for a better interpretation of fossil pollen assemblages.

4.3.1.2 Subsurface sediments

For past climatic inferences, subsurface sediment samples from a trench were collected. A 1-m trench was dug near the Kolhamna Lagoon, 3 km northwest of Ny-Ålesund township (78°56172N–11°50476E; Fig. 4.1B). The sampling site is located on a strandflat (150–300 m wide) and is about 120 m from the shore (Fig. 4.3). The area around the trench is flat, and is crisscrossed by glacial meltwater streams of the Vestre Broggerbreen glacier that terminate into the fjord forming a small lagoon. The trench presents a more or less uniform lithology, comprising of fine- to coarse-grained sand with angular gravels; however, some demarcations in the lithological components can be observed. The top 40 cm shows fine-medium-grained sand; the topmost 10 cm has humus-rich sediments, with some rootlets of the ground vegetation. The middle part (40–55 cm) has

Table 4.1 Surface samples collected from the different geomorphic locations around Ny-Ålesund.

Sample No.	Geomorphic location	Nature of sample	Coordinates (latitude/ longitude)	Altitude (masl)
01	Glacial outwash	Surface soil	78°54′02.6″N, 12°02′20.6″E	34
02	Glacial outwash	Surface soil	78°54′27.0″N, 12°02′03.4″E	19
03	Glacial outwash	Surface soil	78°54′31.7″N, 12°01′37.2″E	19
04	Glacial outwash	Surface soil	78°54′49.4″N, 12°00′47.1″E	09
05	Glacial outwash	Surface soil	78°54′55.3″N, 12°00′10.4″E	09
06	Glacial outwash	Surface soil	78°55′31.9″N, 11°54′03.7″E	13
07	Glacial terrace	Surface soil	78°55′00.8″N, 11°57′12.2″E	40
08	Glacial terrace	Surface soil	78°54′51.6″N, 11°50′39.3″E	44
09	Glacial terrace	Surface soil	78°55′12.7″N, 11°53′56.4″E	28
10	Glacial terrace	Surface soil	78°54′11.7″N, 12°05′52.8″E	17
11	Beach	Surface soil	78°54′04.8″N, 12°09′03.2″E	03
12	Beach	Surface soil	78°54′14.2″N, 12°08′28.3″E	04
13	Beach	Surface soil	78°54′41.6″N, 12°04′43.6″E	05
14	Strandflat	Surface soil	78°54′41.1″N, 12°03′53.3″E	11
15	Strandflat	Surface soil	78°56′26.2″N, 11°49′24.1″E	10
16	Raised shoreline	Surface soil	78°56′36.5″N, 11°48′36.6″E	25
17	Raised shoreline	Surface soil	78°56′23.7″N, 11°46′32.1″E	56
18	Raised shoreline	Surface soil	78°56′58.6″N, 11°45′17.5″E	85
19	Raised shoreline	Surface soil	78°57′20.1″N, 11°42′29.1″E	38
20	Raised shoreline	Surface soil	78°57′52.7″N, 11°41′55.9″E	36

Figure 4.3 (A) The Kolhamna Lagoon and the location of the trench. (B) The trenching site with the Kolhamna Lagoon and Kongsfjord in the background. (C) Dug trench for subsurface sampling.

medium-grained sand, and this zone is further characterized by an intercalation of a brown organic-rich layer, between 41 and 44 cm. The lower part (55–100 cm) has coarse-grained sand, and from 70 cm downward up to the bottom, the sediments also contain some shell pieces. The abundance and size of gravels (2–12 mm) also increase from top to bottom, throughout the trench. Starting from the bottom, 20 sediment samples, about 250 g each, were collected at 5 cm intervals.

4.3.1.3 Polleniferous material

Flowers were collected during field surveys for extracting the pollen and for making reference pollen slides of the respective taxa. The plants bearing the respective flowers were identified with the help of published literature in the field itself and thereafter, pressed dried within paper sheets. These reference slides provide valuable assistance in the proper identification of modern and fossil pollen from the sediments.

4.3.2 Palynological analysis

4.3.2.1 Palynological processing of flowers

The pollen reference slides were made from polleniferous material (flowers), which were collected during field surveys. To extract the pollen from the respective flowers, the method suggested by Erdtman (1943) was followed. The flowers were first soaked in glacial acetic acid. Then they were gently crushed using a glass rod, passed through a 150-mesh sieve, and the filtrate collected and centrifuged. The supernatant liquid was decanted and the residue was treated with freshly formed acetolyzing mixture. It was then placed in a water bath for 3–5 minutes. The material was centrifuged, decanted, and again treated with glacial acetic acid. Thereafter, the material was carefully washed three to four times with distilled water. To prepare the permanent slides, the pure pollen residual material was placed on glass microslides along with glycerin jelly. The slides were then gently heated and sealed with a coverslip.

4.3.2.2 Palynological analysis of surface and subsurface samples (sediments)

The standard procedure of pollen analysis (maceration) as described by Erdtman (1943) was employed to separate the pollen from the sediment. To deflocculate the pollen and spores from the sediments, about 10- to 20-g sample was taken in glass beakers along with 10% KOH solution and boiled for 10 minutes. To separate the large-sized sediments and other impurities, the sample was then passed through a 150-mesh sieve. The filtrate was washed thrice with distilled water for the removal of alkali. Thereafter, the filtrate was treated with 40% HF in plastic jars to remove the silica content. The duration of HF treatment is dependent on the amount of silica present in the sample. For the organic-rich surface samples, the HF treatment took about 2–3 days, whereas for the subsurface samples, the duration of HF treatment was increased for 5–7 days. After the HF treatment, the filtrate was washed with distilled water thrice, so as to remove traces of acids. The filtrate was passed through a 600-mesh sieve and was then centrifuged. Finally, the process of acetolysis was employed using acetolyzing mixture (concentrated sulfuric acid and acetic anhydride in the ratio of 1:9). During the process, the obtained filtrate was first dehydrated using glacial acetic acid and centrifuged. This was followed by treating the samples by acetolyzing mixture. Samples were then heated in a water bath for 10 minutes, followed by centrifuging. This was succeeded by the sequential washes in glacial acetic acid and distilled water, followed by centrifuging after each wash. The obtained residue, having the liberated pollen spores, was transferred in vials containing 50% aqueous glycerin solution. Few drops of phenol were finally added in the macerates to prevent fungal and bacterial growth. For microscopic study, a drop of macerate from the vial was taken on a glass slide, with a drop of glycerin, and mounted with a coverslip. The same set of above samples from the trench were also treated separately for diatoms and foraminifers, but these were found to be totally absent in all the samples.

4.3.2.3 Aeropalynological slide preparation

Aeropalynology is the study of dispersed airborne palynomorphs. To decipher the pattern of pollen dispersal during the flowering season, aeropalynology was attempted. To trap the airborne pollen, a glass slide smeared with glycerin was put in a Burkard air sampler, which was then switched on to suck in the surrounding air. However, even after repeated attempts, it was observed that the Burkard sampler was not able to trap pollen, as the area is a vegetation sparse region with only ephemeral patchy ground vegetation. Thereafter, to increase the probability of trapping airborne palynomorphs, four glass slides, smeared with glycerin, were kept horizontally on the ground, around the Indian Research Station, Himadri, Ny-Ålesund. Additionally, four slides were kept vertically facing the four principal directions to augment the trapping of pollen. The above eight slides were placed for 24 hours and were replaced by fresh slides on the next day. The exposed slides were then placed on a hot plate and were sealed with a coverslip by glycerin jelly.

All the above palynological samples were studied under an Olympus BX61 light microscope with attached CellSens software (for photography).

4.3.3 Quartz grain microtexture and SEM

A total of 5 g of sediments from each sample were processed for quartz grain microtexture analysis following the standard procedures described by Krinsley and Doornkamp (1973). This involved using hydrochloric acid to remove the carbonates and iron oxide; organic matter was removed with hydrogen peroxide. Clay coatings and fine-grained particles were removed from the quartz grains with Calgon solution (sodium hexametaphosphate). After each chemical treatment, the sand was washed thoroughly with deionized water to remove any product remnants that could crystallize on the grain surface when the sample is oven-dried at 60°C. Twenty quartz grains (> 63 μm) from each sample were picked randomly under a binocular stereozoom microscope to study the grain morphology as well as surface microtextures under SEM. Seven quartz grains from each sample were thereby selected and placed on a stub and coated with platinum for SEM analysis.

4.3.4 Magnetic susceptibility measurements

The sediment samples were first air-dried at room temperature and then homogenized manually, using agate mortar and pestle. For magnetic measurements, the sediment samples collected from the trench were packed in 10-cm^3 nonmagnetic plastic cubes. The net weight of the sample at each interval was calculated by subtracting the weight of the empty vials from the weight of the vials packed with respective sample. Magnetic susceptibility (χlf) measurements were performed on Bartington MS2 instrument at the Indian Institute of Geomagnetism, Allahabad, India. Three measurements of low-frequency (LF; 0.47 kHz) magnetic susceptibility were taken and the average was expressed in $m^3\,kg^{-1}$. LF magnetic susceptibility provides an assessment of climatic vicissitudes; the changes are based on the relative transportation and deposition of ferrimagnetic minerals during weathering.

4.3.5 Dating

To arrange the climatic events and the environmental changes in a chronological order, samples from the three lithological units have been dated by ^{14}C AMS method at the Radiocarbon Laboratory of the Institute of Physics, Gliwice, Poland. Two bulk sediment samples and one shell fragment were dated. Two calibrated age ranges with the confidence levels of 68.2% and 95.4% were provided by the laboratory. The average value of the dates with the confidence levels of 68.2% has been taken as it provides a better resolution of the age range of the calendar year (Table 4.2). The three ages from younger to older are 715 ± 25 cal yr B.P. at 0–5 cm (bulk sediment), 8100 ± 60 cal yr B.P. at 40–45 cm (bulk sediment), and 17,294 ± 166 cal yr B.P. at 90–95 cm (shell fragment). With the three available AMS dates, the ages for all the samples have been extrapolated and the age of the bottommost part was fixed at 19,133 cal yr B.P. The sedimentation rates for the complete 1-m trench were also calculated, which are 5.44 $cm\,kyr^{-1}$ at the lower part to 4.74 $cm\,kyr^{-1}$ at the upper part (Fig. 4.4). The ages are robust, as the sedimentation rate is linear, and the last date of the shell provides further reliability to the chronology because of least chances of over-/underestimation by contamination.

Table 4.2 ^{14}C AMS dates of the three trench samples.

S. No.	Depth (cm)	Sample No.	Dating material	Lab. No.	Calibrated age range 68.2%	Calibrated age range 95.4%	Age used for this study (yr B · P.)
1	0–5	P-3/20	Bulk sediment	GdA-2715	740 (68.2%) 690 yr B · P.	790 (95.4%) 680 yr B · P.	715 ± 25
2	35–40	P-3/12	Bulk sediment	GdA-2716	8160 (68.2%) 8040 yr B · P.	8200 (95.4%) 7980 yr B · P.	8100 ± 60
3	85–90	P-3/3	Shell pieces	GdA-2717	17,460 (68.2%) 17,128 yr B · P.	17,580 (95.4%) 16,986 yr B · P.	17,294 ± 166

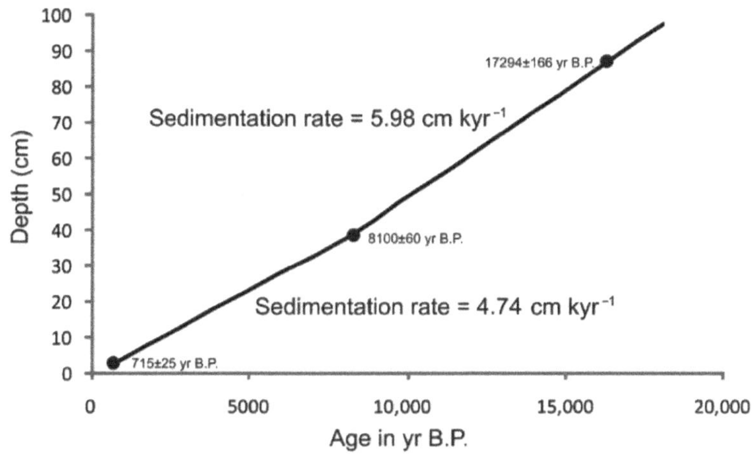

Figure 4.4 The three ^{14}C AMS dates and the age-depth curve showing the sedimentation rate.

4.4 Results

4.4.1 Palynological analysis

4.4.1.1 Pollen from the flowers

One of the important aspects of the palynological work was the collection of polleniferous material from the flowers for preparing reference slides of the pollen taxa. The plants in the Arctic region flower for a very short duration of time, just after the snow melts and therefore, this time phase of the expedition was chosen so as to have the maximum number of plants in their flowering state. The salient points of the work involved the identification of the flowering plants, photography of the plant and flower, and collection of the floral parts for extracting pollen in the laboratory. The list of plants that were identified from around Ny-Ålesund, and their flowers collected for palynological studies are given in Table 4.3. The reference pollen slides that were prepared by extracting the pollen from the flowers (Fig. 4.5) have enabled the accurate identification of respective pollen taxa present in the sediments and air samples.

4.4.1.2 Pollen-vegetation relation

Palynological analysis of surface samples was undertaken to understand the pollen-vegetation relation in the region; that is, how well the pollen of the surrounding vegetation is represented in the sediments. To develop the modern palynological analogues, for the better interpretation of the past

Table 4.3 List of plants identified around Ny-Ålesund, whose flowers were collected for making reference pollen slides.

S. No.	Common name	Botanical name	Family
1	Polar Willow	*Salix polaris*	Salicaceae
2	Mountain Avens	*Dryas octopetala*	Rosaceae
3	Svalbard Poppy	*Papaver dahlianum*	Papaveraceae
4	Arctic Bell-heather	*Cassiope tetragona*	Ericaceae
5	Hairy Lousewort	*Pedicularis hirsuta*	Orobanchaceae
6	Arctic Cottongrass	*Eriophorum scheuchzeri*	Cyperaceae
7	Wood Rush	*Luzula confusa*	Juncaceae
8	Arctic Wood Rush	*Luzula arctica*	
9	Mountain Sorrel	*Oxyria digyna*	Polygonaceae
10	Knot Weed	*Polygonum viviparum*	
11	Pigmy Buttercup	*Ranunculus pygmaeus*	Ranunculaceae
12	Sulfur-colored Buttercup	*Ranunculus sulphureus*	
13	Alpine Meadow-grass	*Poa alpina*	Poaceae
14	Alpine Hair-grass	*Deschampsia alpina*	
15	Cushioned Saltmarsh grass	*Puccinellia phryganodes*	
16	Polar Cress	*Cardamine nymanii*	Brassicaceae
17	Scurvy Grass	*Cochlearia officinalis*	
18	Lapland Whitlow Grass	*Draba lacteal*	
19	Purplish Braya	*Braya purpurascens*	
20	Moss Campion	*Silene acaulis*	Caryophyllaceae
21	Arctic Mouse-ear Chickweed	*Cerastium articum*	
22	Fringed Sandwort	*Arenaria pseudofrigida*	
23	Nodding Lychnis	*Silene wahlbergella*	
24	Hawkweed-leaved Saxifrage	*Saxifraga hieracifolia*	Saxifragaceae
25	Brook Saxifrage	*Saxifraga rivularis*	
26	Tufted Saxifrage	*Saxifraga cespitosa*	
27	Alpine Saxifrage	*Saxifraga nivalis*	
28	Purple Saxifrage	*Saxifraga oppositifolia*	
29	Bog Saxifrage	*Saxifraga hirculus*	
30	Yellow Mountain Saxifrage	*Saxifraga aizoides*	

vegetational changes in the region, 20 surface sediment samples have been palynologically analyzed. The samples were collected in a linear transect, starting from the head of the peninsula (fjord). Sampling was done from the

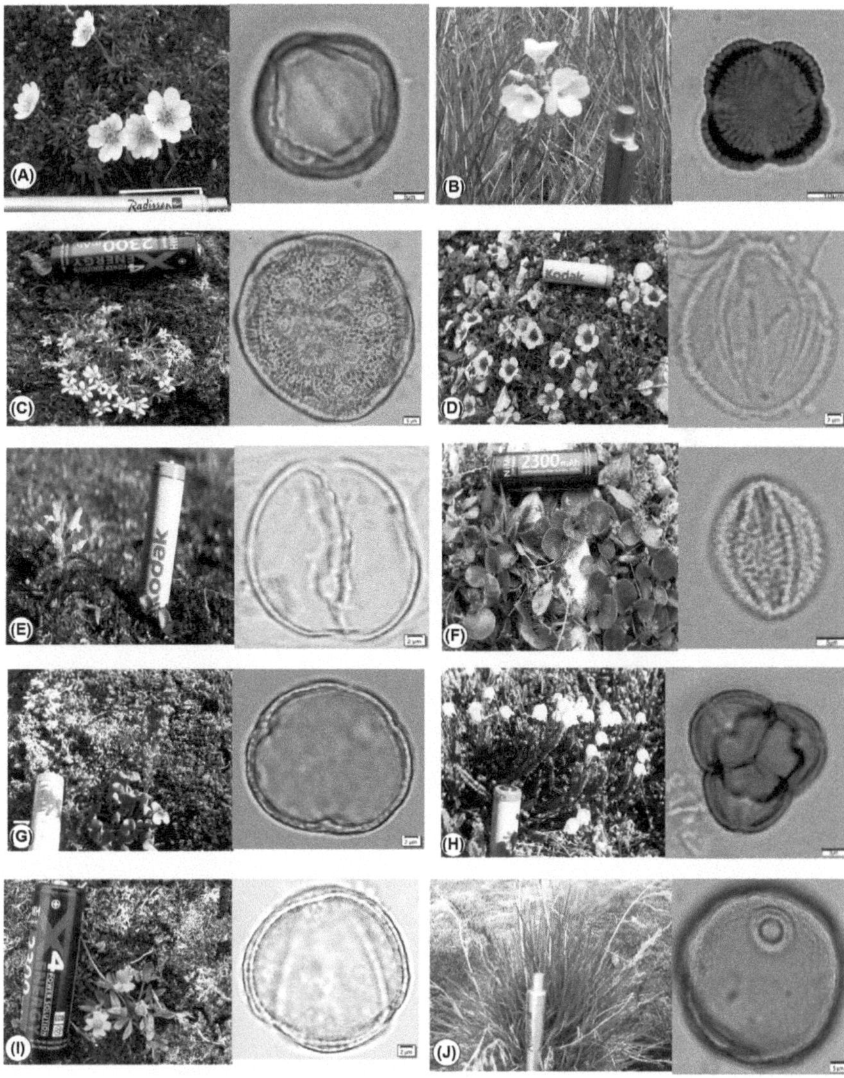

Figure 4.5 Photographs of some flowers in the study area and their respective pollen. (A) *Dryas octopetala* (Rosaceae). (B) *Cardamine nymanii* (Brassicaceae). (C) *Arenaria pseudofrigida* (Caryophyllaceae). (D) *Saxifraga oppositifolia* (Saxifragaceae). (E) *Pedicularis hirsuta* (Orobanchaceae). (F) *Salix polaris* (Salicaceae). (G) *Oxyria digyna* (Polygonaceae). (H) *Cassiope tetragona* (Ericaceae). (I) *Ranunculus pygmaeus* (Ranunculaceae). (J) *Deschampsia alpina* (Poaceae).

various geomorphic locations, such as glacial outwash and terrace, beaches, and raised shoreline, so as to understand the pattern of pollen deposition covering the entire peninsula (Table 4.1; Fig. 4.2). Most of the samples have yielded low numbers of palynomorphs, some even barren, except for a few samples related to particular ecological niches (Fig. 4.6). A list of recovered pollen from each sample is given in Table 4.4.

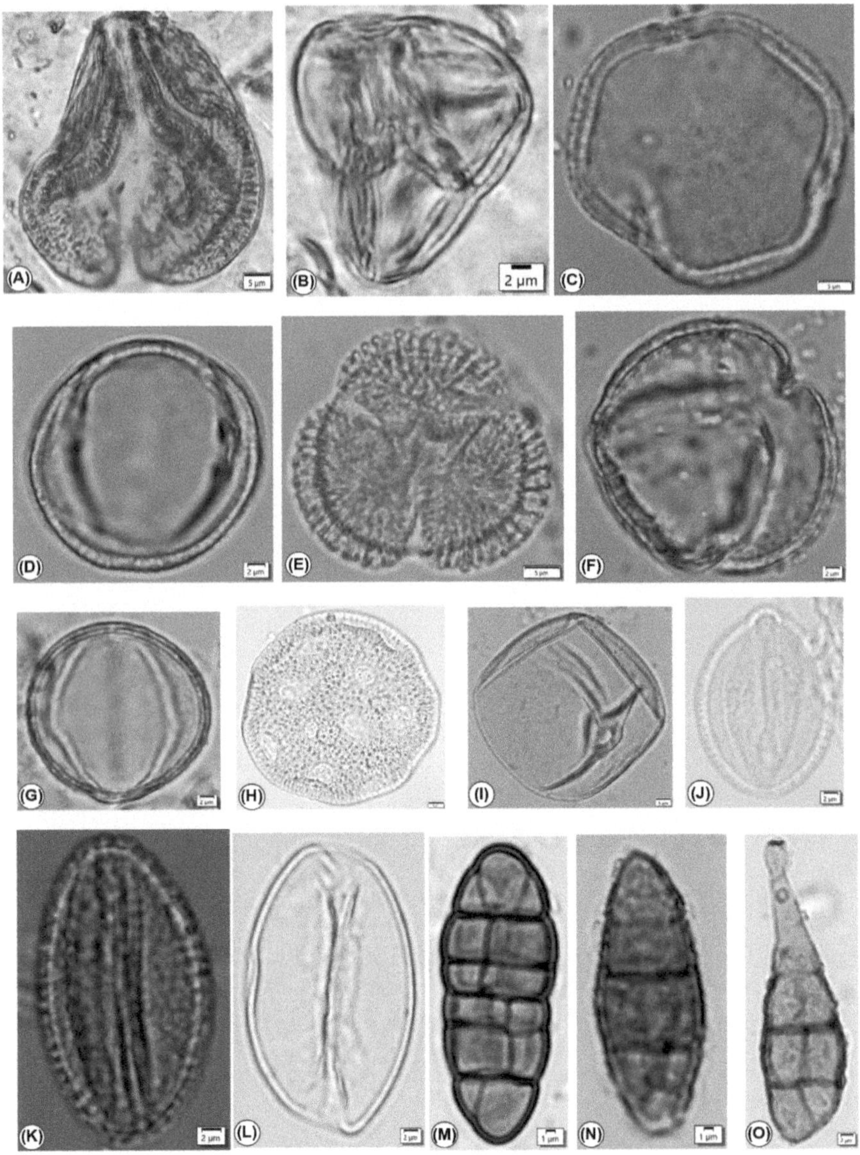

Figure 4.6 Some representative airborne pollen, and palynomorphs recovered from the sediments. (A) *Pinus*. (B) *Cassiope tetragona*. (C) Ranunculaceae. (D) *Oxyria*. (E) Brassicaceae. (F) Papaveraceae. (G) *Dryas octopetala*. (H) Caryophyllaceae. (I) Poaceae. (J) Saxifragaceae. (K) *Salix polaris*. (L) *Pedicularis*. (M—O) Fungal hyphae.

Table 4.4 Number and types of palynomorphs recorded from the surface samples.

Sample No.	Pin.	Sal.	Bet.	Ran.	Bra.	Cyp.	Pap.	Poa.	Car.	Pol.	Sax.	Ros.	Uni.	Fun.	Alg.
SS-01	1	–	–	–	–	–	–	–	1	–	–	–	–	2	–
SS-02	2	–	–	1	–	–	–	–	1	1	–	–	3	2	–
SS-03	–	–	–	–	–	–	1	–	–	–	–	–	–	–	–
SS-04	–	3	–	–	1	–	–	–	1	–	–	–	–	1	1
SS-05	–	–	–	–	–	–	–	–	–	–	–	–	1	–	–
SS-06	–	–	–	–	–	–	–	–	–	–	–	–	–	–	–
SS-07	–	9	–	–	–	–	–	1	3	–	–	2	9	–	–
SS-08	3	38	3	–	1	–	1	2	4	–	–	–	3	23	15
SS-09	–	92	–	3	1	–	–	1	–	–	1	1	4	11	6
SS-10	–	74	–	–	–	–	1	–	1	–	–	–	2	9	3
SS-11	–	–	–	–	–	–	–	–	–	–	–	–	–	1	–
SS-12	–	–	–	–	–	–	–	–	–	–	–	–	–	1	–
SS-13	–	–	–	–	–	–	–	1	–	–	–	–	–	–	1
SS-14	2	75	2	2	1	–	–	7	1	–	1	1	12	15	7
SS-15	1	94	–	–	–	–	–	5	1	–	–	–	2	4	5
SS-16	–	–	–	–	–	–	1	–	1	–	1	–	–	2	2
SS-17	–	–	–	–	1	–	–	–	–	2	–	1	3	1	1
SS-18	3	–	–	–	–	1	–	1	–	1	–	–	4	4	1
SS-19	–	–	1	–	–	2	–	–	2	–	1	1	–	3	2
SS-20	–	–	–	1	–	–	1	–	1	–	–	1	–	1	1

Alg., Algal remains; *Bet.*, Betulaceae; *Bra.*, Brassicaceae; *Car.*, Caryophyllaceae; *Cyp.*, Cyperaceae; *Fun.*, fungal remains; *Pap.*, Papaveraceae; *Pin.*, *Pinus*; *Poa.*, Poaceae; *Pol.*, Polygonaceae; *Ran.*, Ranunculaceae; *Ros.*, Rosaceae; *Sal.*, *Salix*; Sax., Saxifragaceae; *Uni.*, Unidentified pollen.

Six surface sediment samples (SS-1 to SS-6) were collected from the glacial outwash. All the samples show poor yield of pollen. Extra-local *Pinus* pollen is recovered from two samples. *Salix*, Ranunculaceae, Brassicaceae, Caryophyllaceae, and Polygonaceae are recorded in one sample each. Fungal and algal remains are scarce.

Four samples (SS-7 to SS-10) are from a glacial terrace. Because of the relatively flat terrain of this sampling location and formation of soil, a favorable ecological niche has developed, which is marked by the profuse growth of *Salix polaris*. The ground vegetation is faithfully recorded in the pollen assemblage, which has abundant pollen of *Salix*. However, other than *Salix*, the representation of other pollen taxa is still poor, though better than other sampling locations. *Pinus* and Betulaceae pollen are recorded in one sample each. Ranunculaceae, Brassicaceae, Papaveraceae (*Papaver dahlianum*), Poaceae, Caryophyllaceae, Saxifragaceae, and Rosaceae (*Dryas octopetala*) are the terrestrial herbaceous taxa that are recorded. Due to good ground vegetation, fungal remains are quite abundant; as well as the algal remains, reflecting the moist and fertile substrate.

Samples collected along the beach (SS-11 to SS-13) are almost barren in nature. Fungal and algal elements are also negligible in these samples.

Two samples (SS-14 to SS-15) are from the strandflat, near the Kolhamna Lagoon, where the trench for subsurface sampling was dug. Here also, due to the flat terrain and relatively better soil conditions, the location has good ground vegetation, especially *S. polaris*. The pollen assemblage is accordingly dominated by *Salix* pollen, which are encountered in abundance. Poaceae is also present in good amounts, representing its copious presence around the sampling points. Fungal and algal remains are also common. However, the other herbaceous taxa are sporadic, and are represented by Ranunculaceae, Caryophyllaceae, Saxifragaceae, and *D. octopetala*. *Pinus* is recorded in both the samples, whereas, Betulaceae is present in one sample.

Samples collected from the different locations of the raised shoreline (SS-16 to SS-20) have yielded low counts of palynomorphs, nonetheless, the ground vegetation is somewhat represented in the pollen assemblage. Among the herbaceous taxa, Ranunculaceae, Brassicaceae, *P. dahlianum*, Poaceae, Caryophyllaceae, Polygonaceae, Saxifragaceae, and *D. octopetala* are recorded sporadically. Cyperaceae pollen are observed only in these samples and represent the Arctic cottongrass (*Eriophorum scheuchzeri*) growing around the vicinity of the sampling sites. *Pinus* and Betulaceae are observed only in one sample each. Fungal and algal remains are also recorded consistently, though in less numbers.

4.4.1.3 Palynology of subsurface sediments

Though, some amount of pollen spores could be recovered from the topmost sample and the thin organic-rich layer in the middle of the trench, all the other samples were found to be devoid of palynomorphs and other biotic entities (diatoms and foraminifers). Even after repeated maceration, the yield of pollen grains, if at all present in some samples, was extremely sporadic and inadequate for climatic inferences. The topmost sample shows the pollen of the present vegetation growing in the vicinity of the trench. The pollen count is low and is represented by *S. polaris*, Ranunculaceae, Brassicaceae, Caryophyllaceae, Saxifragaceae, and Poaceae. Few fungal spores and hyphae are also present. The organic-rich layer at 41–44 cm depth has also yielded similar pollen taxa as earlier, though with even lesser pollen. Brassicaceae is absent at this level, while *Pinus* pollen are recorded. Good amounts of fungal and algal remains are also observed (Fig. 4.7).

As shown in the results of surface samples, under pollen-vegetation relation, the palynological yield of the soil/sediment samples is poor. Only in the areas of good ground vegetation, in some ecological niches, the pollen count was found to be comparatively better. As the region is

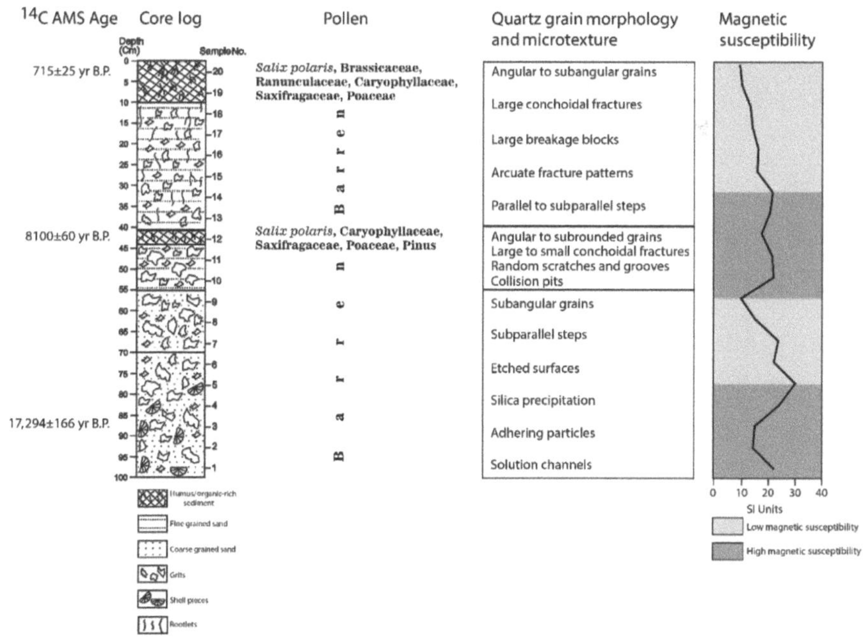

Figure 4.7 Litholog of the trench showing the samples and ^{14}C AMS dates, along with pollen, quartz grain microtexture, and magnetic susceptibility data.

vegetation sparse, with only ephemeral vegetation growing for a short time in isolated patches, it is reflected in the pitiable pollen yield of the sediments. Moreover, pollen spores, which are organic-walled microfossils, get preserved in fine-grained sediments (clay/silt/fine-grained sand) under anaerobic conditions. As the sediments throughout the trench, with the exception of the topmost and the middle organic-rich layer, are coarser in nature, whatever little pollen that might have been deposited during sedimentation seem to have been oxidized, and hence not preserved. Generally, it has been observed, Arctic samples have yielded low pollen counts (Birks, 1991).

4.4.1.4 Aeropalynology

The numbers of the pollen and Non-pollen palynomorphs (NPPs) were not sufficient to construct the pollen percentage diagram, so their actual counts have been taken as the basis to construct the pollen spectra (Fig. 4.8). It was observed that in all the 14 days that aeropalynological sampling was done (August 5, 2009 to August 18, 2009), palynomorphs were recorded, though their counts were low. Altogether, 17 pollen types were identified, along with some algal and fungal remains, and moss spores. Pollen of *S. polaris* (Polar willow) were found to be common. Among the nonarboreals (herbaceous taxa), *D. octopetala* is present in all the 14 days of sampling. *Oxyria, Cassiope tetragona, and Pedicularis* are the

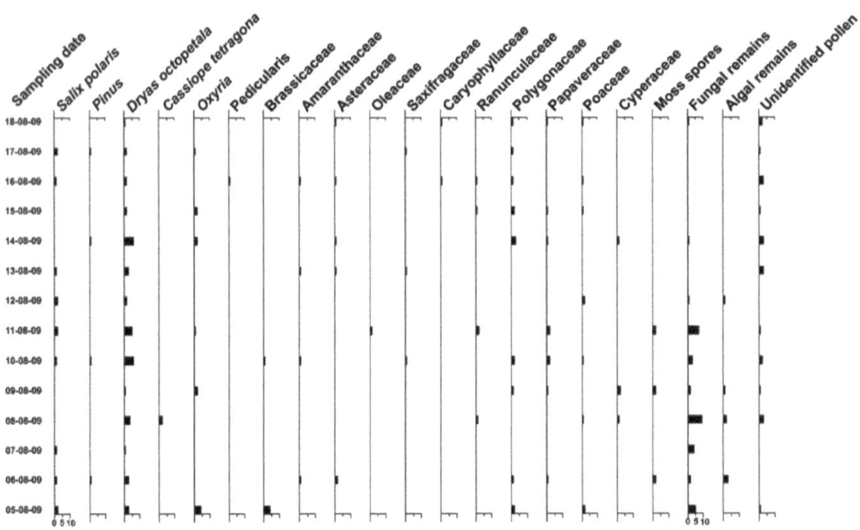

Figure 4.8 The airborne pollen taxa recorded during August 2009.

other main pollen taxa recovered in lower values. Most of the other non-arboreals are identified up to family level—Brassicaceae, Amaranthaceae, Asteraceae, Oleaceae, Saxifragaceae, Caryophyllaceae, Ranunculaceae, Polygonaceae, Papaveraceae, Poaceae, and Cyperaceae are the terrestrial herbaceous taxa, which are recorded sporadically in the aeropalynological assemblages. Moss spores, algal remains and fungal elements, are also recorded sparsely. *Pinus* is the extra-local pollen, which due to its good buoyancy and transporting ability, has been carried by winds from the mainland forests and is recorded, though sporadically.

4.4.2 Quartz grain microtexture

The surface microtextures of detrital quartz grains present in the sediments are formed during the processes of transportation and deposition. In the present study, the quartz grains exhibit the textures of mechanical origin, viz. angular to subangular grains, subrounded to rounded grains, high to low relief, smooth and polished surfaces, breakage blocks, parallel to subparallel steps, arcuate steps, pitted surfaces, collision pits, conchoidal fractures, oriented scratches and grooves, freshly broken smooth surfaces and subparallel linear striations. The textures of chemical origin, such as adhering particles, secondary precipitation, etched surfaces, and solution channels are also noticed on few grains (Fig. 4.9). Combinations of these textures are indicative of glacial action with some eolian and/or aqueous signatures. Depending on these microtextures, the entire section can be divided into three parts (Figs. 4.7 and 4.10).

The topmost part (0–40 cm) exhibits angular to subangular grain morphology (angularity varies from 70% to 90%) with high relief (75% to 90%) and large conchoidal fractures (60% to 75%). Some large breakage blocks (70% to 80%) are also present with some freshly broken smooth surfaces (65% to 75%). Typical arcuate fracture patterns (65% to 75%) along with parallel to subparallel steps (65% to 85%) were also noticed. These morphologies and microtextures are indicative of glacial activities (Figs. 4.7 and 4.10).

The middle part (40–55 cm) carries almost same signatures like the overlying part (angularity varies from 45% to 55%), in addition to few subrounded grains (rounded edges vary from 25% to 40%) along with small conchoidal fractures (45% to 55%) are present. Random scratches and grooves (35% to 45%), collision pits (25% to 35%), and occasional featureless smooth grain surfaces (25% to 40%) are also observed. These

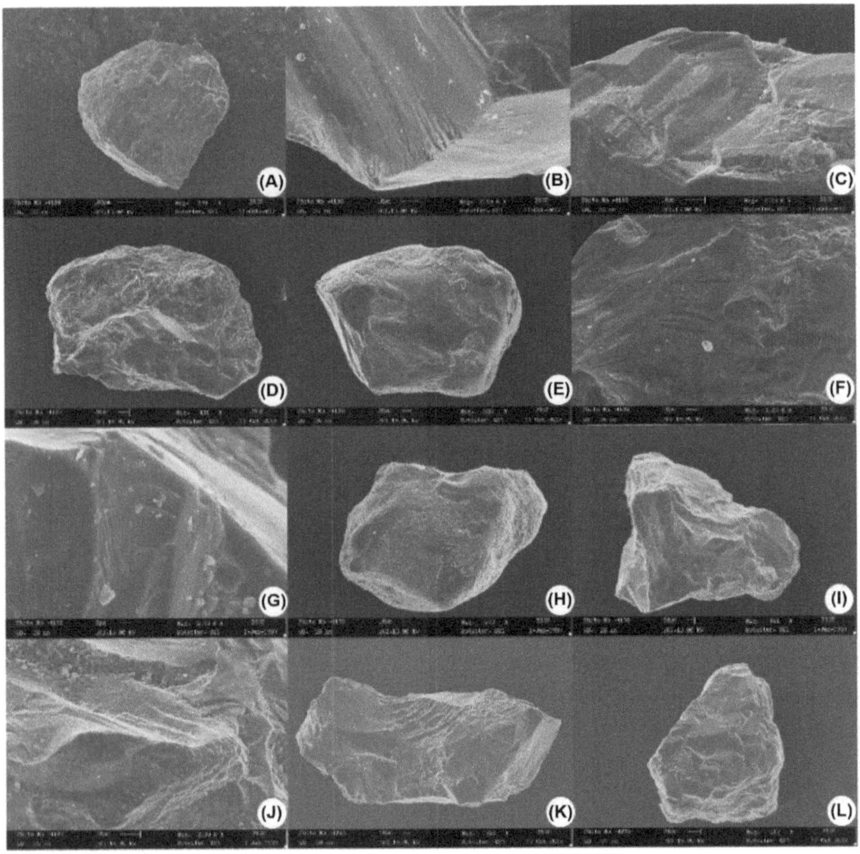

Figure 4.9 SEM photographs showing the representative quartz grain microtextures. (A) Subangular grain. (B) Solution channels and adhering particles. (C) Subparallel steps and silica precipitation. (D) Etched surface. (E) Subrounded grain. (F) Random scratches and grooves. (G) Small conchoidal fractures. (H) Rounded grain with collision pits. (I) Angular grain with arcuate fractures. (J) Parallel steps. (K) Large conchoidal fractures. (L) Large breakage block.

features are representative of some eolian activities within the predominant glacial environment (Figs. 4.7 and 4.10).

The lowermost part (55–100 cm) has mainly evidences of glacial activities (angularity varies from 60% to 75%) with some noticeable amount of etched surfaces (5% to 30%) along with the presence of silica precipitation (5% to 25%), adhering particles (20% to 40%), and solution channels (5% to 15%). These are indicative of glacial activities, with some influence of aqueous interferences (Figs. 4.7 and 4.10).

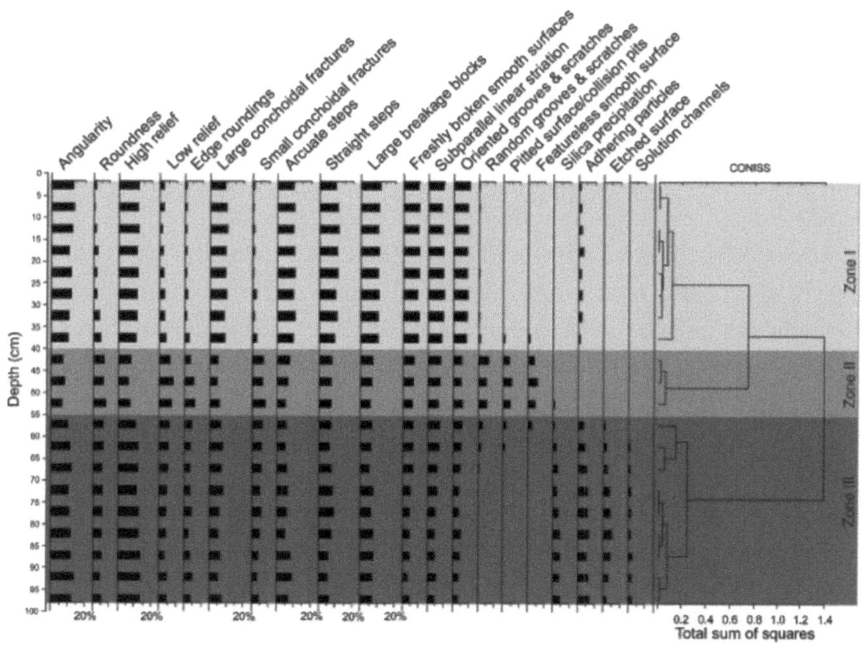

Figure 4.10 CONISS cluster analysis of the trench sediments showing the percentage frequency of characteristic grain morphologies and microtextures.

CONISS cluster analysis of the quartz grain microtextures also exhibits three corresponding zones at the level of total sum of squares as 0.6; Zone I includes 8 samples (0–40 cm), Zone II consists of 3 samples (40–55 cm), and Zone III constitutes a total of 9 samples (55–100 cm) (Fig. 4.10).

4.4.3 Magnetic susceptibility

The magnetic susceptibility (χlf) values show variations across the trench and range between 9.40 and 30.22 SI units. Based on these variations in the magnetic susceptibilities, the trench has been divided into four zones. Zone I (0–30 cm; Recent to 7045 cal yr B.P.) is characterized by the low values of χlf. Zone II (30–55 cm; 7045–11,778 cal yr B.P.) shows a gradual increase in the magnetic susceptibility; however, between 40 and 45 cm, where the organic-rich layer is present, a sharp fall in the concentration of magnetic minerals is seen. Zone III (55–75 cm; 11,778–15,455 cal yr B.P.) is characterized by a considerable decrease in the magnetic susceptibility values. Zone IV (75–100 cm; 15,455–19,132 cal yr B.P.), in the lowest part of the trench, shows a recurrence of increased values of magnetic susceptibility (Fig. 4.7).

4.5 Discussions

Based on the three available ^{14}C AMS dates, which have been extrapolated for all the samples of the trench, and rounded off for ease of presentation, the climatic history in the Ny-Ålesund region can be inferred since the LGM. The late Pleistocene period (19,130–10,860 cal yr B.P.) reveals a marine influence within a predominantly glacial environment. This phase is represented by the lower 100–55 cm of the trench, which is constituted by gravels and coarse sand along with shell pieces. The quartz microtexture also indicates an aqueous influence. This implies that by culmination of the LGM, the glaciers started receding from the fjord, and the sea occupied the vacated area. It has been suggested that there was a sudden warming during 13,000–12,500 yr B.P. in the Svalbard region, as evidenced by rapid glacial retreat (Svendsen and Mangerud, 1992). On the basis of foraminiferal and isotopic studies, a major deglaciation has been observed around the western Svalbard margin during 14–10.8 cal kyr B.P. (Ebbesen et al., 2007). The warm Bølling-Allerød period (c.14,700–12,700 yr B.P.), marked by the transition from perennial sea-ice cover to an ice-free phase, has also been demarcated in the Arctic region (Stranne et al., 2014).

The Early Holocene time span (10,860–8100 cal yr B.P.) provides evidences of continued warming in the region. This phase is represented by the middle 55–40 cm part of the trench, which is characterized by some evidences of eolian activities, as observed by quartz microtexture. The presence of eolian signatures indicates the recession of the glaciers, and that the area got devoid of ice sheet, allowing the eolian activities to get recorded. The presence of a thin organic-rich band, between 41 and 44 cm, dated 8100 cal yr B.P., further confirms the subaerial exposure during this period. The presence of *S. polaris*, Ranunculaceae, Caryophyllaceae, Saxifragaceae, and Poaceae pollen, in this layer, also indicates that the area was exposed for some period of time so as to allow the vegetation to get established and deposit their pollen. Palynological studies from Stuphallet and Blomstrand, around Ny-Ålesund, present a continuous record of pollen over the last 6000 years, indicating that the area has been free from glaciers (Rozema et al., 2006). Paleoenvironmental records from Svalbard indicate that during 9500–4000 yr B.P., summer temperatures were probably higher and glaciated area could have been lesser (Birks, 1991; Svendsen and Mangerud, 1997). From the western margin of Svalbard, a warming period during 10.8–8.8 cal kyr B.P. has been reported (Ebbesen et al., 2007). Based on the study of mollusks, a climate optimum has been recorded in the

Svalbard region between c.8700 and 7700 yr B.P. (Salvigsen et al., 1992). On the basis of sea-ice proxies, Stranne et al. (2014) postulated a prolonged warming phase, from c.10,000 to 6000 yr B.P., with minimum sea ice in the Arctic Ocean. The reduction in sea ice during this period has been explained as a result of the Northern Hemisphere Early Holocene Isolation Maximum (EHIM) (Jakobsson et al., 2010; Polyak et al., 2010; Müller et al., 2012).

During Middle and Late Holocene (8100 cal yr B.P. to the Present), glacial activities prevail in the area, as evidenced by quartz microtexture and the general lithological attributes of the trench sediments. This phase is represented by the upper 40 cm part of the trench and is marked by fluvio-glacial sediments (expect the topmost organic-rich part that reflects the recent vegetation). The area that was encroached by the sea during the late Pliestocene has been taken over by deposits from the meltwater streams emanating from the glaciers. In the North Atlantic region, Late Holocene cooling is an important event (CAPE Project Members, 2001). In the western Svalbard margin as well, an intense cooling period since 8.8 cal kyr B.P. onward has been recorded (Ebbesen et al., 2007). A study involving biotic proxies from the terrestrial deposits of Ny-Ålesund, near to the present study, also reported a cool phase around 8.7 kyr B.P. (Singh et al., 2011). Palynological records from around Ny-Ålesund show a decline in *Salix* pollen, with a corresponding increase in other herbaceous taxa around 3000 yr B.P., reflecting a general cooling trend (Rozema et al., 2006).

The variations in the magnetic susceptibility values observed across the trench characterize the alternate colder and warmer climatic phases. In a predominantly glacial environment, the rapid terrigenous influx along with low organic productivity leads to high susceptibility. During the warmer phases, the high weathering rates may cause dissolution of existing magnetic minerals, thereby resulting in lower values of magnetic susceptibility (Thouveny et al., 1994). Increased biological activities during the warmer periods further tend to reduce the magnetic properties leading to lower susceptibility. A sharp reduction in susceptibility is well displayed by the organic-rich layer at the depth of 41–44 cm, due to an increase in organic content in the sediments.

The lower part of the trench, representing a time span of 19,130–15,450 cal yr B.P., shows high susceptibilities. This corresponds with the post-LGM phase and represents an overall cold glacial environment, which is somewhat compatible to quartz microtexture inferences. Thereafter, the reduced χlf values between 15,450 and 11,780 cal yr B.P.

represent a relatively warm period. A deglaciation from 15,000 to 10,000 yr B.P. in the Arctic region has been observed in the lacustrine records (White et al., 2010). The subsequent phase, between 11,780 and 7050 cal yr B.P., shows high susceptibility and a return to colder conditions. This phase probably represents several cooling events during the Holocene, such as the Pre Boreal Oscillation at 11.3 kyr B.P. (Björck et al., 1997), the Erdalen Event at 9.7 kyr B.P. (Dahl et al., 2002), and the Finse Event at 8.2 kyr B.P. (Klitgaard-Kristensen et al., 1998). From 7050 cal yr B.P. onward, a lowering of the magnetic susceptibility values indicates the initiation of a general warming phase, which could be attributed to the Holocene Climate Optimum. The warming of the present times is reflected by the dominance of *Salix* pollen over the last c.200 years from Ny-Ålesund, post the Little Ice Age (Rozema et al., 2006).

A comparison between the quartz microtexture data and magnetic susceptibility values shows that they are not completely compatible throughout the profile. This is probably due to their different responses to the climatic variations. The magnetic susceptibility reflects the two episodes each of cold and warm events; whereas, the quartz microtexture study reveals two long glacial-dominated phases separated by a short-lived warm event. Nonetheless, within the variations, a broad similarity is also observed. The post-LGM deglaciation and warming, as well as the evidence of Pre Boreal Oscillation are recorded in both the studies. The magnetic minerals seem to be relatively more responsive to the climatic fluctuations, while the quartz grain microtextures generally reflect the depositional environment.

4.6 Conclusions

1. Quartz grain microtextures were useful in reconstructing the paleoenvironmental conditions and reveal a predominance of glacial signatures, which suggests that overall fluvioglacial activities have prevailed since the LGM.
2. Magnetic susceptibility tudies (χlf) were undertaken to decipher the paleoclimatic conditions. Though a broad similarity could be observed between the magnetic susceptibility and quartz grain microtexture inferences, these are not completely synchronous.
3. Due to paucity of palynomorphs in the subsurface sediments, vegetation-based paleoclimatic reconstructions were not possible. Nonetheless, pollen spores from the surface sediments, though scarce,

have helped in establishing the pollen-vegetation relation, which may be used for reference in future palynological studies from the region.

4. After the culmination of the LGM, during late Pleistocene (19,130–10,860 cal yr B.P.), a warming phase is observed with evidences of glacial retreat.
5. As the glaciers receded, the deglaciated area was taken over by the sea. This is well recorded by the presence of shell pieces in the lower part of the trench (70–100 cm). Aqueous markers in the quartz microtextures also provide evidences of marine incursion.
6. The Early Holocene period (10,860–8100 cal yr B.P.) is marked by continuing warming. The further recession of glaciers resulted in an overall influx of terrigenous sediments by glacial meltwater streams, which were deposited over the marine sediments.
7. The recession of the glaciers is also marked by the presence of eolian markers. As the area got devoid of ice sheet, the eolian activities too got registered in the depositional regime. The presence of a thin organic-rich band in the trench, dated 8100 cal yr B.P., further confirms the subaerial exposure during this period.
8. During Middle to Late Holocene (8100 cal yr B.P. to the Present), a general fluvioglacial environment in and around the study area is observed, with glacial meltwater streams crisscrossing the strandflat and flowing into the fjord.

Acknowledgments

We are thankful to Dr. Vandana Prasad, Director, BSIP for providing the necessary working facilities and permission to publish the paper. The logistic and financial support for the Arctic Expedition by the National Centre for Polar and Ocean Research, Ministry of Earth Sciences, India is gratefully acknowledged. RK thanks Dr. Neloy Khare for the invitation to contribute a chapter in the book—*Understanding Present and Past Arctic Environments.*

References

Ahlmann, H.W., 1953. Glacier variations and climatic fluctuations. Am. Geograph. Soc. Bowman Memorial Lectures, Ser 3 New York.

Amstrong-Altrin, J.S., Natalhy-Pineda, O., 2014. Microtextures of detrital sand grains from the Tecolutla, Nautla, and Veracruz beaches, western Gulf of Mexico, Mexico: implications for depositional environment and paleoclimate. Arab. J. Geosci 7, 4321–4333.

Anderson, P.M., 1985. Late Quaternary vegetational change in the Kotzebue Sound area, northwestern Alaska. Quarternary Res. 24, 307–321.

Arctic Climate Impact Assessment (ACIA), 2005a. Arctic Climate Impact Assessment. Cambridge University Press, Cambridge.

Arctic Climate Impact Assessment (ACIA), 2005b. Arctic Climate Impact Assessment. Cambridge University Press, Cambridge.

Bajpai, R., Kar, R., 2018. Modern pollen deposition in glacial settings in the Himalaya (India): abundance of *Pinus* pollen and its significance. Palynology . Available from: https://doi.org/10.1080/01916122.2017.1407835.

Benestad, R.E., Hanssen-Bauer, I., Skaugen, T.E., Forland, E.J., 2002. Associations Between Sea-ice and the Local Climate on Svalbard. Norwegian Meteorological Institute, Norway, Report No. 07/02. KLIMA, Oslo.

Birks, H.H., 1991. Holocene vegetational history and climatic change in west Spitsbergen – plant macrofossils from Skardtjørna, an Arctic lake. Holocene 13, 209–218.

Birks, H.J.B., 1995. Quantitative palaeoenvironmental reconstructions. In: Maddy, D., Brew, J.S. (Eds.), Statistical Modelling of Quaternary Science Data. Technical Guide, vol. 5. Quaternary Research Association, Cambridge, pp. 161–254.

Birks, H.H., Birks, H.J.B., 2005. Reconstructing Holocene climates from pollen and plant macrofossils. In: Mackay, A., Battarbee, R.W., Birks, H.J.B., Oldfield, F. (Eds.), Global Change in the Holocene. Hodder Arnold, London, pp. 342–357.

Björck, S., Rundgren, O., Ingolfsson, O., Funder, S., 1997. The preboreal oscillation around the nordic seas: terrestrial and lacustrine responses. J. Quat. Sci. 12, 455–465.

Boitsov, V.D., Karsakov, A.L., 2012. Atlantic water temperature and climate in the Barents Sea in 2000–2009. J. Mar. Sci. 69, 833–840.

CAPE Project Members, 2001. Holocene palaeoclimate data from the Arctic: testing models of global climate change. Quat. Sci. Rev. 20, 1275–1287.

Comiso, J.C., Parkinson, C.L., 2004. Satellite-observed changes in the Arctic. Phys. Today 57, 38–44. Available from: https://doi.org/10.1063/1.1801866.

Comiso, J.C., Parkinson, C.L., Gersten, R., Stock, L., 2008. Accelerated decline in the Arctic sea ice cover. Geophys. Res. Lett. 35 (L01703). Available from: https://doi.org/10.1029/2007GL031972.

Dahl, S.O., Nesje, A., Lie, O., Fjordheim, K., Matthews, J.A., 2002. Timing, Equilibrium line altitudes and climatic implications of two early Holocene glacier readvances during the Erdalen Events. Holocene 12, 17–25.

Dallmann, W.K., 1999. Lithostratigraphic Lexicon of Svalbard. Upper Palaeozoic to Quaternary Bedrock. Review and Recommendations for Nomenclature Use, Committee on the Stratigraphy of Svalbard, Norsk Polarinstitutt.

Ebbesen, H., Hald, M., Eplet, T.H., 2007. Lateglacial and early Holocene climatic oscillations on the western Svalbard margin, European Arctic. Quat. Sci. Rev. 26, 1999–2011.

Erdtman, G., 1943. An Introduction to Pollen Analysis. Chronica Botanica, MA.

Erdtman, G., 1952. Pollen morphology and plant taxonomy; angiosperms. An Introduction to Palynology, I. Almqist & Wiksell, Stockholm.

Eriksen, E., Ingvaldsen, R.B., Nedreaas, K., Prozorkevich, D., 2015. The effect of recent warming on polar cod and beaked redfish juveniles in the Barents Sea. Reg. Studies Mar. Sci 2, 105–112.

Faegri, K., Iversen, J., 1989. Textbook of Pollen Analysis, fourth ed. John Wiley & Sons, Chichester, revised by Fægri, K., Kaland, P.E., Krzywinski, K.

Gurney, S.D., White, C.K., 2005. Sediment magnetic properties of glacial till deposited since the Little Ice Age Maximum for selected glaciers at Svartisen and Okstindan northern Norway. Boreas 34, 75–83.

Guthrie, R.D., 2001. Origins and causes of the mammoth steppe: a story of cloud cover, woolly mammoth tooth pits, buckles, and inside-out Beringia. Quat. Sci. Rev. 20, 549–574.

Hart, J.K., 2006. An investigation of subglacial processes at the microscale from Briksdalsbreen, Norway. Sedimentol 53, 125–146.

Hassol, S.J., 2004. Impact of a Warming Arctic. Arctic Climate Impact Assessment. Cambridge Universiry Press, p. 139.
Houghton, J.T., Ding, Y., Griggs, D.J., Noguer, M., van der Linden, P.J., Dai, X., et al., 2001. Climate Change 2001: The Scientific Basis. Contribution of Working Group I to the Third Assessment Report of the Intergovernmental Panel on Climate Change. Cambridge University Press, Cambridge.
Humlum, O., Elberling, B., Hormes, A., Fjordheim, K., Hansen, O.H., Heinemeier, J., 2005. Late-Holocene glacier growth in Svalbard - documented by subglacial relict vegetation and living soil microbes. Holocene 15, 396–407.
IPCC, 2001. Report of the Intergovernmental Panel on Climate Change.
Jakobsson, M., Long, A., Ingóllfsson, Ó., Kjær, K.H., Spielhagen, R.F., 2010. New insights on Arctic Quaternary climate variability from palaeo-records and numerical modelling. Quat. Sci. Rev. 29, 3349–3358.
Kar, R., Quamar, M.F., 2019. Pollen-based Quaternary palaeoclimatic studies in India: an overview of recent advances. Palynology 43, 76–93.
Kar, R., Mazumder, A., Mishra, K., Patil, S.K., Ravindra, R., Ranhotra, P.S., et al., 2018. Climatic history of Ny-Alesund region, Svalbard, over the last 19,000 yr: insights from quartz grain microtexture and magnetic susceptibility. Polar Sci. 18, 189–196.
Klitgaard-Kristensen, D., Sejrup, H.P., Haflidason, H., Johnson, S., Spunk, M., 1998. A regional 8200 cal yr BP cooling event in the northwest Europe, induced by final stages of the Laurentide ice-sheet deglaciation? J. Quat. Sci. 13, 165–169.
Krinsley, D.H., Donahue, J., 1968. Environmental interpretation of sand grain surface textures by electron microscopy. Geol. Soc. Am. Bull. 79, 743–748.
Krinsley, D.H., Doornkamp, J.C., 1973. Atlas of Quartz Sand Surface Textures. Cambridge University Press, Cambridge.
Lamb, H.H., 1977. Climate Present, Past and Future, vol. 2. Climatic History and the Future, Methuen.
Liu, Q., Roberts, A.P., Larrasoana, J.C., Banerjee, S.K., Guyodo, Y., 2012. Environmental magnetism: principles and applications. Rev. Geophys. 50, 1–50.
Mahaney, W.C., Claridge, G., Campbell, I., 1996. Microtextures on quartz grains in tills from Antarctica. Palaeogeogr. Palaeoclimatol. Palaeoecol. 121, 89–103.
Manley, T.O., 1995. Branching of Atlantic Water within the Greenland-Spitsbergen passage: an estimate of recirculation. J. Geophys. Res. 100, 20627–20634.
Margolis, S.V., Kennett, J.P., 1971. Cenozoic palaeoglacial history of Antarctica recorded in subantarctic deep sea cores. Am. J. Sci. 271, 1–36.
Margolis, S.V., Krinsley, D.H., 1974. Process of formation and environmental occurrence of microfeatures on detrital quartz grains. Am. J. Sci. 274, 449–464.
Mathur, A.K., Mishra, V.P., Singh, J., 2009. Study of quartz grain surface texture by electron microscopy—a tool in evaluating palaeoglacial sediments in Uttarakhand. Curr. Sci. 96, 1377–1382.
Meier, W., Stroeve, J., Fetterer, F., Knowles, K., 2005. Reductions in Arctic sea ice cover no longer limited to summer. EOS Trans. Am. Geophys. Union 86, 326–327.
Müller, J., Werner, K., Stein, R., Fahl, K., Moros, M., Jansen, E., 2012. Holocene cooling culminates in sea ice oscillations in Fram Strait. Quat. Sci. Rev. 47, 1–14.
Nolan, S.R., Bloemendal, J., Boyle, J.F., Jones, R.T., Oldfield, F., Whitney, M., 1999. Mineral magnetic and geochemical records of late Glacial climatic changes from two northwest European carbonate lakes. J. Paleolimnol. 22, 97–107.
Pelto, M.S., Higgins, S.M., Hughes, T.J., Fastook, J.L., 1990. Modeling mass-balance changes during a glaciation cycle. Ann. Glaciol. 14, 238–241.
Polyak, L., Alley, R.B., Andrews, J.T., Brigham-Grette, J., Cronin, T.M., Darby, D.A., et al., 2010. History of sea ice in the Arctic. Quat. Sci. Rev. 29, 1757–1778.

Rigor, I.G., Wallace, J.M., 2004. Variations in the age of Arctic sea-ice and summer sea-ice extent. Geophys. Res. Lett. 31, L09401. Available from: http://dx.doi.org/10.1029/2004GL019492.

Rozema, J., Boelen, P., Doorenbosch, M., Bohncke, S., Blokker, P., Boekel, C., et al., 2006. A vegetation, climate and environment reconstruction based on palynological analyses of high arctic tundra peat cores (5000—6000 years BP) from Svalbard. Plant Ecol. 182, 155—173.

Salvigsen, O., Forman, S.L., Miller, G.H., 1992. Thermophilous molluscs on Svalbard during the Holocene and their paleoclimatic implications. Pol. Res. 11, 1—10.

Schweger, C.E., 1982. Late Pleistocene vegetation of eastern Beringia: pollen analysis of dated alluvium. In: Hopkins, D.M., Matthews, J.V., Schweger Jr, C.E., Young, S.B. (Eds.), Paleoecology of Beringia. Academic Press, New York, pp. 95—112.

Serreze, M.C., Walsh, J.E., Chapin III, F.S., Ostercamp, T., Dyurgerov, M., Romanovsky, V., 2000. Observational evidence of recent change in the northern high-latitude environment. Clim. Change 46, 159—207. Available from: https://doi.org/10.1023/A:1005504031923.

Serreze, M.C., Maslanik, J., Scambos, T.A., Fetterer, F., Stroeve, J., Knowles, K., et al., 2003. A record minimum Arctic sea ice extent and area in 2002. Geophys. Res. Lett. 30, 1110—1113.

Singh, D.S., Ravindra, R., 2011. Control of glacial and fluvial environments in the Ny-Alesund region. Arctic. Mausam 62, 641—646.

Singh, V., Farooqui, A., Mehrotra, N.C., Singh, D.S., Tewari, R., Jha, N., et al., 2011. Late pleistocene and early holocene climate of Ny-Alesund, Svalbard (Norway): a study based on biological proxies. J. Geol. Soc. Ind. 78, 109—116.

Stranne, C., Jakobsson, M., Björk, G., 2014. Arctic Ocean perennial sea ice breakdown during the early holocene isolation Maximum. Quat. Sci. Rev. 92, 123—132.

Stroeve, J., Serreze, M., Drobot, S., Gearheard, S., Holland, M., Maslanik, J., et al., 2008. Arctic sea ice extent plummets in 2007. EOS Trans. Am. Geophys. Union 89, 13—20. Available from: https://doi.org/10.1029/2008EO020001.

Svendsen, J.I., Mangerud, J., 1992. Palaoclimatic inferences from glacial fluctuations on Svalbard during the last 20000 years. Clim. Dyn. 6, 213—220.

Svendsen, J.I., Mangerud, J., 1997. Holocene glacial andclimatic variations on Spitsbergen, Svalbard. Holocene 7, 45—57.

Thouveny, N., Beaulieu, J.L., Bonifay, E., Creer, K.M., Guiot, J., Icole, M., et al., 1994. Climate variation in Europe over the past 140 kyrs deduced from rockmagnetism. Nature 371, 503—506.

Walden, J., Ballantyne, C.K., 2002. Use of environmental magnetic measurements to validate the vertical extent of ice masses at the Last Glacial maximum. J. Quat. Sci. 17, 193—200.

Whalley, W.B., Krinsley, D.H., 1974. A scanning electron microscope study of surface textures of quartz grains from glacial environments. Sedimentology 21, 87—105.

White, J.W.C., Alley, R.B., Brigham-Grette, J., Fitzpatrick, J.J., Jennings, A.E., Johnson, S.J., et al., 2010. Past rates of climate change in the Arctic. Quat. Sci. Rev. 29, 1716—1727.

CHAPTER 5

Ocean circulation in high northern latitudes and its influence on Arctic climate through the ages

Ashutosh K. Singh[1], Devesh K. Sinha[1], Kirtiranjan Mallick[2], Vikram Pratap Singh[3], Ankush Shrivastava[4] and Tushar Kaushik[5]
[1]Department of Geology, Centre of Advanced Studies, University of Delhi, Delhi, India
[2]Department of Geology, Utkal University, Bhubaneswar, India
[3]Department of Geology, Indira Gandhi National Tribal University, Amarkantak, India
[4]Department of Geology, Mohanlal Sukhadia University, Udaipur, India
[5]Biodiversity and Palaeobiology, Agharkar Research Institute, Pune, India

5.1 Introduction

The high northern latitude oceans are of immense significance because of their direct and indirect influences on the global climate through the formation of permanent and seasonal ice covers, the transfer of sensible and latent heat to the atmosphere, deepwater rejuvenation, and ventilation by thermohaline circulation, all of which influence both oceanic and atmospheric properties (Thiede et al., 1998). The Arctic is an ocean basin surrounded by land masses, while its counterpart, Antarctica, is a landmass surrounded by oceans (Fig. 5.1). In terms of simple physics, the high specific heat material (sea) is surrounded by low specific heat material (land) in the North, while the opposite condition prevails in the South. This contrast in the land—sea distribution has played a significant role in the amplitudes, rates, patterns, and intensity of climatic variation at the two poles during the geological past. Debates have been centered on the issue of synchronous (in-phase) or asynchronous (out-of-phase) climate changes in the Arctic and Antarctic through the ages (Bender et al., 1994; Severinghaus, 2009; deConto et al., 2008; Tripati et al., 2005, 2008). If one considers the climate forcing factors like solar insolation, orbital parameters, and greenhouse gas concentration, then one should expect synchronous climate changes at the two poles in the geological past. However, the paleo-records based on reliable proxy indicators reveal that

Understanding Present and Past Arctic Environments
DOI: https://doi.org/10.1016/B978-0-12-822869-2.00017-7

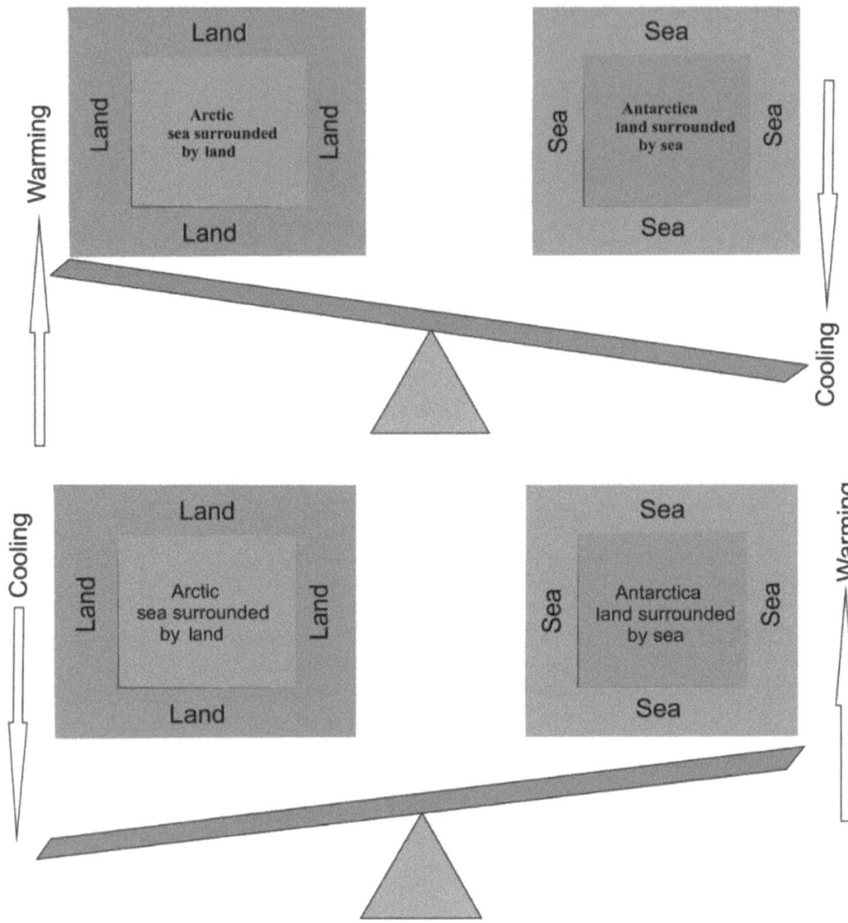

Figure 5.1 Contrasting geographical set ups of the Arctic and Antarctic, and the see-saw.

during many intervals of the geological past, the North and South have responded to the external forcing in opposite ways, termed as a bipolar see-saw (Severinghaus, 2009). One of the reasons for asynchronous climatic changes at the two poles may be due to the contrasting geography of the Arctic and Antarctic (Fig. 5.1). At both the poles, the climate changes have been influenced mainly by changing ocean circulation through the ages besides various forcing factors, including solar insolation, orbital parameters, and greenhouse gas concentration. The Arctic is more vulnerable because of the presence of the ocean at the North Pole.

Large-scale ocean circulation is, to a great extent, responsible for the climate teleconnection between the Arctic and Antarctica (Bender et al., 1994). Thus it is of interest to study the ocean circulation's influence on the high northern latitude climate in general and the Arctic in particular. In simple terms, we need first to discuss the factors which affect the Arctic climate.

5.2 The Arctic climate

5.2.1 Climate variability and ocean circulation

The Arctic climate has been changing with respect to the sea ice, precipitation, and temperature for the last several decades (Graham et al., 2017; Landrum and Holland, 2020). The satellite measurements have revealed that from the late 1970s to the mid-2000s, the Arctic summer sea ice area has declined by more than 10% per decade (Stroeve et al., 2011). Projecting this trend into the future, one cannot rule out the possibility of the Arctic becoming ice-free in summer for the first time in the next few decades. In general, one is inclined to believe that this warming is probably attributed to the current trend in the increase in greenhouse gases, mainly CO_2; however, this is not as simple as it appears. If greenhouse forcing is to be blamed for all the Arctic warming, then it would be a linear relation to be expected in the past and the ice extent could be related simply to the concentration of CO_2 in the atmosphere. It is worthwhile mentioning that greenhouse forcing is related to the cooling/warming of the climate, but sea ice/glaciations are linked to both the temperature and moisture supply. Gildor (2003), in a classic paper, compared the situation with a freezer door left slightly ajar for a few days when the owner goes on vacation and on return, enough ice is found in the freezer, teaching us an important lesson that ice does not form when conditions are merely cold, but when there is a supply of warm, moist air to a sufficiently cold environment. Many workers in the past have advocated the necessity of a strong hydrological cycle for any glaciations to occur (Gildor, 2003; Miller and de Vernal, 1992; Gildor and Tziperman, 2000; van der Veen, 2002). Another critical issue is our poor understanding of the long-term behavior of the North Atlantic—Arctic Sea Ice system. For example, in 2019, Jakobshavn Isbrae, Greenland's fastest flowing and thinning glacier, was found to grow and become thicker despite the ever-increasing concentration of CO_2 in the atmosphere. Investigations led to the conclusion that this was due to the North Atlantic Oscillation (NAO), which plays a vital role in modulating various ocean properties, including North Atlantic overturning. Such an example demonstrates that several forcing factors operate in the climate system and there is a complex interconnectivity amongst the forcing factors themselves.

Scientists have been using numerous models to explain any future variations in the Arctic climate, including precipitation and sea ice cover. Still, all such models can be tested only over time. Any model, if it is able to predict no sea ice in the Arctic by the end of this century, can be validated only once we reach the end of the century. Model simulations of many climate phenomena remain highly uncertain, despite scientific advances and huge amounts of data (Carslaw et al., 2018). The uncertainties in various models are to be addressed through input from the paleoproxy data.

The paleoclimatologists and paleoceanographers are at some advantage as their interpretations are based on actual data from proxy records for the last several million years at different time resolutions that help understand the behavior of climate. With this background, the role of ocean circulation, mostly related to the Atlantic Meridional Overturning Circulation (AMOC) and its effect on the high northern latitude climate, is intriguing (Srokosz and Bryden, 2015). The long-term AMOC decline has been conjectured to cause the so-called "warming hole" persisting in the subpolar North Atlantic (Rahmstorf et al., 2015; Drijfhout et al., 2012). In addition, the Arctic sea ice variation has a significant influence on the strength of the AMOC (Sévellec et al., 2017). Thus both can form a negative-feedback system (Fig. 5.2), where the decline in Arctic sea ice can cause AMOC to slow down due to fresh meltwater forcing. The slowing down of the AMOC will result in reduced poleward heat supply and cooling, as conceptualized by Sévellec et al. (2017). It is difficult to understand such complex systems by applying present-day models. The paleoclimatic data can be used to test such feedback systems.

Further, we need to understand the number of components to be considered while understanding the ocean circulation's role in high northern latitude climates. If we consider only the AMOC and Arctic sea ice cover as two members of a negative-feedback system (Fig. 5.2), the explanations, as discussed earlier, are relatively simple. However, the reality is different. There are many components involved in this system and that makes the relationship complex. For example, the moisture supply for the Arctic sea ice needs a robust hydrological cycle. It is a fact that the AMOC, an essential element of oceanic circulation monitored over the

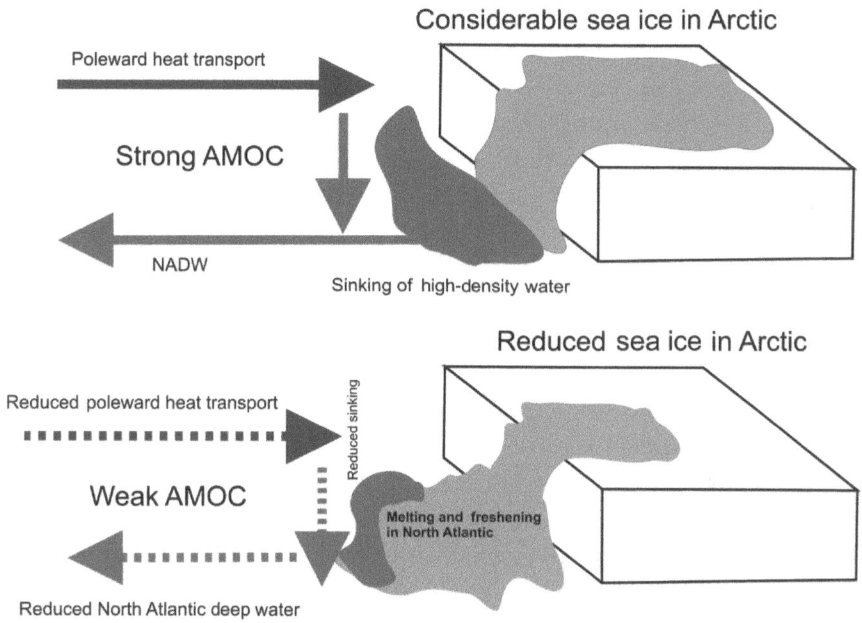

Figure 5.2 A scenario of negative feedback involving two components, the AMOC and Arctic sea ice. Reduced sea ice and freshening of the area of deepwater formation will result in reduced sinking and weakening of the AMOC and thus reduced poleward heat supply, which may again result in cooling and restoration of sea ice. The moisture supply has not been considered.

past decade, is possibly slowing down at a rate as high as 0.4 Sv per year (Sévellec et al., 2017; McCarthy et al., 2012; Smeed et al., 2014).

However, is the Arctic sea ice cover responsible for the entire process of slowing down of the AMOC, or are there other factors responsible also (Jackson et al., 2016; Zhao and Johns, 2014)? One such factor is oceanic natural decadal variability (Buckley and Marshall, 2016). Applying the Huttonian principle of "the present is the key to the past" has limitations because much anthropogenic forcing in the last several decades might have resulted in changing boundary conditions.

5.2.2 Analog in the past and the problem of threshold

One approach to understanding the climatic variability at high northern latitudes is to study a geological time interval when the concentration of greenhouse gases was similar to the present one and see how did the North Atlantic and Arctic climate change from a naturally warm state to a colder state. However, this is not as simple as it looks. Any previous warming due to similar greenhouse gas levels was natural and the rate of increase was quite gradual. However, in the last century, the increase in greenhouse gas concentration has been very steep with anthropogenic influence. Thus any analog from the past for the present climate condition may not correctly predict the future scenario.

One peculiarity observed in the high northern latitude climate changes through the ages has been its abruptness and instability during pre-Holocene time compared to the South where the climate change has been slow and gradual (Blunier, 2001). This abruptness and instability have been mainly attributed to the conveyor circulation in the Atlantic, which has a history of abruptly switching on, causing rapid warming of the high northern latitude climate (Broecker and Denton, 1989). Also, the threshold defined by the specific numerical values may not hold well in the current circumstances. For example, can we have a threshold value of carbon dioxide and related warming when the Earth's climate can go back to or toward a cooling/glacial stage? Can we predict the end of the Holocene Interglacial? Can we predict the slowing down of the AMOC in the next 1000 years? Even Wally Broecker, the author of *Great Ocean Conveyor Belt*, didn't believe in a simplified scenario of shutting down of the conveyor belt, resulting in North America and Europe going into an ice age. Broecker (1991) argued that the conditions like Younger Dryas (YD) when the conveyor belt was considerably slowed down cannot find a perfect match in the present global warming scenario. Broecker (2012), while discussing his landmark paper of 1987, "The Biggest Chill," published in the *Natural History* magazine (Broecker, 1987) mentions how, to his dismay, the *Natural History* magazine placed a "stimulator" on the issue's cover which stated, "Europe beware; the big chill may be coming." Many theoretical models suggest that if the strength of the AMOC drops below a certain threshold, it could suddenly shut down and result in a horrible situation of an ice-covered North America and Europe as shown in the 2004 disaster movie "The Day After Tomorrow."

A better understanding of the thresholds under which a switchover from glacial to interglacial and vice versa occurs at high northern latitudes is warranted. The Earth system's behavior is full of strong nonlinearities (Rial et al., 2004). One finds from the paleo-proxy data that the system is not characterized by several stable equilibrium conditions separated by specific periodicities. Also, very little perturbations in the system, particularly in the climate forcing factors, can make the system reach and cross a threshold, which may result in abrupt changes in the behavior of the system. Many examples from the paleoclimatic records testify to this aspect, like the ice age terminations, the conspicuous and sudden cooling/warming and events in the high northern latitudes. Even the carbon dioxide threshold values for glaciation to occur in the Arctic and the Antarctic differ (DeConto et al., 2008). The lower threshold for CO_2 for the high northern latitude glaciations is because of their greater seasonality and warmer summers (DeConto et al., 2008). Can we get some clues from the past? Can we have an analog in the past representing the present climate?

The Last Interglacial (LIG), a warmer period 130,000–116,000 years before the present, is a potential analog for future climate change (Guarino et al., 2020). How did the climate return to a cooler state from this warm state when the stronger LIG summertime insolation at high northern latitudes drove Arctic land summer temperatures 4°C–5°C higher than in the preindustrial times (Guarino et al., 2020)? The LIG interval came to an end by the commencement of a long, progressive glaciation attributed to astronomically influenced changes in the seasonal distribution of sunlight over the Earth (Mokeddem et al., 2014). Was the orbital forcing the only mechanism that caused the initiation of the last glaciation, or were there other factors responsible? Based on foraminiferal abundance and stable isotope ratios in deep-sea sediments from Ocean Drilling Program (ODP) site 984 south of Iceland, Mokeddem et al. (2014) showed that a crucial early step in glacial inception during the last Glacial involved the weakening of the subpolar gyre circulation of the North Atlantic Ocean. Thus ocean circulation and ocean–atmosphere reorganization play a vital role in the initiation or termination of the glacial periods (Broecker and Denton, 1989).

The present situation differs from the LIG in terms of anthropogenic forcing. Greenhouse forcing is an obvious reason for increasing warming. Orbital parameters can be ignored for the near future. The only possible candidate that remains is ocean–atmosphere reorganization, which includes the changing strength of the AMOC (Sévellec et al., 2017; McCarthy et al., 2012; Smeed et al., 2014) and long-term changes in Hadley and Walker circulation (Hu et al., 2018; Gastineau et al., 2009). These changes, to some extent, are perceptible. Should we expect the climate to go back to the glacial stage the way it did in the past after LIG? The obvious answer is negative. The reason is the rate of increase in greenhouse gas and the rate at which climate sustainability works differ from the present and LIG. Under the present anthropogenic forcing, can we expect the climate to have the same thresholds to switch from interglacial to glacial stages? The Arctic condition is different from the Antarctic because of the presence of sea ice. The sea ice cover is quite vulnerable to small perturbations in climate. Thus, with the earlier discussion, we are faced with intriguing questions about the climate forcing factors that controlled the Arctic climate in the past. As per the document from the National Snow and Ice Data Centre (NSIDC), the factors affecting the Arctic climate are latitude and sunlight, atmospheric pressure, wind, temperature, humidity, cloud precipitation and seasonal sea ice cover, ocean circulation, and land–sea distribution.

We shall discuss the factors and focus then on the changing ocean circulation and its effect on the high northern latitude climate, including the Arctic through the ages. How can we envisage the future of the Arctic climate in the present global warming scenario? How did the past ocean circulation affect the Arctic climate? How did the various forcing factors interact in the past? Related to this issue is an understanding of the climate system behavior in the past, which caused the climate to go back to the cooler mode from the warm mode, though without anthropogenic influence. Besides ocean circulation, some of the factors which affect high northern latitude climate are albedo and associated feedback, and Arctic amplification.

5.2.3 Albedo and feedback

The high northern latitudes, including the Arctic Ocean and North Atlantic, Greenland, and Norwegian Iceland seas, have recently attracted the attention of climatologists because of their vulnerability to climate change. While the south pole is occupied by a continent as large as Antarctica covered with thick ice caps, the north pole is an ocean where sea ice is not a permanent feature and has been subject to variation in global climate through the geological past. Another feature that makes the north pole more dynamic in terms of its participation in climate change is frequent changes in albedo from ice to sea and from sea to ice, which is relatively high as compared to the south pole, where melting results in changes from land to ice and ice to land, particularly in east Antarctica. Sea ice has a very high albedo compared to surrounding oceans. In general, the ocean albedo is 0.06, while that of sea ice is 0.5–0.7 and that of snow is almost 0.9. Thus the sea will reflect only 6% of the incoming solar radiation and absorb the rest, while sea ice will reflect 50%–70% of the incoming solar radiation. Thus when the Arctic sea ice cover diminishes, the seawater is exposed to sunlight and this causes heat to be absorbed. This warm seawater adds to Arctic ice melting. The change in albedo is almost from 70% to 7%, resulting in positive feedback for a warming scenario (Fig. 5.3). The surface albedo feedback (SAF) has been identified as the primary driver of polar amplification in several studies (Hall, 2004; Manabe and Wetherald, 1975; Manabe and Stouffer, 1980; Screen and Simmonds, 2010; Taylor et al., 2013). The surface albedo of the Arctic sea-ice zone is a crucial component in the energy budget of the Arctic region (Hall, 2004; Perovich et al., 2007). Some innovative work has suggested that artificially increasing high-latitude ocean albedo could restore sea ice, but the climate impacts of such a strategy have not been previously explored (Cvijanovic et al., 2015). Thus, of several mechanisms that may contribute to the deteriorating Arctic climate that have been identified, the most well-known is the SAF that occurs due to the higher reflectivity of ice compared to open water (Beer et al., 2020). Due to complex feedback processes (collectively known as "polar amplification"),

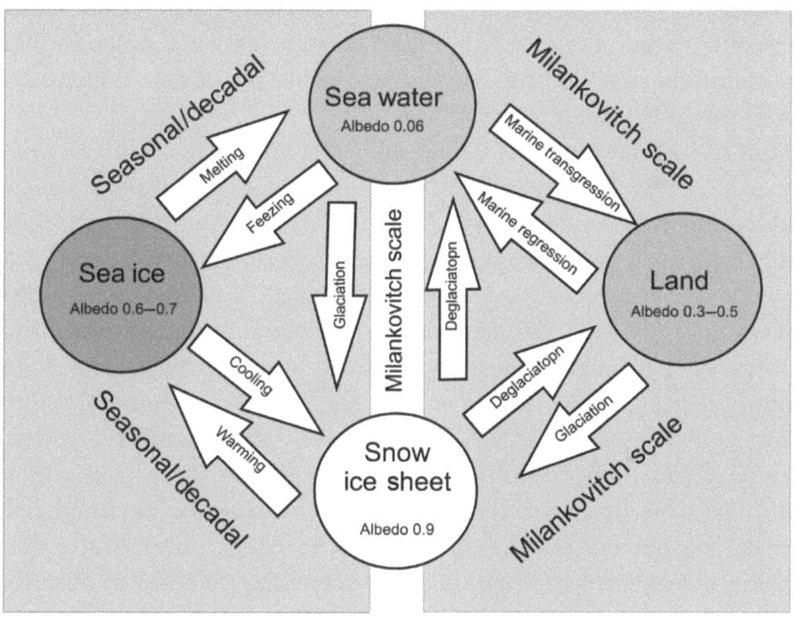

Figure 5.3 Feedback mechanism, interaction, and times scale at which various components of the high-latitude climate system work. The left rectangle mostly depicts the interaction at high northern latitudes, while the right rectangle shows interaction prevalent at and near Antarctica.

the Arctic is both a contributor to climate change and a region that will be most affected by global warming (Serreze and Barry, 2011; Stocker et al., 2013; Stein, 2019).

5.2.4 Arctic amplification

Polar amplification is commonly defined as the ratio of polar warming to tropical warming. Over the past few decades, the warming in the Arctic has been almost twice the rate of the entire globe, a phenomenon known as Arctic amplification. Many studies have concluded that in the last few decades Arctic warming has been unprecedented (Cowtan and Way, 2014; Hartmann et al., 2019; Holland and Bitz, 2003). Scientists generally believe that Arctic amplification is a result of anthropogenic activities. Arctic amplification can have a significant influence on the pathways of severe storms, the jet stream, and other planetary waves and their associated energy propagation (Stjern et al., 2019; Cohen et al., 2014; Coumou et al., 2018) and thus affect the mid latitude weather. Besides, several studies have pointed out teleconnectivity between Arctic amplification and Asian monsoon (Gong et al., 2001; Stjern et al., 2019). Increases in poleward atmospheric heat transport tend to enhance polar amplification, although this can be reduced by other feedbacks (Alexeev and Jackson, 2012; Beer et al., 2020).

Thus the Arctic region is warming at a faster rate than the rest of the globe. Also, there is a possibility of Arctic warming due to the heat reaching the surface from deep layers, because in the Arctic Ocean, a large amount of heat is stored in relatively warm waters at depth, with a cold layer of water and sea ice cover above (Beer et al., 2020). The results of the study by Beer et al. (2020) revealed that the amount of heat from this warm water that reaches the sea ice cover would increase under global warming, enhancing the rate of warming in the Arctic region and further adding to Arctic amplification. There have been considerable debates regarding gradual versus abrupt climate change and the apparent reasons behind this have been our anthropocentric thinking. As discussed earlier, the main fear behind such thought is a poor understanding of the threshold value. Thus there is a need to understand the mechanism controlling the climate of the high northern latitudes and several studies point toward the role of ocean circulation changes as major controlling factor. In addition, the feedback mechanisms, as discussed earlier (Fig. 5.1), also play a crucial role in accelerating climate change. Theoretical modeling combined with the proxy records from the oceanic sediments (Polyak and Jakobsson, 2015; IODP Expedition 302) and Greenland ice cores (Alley, 2000) indicates that the rate of climate change at high northern latitudes is unprecedented. The deep-sea cores obtained through Deep Sea Drilling Project (DSDP) and ODP cruises present a relatively complete record of the commencement and variability of circum-Arctic continental glaciation, the formation of sea-ice covers, and the interaction of the glaciated continental hinterland with the adjacent ocean basin (Thiede et al., 1998).

5.3 Ocean circulation at high northern latitudes

5.3.1 Modern ocean circulation in high northern latitudes

The climate of the Arctic, to a great extent, is controlled by the intensity of inward (warm) and outward (cold) surface currents. The North Atlantic Drift (North Atlantic Current), originating from the Gulf Stream system in the central North Atlantic basin, passes through the Greenland–Scotland Ridge's (GSR) eastern flank. The current crosses Fram Strait into the Arctic Ocean. Its warmer waters can be traced to the north of Svalbard (Langehaug and Falck, 2012), representing the primary path of heat advection into the Arctic (Spielhagen et al., 2011) (Fig. 5.4). The warm waters entering the Arctic are partly balanced by the cold water leaving the Arctic through the East Greenland Current, which carries cold, brackish, partly and seasonally highly variable ice-covered Arctic surface water from the Arctic Ocean (Thiede et al., 1998) (Fig. 5.4). The deep waters of the Norwegian–Greenland Sea are produced by sinking of very cold, saline, and dense oxygenated bottom waters between the Polar and Arctic fronts in the open Greenland Sea (Koltermann and Lüthje, 1989). The production of dense brines due to variable sea-ice formation on the shelves that flow across the continental margins into the adjacent deep-sea basins also adds to these deep waters (Thiede et al., 1998; Quadfasel et al., 1987)

These cold, deep waters fill the Norwegian–Greenland Sea quite above the sill depth of the southern gateway and overflow through the Denmark Strait, Iceland–Faeroe Ridge, and Faeroe–Shetland Channel into the deep North Atlantic to feed the North Atlantic deep water (NADW) (Thiede et al., 1998). The overflow of cold, saline, oxygenated deep water from the Norwegian–Greenland Sea into the North Atlantic Ocean (Meincke, 1983) and its variability through geological time as documented in the ocean sediments in the form of water mass sensitive planktic foraminifera and other geochemical proxies, is very significant

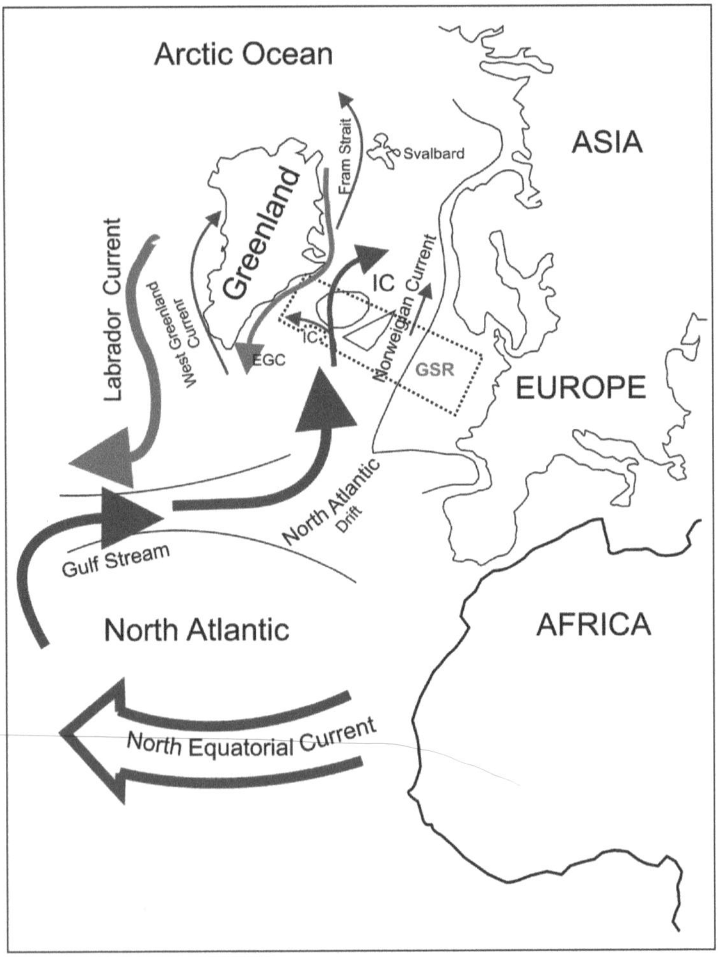

Figure 5.4 Circulation pattern in the North Atlantic and Arctic Oceans (not to the scale). *EGC*, East Greenland Current; *GSR*, Greenland–Scotland Ridge; *IC*, Irminger Current.

and is one of the engines of the great oceanic conveyor belt (Broecker, 1991). Also, it is very challenging to study the variability of NADW, slowing down of the conveyor belt, and its response to any climate change in the high northern latitudes in the future. Thus the water mass dynamics of the upper water column in the Norwegian–Greenland Sea will have a global impact through the conveyor belt and essential feedback and linkages can be observed between the mode of Northern Hemisphere climates, and atmospheric and oceanic circulation patterns (Thiede et al., 1998).

The production of deepwater in the North Atlantic Ocean plays a vital role in maintaining the global meridional overturning circulation (MOC). NADW, the lower branch of the Atlantic part of the MOC (AMOC), forms in the Labrador and Nordic seas as surface waters cool and become dense (Coxall et al., 2018). The sinking of dense water is the most significant part of the AMOC. This is primarily controlled by water mass stratification at the sinking sites (Broecker, 1998). The transport of warm salty water from low latitudes provides sources to the sinking waters. The sinking is also affected by cold freshwater flow from the Arctic Ocean and local heat and freshwater fluxes (Coxall et al., 2018), and wind-driven upwelling in the Southern Ocean (de Boer et al., 2008), which returns deepwater to the surface (Toggweiler and Samuels, 1998; de Boer et al., 2008).

An important feature of the present AMOC is the formation of the NADW by water flowing over the GSR from the Nordic seas (Abelson and Erez, 2017). The North Atlantic forces the entrance of warm saline surface water with high evaporation-induced salinity, which later turns into cold, dense, deepwater that sinks after heat loss to the atmosphere and then flows southward. The total flux of the deepwater flowing over the GSR is around 5 Sv (Abelson and Erez, 2017; Dickson and Brown, 1994). The total ~15 Sv of the southward flowing NADW is obtained from the 2–4 Sv of the Labrador Sea and additional entrainment of waters along the currents leaving the northern basins (Kuhlbrodt et al., 2007). This seawater exchange between the Nordic seas and the North Atlantic can be described as a system of antiestuarine circulation, which could have been regulated by depth variations of the GSR sill (Abelson and Erez, 2017; Abelson et al., 2008; Wright and Miller, 1996).

5.3.2 Early plate movements and ocean circulation changes at high northern latitudes

The paleoceanography and the modern ocean circulation in the North Atlantic and Arctic region owe their origin to the opening of the North Atlantic due to seafloor spreading between Greenland and North America, which started in the Late Mesozoic and continued during the Cenozoic (Oakey and Chalmers, 2012). Such plate tectonic movements resulted in the formation of several deep basins like the North Atlantic basin in the south, and the Norwegian–Greenland Sea and Eastern Arctic basin in the North. Water mass exchange continued to take place amongst these basins through the North Atlantic–Arctic gateways. The GSR constricted the flow between north and south; however, shallow passages existed in the form of the Denmark Strait in the west and the Faeroe–Shetland channel in the east. The Denmark Strait lies between Greenland to its northwest and Iceland to its southeast, while the Faeroe–Shetland channel lies between two Island groups of the Shetland and Faroe Backspace Islands. In the Paleocene, seafloor spreading first occurred between Canada and Greenland and later in the Eocene between Greenland and Scandinavia, forming the Reykjanes-Iceland-Jan Mayen-Spitzbergen Ridge with the Faeroes and Rockall Plateau splitting off the European continent (Pomerol, 1982). The Fram Strait lies between the continental margins of northeast Greenland and Svalbard. These connecting passages between the North Atlantic and Arctic developed during the Late Cenozoic have resulted in significant changes in surface and bottom water circulations, which have great climatic implications. One such is the deepening of the Greenland–Scotland Ridge, which allowed initiation of the North Component Water (NCW), later known as the NADW (Abelson and Erez, 2017).

5.4 Ocean circulation and climate at high northern latitudes

The tectonic movements probably impacted the early Cenozoic Meridional Overturning Circulation state, when Atlantic bathymetry and ocean gateways were different and global temperatures were warmer than today (Abelson and Erez, 2017; Cramer et al., 2009). However, to resolve their interplay at the onset of NADW production, referred to here as its paleo precursor Northern Component Water (NCW), is challenging because the early history of the AMOC remains poorly constrained.

Besides the direct effect of an increase in the greenhouse gas concentration, the meridional heat transport through ocean and atmosphere have been major contributing factors to northern high-latitude climate, including the Arctic (Fedorov et al., 2004; Manabe and Stouffer, 1995, 1999; Rahmstorf, 1995; Stocker and Schmittner, 1997; Alley, 2003; Seidov et al., 2005). Though theoretical models convincingly demonstrate that a slowing down of the global ocean conveyor belt and the resultant decrease in meridional heat transport would result in a cooling of high northern latitudes, the debates still revolve around the threshold conditions (Hand, 2016). The meridional heat transport to the high northern latitudes also occurs through the atmosphere (Masuda, 1988). Another critical issue is the response of the high-latitude climate to the variation in the NADW formation (Muschitiello et al., 2019), intimately linked to the MOC. Using a continuous record of deepwater ventilation from the Nordic seas, Muschitiello et al. (2019) found a ~400-year lead of changes in high-latitude NADW formation ahead of abrupt climate changes recorded in Greenland ice cores at the onset and end of the Younger Dryas Stadial (YDS), which likely occurred in response to gradual changes in temperature and wind-driven freshwater transport. The YDS (~12,900–11,600 years ago) in the Northern Hemisphere is a classic example of abrupt cooling due to changes in meridional heat transport. Although this event involved a global reorganization of atmospheric and oceanic circulation (Denton et al., 2005), the magnitude, seasonality, and geographical footprint of YDS cooling remain unresolved and pose a challenge to our understanding of abrupt climate change in the high northern latitudes (Gordon et al., 2014). The AMOC has both direct and indirect effects on the high-latitude climate. The AMOC also plays a vital role in removing carbon from the atmosphere and storing it in the deep ocean (Takahashi et al., 2009; Perez et al., 2013).

5.5 Initiation of the Atlantic Meridional Overturning Circulation

5.5.1 Eocene-Oligocene transition

Let us first discuss the initiation of the AMOC or AMOC-like circulation. There are two fundamental processes causing ocean overturning. One is the sinking of high-density water at high latitudes, which results in dragging of surface waters from low latitudes. The other is large-scale upwelling, which removes deep waters from the ocean's interior and brings them to the surface. Some authors, such as Cramer et al. (2009), have placed more emphasis on the process of upwelling and considered deepwater production as a second-order characteristic of ocean overturning. The first-order ocean-mixing processes were attributed to intensification of the wind-driven circulation such as the Antarctic Circumpolar Current (ACC). The kinetic energy imparted by the wind to the ocean surface layers and tidal mixing (Wunsch, 1998) were considered as candidates for the first-order reorganization of ocean circulation, and this process was believed to have controlled the ocean overturning. Wind-driven ACC is associated with large-scale upwelling and is the dominant mode of water transport from the ocean interior to the surface in the modern ocean (Cramer et al., 2009). The ACC is primarily responsible for density stratification of the ocean from southern to high northern latitudes (Toggweiler and Samuels, 1998). The ACC also creates a thermal barrier for Antarctica, preventing warm ocean currents from reaching the icy continent. The characteristic isotopic values of oxygen and carbon isotope ($\delta^{18}O$ and $\delta^{13}C$) dependent on the density/depth and nutrients of the NADW and AABW are also a result of the ACC (Cramer et al., 2009; Toggweiler and Bjornsson, 2000). Cramer et al. (2009) believed that the ACC is the driving force in organizing features of the modern ocean, responsible for the density structure, deep circulation, NCW, and Southern Component Water. Thus the chronology of the opening of the Drake Passage and Tasmanian Seaway and the resultant development of ACC may have some uncertainties, but logically speaking, the ACC was the main driver of North Atlantic Circulation at high northern latitudes during the Eocene-Oligocene transition (EOT).

Another point of debate regarding whether the ACC has been the driver of NCW in the EOT is the geochemical signature of the NCW. Coxall et al. (2018) discussed the geochemical fingerprint of the NCW. During the EOT, the northwestern Atlantic was well stratified, with nutrient-rich, low-salinity bottom waters (Coxall et al., 2018). Later the basin witnessed inflow of the initiation of NCW flowing toward the south and this event was about a million years before the initial Antarctic glaciation (Coxall et al., 2018). Instead of attributing the initiation of NCW to ACC, these authors believed that the probable trigger for the initiation of NCW was plate tectonic movements in seas around the Arctic that enabled an increased exchange of waters across the GSR (Coxall et al., 2018). There was a substantial climatic implication of the initiation of the NCW due to tectonic readjustments of the Arctic seas. The Arctic released a vast amount of fossil carbon as indicated by its carbon isotopic signature from modern values. Export of this water with enough nutrients pushed a temporary CO_2 input to the atmosphere and hydrosphere, which probably resulted in a short-lived warming phase. In contrast, the long-term effect of an increase in NCW formation resulted in enhanced northward heat transport that cooled Antarctica (Coxall et al., 2018).

As discussed in Section 5.1, one of the major factors affecting high northern latitude and Arctic climate is the AMOC and associated heat supply to the high northern latitudes. Thus it is worthwhile discussing the onset of the AMOC and how it affected the high northern latitude and Arctic climate. As stated earlier, the reverse is also true, that is, the Arctic climate also affects the AMOC through a negative-feedback mechanism. AMOC and the Arctic climate are connected by both positive- and negative-feedback mechanisms depending on many other factors. Thus we discuss now the major ocean circulation changes that profoundly affected the high northern latitude climate and the primary candidate is the AMOC. It is intriguing to note that the many authors beleive that the ACC has been the main driver of the AMOC (Cox, 1989; Toggweiler and Samuels, 1995; Toggweiler and Bjornsson, 2000; Kuhlbrodt et al., 2007).

Interestingly, the history of the onset and development of the AMOC and related circulation in the high northern latitudes owe their origin to the onset and development of ACC attributed to the opening of the Drake Passage and Tasmanian Seaway. A shallow connection through the Drake Passage opened at the late Eocene (41 Ma), based on the spreading history of the Scotia Sea (Eagles et al., 2005; Livermore et al., 2007) and Nd isotopes from ODP Site 1090 in the Southern Ocean. Another event is the onset of the ACC flow through the Tasman Gateway at ~30 Ma, as seen by Nd isotopes (Scher et al., 2015). These two gateways opening resulted in the full development of the ACC and thermal isolation of Antarctica (Kennett, 1977). A full ACC with modern characteristics is seen from sedimentary records only by the late Oligocene at about 23–25 Ma (Barker and Thomas, 2004; Pfuhl and McCave, 2005; Lyle et al., 2007; Abelson and Erez, 2017).

Should we then believe that ACC is the primary driver of the AMOC and the AMOC onset also is linked to the opening of these two gateways in the south? The logical answer would be yes, but whether it could have initiated the interhemispheric circulation of the AMOC remains uncertain (Sijp and England, 2004; Sijp et al., 2011). Let us look at the evidence for the onset of the AMOC. Abelson and Erez (2017) discussed the role of the opening of the Nordic seas gateway, forming intensified formation of deepwater as a critical factor in the interhemispheric pattern of the overturning circulation. With the integration of the variety of ocean circulation proxies into a common scheme, Abelson and Erez (2017) showed that the EOT marks the transition to a world with an AMOC-like circulation, that is, the onset of interhemispheric northern-sourced circulation cell.

The EOT was marked by a major reorganization of the deep ocean circulation worldwide (Miller et al., 2009; Cramer et al., 2009). Many authors, such as Moore et al. (1978), Miller and Tucholke (1983),

Mountain and Miller (1992), Rea and Lyle (2005), and Miller et al. (2009) reported worldwide deep-sea hiatuses in the sedimentary record, probably caused by intensified bottom water circulation. This vigorous deep circulation was also present in the north Atlantic at high northern latitudes and probably caused a striking intensification in NCW (analogous to the present NADW) seen around that time (Davies et al., 2001; Miller et al., 2009). Also, the onset of the North Atlantic drift sedimentation is established during this time as represented by Feni drift, which is an elongate and mounded contourite deposit located along the western margin of the Rockall Trough in the NE Atlantic Ocean (McCave and Tucholke, 1986; Miller et al., 2009). In addition, the evidence from carbon and oxygen isotopes along the Atlantic Ocean indicates that a major reorganization of the ocean circulation started much earlier (40 Ma) and also later (30 Ma) than the EOT. This means that the ACC may have had little role as a driver for the initiation of NCW (Cramer et al., 2009). Coxall et al. (2018) believed that during Late Eocene, the depth of the GSR increased, for which there is geological evidence (Abelson and Erez, 2017; Stärz, et al., 2017). This increase in the ridge's bathymetry would have increased the Nordic overflows (Fig. 5.5) and thus strengthened NCW production.

Various models run with increasing depth of the GSR indicate that increasing bathymetry of the GSR beyond a critical depth of 50 m initiates a threshold switch from lagoonal to estuarine circulation that salinifies the Nordic seas sufficiently to intensify the northern deepwater production (Abelson and Erez, 2017; Stärz et al., 2017).

Geological evidence suggests that the Barents Sea—Arctic passageway shoaled in the latest Eocene (Musatov and Pogrebitskij, 2000; Kharin, and Lukashina, 2010) and those relative sea-level variations in the Arctic were decoupled from global trends during the late Eocene to the early Miocene (Hegewald, and Jokat, 2013). This paleogeographic Arctic isolation enhanced the salinification in the Nordic seas as brackish Arctic outflows were gradually cut off (Coxall et al., 2018). Now the question is how far this initiation of the AMOC around EOT affected the high northern latitude climate?. As discussed in the previous section, the initiation of the NCW (proto-NADW) could have resulted in increased poleward heat supply and moisture and could have enhanced precipitation resulting in ice formation as also envisaged by Haug and Tiedemann (1998) and Driscoll and Haug (1998). However, all such hypotheses can be debated and some authors beleive that the strongest candidate for the

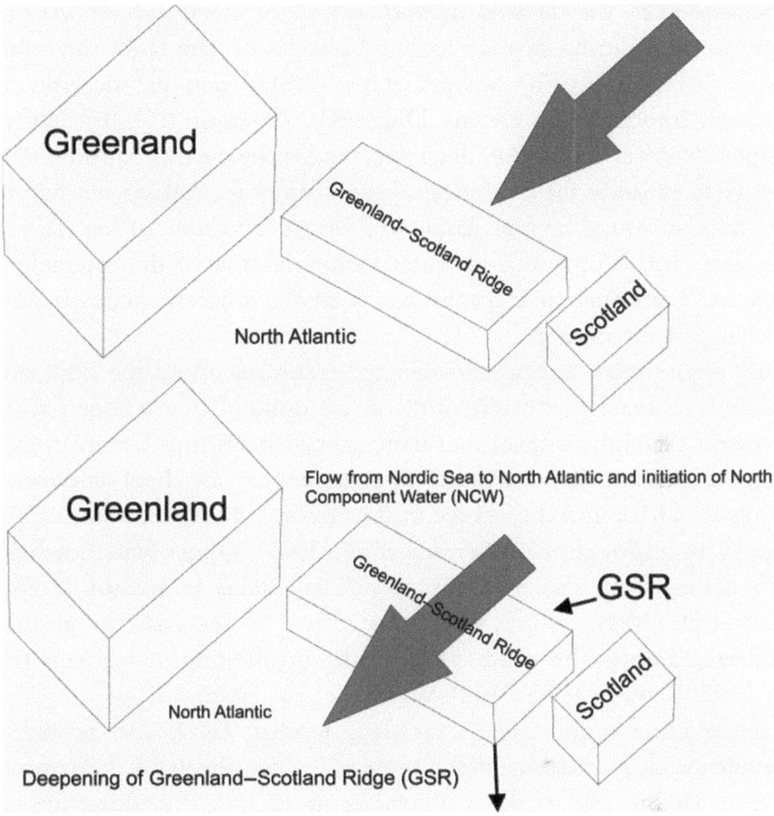

Figure 5.5 Greenland–Scotland Ridge and Nordic Sea overflow leading to initiation of NCW. *Concept from Abelson, M., Erez, J., 2017. The onset of modern-like Atlantic meridional overturning circulation at the Eocene-Oligocene transition: evidence, causes, and possible implications for global cooling. Geochem. Geophys. Geosyst. 18 (6), 2177–2199. https://doi.org/10.1002/2017gc006826.*

cooling of the northern high latitude at the EOT is greenhouse forcing (Tripati et al., 2005). Though consequent to the initial glaciations, the high northern latitude climate during Miocene and later times has been much affected by ocean circulation changes, including the strength of the AMOC (Bryden et al., 2005; Roemmich and Wunsch, 1985; Caesar et al., 2019).

5.5.2 Earliest evidence of ice at high northern latitudes

Land-based paleo-proxy records of the Northern Hemisphere Glaciation (NHG) through the ages are inconsistent and imperfect due to strong erosion caused by the vigorous movement of ice sheets, which were temporarily formed during several cooling episodes on the lands surrounding the Arctic. However, the advent of the DSDP and its successors, the ODP and International Ocean Discovery Program (IODP), retrieved high-quality cores from the deep sea, and provided an unprecedented opportunity to study the evidence of initiation of glaciations on the continental mass surrounding the Arctic and also the history of sea ice cover. These deep-sea records also provide documentation of the interaction of the glaciated continental surroundings with the adjacent ocean basin and their changing circulation.

One of the most critical and debated issues regarding the high northern latitude climate is the initiation of the buildup of ice sheets and the subsequent establishment of permanent ice sheets on Greenland. The northern high-latitude ice consists of a permanent ice sheet at Greenland (Alley et al., 2010) and the sea ice in the Arctic, which has been variability governed by global climate (Serreze et al., 2003). Direct evidence for glaciation comes from the presence of detrital grains in marine sediments (Tripati et al., 2008), which were later carried to the coast by glacier ice and released during the production, drift, and melting of icebergs (Bond et al., 1992; Krissek, 1995; Wolf-Welling et al., 1996).

Such sediments are called ice-rafted detritus (IRD) and provide the best evidence of glaciations or the presence of ice sheets of the continents (Fig. 5.6). Before the work of Tripati et al. (2008), the oldest ice-rafted debris (IRD) documented in the high northern latitudes was found in sediments ranging in age from 2 to 15 Ma (Krissek, 1995; Wolf-Welling et al., 1996; Helland and Holmes, 1997; St. John, 1999; St. John and Krissek, 1999). Eldrett et al. (2007) and Tripati et al. (2008) reported the presence of ice-rafted debris (IRD) in ~44 to 30 Ma sediments from Site 913 in the Greenland Sea and considered it as evidence for glaciation in the North Atlantic during the Middle Eocene to Early Oligocene.

Their sedimentological studies (Tripati et al., 2008) indicated that glaciers extended to sea level in the region, allowing icebergs to be produced. Wright and Miller (1996), Eldrett et al. (2004), and Moran et al. (2006) pointed out that the reason for the rarity of older IRD layers in high northern latitudes was because of hiatuses resulting from vigorous bottom current activity and erosion during the Oligocene and earlier in most sediment cores from the subpolar oceans. The paleoclimatic studies are often confirmed from more than one proxy record. Tripati et al. (2008) showed that the peaks in ice-rafting occur at 42–40 Ma,

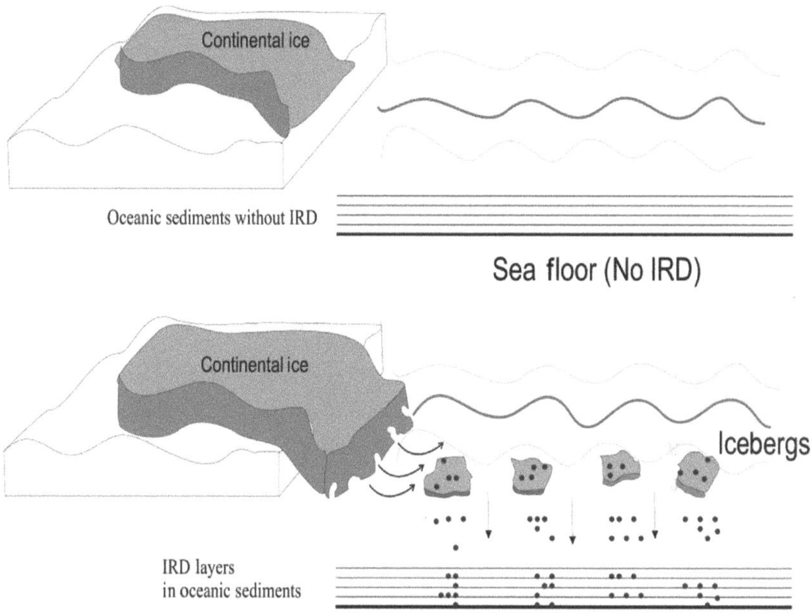

Figure 5.6 Mechanism of the formation of the IRD layers which are indicative of glacial erosion by continental ice sheets and deposition due to later melting of the icebergs.

39–37 Ma, and from 34 to 30 Ma, approximately coincident with previously reported major glacial events inferred from deepenings of the carbonate compensation depth and records of seawater $\delta^{18}O$ for Site 1218 in the deep Pacific (Tripati et al., 2005). Thus when the IRD peaks at high northern latitudes are coincident with seawater $\delta^{18}O$ peaks, the interpretation about the presence of ice becomes more robust because of the ice volume effect on $\delta^{18}O$ of carbonate microfossils. However, the attribution of the $\delta^{18}O$ peak to both hemispheres cannot be emphatically stated. There have been considerable debates about the bipolar or unipolar (Antarctica) glaciations during the Oligocene. DeConto et al. (2008) disagreed with the concept of bipolar glaciations and attributed all the $\delta^{18}O$ peaks to Antarctic glaciations during the Oligocene (Oi-1). Some later IRD pulses before the onset of NHG are the composition of the coarse fraction from Site 909 in the Fram Strait, which indicates that the oldest IRD pulses are documented around 14 Ma (Middle Miocene) and around 12.6 Ma on the Voring Plateau (Leg 104, Site 642) in the Norwegian–Greenland Sea (Thiede et al., 1998).

Nevertheless, the evidence of IRD layers during 43–30 Ma in the North Atlantic does point toward the presence of ice sheets at high northern latitudes. It is worthwhile mentioning that generally there is agreement that the permanent ice sheets in Antarctica and high northern latitudes formed at different times, with the north lagging behind the south by over 30 million years. The ocean circulation changes due to tectonic movements have played a crucial role in both hemispheres in developing permanent ice sheets, as discussed earlier.

5.6 Neogene-quaternary high northern latitude circulation and climate

As stated earlier, Tripati et al. (2008) reported the earliest IRD layers dating back to 44 Ma in the Greenland Sea, indicating the presence of continental ice on the land surrounding the Arctic. The Neogene record of the NHGs is much more prevalent. The Miocene to Pleistocene records of the NHGs are quite robust, and there have been many studies covering different time slices. The most conspicuous feature of the Miocene has been the Miocene climatic optimum (MCO), an intriguing climatic period due to its above-modern temperatures in mid-to-high latitudes in the presence of close-to-modern CO_2 concentrations, and Late Miocene Cooling and closing of the strait of Gibraltar associated with Messinian Salinity Crisis. As far as the ocean circulation at high northern latitudes and its effect on Arctic climate is concerned, the SST records from the North Atlantic DSDP and ODP sites provide important clues. Based on the TEX86 and Uk'37 proxy estimates at Deep Sea Drilling Project Site 608 (52°N), north of the Azores Front, and Ocean Drilling Program Site 982 (57°N) under the North Atlantic Current in a north–south transect, Super et al. (2020) estimated a sea surface temperature gradient of ~4°C between the two sites before the Early MCO (>17 Ma). Super et al. (2020) ruled out the existence of a broad connection to the Arctic through high-latitude oceanic gateways. Thus the North Atlantic Current (NAC) couldn't extend as far northward as today and North Atlantic waters may have been so warm as to prevent buoyancy-driven deep overturning and NCW formation. Alternatively, an early precursor of NCW may have formed further to the south (Coxall et al., 2018) because the NAC failed to reach far northern latitudes and thus, the overturning circulation might not have been possible as today. However, this is not in conformitywith the interpretation of Abelson and Erez (2017) about the initiation of NCW at EOT due to the subsidence of the GSR. All such events would have been temporary until forming a permanent ice sheet in the Northern Hemisphere in the Late Pliocene. Another tectonic event that would have affected the high northern latitude climate, including the Arctic, was the probable opening of Fram Strait between the Nordic seas and the Arctic at around 17 Ma in the Early Miocene (Jakobsson et al., 2007). This passage may have contributed to the MCO by allowing the flow of warmer surface waters to enter the Arctic basin, thereby increasing Arctic SSTs and decreasing seasonal Arctic sea ice (St John, 2008). As discussed earlier, the positive feedback related to reduced Arctic sea ice cover might have contributed to further warming of the oceans coincident with the MCO (Moon et al., 2019; Jakobsson et al., 2007). Near the end of the Early MCO, the overturning circulation enhanced, causing increased input of warm saline waters to the Nordic seas by the North Atlantic current. Super et al. (2020) suggested that during the MCO, the warming of the North Atlantic surface waters was not related to the AMOC. This was probably associated with globally warm conditions or increased regional surface ocean stratification due to an enhanced hydrological cycle, though it can still be debated.

Is there enough record of the IRD during the Miocene interval discussed earlier? The clear evidence of the IRD is reported later than the MCO. The MCO was followed by the Middle Miocene Climatic Transition (MMCT). As stated by Super et al. (2020), the end of the MCO was marked by enhanced MOC-like circulation and poleward heat transport. That might have introduced the sinking of warm saline waters further north and AMOC-like circulation to establish. The ocean circulation and sea surface temperature data point toward an ice-free Arctic or one with reduced ice. Wolf-Welling et al. (1996) reported occurrences of ice-rafted debris (IRD) at ~14 Ma, possibly reflecting the floating of icebergs in the central Fram Strait during the middle Miocene. In addition, Knies and Gaina (2008) showed that after an ice-free period during the MCO (17–15 Ma), glacially eroded materials from the uplifted northern Barents Sea were transported by iceberg flotillas toward the Fram Strait. Opening of the North Atlantic–Arctic gateway allowed transportation of icebergs from the North. The expansive ice growth is probably induced by both large-scale changes in ocean circulation due to enhanced flow of Atlantic water into the Arctic Ocean during the opening of the gateway and concurrent global cooling during the MMCT. This further testifies that the entry of warm water into the Arctic and strong AMOC could have led to establishing a strong hydrological effect, leading to ice growth as proposed conceptually by Gildor (2003).

The MMCT was characterized by major Antarctic ice-sheet expansion and global cooling during the interval ~15–13 Ma (Frigola et al., 2018). However, the high northern latitudes too witnessed dramatic changes in ocean circulation patterns resulting in a change in sedimentation from a biosiliceous to calcareous-rich sediment pattern in the Norwegian Sea (Bohrmann et al., 1990; Poole and Vorren, 1993; Flower and Kennett, 1994). By this time, the modern circulation pattern, including the ice drift pathways and water mass exchange between the Arctic and the Nordic seas, was also established. The MMCT witnessed a proto thermohaline circulation in the Norwegian–Greenland seas with southward flowing cold Arctic waters and northward intrusion of Atlantic water. Wright et al. (1992) and Woodruff and Savin (1989) envisaged intensification of NADW production during the middle Miocene. As discussed earlier, based on the occurrence of the IRD layers, the initiation of deepwater production and circulation from the Norwegian Sea into the North Atlantic was supposed to have started within the early Oligocene (35 Ma) or EOT, or even in the Eocene around 44 Ma (Tripati et al., 2008; Davies et al., 2001; Via and Thomas, 2006). Thus the relationship between deepwater production and climate change during the middle Miocene may be an analog to the early Oligocene (Knies and Gaina, 2008). Peak production of proto-NADW around 17 Ma (Wright et al., 1992) is closely followed by the buildup of glacial ice in the northern Barents Sea (Knies and Gaina, 2008). Thus the opening of the ocean gateways (Fram Strait) played a significant role in causing the expansion of the Barents Sea glaciations. This also resulted in the onset of water mass exchange between the Arctic and the Nordic seas, coupled with a general global cooling trend during the MMCT (Knies and Gaina, 2008). It is intriguing to note that the Middle Miocene expansion of the Antarctic ice sheet (14 Ma) is also at the same time, the reason for which has been attributed to a trigger by a combination of low summer insolation and declining atmospheric carbon dioxide (Holbourn et al., 2005) and orbitally paced ocean circulation changes (Shevenell et al., 2004). The generation of positive feedbacks in the ocean–atmosphere system has promoted the build-up of ice sheets in both hemispheres during the middle Miocene global cooling; a scenario that is comparable with recent findings of ice storage in both hemispheres during the late Eocene to early Oligocene (Eldrett et al., 2007; Knies and Gaina, 2008).

Warnke et al. (1977) reported IRD layers from DSDP Leg 38 (Norwegian–Greenland seas). The IRD layer distribution at Sites 336 and 344 of Leg 38 suggested that the Norwegian Current was intermittently active throughout the time of deposition (Warnke et al., 1977). However, due to poor biostratigraphic control, these authors were not sure of the ages of these IRD layers, which indicated the presence of glaciers in the hinterlands. By using rates of sedimentation based on available paleontological information and a hypothetical Plio–Pleistocene boundary in the cores, Warnke et al. (1977) assigned an apparent age of 7.4 Ma (Late Miocene) for the basal portions of these sediments at both sites. Wolf and Thiede (1991) and Larsen et al. (1994) indicated a middle to late Miocene onset of glaciation on Greenland. Thus from 44 Ma onward, there have been reports of the IRD layers in the North Atlantic and therefore, the possibility of early Arctic glaciations cannot be ruled out.

5.7 Development of a permanent ice sheet (Northern Hemisphere Glaciation)

There have been diverse views on the development of a permanent ice sheet on Greenland. The debate has been both on the timing and the causative factors for the formation of a permanent ice sheet. As discussed in previous sections, the IRD in the oceanic sediments dates back to 43 Ma (Tripati et al., 2008). Still, large-scale Arctic sea ice only appeared since the Late Pliocene ($\sim$3.2 Ma), when a major reorganization of the ocean–climate system took place (Bartoli et al., 2005). In addition to the oldest record of the IRD at 43 Ma, several workers claim that the NHG started as early as in the Late Miocene (Fronval and Jansen, 1996) based on ODP Leg 151 results or even earlier. Minor amounts of IRDs are observed in the Norwegian seas as early as 5.45 Ma (Jansen and Sjøhølm, 1991). However, the climate system was unable to generate a permanent ice sheet in the Northern Hemisphere until the Late Pliocene. Thus the presence of glaciers in the late Miocene cannot be ruled out as discussed in the previous section. Many interesting theories have emerged regarding the onset of NHGs. These studies pertain to either advocating a very gradual or abrupt process leading to NHGs.

5.7.1 Role of ocean circulation in Northern Hemisphere Glaciation

Surface ocean circulation largely depends on the wind stress and ocean–continent geometry, besides indirect dragging of low-latitude surface waters toward poles due to thermohaline circulation. The kinetic energy imparted to the surface ocean by the winds is the primary driver of the surface circulation. The Coriolis force modifies the surface circulation causing spectacular gyres, which are clockwise in the Northern Hemisphere and anticlockwise in the Southern Hemisphere. Many important surface currents owe their origin to a combination of wind stress, Coriolis force, and ocean–continent geometry. The North and South Equatorial Currents flowing around the entire world ocean follow the trade wind pattern. These currents, under the influence of ocean–continent geometry and Coriolis force, intensify Western boundary currents, which carry tropical heat toward poles and modify the climate at high northern latitudes.

Interestingly, ocean circulation changes in the geological past have been primarily due to the shift in ocean–continent geometry and opening and closing of the ocean gateways due to plate tectonic processes (Srinivasan and Sinha, 1998, 2000). As discussed in the previous section, one of the major causes of the development of permanent ice sheets on Antarctica has been the opening of the Drake Passage and Tasmanian Seaway, which led to the establishment of a circum Antarctic psychrospheric circulation and thermal isolation of Antarctica (Kennett, 1977). In the Northern Hemisphere, the deepening of the GSR has been a major cause for the initiation of the proto-NADW, often called the NCW. The passage provided by the Fram Strait is an important gateway for the relatively warm North Atlantic current to enter the Arctic. Thus, in the Northern Hemisphere also, the major changes in ocean circulation have been in response to plate movements.

5.7.2 Closing of the Central American Seaway, changing ocean circulation, and high northern latitude climate

While the deepening of the GSR (Abelson and Erez, 2017) and opening of Fram Strait (Jakobsson et al., 2007) have been the major tectonic processes allowing the exchange of deeper water between the Arctic and North Atlantic and subsequent change in ocean circulation and climate, the closing of the Central American Seaway (CAS) has been another major candidate for a major shift in ocean circulation at high northern latitudes. Though there have been debates about the abrupt versus gradual closing of the CAS, there is a general agreement about the strengthening of poleward heat transport and resultant climate impact. Keigwin (1982), Maier-Reimer et al. (1990), Tiedemann and Franz (1997), Haug and Tiedemann (1998), Rea et al. (1998), and many others were of the opinion that the gradual closing of the CAS and emergence of Panama Isthmus in the late Miocene and early Pliocene led to the major intensification of NHGs that occurred between 3.2 and 2.7 Ma. A significant part of the Gulf Stream and North Equatorial Current used to leak into the Pacific because the CAS provided a passage. After the closure of the CAS, the northward transport of the warm saline waters through the Gulf Stream were strengthened and there was an enhancement in moisture supply to high northern latitudes as a result of evaporation. The high salinity water, once cooled, started sinking, which in turn increased the formation of NADW (Tiedemann and Franz, 1997; Haug and Tiedemann, 1998). Increased evaporation at high latitudes would have also provided additional moisture, an essential requirement for ice-sheet growth. Gildor (2003) strongly advocated the need for a strong hydrological cycle to initiate any ice-sheet growth besides cooling. The hydrological cycle requires a warm region for the origin of moisture, which in this case was the vast area covered with warm waters having enhamced sea surface temperature and cooling at the site of deposition, at the high northern latitude provided by the hinterland of the Arctic ocean (Fig. 5.7).

Many authors have discussed the evidence of positive excursion of benthic foraminiferal stable isotope of oxygen as evidence for the cooling and glaciations. That is true, but the positive excursion only occurs where a large quantity of seawater has evaporated to be deposited as ice sheets making seawater depleted in the heavier isotope. This strong positive excursion in the oxygen isotope also supports a robust hydrological cycle as a precursor for any glaciations. This increased moisture supply led previous researchers to propose that the closing of the Panamanian isthmus caused NHGs (Keigwin, 1982; Rea et al., 1998). However, Haug and Tiedemann (1998) argued that the altered ocean circulation would also increase heat transport to the higher latitudes, as evidenced by a slight warming trend in the Northern Hemisphere at approximately 4.6 Ma. Enhanced Atlantic Ocean thermohaline circulation and the associated heat transport to high latitudes thus seem a bit contradictory with the formation of large ice sheets in the Northern Hemisphere (Driscoll and Haug, 1998).

5.7.3 Ocean circulation change, freshening of the Arctic relation, and relation to NHG

The closing of the Panama isthmus stopped leakage of the Atlantic surface circulation into the Pacific. The enhanced northward flow of warm

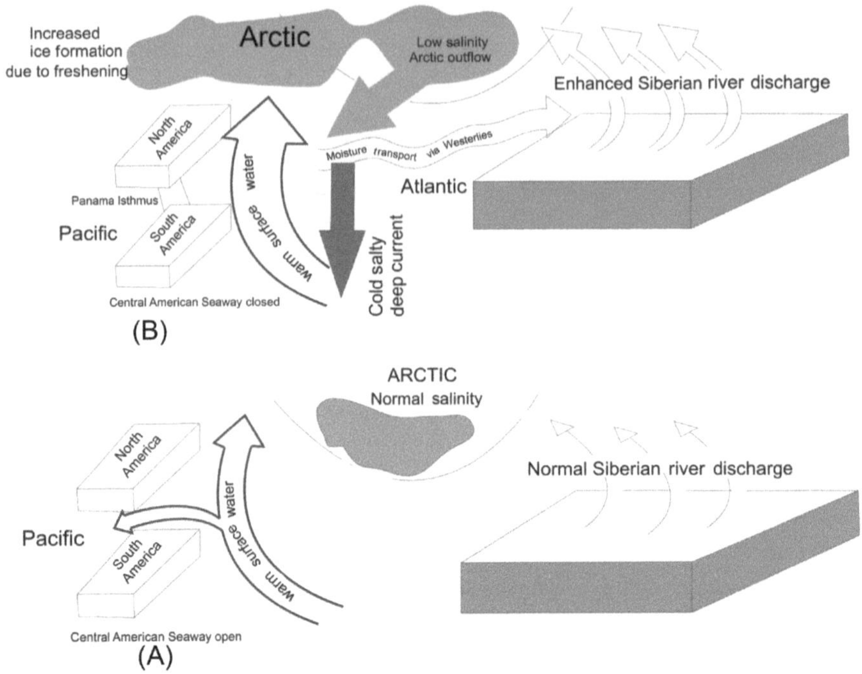

Figure 5.7 A. The cartoon shows the condition when Central Americal seaway was open. B. The Cartoon shows the effect of closing of the Central American Seaway and related enhanced moisture transport associated with NHGs and the freshening of the Arctic and the subsequent impact on ice cover. *Concept from Driscoll N.W., Haug, G.H., 1998. A short circuit in thermohaline circulation: a cause for Northern Hemisphere Glaciation? Science 282 436–438.*

Atlantic surface circulation resulted in increased moisture delivery to the Eurasian continent via atmospheric transport by the westerlies. This change is recorded in drill cores recovered from Lake Baikal (Wennrich et al., 2014). From 3.58 to 3.3 Ma (Nowaczyk et al., 2013) a high flux of clastic material into the lake was recorded, indicating probably enhanced precipitation and the absence of permafrost in the catchment (Brigham-Grette et al., 2013; Tarasov et al., 2013). Driscoll and Haug (1998) argued that diatom abundances and assemblages, spore and pollen data, magnetic susceptibility, and sedimentological evidence indicate, as expected, that the climate was warmer in the early Pliocene and that Siberian humidity increased around 4.5 Ma. As most of the drainage from Siberia empties into the Arctic, with such increased moisture supply, subsequent precipitation would have led to the Arctic freshening (Fig. 5.7). The salty water with an average salinity of 32 psu would freeze at −1.8°C. Thus freshening of seawater in the Arctic would have caused the formation of extensive sea ice by allowing the water to freeze quickly. The hypothesis of Driscoll and Haug (1998) is simple and supported by the simple physical properties of water. The quick formation of sea ice would have resulted in an enormous increase in the surface albedo at the high northern latitudes and would have created positive feedback by insulating the atmosphere from the high heat capacity of the ocean. Driscoll and Haug (1998) compared this situation as if the Arctic had a landmass (ice) with high albedo.

5.8 Conclusions

The high northern latitude climate is vulnerable to the current climatic deterioration due to anthropogenic forcing; however, the geological past has witnessed waxing and waning of the Greenland ice sheet and variation in Arctic sea ice cover due to more than one reason. The high northern latitude climate changes are a definite sign of concern because of Arctic amplification. Ocean circulation has been a primary driver of the high northern latitude climate, including the Arctic, besides another forcing like greenhouse gas concentration and solar insolation variations.

The earliest evidence of NHGs has been primarily based on the ice-rafted sediment records from the deep ocean basins. Though the IRDs date back to as early as 44 Ma, the development of a permanent ice sheet in the Northern Hemisphere only occurred in the Late Pliocene.

The significant changes in ocean circulation at a high northern latitude have been the connection of Arctic and Nordic seas with the North Atlantic, which allowed early NCWs to enter Atlantic and Atlantic waters and then to enter the Arctic. Two major tectonic events made this possible. One was the deepening of the GSR and the other was the opening of the Fram Strait.

The high northern latitude climate has been very much affected by the AMOC. Cenozoic plate movements, including the closing of the CAS, have been a major cause of enhanced AMOC and the transport of warm waters toward the pole. This was attributed as a major cause of the supply of moisture and development of permanent ice sheets in the Northern Hemisphere in the late Pliocene.

Acknowledgments

DKS and AKS thank the Delhi School of Climate Change and Sustainability (DSCCS) of the Institution of Eminence (IoE) and the Department of Geology, University of Delhi, for providing infrastructural support. Financial assistance to AKS and DKS from the Ministry of Earth Sciences, Government of India (Sanction No. MoES/CCR/Paleo-4/2019) is thankfully acknowledged. Other authors (KM, VPS, AS, TK) thank their respective institutions for logistic help.

References

Abelson, M., Erez, J., 2017. The onset of modern-like Atlantic meridional overturning circulation at the Eocene-Oligocene transition: evidence, causes, and possible implications for global cooling. Geochem. Geophys. Geosyst. 18 (6), 2177–2199. Available from: https://doi.org/10.1002/2017gc006826.

Abelson, M., Agnon, A., Almogi-Labin, A., 2008. Indications for control of the Iceland plume on the Eocene-Oligocene 'greenhouse-icehouse' climate transition. Earth Planet. Sci. Lett. 265, 33–48.

Alexeev, V.A., Jackson, C.H., 2012. Polar amplification: is atmospheric heat transport important? Clim. Dyn. 41 (2), 533–547. Available from: https://doi.org/10.1007/s00382-012-1601-z.

Alley, R.B., 2000. Ice-core evidence of abrupt climate change. PNAS 97 (4), 1331–1334. Available from: https://doi.org/10.1073/pnas.97.4.1331.

Alley, R.B., 2003. Abrupt climate change. Science 299 (5615), 2005–2010. Available from: https://doi.org/10.1126/science.1081056.

Alley, R.B., Andrews, J.T., Brigham-Grette, J., Clarke, G.K.C., Cuffey, K.M., Fitzpatrick, J.J., et al., 2010. History of the Greenland Ice Sheet: paleoclimatic insights. Quat. Sci. Rev. 29 (15–16), 1728–1756. Available from: https://doi.org/10.1016/j.quascirev.2010.02.007.

Barker, P.F., Thomas, E., 2004. Origin, signature and palaeoclimatic influence of the Antarctic Circumpolar Current. Earth-Sci. Rev. 66 (1–2), 143–162. Available from: https://doi.org/10.1016/j.earscirev.2003.10.003.

Bartoli, G., Sarnthein, M., Weinelt, M., Erlenkeuser, H., Garbe-Schönberg, D., Lea, D. W., 2005. Final closure of Panama and the onset of northern hemisphere glaciation. Earth Planet. Sci. Lett. 237 (1–2), 33–44. Available from: https://doi.org/10.1016/j.epsl.2005.06.020.

Beer, E., Eisenman, I., Wagner, T.J.W., 2020. Polar amplification due to enhanced heat flux across the Halocline. Geophys. Res. Lett. 47 (4). Available from: https://doi.org/10.1029/2019gl086706.

Bender, M., Sowers, T., Dickson, M.-L., Orchardo, J., Grootes, P., Mayewski, P.A., et al., 1994. Climate correlations between Greenland and Antarctica during the past 100,000 years. Nature 372 (6507), 663–666. Available from: https://doi.org/10.1038/372663a0.

Blunier, T., Brook, E.J., 2001. Timing of millennial-scale climate change in Antarctica and Greenland during the Last Glacial Period. Science 291 (5501), 109–112. Available from: https://doi.org/10.1126/science.291.5501.109.

Bohrmann, G., Henrich, R., Thiede, J., 1990. Miocene to Quaternary palaeoceanography in the northern North Atlantic: variability in carbonate and biogenic opal accumulation. In: Bleil, U., Thiede, J. (Eds.), Geological History of the Polar Oceans: Arctic Versus Antarctic. Kluwer Acad., Norwell, MA, pp. 647–675.

Bond, G., Heinrich, H., Broecker, W., Labeyrie, L., McManus, J., Andrews, J., et al., 1992. Evidence for massive discharges of icebergs into the North Atlantic ocean during the last glacial period. Nature 360, 245–249.

Brigham-Grette, J., Melles, M., Minyuk, P., Andreev, A., Tarasov, P., DeConto, R., et al., 2013. Pliocene warmth, polar amplification, and stepped Pleistocene cooling recorded in NE Arctic Russia. Science 340 (6139), 1421–1427. Available from: https://doi.org/10.1126/science.1233137.

Broecker, W.S., 1987. The biggest chill. Nat. Hist. Mag. 97, 74–82.

Broecker, W.S., 1991. The Great Ocean conveyor. Oceanography 4 (2), 79–89.

Broecker, W.S., 1998. Paleocean circulation during the last deglaciation: a bipolar seesaw? Paleoceanography 13, 119–121.

Broecker, W.S., 2012. The carbon cycle and climate change: memoirs of my 60 years in science. Geochem. Perspect. 1 (2), 221–340.

Broecker, W.S., Denton, G.H., 1989. The role of ocean-atmosphere reorganizations in glacial cycles. Geochim. Cosmochim. Acta 53 (10), 2465–2501. Available from: https://doi.org/10.1016/0016-7037(89)90123-3.

Bryden, H.L., Longworth, H.R., Cunningham, S.A., 2005. Slowing of the Atlantic meridional overturning circulation at 25° N. Nature 438 (7068), 655–657. Available from: https://doi.org/10.1038/nature04385.

Buckley, M.W., Marshall, J., 2016. Observations, inferences, and mechanisms of the Atlantic Meridional Overturning Circulation: a review. Rev. Geophys. 54, 5–63.

Caesar, L., et al., 2019. Observed fingerprint of a weakening Atlantic Ocean overturning circulation. Nature . Available from: https://doi.org/10.1038/s41586-018-0006-5.

Carslaw, K.S., Lee, L.A., Regayre, L.A., Johnson, J.S., 2018. Climate models are uncertain, but we can do something about it. Eos 99. Available from: https://doi.org/10.1029/2018EO093757.

Cohen, J., Screen, J.A., Furtado, J.C., Barlow, M., Whittleston, D., Coumou, D., et al., 2014. Recent Arctic amplification and extreme mid-latitude weather. Nat. Geosci. 7 (9), 627–637. Available from: https://doi.org/10.1038/ngeo2234.

Coumou, D., Di Capua, G., Vavrus, S., Wang, L., Wang, S., 2018. The influence of Arctic amplification on mid-latitude summer circulation. Nat. Commun. 9 (1). Available from: https://doi.org/10.1038/s41467-018-05256-8.

Cowtan, K., Way, R.G., 2014. Coverage bias in the HadCRUT4 temperature series and its impact on recent temperature trends. Q. J. R. Meteorol. Soc. Available from: https://doi.org/10.1002/qj.2297.

Cox, M.D., 1989. An idealized model of the world ocean. Part I: The global-scale water. Masses. Journal of Physical Oceanography, 19(11), 1730–1752. Available from: https://doi.org/10.1175/1520-0485(1989)019 < 1730:aimotw > 2.0.co;2.

Coxall, H.K., Huck, C.E., Huber, M., Lear, C.H., Legarda-Lisarri, A., O'Regan, M., et al., 2018. Export of nutrient-rich Northern Component Water preceded early Oligocene Antarctic glaciation. Nat. Geosci. 11 (3), 190–196. Available from: https://doi.org/10.1038/s41561-018-0069-9.

Cramer, B.S., Toggweiler, J.R., Wright, J.D., Katz, M.E., Miller, K.G., 2009. Ocean overturning since the Late Cretaceous: inferences from a new benthic foraminiferal isotope compilation. Paleoceanography 24 (4). Available from: https://doi.org/10.1029/2008pa0016.

Cvijanovic, I., Caldeira, K., MacMartin, D.G., 2015. Impacts of ocean albedo alteration on Arctic sea ice restoration and Northern Hemisphere climate. Environ. Res. Lett. 10 (4), 1.

Davies, R., Cartwright, J., Pike, J., Line, C., 2001. Early Oligocene initiation of North Atlantic Deep Water formation. Nature 410, 917–920. Available from: https://doi.org/10.1038/35073551.

de Boer, A.M., Toggweiler, J.R., Sigman, D.M., 2008. Atlantic dominance of the meridional overturning circulation. J. Phys. Oceanogr. 38, 435–450.

DeConto, R.M., Pollard, D., Wilson, P.A., Pälike, H., Lear, C.H., Pagani, M., 2008. Thresholds for Cenozoic bipolar glaciation. Nature 455 (7213), 652–656. Available from: https://doi.org/10.1038/nature07337.

Denton, G., Alley, R., Comer, G., Broecker, W., 2005. The role of seasonality in abrupt climate change. Quat. Sci. Rev. 24 (10–11), 1159–1182. Available from: https://doi.org/10.1016/j.quascirev.2004.12.002.

Dickson, R.R., Brown, J., 1994. The production of North Atlantic Deep Water: sources, rates, and pathways. J. Geophys. Res. 99 (C6), 12319. Available from: https://doi.org/10.1029/94jc00530.

Drijfhout, S., van Oldenborgh, G.J., Cimatoribus, A., 2012. Is a decline of AMOC causing the warming hole above the North Atlantic in observed and modeled warming patterns? J. Clim. 25, 8373–8379.

Driscoll, N.W., Haug, G.H., 1998. A short circuit in thermohaline circulation: a cause for Northern Hemisphere Glaciation? Science 282, 436–438.

Eagles, G., Livermore, R.A., Fairhead, J.D., Morris, P., 2005. Tectonic evolution of the west Scotia Sea. J. Geophys. Res.: Solid Earth 110 (B2). Available from: https://doi.org/10.1029/2004jb003154.

Eldrett, J.S., Harding, I.C., Firth, J.V., Roberts, A.P., 2004. Magnetostratigraphic calibration of Eocene–Oligocene dinoflagellate cyst biostratigraphy from the Norwegian-Greenland Sea. Mar. Geol. 204, 91–127.

Eldrett, J.S., Harding, I.C., Wilson, P.A., Butler, E., Roberts, A.P., 2007. Continental ice in Greenland during the Eocene and Oligocene. Nature 446, 176–179.

Fedorov, A.V., Pacanowski, R.C., Philander, S.G., Boccaletti, G., 2004. The effect of salinity on the wind-driven circulation and the thermal structure of the Upper Ocean. J. Phys. Oceanogr. 34 (9), 1949–1966. Available from: https://doi.org/10.1175/1520-0485(2004)034<1949:teosot>2.0.co;2.

Flower, B.P., Kennett, J.P., 1994. The middle Miocene climatic transition: East Antarctic ice sheet development, deep ocean circulation and global carbon cycling. Palaeogeogr. Palaeoclimatol. Palaeoecol. 108, 537–555. Available from: https://doi.org/10.1016/0031-0182(94)90251-8.

Frigola, A., Prange, M., Schulz, M., 2018. Boundary conditions for the Middle Miocene Climate Transition (MMCT v1.0). Geosci. Model. Dev. 11 (4), 1607–1626. Available from: https://doi.org/10.5194/gmd-11-1607-2018.

Fronval, T., Jansen, E., 1996. Late Neogene palaeoclimates and palaeoceanography in the Iceland-Norwegian Sea: evidence from the Iceland and Vøring Plateaus. Proc. Ocean. Drill. Program. Sci. Results 151, 455–468.

Gastineau, G., Li, L., Le Treut, H., 2009. The Hadley and Walker Circulation changes in global warming conditions described by idealized atmospheric simulations. J. Clim. 22 (14), 3993–4013. Available from: https://doi.org/10.1175/2009jcli2794.1.

Gildor, H., 2003. When Earth's freezer door is left ajar. EOS Trans. Am. Geophys. Union 84 (23), 215. Available from: https://doi.org/10.1029/2003eo230005.

Gildor, H., Tziperman, E., 2000. Sea ice as the glacial cycles climate switch: role of seasonal and Milankovitch forcing. Paleoceanography 15 (6), 605–615.

Gong, D.-Y., Wang, S.-W., Zhu, J.-H., 2001. East Asian Winter Monsoon and Arctic Oscillation. Geophys. Res. Lett. 28 (10), 2073–2076. Available from: https://doi.org/10.1029/2000gl012311.

Gordon, R.M.B., Putnam, A.E., Rademaker, K.M., Lowell, T.V., Schaefer, J.M., Hall, B., et al., 2014. Younger Dryas deglaciation of Scotland driven by warming summers. Proc. Natl. Acad. Sci. 111 (17), 6215–6219. Available from: https://doi.org/10.1073/pnas.1321122111.

Graham, R.M., Cohen, L., Petty, A.A., Boisvert, L.N., Rinke, A., Hudson, S.R., et al., 2017. Increasing frequency and duration of Arctic winter warming events. Geophys. Res. Lett. 44 (13), 6974–6983. Available from: https://doi.org/10.1002/2017gl073395.

Guarino, M.-V., Sime, L.C., Schröeder, D., Malmierca-Vallet, I., Rosenblum, E., Ringer, M., et al., 2020. Sea-ice-free Arctic during the Last Interglacial supports fast future loss. Nat. Clim. Change. Available from: https://doi.org/10.1038/s41558-020-0865-2.

Hall, A., 2004. The role of surface albedo feedback in climate. J. Clim. 17, 1550–1568.

Hand, E., 2016. New scrutiny for a slowing Atlantic conveyor. Science 352 (6287), 751–752. Available from: https://doi.org/10.1126/science.352.6287.751.

Hartmann, M., Blunier, T., Brugger, S.O., Schmale, J., Schwikowski, M., Vogel, A., et al., 2019. Variation of ice nucleating particles in the European Arctic over the last centuries. Geophys. Res. Lett. 46, 1–10.

Haug, G.H., Tiedemann, R., 1998. Effect of the formation of the Isthmus of Panama on Atlantic Ocean thermohaline circulation. Nature 393673–393676.

Hegewald, A., Jokat, W., 2013. Tectonic and sedimentary structures in the northern Chukchi region, Arctic Ocean. J. Geophys. Res.: Solid Earth 118 (7), 3285–3296. Available from: https://doi.org/10.1002/jgrb.50282.

Helland, P.E., Holmes, M.A., 1997. Surface textural analysis of quartz sand grains from ODP Site 918 off the southeast coast of Greenland suggests glaciation of southern Greenland at 11 Ma. Palaeogeogr. Palaeoclimatol. Palaeoecol. 135, 109–121.

Holbourn, A., Kuhnt, W., Schulz, M., Erlenkeuser, H., 2005. Impacts of orbital forcing and atmospheric carbon dioxide on Miocene ice-sheet expansion. Nature 438, 483–487.

Holland, M.M., Bitz, C.M., 2003. Polar amplification of climate change in coupled models. Clim. Dyn. 21 (3–4), 221–232. Available from: https://doi.org/10.1007/s00382-003-0332-.

Hu, Y., Huang, H., Zhou, C., 2018. Widening and weakening of the Hadley circulation under global warming. Sci. Bull. 63 (10), 640–644.

Jackson, L.C., Peterson, K.A., Roberts, C.D., Wood, R.A., 2016. Recent slowing of Atlantic Overturning Circulation as a recovery from earlier strengthening. Nat. Geosci. 9, 518522.

Jakobsson, M., Backman, J., Rudels, B., Nycander, J., Frank, M., Mayer, L., et al., 2007. The early Miocene onset of a ventilated circulation regime in the Arctic Ocean. Nature 447 (7147), 986–990. Available from: https://doi.org/10.1038/nature05924.

Jansen, E., Sjøhølm, J., 1991. Reconstruction of glaciation over the past 6 Myr from ice-borne deposits in the Norwegian Sea. Nature 349, 600–603.

Keigwin, L.D., 1982. Isotope paleoceanography of the Caribbean and east Pacific: role of Panama uplift in late Neogene time. Science 217, 350–353.

Kennett, J.P., 1977. Cenozoic evolution of Antarctic glaciation, the circum-Antarctic Ocean, and their impact on global paleoceanography. J. Geophys. Res. 82 (27), 3843–3860. Available from: https://doi.org/10.1029/jc082i027p03843.

Kharin, G.S., Lukashina, N.P., 2010. Paleogeography of the Norwegian-Greenland and Northwestern European Sea Basins in the Paleogene. Oceanology 50 (2), 226–239. Available from: https://doi.org/10.1134/s0001437010020086.

Knies, J., Gaina, C., 2008. Middle Miocene ice sheet expansion in the Arctic: views from the Barents Sea. Geochem. Geophys. Geosyst. 9 (2). Available from: https://doi.org/10.1029/2007gc001824.

Koltermann, K.P., Lüthje, H., 1989. Hydrographic Atlas of the Greenland and Northern Norwegian Seas (1979–1987). Deutsches Hydrographisches Institut, Hamburg, No. 2328.

Krissek, L.A., 1995. Late Cenozoic ice-rafting records from Leg 145Sites in the North Pacific: Late Miocene onset, Late Pliocene intensification, and Plio-Pleistocene Events. In: Rea, D.K., Basov, I.A., Scholl, D.W., Allan, J.F. (Eds.), Proc. ODP, Sci. Results. Ocean Drilling Program, College Station, TX, pp. 179–194.

Kuhlbrodt, T., Griesel, A., Montoya, M., Levermann, A., Hofmann, M., Rahmstorf, S., 2007. On the driving processes of the Atlantic meridional overturning circulation. Rev. Geophys. 45 (2). Available from: https://doi.org/10.1029/2004rg000166.

Landrum, L., Holland, M.M., 2020. Extremes become routine in an emerging new Arctic. Nat. Clim. Change. Available from: https://doi.org/10.1038/s41558-020-0892-z.

Langehaug, H.R., Falck, E., 2012. Changes in the properties and distribution of the intermediate and deep waters in the Fram Strait. Progr. Oceanogr. 96, 57–76. Available from: https://doi.org/10.1016/j.pocean.2011.10.002.

Larsen, H.C., Saunders, A.D., Clift, P.D., Beget, J., Wei, W., Spezzaferri, S., 1994. Seven million years of glaciation in Greenland. Science 264 (5161), 952–955. Available from: https://doi.org/10.1126/science.264.5161.952.

Livermore, R., Hillenbrand, C.-D., Meredith, M., Eagles, G., 2007. Drake Passage and Cenozoic climate: an open and shut case? Geochem. Geophys. Geosyst. 8 (1). Available from: https://doi.org/10.1029/2005gc001224.

Lyle, M., Gibbs, S., Moore, T.C., Rea, D.K., 2007. Late Oligocene initiation of the Antarctic Circumpolar Current: evidence from the South Pacific. Geology 35 (8), 691. Available from: https://doi.org/10.1130/g23806a.1.

Maier-Reimer, E., Mikolajewicz, U., Crowley, T.J., 1990. Ocean general circulation model sensitivity experiment with an open American Isthmus. Paleoceanography 5, 349–366.

Manabe, S., Stouffer, R.J., 1980. Sensitivity of a global climate model to an increase of CO_2 concentration in the atmosphere. J. Geophys. Res. 85, 5529–5554.

Manabe, S., Stouffer, R.J., 1995. Simulation of abrupt climate change induced by freshwater input to the North Atlantic Ocean. Nature 378 (6553), 165–167. Available from: https://doi.org/10.1038/378165a0.

Manabe, S., Stouffer, R.J., 1999. Are two modes of thermohaline circulation stable? Tellus 51A, 400–411.

Manabe, S., Wetherald, R.T., 1975. The effect of doubling the CO_2 concentration on the climate of a general circulation model. J. Atm. Sci. 32, 3–15.

Masuda, K., 1988. Meridional heat transport by the atmosphere and the ocean: analysis of FGGE data. Tellus A 40A (4), 285–302. Available from: https://doi.org/10.1111/j.1600-0870.1988.tb00348.x.

McCarthy, G.D., Frajka-Williams, E., Johns, W.E., et al., 2012. Observed interannual variability of the Atlantic meridional overturning circulation at 26.5°N. Geophys. Res. Lett. 39, L19609.

McCave, I.N., Tucholke, B.E., 1986. Deep current controlled sedimentation in the western North Atlantic. In: The Geology of North America, M. The Western North Atlantic Region. Geol. Soc. Am. Bull., Boulder, CO, pp. 451–468.

Meincke, J., 1983. The modern current regime across the Greenland-Scotland Ridge. In: Bott, M.H.P., Saxov, S., Talwani, M., Thiede, J. (Eds.), Structure and Development of the Greenland 206 Quaternary Science Reviews: Volume 17 Scotland Ridge: New Methods and Concepts. Plenum, New York, pp. 637–650.

Miller, K.G., Tucholke, B.E., 1983. Development of Cenozoic abyssal circulation south of the Greenland-Scotland Ridge. In: Bott, M., et al., (Eds.), Structure and Development of the Greenland-Scotland Ridge: New Methods and Concepts, 8. NATO Conf. Ser. IV, pp. 549–589.

Miller, G.H., and A. de Vernal, Will greenhouse warming lead to Northern Hemisphere ice-sheet growth?, 1992 *Nature*, 355, 244–246.

Miller K.G., K.G., Wright, J.D., Katz, M.E. Wade, B.S. Browning, J.V., Cramer, B., Rosenthal, Y., 2009. Climate threshold at the Eocene-Oligocene transition: AIS influence on ocean circulation. Special Paper of the Geological Society of America 452:169–178. Available from: https://doi.org/10.1130/2009.2452(11).

Mokeddem, Z., McManus, J.F., Oppo, D.W., 2014. Oceanographic dynamics and the end of the last interglacial in the subpolar North Atlantic. Proc. Natl. Acad. Sci. 111 (31), 11263–11268. Available from: https://doi.org/10.1073/pnas.1322103111.

Moon, T.A., Overeem, I., Druckenmiller, M., Holland, M., Huntington, H., Kling, G., et al., 2019. The expanding footprint of rapid Arctic change. Earth's Future 7, 212–218.
Moore Jr., T.C., Van Andel, Tj. H., Sancetta, C., Pisias, N., 1978. Cenozoic hiatuses in pelagic sediments. Micropaleontology 24, 113–138.
Moran, K., Backman, J., Brinkhuis, H., Clemens, S.C., Cronin, T., Dickens, G.R., Kristoffersen, Y., 2006. *The Cenozoic palaeoenvironment of the Arctic Ocean*. Nature 441 (7093), 601–605. Available from: https://doi.org/10.1038/nature04800.
Mountain, G.S., Miller, K.G., 1992. Seismic and geologic evidence for early Paleogene deepwater circulation in the western North Atlantic. Paleoceanography 7 (4), 423–439. Available from: https://doi.org/10.1029/92PA01268.
Musatov, E.E.L., Pogrebitskij, Y.E., 2000. Late Mesozoic-Cenozoic Evolution of the Barents Sea and Kara Sea Continental Margins. Polarforschung 68, 283–290.
Muschitiello, F., D'Andrea, W.J., Schmittner, A., Heaton, T.J., Balascio, N.L., deRoberts, N., et al., 2019. Deep-water circulation changes lead North Atlantic climate during deglaciation. Nat. Commun. 10 (1). Available from: https://doi.org/10.1038/s41467-019-09237-3.
Nowaczyk, N.R., Haltia, E.M., Ulbricht, D., Wennrich, V., Sauerbrey, M.A., Rosen, P., et al., 2013. Chronology of Lake El'gygytgyn sediments. Clim. Past Discuss. 9, 3061–3102. Available from: https://doi.org/10.5194/cpd-9-3061-2013.
Oakey, G.N., Chalmers, J.A., 2012. A new model for the Paleogene motion of Greenland relative to North America: plate reconstructions of the Davis Strait and Nares Strait regions between Canada and Greenland. J. Geophys. Res.: Solid Earth 117 (B10). Available from: https://doi.org/10.1029/2011jb008942.
Perez, F.F., Mercier, H., Vazquez-Rodriguez, M., L'herminier, P., Velo, A., Pardo, P., et al., 2013. Atlantic Ocean CO_2 uptake reduced by weakening of the meridional overturning circulation. Nat. Geosci. 6, 146–152. Available from: https://doi.org/10.1038/ngeo1680.
Perovich, D.K., et al., 2007. Increasing solar heating of the Arctic Ocean and adjacent seas, 1979–2005: attribution and role in the ice-albedo feedback. Geophys. Res. Lett. 34, L19505.
Pfuhl, H.A., McCave, I.N., 2005. Evidence for late Oligocene establishment of the Antarctic Circumpolar Current. Earth Planet. Sci. Lett. 235 (3–4), 715–728. Available from: https://doi.org/10.1016/j.epsl.2005.04.025.
Polyak, L., Jakobsson, M., 2015. Quaternary sedimentation in the Arctic Ocean: recent advances and further challenges. Oceanography 24 (3), 53–64.
Pomerol, C., 1982. The Cenozoic Era: Tertiary and Quaternary. Ellis Horwood for the British Micropalaeontological Society, Chichester, 272 pp., numerous illustrations.
Poole, D.A.R., Vorren, T.O., 1993. Miocene to Quaternary paleoenvironments and uplift history on the mid Norwegian shelf. Mar. Geol. 115, 173–205.
Quadfasel, D., Gascard, J.-C., Koltermann, P.K., 1987. Large-scale oceanography in Fram Strait during the 1984 Marginal Ice Zone experiment. J. Geophys. Res. 92, 6719–6728.
Rahmstorf, S., 1995. 'Bifurcations of the Atlantic thermohaline circulation in response to changes in the hydrological cycle. Nature 378, 145–149.
Rahmstorf, S., Box, J.E., Feulner, G., Mann, M.E., Robinson, A., Rutherford, S., et al., 2015. Exceptional twentieth-century slowdown in Atlantic Ocean Overturning Circulation. Nat. Clim. Change 5, 475–480.
Rea, D.K., Lyle, M.W., 2005. Paleogene calcite compensation depth in the eastern subtropical Pacific: answers and questions. Paleoceanography 20. Available from: https://doi.org/10.1029/2004PA001064. PA1012.
Rea, DK, Snoeckx, H, Joseph, LH, 1998. Late Cenozoic Eolian deposition in the North Pacific: Asian drying, Tibetan uplift, and cooling of the Northern Hemisphere. Paleoceanography 13, 215–224.

Rial, J.A., Pielke, R.A., Beniston, M., et al., 2004. Nonlinearities, feedbacks and critical thresholds within the Earth's climate system. Clim. Change 65, 11–38. Available from: https://doi.org/10.1023/B:CLIM.0000037493.89489.3f.

Roemmich, D., Wunsch, C., 1985. Two transatlantic sections: meridional circulation and heat flux in the subtropical North Atlantic Ocean. Deep-Sea Res. 32, 619–664.

Scher, H.D., Whittaker, J.M., Williams, S.E., Latimer, J.C., Kordesch, W.E.C., Delaney, M.L., 2015. Onset of Antarctic Circumpolar Current 30 million years ago as Tasmanian Gateway aligned with westerlies. Nature 523 (7562), 580–583. Available from: https://doi.org/10.1038/nature14598.

Screen, J.A., Simmonds, I., 2010. The central role of diminishing sea ice in recent Arctic temperature amplification. Nature 464 (7293), 1334–1337. Available from: https://doi.org/10.1038/nature09051.

Seidov, D., Stouffer, R.J., Haupt, B.J., 2005. Is there a simple bi-polar ocean see-saw? Glob. Planet. Change 49 (1–2), 19–27. Available from: https://doi.org/10.1016/j.gloplacha.2005.05.001.

Serreze, M.C., et al., 2003. A record minimum arctic sea ice extent and area in 2002. Geophysical Research Letters 30 (3). Available from: https://doi.org/10.1029/2002gl016406.

Serreze, M.C., Barry, R.G., 2011. Processes and impacts of Arctic amplification: a research synthesis. Glob. Planet. Change 77 (1–2), 85–96. Available from: https://doi.org/10.1016/j.gloplacha.2011.03.004.

Sévellec, F., Fedorov, A.V., Liu, W., 2017. Arctic sea-ice decline weakens the Atlantic Meridional Overturning Circulation. Nat. Clim. Change 7 (8), 604–610. Available from: https://doi.org/10.1038/nclimate3353.

Severinghaus, J.P., 2009. Southern see-saw seen. Nature 457 (7233), 1093–1094. Available from: https://doi.org/10.1038/4571093a.

Shevenell, A.E., Kennett, J.P., Lea, D.W., 2004. Middle Miocene Southern Ocean cooling and Antarctic cryosphere expansion. Science 305, 1766–1770.

Sijp, W.P., England, M.H., 2004. Effect of the Drake Passage throughflow on global climate. J. Phys. Oceanogr. 34 (5), 1254–1266. Available from: https://doi.org/10.1175/1520-0485(2004)034 < 1254:eotdpt > 2.0.co;2.

Sijp, W.P., England, M.H., Huber, M., 2011. Effect of the deepening of the Tasman Gateway on the global ocean. Paleoceanography 26 (4). Available from: https://doi.org/10.1029/2011pa002143.

Smeed, D.A., McCarthy, G.D., Cunningham, S.A., Frajka-Williams, E., Rayner, D., Johns, W.E., et al., 2014. Observed decline of the Atlantic meridional overturning circulation 2004–2012. Ocean Sci. 10 (1), 29–38. Available from: https://doi.org/10.5194/os-10-29-2014.

Spielhagen, R.F., Werner, K., Sorensen, S.A., Zamelczyk, K., Kandiano, E., Budeus, G., et al., 2011. Enhanced modern heat transfer to the Arctic by warm Atlantic Water. Science 331 (6016), 450–453. Available from: https://doi.org/10.1126/science.1197397.

Srinivasan, M.S., Sinha, D.K., 1998. Early Pliocene closing of the Indonesian Seaway: evidence from northeast Indian Ocean and southwest Pacific deep sea cores. J. Southeast Asian Earth Sci. 16 (1), 29–44. Pergamon Press, UK.

Srinivasan, M.S., Sinha, D.K., 2000. Ocean circulation changes in the Indo-Pacific during 5.6 to 4.2 Ma: Planktic foraminiferal and isotopic evidences. Earth Planet. Sci. Indian Acad. Sci 109 (3), 315–328.

Srokosz, M., Bryden, H.L., 2015. Observing the Atlantic Meridional Overturning Circulation yields a decade of inevitable surprises. Science 348, 1255575.

St. John, K. 2008. Cenozoic ice-rafting history of the central Arctic Ocean: Terrigenous sands on the Lomonosov Ridge. Paleoceanography, 23(1), n/a–n/a. Available from: https://doi.org/10.1029/2007pa001483.

St. John, K.E.K., 1999. Site 918 IRD mass accumulation rate record, late Miocene–Pleistocene. Ocean Drill. Program. Proc., Sci. Results 163, 163–166.
St. John, K.E.K., Krissek, L.A., 1999. Regional patterns of Pleistocene ice-rafted debris flux in the North Pacific. Paleoceanography 14, 653–662.
Stärz, M., Jokat, W., Knorr, G., Lohmann, G., 2017. Threshold in North Atlantic-Arctic Ocean circulation controlled by the subsidence of the Greenland-Scotland Ridge. Nat. Commun. 8, 15681. Available from: https://doi.org/10.1038/ncomms15681.
Stein, R., 2019. The late Mesozoic-Cenozoic Arctic Ocean climate and sea ice history: a challenge for past and future scientific ocean drilling. Paleoceanogr. Paleoclimatol. Available from: https://doi.org/10.1029/2018pa003433.
Stjern, C.W., Lund, M.T., Samset, B.H., Myhre, G., Forster, P.M., Andrews, T., et al., 2019. Arctic amplification response to individual climate drivers. J. Geophys. Res.: Atmos. Available from: https://doi.org/10.1029/2018jd029726.
Stocker, T., Schmittner, A., 1997. Influence of CO_2 emission rates on the stability of the thermohaline circulation. Nature 388, 862–865.
Stocker, T., et al., 2013. Climate Change (2013): The Physical Science Basis. Contribution of Working Group I to the Fifth Assessment Report of the Intergovernmental Panel on Climate Change. Cambridge University Press, Cambridge and New York.
Stroeve, J.C., Serreze, M.C., Holland, M.M., Kay, J.E., Malanik, J., Barrett, A.P., 2011. The Arctic's rapidly shrinking sea ice cover: a research synthesis. Climatic Change 110 (3–4), 1005–1027. Available from: https://doi.org/10.1007/s10584-011-0101-1.
Super, J.R., Thomas, E., Pagani, M., Huber, M., O'Brien, C.L., Hull, P.M., 2020. Miocene evolution of North Atlantic sea surface temperature. Paleoceanogr. Paleoclimatol Available from: https://doi.org/10.1029/2019pa003748e2019PA003748.
Takahashi, T., Sutherland, S.C., Wanninkhof, R., Sweeney, C., Feely, R.A., Chipman, D. W., et al., 2009. Climatological mean and decadal change in surface ocean pCO_2, and net sea-air CO_2 flux over the global oceans. Deep. Sea Res. Part II: Topical Stud. Oceanogr. 56 (8–10), 554–577. Available from: https://doi.org/10.1016/j.dsr2.2008.12.009.
Tarasov, P.E., Andreev, A.A., Anderson, P.M., Lozhkin, A.V., Leipe, C., Haltia, E., et al., 2013. A pollen-based biome reconstruction over the last 3.562 million years in the Far East Russian Arctic – new insights into climate–vegetation relationships at the regional scale. Clim. Past. 9, 2759–2775.
Taylor, P.C., Cai, M., Hu, A., Meehl, J., Washington, W., Zhang, G.J., 2013. A decomposition of feedback contributions to polar warming amplification. J. Clim. 26 (18), 7023–7043. Available from: https://doi.org/10.1175/jcli-d-12-00696.1.
Thiede, J., Winkler, A., Wolf Welling, T., Eldholm, O., Myhre, A.M., Baumann, K.-H., et al., 1998. Late Cenozoic History of the polar North Atlantic: results from the Ocean Drilling. Quat. Sci. Rev. 17 (1–3), 185–208. Available from: https://doi.org/10.1016/s0277-3791(97)00076-0.
Tiedemann, R., Franz, S.O., 1997. Deep-water circulation, chemistry, and terrigenous sediment supply in the equatorial Atlantic during the Pliocene, 3.3–2.6 Ma and 5–4.5 Ma. Proc. ODP Sci. Res. 154, 299–318.
Toggweiler, J.R., Bjornsson, H., 2000. Drake Passage and palaeoclimate. J. Quat. Sci. 15 (4), 319–328. Available from: https://doi.org/10.1002/1099-1417(200005)15:4 < 319::AID-JQS545 > 3.0.CO;2-C.
Toggweiler, R., Samuels, B., 1995. Effect of Drake Passage on the global thermohaline circulation. Deep Sea Res. I 42, 477–500.
Toggweiler, J.R., Samuels, B., 1998. On the ocean's large-scale circulation near the limit of no vertical mixing. J. Phys. Oceanogr. 28, 1832–1852.
Tripati, A., Backman, J., Elderfield, H., Ferretti, P., 2005. Eocene bipolar glaciation associated with global carbon cycle changes. Nature 436 (7049), 341–346. Available from: https://doi.org/10.1038/nature03874.

Tripati, A.K., Eagle, R.A., Morton, A., Dowdeswell, J.A., Atkinson, K.L., Bahé, Y., et al., 2008. Evidence for glaciation in the Northern Hemisphere back to 44 Ma from ice-rafted debris in the Greenland Sea. Earth Planet. Sci. Lett. 265 (1–2), 112–122. Available from: https://doi.org/10.1016/j.epsl.2007.09.045.

van der Veen, C.J., 2002. Polar ice sheets and global sea level: how well can we predict the future? Glob. Planet. Change 32, 165–194.

Via, R.K., Thomas, D.J., 2006. Evolution of Atlantic thermohaline circulation: Early Oligocene onset of deepwater production in the North Atlantic. Geology 34, 441–444.

Warnke, D.A., Hansen, M.E., Hayward/Calif, 1977. Sediments of Glacial Origin in the Area of Operations of D.S.D.P. Leg 38 (Norwegian—Greenland Seas): Preliminary Results from Sites 336 and 344 Ber. Naturf. Ges. Freiburg i. Br. — Pfannenstiel Gedenkband — 67 S. 371—392 Freiburg.

Wennrich, V., Minyuk, P.S., Borkhodoev, V. Ya, Francke, A., Rit-ter, B., Nowaczyk, N., et al., 2014. Pliocene to Pleistocene climate and environmental history of Lake El'gygytgyn, Far East Russian Arctic, based on high-resolution inorganic geochemistry data. Clim. Past 10, 1381–1399.

Wolf, T.C., Thiede, J., 1991. History of terrigenous sedimentation during the past 10 m. y. in the North Atlantic (ODP Legs 104 and 105 and DSDP Leg 81). Mar. Geol. 101 (1–4), 83–102. Available from: https://doi.org/10.1016/0025-3227(91)90064-b.

Wolf-Welling, T.C.W., Cremer, M., O'Connell, S., Winkler, A., Thiede, J., 1996. Cenozoic Arctic Gateway Paleoclimate variability: indications from changes in coarse-fraction composition. In: Thiede, J., Myhre, A.M., Firth, J.V., Johnson, G.L., Ruddiman, W.F. (Eds.), Ocean Drill. Program Proc., Sci. Results, 151. ODP, College Station, TX, pp. 515–568.

Woodruff, F., Savin, S.M., 1989. Miocene deepwater oceanography. Paleoceanography 4, 87–140.

Wright, J.D., Miller, K.G., 1996. Control of North Atlantic Deep Water Circulation by the Greenland-Scotland Ridge. Paleoceanography 11 (2), 157–170. Available from: https://doi.org/10.1029/95pa03696.

Wright, J.D., Miller, K.G., Fairbanks, R.G., 1992. Miocene stable isotopes: implications for deepwater circulation and climate. Paleoceanography 7, 357–389.

Wunsch, C., 1998. The work done by the wind on the oceanic general circulation. J. Phys. Oceanogr. 28, 2332–2340.

Zhao, J., Johns, W., 2014. Wind-forced interannual variability of the Atlantic Meridional Overturning Circulation at 26.5_N. J. Geophys. Res. 119, 2403–2419.

CHAPTER 6

Teleconnection between Arctic climate and tropical Indian monsoon during the Holocene

Anil K. Gupta[1], Som Dutt[2], Moumita Das[3] and Raj K. Singh[4]
[1]Department of Geology and Geophysics, Indian Institute of Technology Kharagpur, Kharagpur, India
[2]Wadia Institute of Himalayan Geology, Dehradun, India
[3]Mahila Mahavidyalaya, Banaras Hindu University, Varanasi, India
[4]School of Earth, Ocean and Climate Sciences, Indian Institute of Technology, Bhubaneswar, India

6.1 Introduction

The agriculture-based economy of south and southeast Asia, including the Indian subcontinent, is primarily dependent on precipitation contributed by the Indian monsoon, which is characterized by seasonal reversals in wind directions. Monsoon winds are suggested as a manifestation of the land–sea pressure gradient and latitudinal positioning of the Inter Tropical Convergence Zone (ITCZ) (Halley, 1686; Quade et al., 1995; Fleitmann et al., 2003; Gadgil, 2003). During summer, a pressure gradient develops between the ocean and Indian landmass due to their differential heating, and winds start blowing from the ocean to the continent, generally associated with large-scale precipitation on continental areas. The direction reverses in winter and winds blow from the northeast to the southwest, that is, from the continent to the ocean. The Asian monsoon system comprising of the Indian summer monsoon (ISM) and East Asian summer monsoon plays a pivotal role in large-scale climate variability in terms of global heat and moisture transport from low to high latitudes (Wang et al., 2001, 2017; Tada et al., 2018). During the northern hemisphere summers, moisture-laden winds from the Indian Ocean blow toward the Indian subcontinent, bringing abundant rainfall across major parts of the Indian subcontinent between June and September (Parthasarathy et al., 1993; Quade et al., 1995; Gadgil, 2003; Shewale and Kumar, 2005). Winter or northeast monsoon winds, in contrast, blow from the continent to the Indian Ocean and are relatively dry, supplying moisture to a small part of southern India.

Understanding Present and Past Arctic Environments
DOI: https://doi.org/10.1016/B978-0-12-822869-2.00001-3

Instrumental as well as proxy records from natural archives including oceans, lakes, speleothems, peat bogs, rivers, tree rings, and ice cores suggest that ISM rainfall is highly variable and influenced by several factors. On a sub-annual scale, the strength of the ISM is influenced by an active-break cycle, while other factors such as solar insolation, temperature of the northern hemisphere, latitudinal positioning of the ITCZ, El Niño-Southern Oscillation (ENSO), Indian Ocean Dipole (IOD), snow conditions in the Himalaya and western Europe, temperature conditions in the Arctic and associated sea ice conditions in the North Atlantic as well as North Atlantic Meridional Oscillation (AMOC) play important roles at multiple time scales ranging from subdecadal to millennial and orbital time scales (Bamzai and Shukla, 1999; Ashok et al., 2001; Neff et al., 2001; Fleitmann et al., 2003; Gupta et al., 2003, 2005; Kumar et al., 2006; Gupta et al., 2015; Sinha et al., 2015; Dutt et al., 2018). Numerous studies have reported decadal to centennial-scale changes in monsoon precipitation since the Last Glacial Maximum (LGM) particularly during the Holocene (Gupta et al., 2003, 2005, 2013; Dutt et al., 2015; Das et al., 2017, etc.). The abrupt climatic events such as Bølling-Ållerød and Younger Dryas (YD) also had pronounced effects on monsoon precipitation (Dutt et al., 2015; Partin et al., 2015). The mechanism that drives these decadal- to centennial-scale or abrupt changes in monsoon precipitation is either linked to solar variability (Gupta et al., 2005, 2013; Dutt et al., 2015; Das et al., 2017) or movement of the ITCZ (Fleitmann et al., 2003). Changes in the AMOC have also been reported as a crucial factor for monsoon variability (Partin et al., 2015; Schlolaut et al., 2017). This study summarizes tropical monsoon variability during the Holocene as recorded in various marine and continental proxies and its linkages with climatic variability over the Arctic region (Fig. 6.1). We would also like to understand the influence of monsoon variability on South Asian societies and their response to abrupt monsoon shifts.

6.2 Holocene Indian summer monsoon variability

The Holocene period [11.7 kyr (thousand years) to the Present] was earlier considered as a time interval of warm climate with secular variations. However, the availability of high-resolution proxy records and recent modeling approaches indicate that Holocene climate also varied considerably with repeated occurrences of centennial- to millennial-scale warm and cold intervals in the North Atlantic, which are aligned with strong

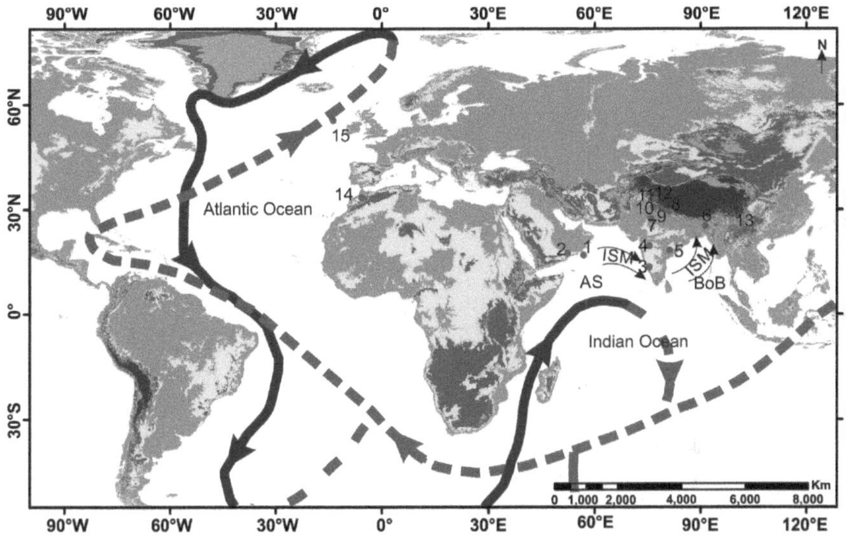

Figure 6.1 Locations of the paleoclimatic proxy records on DEM (GTOPO 30) images (https://earthexplorer.usgs.gov/) for the Indian monsoon and compared with records of North Atlantic sites (round dots). (1) Hole 723A, Arabian Sea (Gupta et al., 2005, 2013; Das et al., 2017), (2) Qunf cave, Oman (Fleitmann et al., 2003), (3) Shanti Sagara Lake, southern India (Sandeep et al., 2017), (4) Lonar Lake, Central India (Menzel et al., 2014), (5) Kotumsar cave, central-eastern India (Band et al., 2018), (6) Mawmluh cave, northeastern India (Berkelhammer et al., 2012; Dutt et al., 2015), (7) Kotla Dahar paleolake, north-western India (Dixit et al., 2014), (8) Lake Rara, Nepal (Nakamura et al., 2016), (9) Benital Lake, Uttarakhand (Bhushan et al., 2018), (10) Sahiya cave, Uttarakhand (Kathayat et al., 2017), (11) Tso Moriri Lake, Ladakh (Mishra et al., 2015), (12) Chandra peat bog, Himachal (Rawat et al., 2015), (13) Xiaobailong cave, China (Cai et al., 2015), (14) Charra cave (Brahim et al., 2019), and (15) MC52-VM29–191 from the North Atlantic (Bond et al., 2001). Thin curve lines are Indian summer monsoon at Arabian Sea (AS) and Bay of Bengal (BoB). A generalised schematic diagram of the oceanic thermohaline circulation is shown by thick lines - dashed line indicates warm surface flow and continuous line indicate cold subsurface flow.

and weak monsoon conditions, respectively, in the tropical region. Therefore, climatic shifts in the North Atlantic have potential effects on the tropical climate in the Indian Ocean (Fig. 6.1).

Proxy records of the ISM suggest significant variability in rainfall in the Indian subcontinent during the Holocene. The monsoon was strong, and high rainfall conditions prevailed in the Indian subcontinent during the early Holocene (Greenlandian)—a time often known as the Holocene wet phase or Holocene Climatic Optimum, followed by drier conditions in the late Holocene as demonstrated by marine, lacustrine, and speleothem records across south Asia (Fig. 6.1).

6.2.1 Marine proxies

The ocean proxies of ISM variability are well preserved in the Arabian Sea and the Bay of Bengal (BoB). Two recently concluded International Ocean Discovery Program (IODP) expeditions in the BoB (Clemens et al., 2016) and Arabian Sea (Pandey et al., 2016) were aimed at understanding ISM variability on multiple time scales. These two IODP expeditions were aimed at complementing the results of previously drilled Ocean Drilling Program Leg 117 in the western Arabian Sea, which was earlier drilled with the main focus of unraveling the history of the Indian monsoon (Prell et al., 1992).

In each summer, seasonal reversal of winds that blow across the Arabian Sea causes upwelling adjacent to the Oman margin. The upwelling cools the sea surface and brings nutrients to the surface, leading to blooming of distinct fauna and flora including planktic foraminifera *Globigerina bulloides* (Fig. 6.1; Prell et al., 1992; Anderson et al., 2002; Gupta et al., 2003, 2005, 2013). Gupta et al. (2003) reconstructed the history of monsoon variability during the Holocene and found that during the cold intervals, the ISM exhibited weakening with the repeated occurrence of millennial-scale abrupt shifts, which were aligned to millennial-scale cold phases in the Arctic and North Atlantic (Fig. 6.2). The multidecadal to centennial time scale changes in the ISM appear to be closely associated with solar activity (Gupta et al., 2005; Gupta et al., 2013).

The ISM-induced biogenic blooms cause intensification of the oxygen minimum zone (OMZ) in the western Arabian Sea that spreads across the entire length and width of the Arabian Sea. Changes in the intensity of the Arabian Sea OMZ have been closely correlated with the ISM strength (denDulk et al., 2000), changes in intermediate water circulation (Fig. 6.2; Das et al., 2017), and fluctuations in the AMOC during the LGM (Böning and Bard, 2009). Das et al. (2017) suggested that intensification of the Arabian Sea OMZ could be associated with the replacement of oxygen-rich southern source sub-Antarctic mode and Antarctic Intermediate Waters with oxygen-poor Red Sea water during the Northgrippian time. Various other authors have proposed solar forcing, El-Nino Southern Oscillation, and IOD as triggers for ISM variability on multi-decadal to centennial time scales. However, the millennial-scale

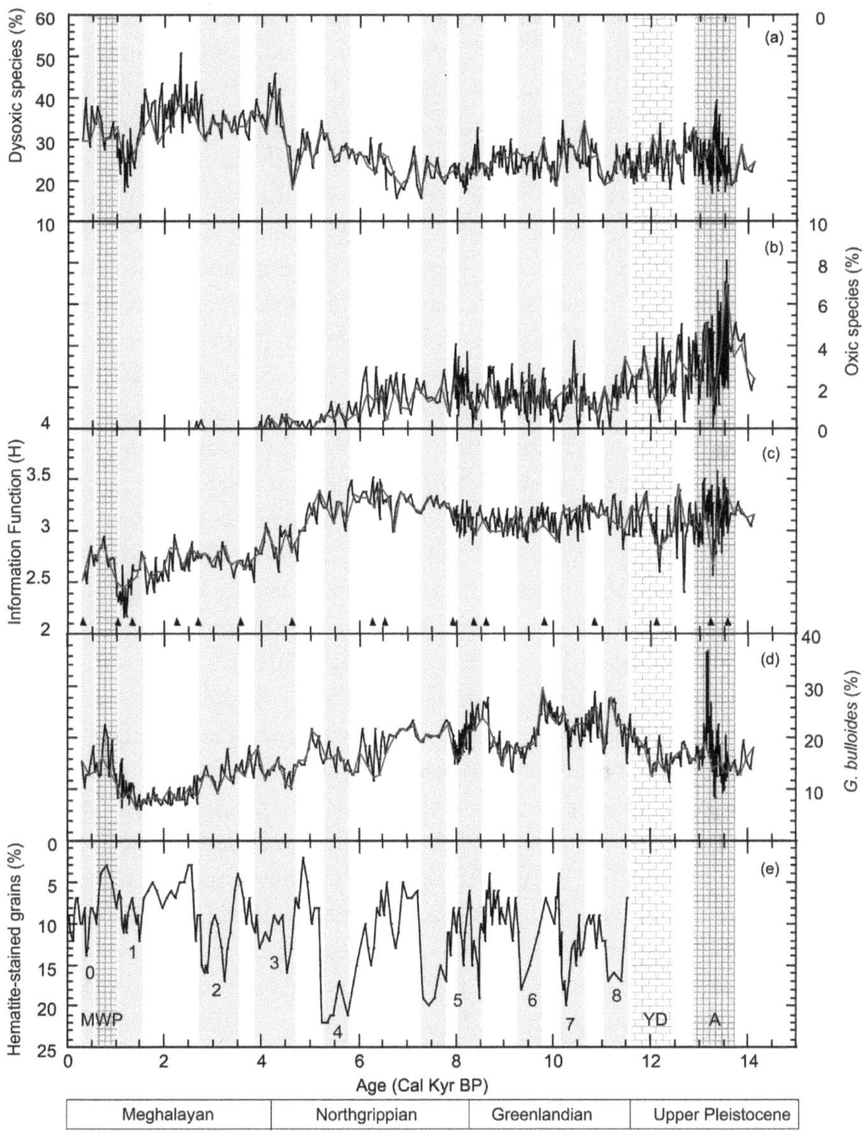

Figure 6.2 Benthic foraminifera (a) Dysoxic, (b) Oxic species (%) abundance (c) diversity in terms of Information Function (H), and (d) Planktic foraminifera *Globigerina bulloides* (%) abundance at Hole 723A, from the Arabian Sea (Gupta et al 2005; Gupta et al 2013; Das et al 2017) and (e) Hematite-stained grains (%) in core MC52-VM29-191 from the North Atlantic are plotted against numerical ages and Bond events labelled from 0–8 (Bond et al., 2001). Grey bars mark the Bond events, Square filled bars show a relatively warm period (Ållerød Period and Medieval Climate Anomaly) and bricks filled bar marks the Younger Dryas (YD). Triangles mark the AMS ^{14}C dates used for age interpolation.

variability in the ISM is linked with changes in the North Atlantic and Arctic climates (Bond et al., 2001; Agnihotri et al., 2003; Gupta et al., 2005; Tiwari et al., 2010; Saraswat et al., 2013; Satpathy et al., 2019).

The BoB receives 8% of the total riverine supply of the world oceans contributed by Himalayan and Peninsular rivers (Milliman and Syvitski, 1992). Peak changes in the BoB occur during the southwest monsoon season, which stratifies the near-surface water affecting marine productivity and fauna. The increased freshwater influx significantly influences the surface water salinity and sea surface temperature (SST), as observed in the oxygen isotope and Mg/Ca-derived SST records (Clemens et al., 2016; Naidu and Govil, 2017; The Expedition 353 Scientists, 2016). The Holocene weakening/strengthening of the ISM causes low/high influx of freshwater, sediment discharge, and associated chemical changes in the northern BoB (Kudrass et al., 2001; Govil and Naidu et al., 2011; Sijinkumar et al., 2016; Naik et al., 2019). The glacial—deglacial climatic conditions in the Himalayan and Peninsular region influenced the sediment supply to the BoB with reduced sediment supply during the LGM due to the combined effect of glaciations in the Himalaya and reduced influence of the monsoon (Tripathy et al., 2011, 2014).

During the Greenlandian, sediment discharge significantly increased from the Himalayan and Peninsular rivers with the strengthening of the ISM. Various factors influenced decadal- to centennial-scale changes in the BoB, but millennial-scale changes in the BoB and Andaman Sea were associated with North Atlantic and Arctic climates (Clemens et al., 2016; Sijinkumar et al., 2016; Miriyala et al., 2017). The various isotopic, geochemical, and foraminifer-based proxies that respond to large-scale hydrological changes in the BoB on a millennial scale suggest strong teleconnection with the North Atlantic and Arctic Climate (Kudrass et al., 2001; Marzin et al., 2013; Sijinkumar et al., 2016; Miriyala et al., 2017; Nilsson- Kerr et al., 2019). Overall the ISM was strong during the Greenlandian, and subsequently weakened during the Northgrippian. During the Meghalayan Age significant monsoon shifts were observed during the Little Ice Age (LIA), Medieval Climate Anomaly (MCA), and Current Warm Period (CWP).

6.2.2 Speleothem proxies

Speleothems are widely used to reconstruct high-resolution records of past monsoon variability in the South Asian region using stable oxygen isotope ratios. Speleothems can be dated precisely using a U-Th disequilibrium series dating method. Oxygen isotope ratios in speleothems from the tropical regions reflect the changes in meteoric rainfall, which in itself depends on rainfall amount, the distance of moisture traveled, and moisture sources such as ISM (the BoB or Arabian Sea branch) and mid-latitude westerlies, etc. (Fleitmann et al., 2003; Breitenbach et al., 2010). Generally, a negative excursion in $\delta^{18}O$ ratios for a speleothem from a typical Indian monsoon domain implies more time for a summer monsoon rainfall (early onset and late withdrawal) and long moisture travel, indicating higher rainfall in the Indian subcontinent and vice versa (Berkelhammer et al., 2012; Dutt et al., 2015; Lechleitner et al., 2017; Kaushal et al., 2018). Other proxies such as lamina counting and width estimation, fluorescence, petrology, and elemental concentration are also used in speleothem study for reconstructing past climates (Fairchild et al., 2006).

Speleothem records of ISM variability show highly negative $\delta^{18}O$ ratios during the Greenlandian (Fig. 6.3), reflecting strong ISM conditions during 11.7–8.3 kyr BP, which can be ascribed to the warming of the global climate following the YD interval. It led to increased evaporation from the oceans and more convective activities. The summer monsoon strengthened gradually during the Greenlandian age, with the strongest conditions observed during ~10–8 kyr BP. During the Northgrippian (8.3–4.2 kyr BP; Fig. 6.3), the ISM records from northeastern India show highly variable climatic conditions with centennial-scale strong and weak ISM events and a general declining trend toward the end of the Northgrippian (Berkelhammer et al., 2012; Dutt et al., 2015). The proxy record from the Kotumsar cave, in central-eastern India, also indicates a significant dry spell in the region between 7 and 6 kyr BP (Band et al., 2018).

The speleothem records for the Meghalayan Age (4.2 kyr BP to the Recent; Fig. 6.3) are fragmentary except a record by Kathayat et al. (2017). Berkelhammer et al. (2012) suggest a 200 year long and abrupt dry spell and weak ISM in northeastern India, which began at around 4.0 kyr BP, popularly known as the 4.2 ka event. Recently, Kathayat et al. (2018) suggested several episodes of multidecadal weakening of the summer monsoon during the same event. Nevertheless, cave records from Uttarakhand, northern India, show a gradual weakening in ISM strength after 4.2 kyr BP with no abrupt shift. Cave records from central-eastern and northeastern India suggest a warm climate in the Indian subcontinent

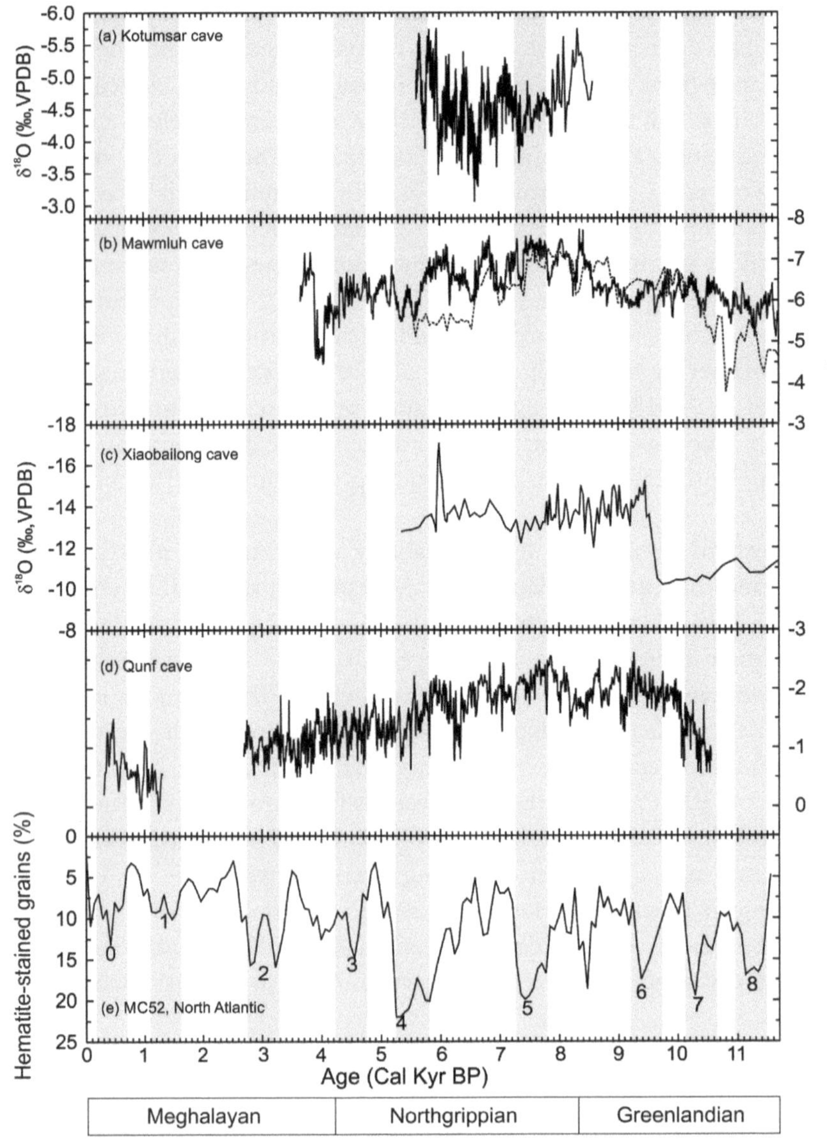

Figure 6.3 Comparison of ISM variability through speleothem proxy records with that of North Atlantic climate changes. (a) $\delta^{18}O$ time series of ISM strength from the Kotumsar cave, central-eastern India (Band et al., 2018), (b) $\delta^{18}O$ time series from the Mawmluh cave, northeastern India (Berkelhammer et al., 2012; Dutt et al., 2015), (c) $\delta^{18}O$ time series from the Xiaobailong cave, China (Cai et al., 2015), (d) $\delta^{18}O$ time series of ISM strength from the Qunf cave, Oman (Fleitmann et al., 2003), and (e) Hematite-stained grains (%) in core MC52 in the North Atlantic (Bond et al., 2001). Gray bars and numbers (0–8) indicate Bond events (Bond et al., 2001). *ISM*, Indian summer monsoon.

during the MCA (950–1250 CE) and CWP (following 1850 CE), whereas there was a weak summer monsoon during the cold interval of the LIA. Recently, Gupta et al. (2019) identified multiple episodes of long weakening and strengthening of the ISM during the LIA along with two major flood events across the MCA–LIA transition. The beginning and termination of the LIA were accompanied by a very weak summer monsoon in the Indian subcontinent that adversely affected agriculture and peasants, thus impacting the economy of the region.

6.2.3 Lake proxies

Paleo records of ISM variability from freshwater lakes are widely available as numerous freshwater lakes are present across the Indian subcontinent with easy access. Further, their depositional environment and lithologic variations are sensitive to climate change. Lake deposits can be dated with AMS ^{14}C and optically stimulated luminescence dating techniques. Grain size characteristics of lake sediments suggest the mode and energy of transport. Other proxies such as environmental magnetism, elemental concentration, mineral assemblages, stable isotopic ratios (C, N, O, P, S, etc.), and faunal and floral remains indicate changes in regional temperature, hydrology, vegetation, and rainfall conditions. The size and depth of lakes also affect the response of proxies and need utmost care in their interpretation for climate change studies.

Lake records from the Indian subcontinent indicate a more unstable climate and highly fluctuating rainfall in the region, temporally as well as spatially. Proxy records from the Tso Kar and Tso Moriri lakes, Ladakh (Wünnemann et al., 2010; Mishra et al., 2015; Dutt et al., 2018); Lonar Lake, Central India (Menzel et al., 2014), and Shanti Sagara Lake, southern India (Sandeep et al., 2017) indicate a warm regional climate accompanied by high rainfall during the Greenlandian (Fig. 6.4). Chandra peat deposits also show a high percentage in the abundance of broad-leaved plants and a warmer climate than the Recent in the western Himalaya during the Greenlandian (Rawat et al., 2015). These records suggest that the highest ISM rainfall conditions occurred between 11 and 10 kyr BP and then rainfall decreased in the younger interval. However, other records, such as those from the Benital Lake, Uttarakhand Himalaya (Bhushan et al., 2018), hint at low-precipitation conditions in parts of the Himalaya during the beginning of the Greenlandian that strengthened later at around 8 kyr BP as indicated by the speleothem records

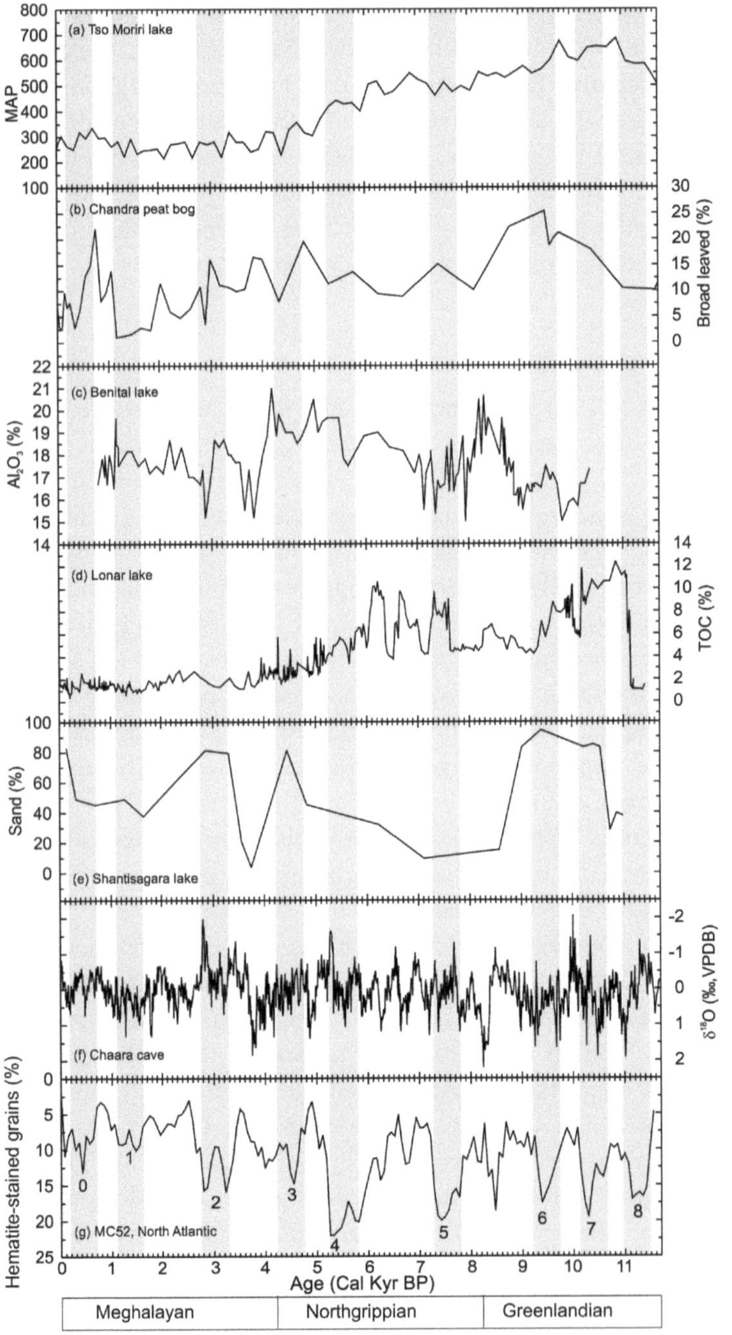

(a) Tso Moriri lake
MAP
(b) Chandra peat bog
Broad leaved (%)
(c) Benital lake
Al_2O_3 (%)
(d) Lonar lake
TOC (%)
(e) Shantisagara lake
Sand (%)
(f) Chaara cave
$\delta^{18}O$ (‰,VPDB)
(g) MC52, North Atlantic
Hematite-stained grains (%)
Age (Cal Kyr BP)
Meghalayan
Northgrippian
Greenlandian

(Berkelhammer et al., 2012; Dutt et al., 2015). Similar asynchronicity is visible in ISM reconstruction during the Northgrippian from lake records from different parts of the Indian subcontinent. The lake records from the western Himalaya and southern India show gradually declining precipitation conditions with weakening of the ISM, and no major abrupt event was observed, whereas the Lonar Lake record indicates highly fluctuating monsoon conditions during the Northgrippian with centennial-scale repeated strong and weak events (Leipe et al., 2014; Menzel et al., 2014; Mishra et al., 2015). The Benital Lake data indicate increasing rainfall during the Northgrippian (Bhushan et al., 2018). Similarly, a warm climate with high precipitation has been deduced through multiple proxy records from the Kedarnath peat deposits (Srivastava et al., 2017).

The majority of lake records suggest weak ISM conditions during the Meghalayan time. The sudden occurrence of aridity is evident in the Indian subcontinent which began any time between 4.4 and 4.2 kyr BP and lasted for 500–1100 years (Fig. 6.5; Dixit et al., 2014; Mishra et al., 2015; Nakamura et al., 2016; Dutt et al., 2018). This phase has been suggested to have caused low agricultural output in the Indus valley area, triggering south-eastward population migration and transformation of a large urban culture to small urban or village communities (Dutt et al., 2019). During the last 1500 years, the ISM has varied considerably, marked by pronounced flood and drought events. The climate was warm and humid, and the ISM was moderate during the MCA, weak during the LIA, and has strengthened following the LIA with frequent occurrences of extreme precipitation events over the last four centuries, complementing the marine records (Dutt et al., 2018; Gupta et al., 2019).

Figure 6.4 Comparison of ISM variability through lake proxy records with that from North Atlantic climate changes. (a) Mean annual precipitation reconstruction from the Tso Moriri Lake, Ladakh (Leipe et al., 2014), (b) broad-leaved (%) from the Chandra peat bog (Rawat et al., 2015), (c) Al_2O_3 from the Benital Lake, Uttarakhand Himalaya (Bhushan et al., 2018), (d) TOC (%) from the Lonar Lake, central India (Menzel et al., 2014), (e) sand (%) from the Shanti Sagara Lake, southern India (Sandeep et al., 2017), (f) $\delta^{18}O$ time series from the Charra cave (Brahim et al., 2019), (g) Hematite-stained grains (%) in core MC52 in the North Atlantic (Bond et al., 2001). Gray bars and numbers (0–8) indicate the Bond events (Bond et al., 2001). *ISM*, Indian summer monsoon.

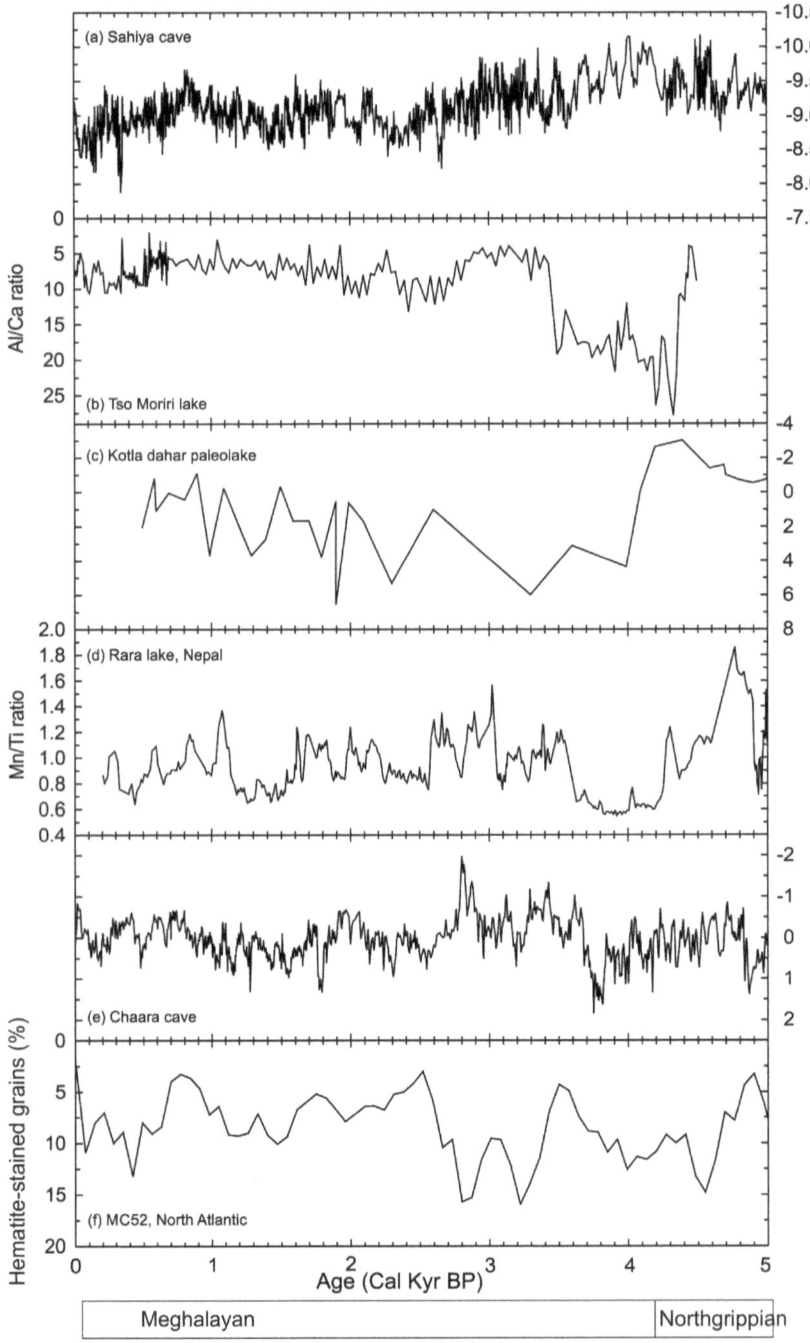
(a) Sahiya cave
-10.5
-10.0
-9.5
-9.0
-8.5
-8.0
-7.5
δ18O (‰, VPDB)
(b) Tso Moriri lake
Al/Ca ratio
0
5
10
15
20
25
(c) Kotla dahar paleolake
-4
-2
0
2
4
6
8
δ18O (‰, VPDB)
(d) Rara lake, Nepal
Mn/Ti ratio
2.0
1.8
1.6
1.4
1.2
1.0
0.8
0.6
0.4
(e) Chaara cave
-2
-1
0
1
2
δ18O (‰, VPDB)
(f) MC52, North Atlantic
Hematite-stained grains (%)
0
5
10
15
20
0
1
2
3
4
5
Age (Cal Kyr BP)
Meghalayan
Northgrippian

6.3 Teleconnections between Arctic climate and tropical monsoon variability

The modern climate over the Arctic region is characterized by long freezing winters and short summers, experiencing extreme solar radiation in both summer and winter seasons. The average temperature during the summer ranges from −10°C to 10°C and during the winter from −34°C to 0°C (en.wikipedia.org/wiki/Climate_of_the_Arctic). Past climate variability over the Arctic region is mainly deduced from the $\delta^{18}O$ record of the Greenland Ice Core Project (Dansgaard et al., 1993) and Greenland Ice Sheet Project, as well as proxy records from the North Atlantic Ocean. Variability in $\delta^{18}O$ of the ice core shows nearly 25 large abrupt climate perturbations since the LGM and a number of other high-frequency climate oscillations with 1000-, 1450-, and 3000-year cyclicities known as Dansgaard–Oeschger cycles or D-O cycles, Heinrich events associated with sudden iceberg discharge into the North Atlantic and Bond events of the Holocene, which are comparable to D-O cycles during the Holocene (Heinrich, 1988; Bond et al., 1993; Dansgaard et al., 1993).

Further, alteration of warm and cold periods, also known as stadials and interstadials that lasted up to millennia are observed from different parts of the world. The North Atlantic and the southern Arctic originated thermohaline circulation known as North Atlantic Deep Water (NADW) acts as a major ocean conveyor belt that may be the primary mechanism that regulates the global climate (Fig. 6.1; Tamsitt et al., 2017). The formation of the NADW through the sinking of warm saline waters and release of heat energy during cold conditions led to decreased production or complete shutdown of NADW formation by the increased influx of freshwater due to the surge of ice sheets from northern Europe (Fanning and Weaver, 1997; Grousset et al., 2000; Álvarez-Solas et al., 2011). The thermohaline conveyor belt carried and distributed the signals of

Figure 6.5 Proxy records of ISM variability for the last 5000 years correlated with the North Atlantic records. (a) $\delta^{18}O$ time series from the Sahiya cave, Uttarakhand Himalaya (Kathayat et al., 2017), (b) Al/Ca ratio from the Tso Moriri Lake, Ladakh (Dutt et al., 2018), (c) $\delta^{18}O$ time series from the Kotla Dahar paleolake (Dixit et al., 2014), (d) Mn/Ti ratio from the Lake Rara, Nepal (Nakamura et al., 2016), (e) $\delta^{18}O$ time series from the Chaara cave (Brahim et al., 2019), (f) Hematite-stained grains (%) in core MC52 in the North Atlantic (Bond et al., 2001). *ISM*, Indian summer monsoon.

warm/cold wave conditions of the Arctic across the globe. Their imprints were preserved in the global climate proxies as evidence of D-O cycles, Bond cycles, and Heinrich events.

The Indian monsoon proxies from the marine, lake, and speleothem records suggest pronounced variability in monsoon precipitation and its linkage with the Heinrich events, D-O cycles, and Bond events since the LGM including the Holocene. All the proxy records are concurrent with a decreased ISM precipitation during the YD due to AMOC weakening, and a strengthened ISM thereafter during the Greenlandian age (Figs. 6.2 and 6.3; Gupta et al., 2003; Berkelhammer et al., 2012; Leipe et al., 2014; Dutt et al., 2015 and references therein). The meridional shifts of the ITCZ through tropospheric–stratospheric connectivity could have been interlinked with changes in meltwater discharge, AMOC circulation, North Sea ice coverage, etc. The apparent similarities in millennial-scale variability in the North Atlantic climate and tropical monsoon as evident in the land (speleothem) and marine (*G. bulloides*) proxies strongly advocate a mechanistic link between tropical monsoon and Arctic climate, with a possible linkage through the mid-latitude westerlies (Bond et al., 2001; Wang et al., 2001; Gupta et al., 2003, 2005; Dutt et al., 2015; Duan et al., 2018). During low atmospheric temperature, sea ice expands in the North Atlantic, causing further cooling, and that signal is transported to the tropics through the mid-latitude westerlies leading to low precipitation conditions. Ice rafted events in the Atlantic ocean during the Holocene (Bond events) are also linked to the reduced ISM strength because of substantial freshwater input due to ice melting and subsequently weakening of the AMOC (Bond et al., 2001). The signal of Bond events is better preserved in marine archives of the ISM than that of the continental records and solar-triggered reduction in the NADW formation was accompanied by enhanced sea-ice flux into the Nordic sea (Figs. 6.2–6.4; Bond et al., 2001; Fleitmann et al., 2003; Gupta et al., 2003; Gupta et al., 2005; Berkelhammer et al., 2012; Leipe et al., 2014). The continental records are not well-correlated with the bond events and are thus more influenced by long-term solar insolation, El-Niño Southern Oscillation, and volcanism during the Holocene (Figs. 6.3–6.5; Dutt et al., 2018; Gupta et al., 2019). The lake records have preserved the local precipitation signal and may or may not coincide with the global climate signal association with climatic variability over the Arctic region.

6.4 Summary

The ISM witnessed abrupt shifts during the Holocene driven by both internal and external forcing factors over different time scales. The monsoon proxies are very well preserved in sediments of the Arabian Sea and BoB, as well as stalagmites and freshwater lake deposits from the Indian landmass. These proxies have enabled deduction of signals of ISM variability and their societal impacts. The monsoon was weak during the cold YD and began to strengthen during the Greenlandian Age. All the proxies suggest stronger monsoon during the Greenlandian Age followed by a gentle declining trend during the Northgrippian Age with more considerable variability during the stadial. The ISM was very weak at the beginning of the Meghalayan Age and strengthened during the MCA and CWP. The proxy record suggests a weak ISM during the LIA. Subdecadal to centennial-scale changes in the ISM are linked with the solar insolation, ENSO, and IOD, but millennial-scale changes are linked with Arctic and North Atlantic climate variability. All ISM proxies suggest a strong teleconnection between the ISM strength and Arctic climate anomaly, but this is more pronounced in the marine archives. The mechanistic link is associated with NADW formation and AMOC circulation, which is directly related to climate variability over the Arctic region.

Acknowledgments

AKG acknowledges IIT Kharagpur for the necessary facilities and Department of Science and Technology, Govt. of India for the JC Bose Fellowship. SD acknowledges Wadia Institute of Himalayan Geology for providing the necessary infrastructure facilities. MD thanks the Principal of Mahila Mahavidyalaya, Banaras Hindu University, Varanasi. RKS acknowledges IIT Bhubaneswar for providing the necessary infrastructure facilities during this work.

References

Agnihotri, R., Sarin, M.M., Somayajulu, B.L.K., 2003. Late-quaternary biogenic productivity and organic carbon deposition in the eastern Arabian Sea. Palaeogeogr. Palaeoclim. Palaeoecol. 197 (1-2), 43–60.

Álvarez-Solas, J., Montoya, M., Ritz, C., Ramstein, G., Charbit, S., Dumas, C., et al., 2011. Heinrich event 1: an example of dynamical ice-sheet reaction to oceanic changes. Clim. Past. 7 (4), 1297–1306.

Anderson, D.M., Overpeck, J.T., Gupta, A.K., 2002. Increase in the Asian southwest monsoon during the past four centuries. Science 297, 596–599.

Ashok, K., Guan, Z., Yamagata, T., 2001. Impact of the Indian Ocean dipole on the relationship between the Indian monsoon rainfall. Geophys. Res. Lett. 28 (23), 4499–4502.

Bamzai, A.S., Shukla, J., 1999. Relation between Eurasian snow cover, snow depth, and the Indian summer monsoon: an observational study. J. Clim. 12 (10), 3117–3132.

Band, S., Yadava, M.G., Lone, M.A., Shen, C.-C., Sree, K., Ramesh, R., 2018. High-resolution mid-Holocene Indian Summer Monsoon recorded in a stalagmite from the Kotumsar Cave, Central India. Q. Int. 479, 19–24.

Berkelhammer, M., Sinha, A., Stott, L., Cheng, H., Pausata, F.S.R., Yoshimura, K., 2012. An abrupt shift in the Indian monsoon 4000 years ago. In: InGiosan, L., Fuller, D.Q., Nicoll, K., Flad, R.K., Clift, P.D. (Eds.), Climates, Landscapes, and Civilizations, 198. American Geophysical Union Geophysical Monograph Series, pp. 75–87.

Bhushan, R., Sati, S.P., Rana, N., Shukla, A.D., Mazumdar, A.S., Juyal, N., 2018. High-resolution millennial and centennial-scale Holocene monsoon variability in the Higher Central Himalayas. Palaeogeogr. Palaeoclim. Palaeoecol. 489, 95–104.

Bond, G., Broecker, W., Johnsen, S., McManus, J., Labeyrie, L., Jouzel, J., et al., 1993. Correlations between climate records from North Atlantic sediments and Greenland ice. Nature 365 (6442), 143. Available from: https://www.nature.com/articles/365143a0.

Bond, G., Kromer, B., Beer, J., Muscheler, R., Evans, M.N., Showers, W., et al., 2001. Persistent solar influence on North Atlantic climate during the Holocene. Science 294 (5549), 2130–2136.

Böning, P., Bard, E., 2009. Millennial/centennial-scale thermocline ventilation changes in the Indian Ocean as reflected by aragonite preservation and geochemical variations in Arabian Sea sediments. Geochim. Cosmochim. Acta 73 (22), 6771–6788.

Brahim, Y.A., Wassenburg, J.A., Sha, L., Cruz, F.W., Deininger, M., Sifeddine, A., et al., 2019. North Atlantic ice-rafting, ocean and atmospheric circulation during the Holocene: insights from Western Mediterranean speleothems. Geophys. Res. Lett. 46, 7614–7623.

Breitenbach, S.F.M., Adkins, J.F., Meyer, H., Marwan, N., Kumar, K.K., Haug, G.H., 2010. Strong influence of water vapor source dynamics on stable isotopes in precipitation observed in Southern Meghalaya, NE India. Earth Planet. Sci. Lett. 292 (1–2), 212–220.

Cai, Y., Fung, I.Y., Edwards, R.L., An, Z., Cheng, H., Lee, J.-E., et al., 2015. Variability of stalagmite-inferred Indian monsoon precipitation over the past 252,000 y. Proc. Natl. Sci. Acad. U. S. A. 112 (10), 2954–2959.

Clemens, S.C., Kuhnt, W., LeVay, L.J., The Expedition 353 Scientists, 2016. Indian monsoon rainfall. Proceedings of the International Ocean Discovery Program, 353: College Station, TX (International Ocean Discovery Program). https://doi.org/10.14379/iodp.proc.353.2016

Dansgaard, W., Johnsen, S.J., Clausen, H.B., Dahl-Jensen, D., Gundestrup, N.S., Hammer, C.U., et al., 1993. Evidence for general instability of past climate from a 250-kyr ice-core record. Nature 364 (6434), 218.

Das, M., Singh, R.K., Gupta, A.K., Bhaumik, A.K., 2017. Holocene strengthening of the Oxygen Minimum Zone in the northwestern Arabian Sea linked to changes in intermediate water circulation or Indian monsoon intensity? Palaeogeogr. Palaeoclim. Palaeoecol. 483, 125–135.

den Dulk, M., Reichart, G.J., van Heyst, S., Zachariasse, W.J., Zwaan, G.J.V., 2000. Benthic foraminifera as proxies of organic matter flux and bottom water oxygenation? A case history from the northern Arabian Sea. Palaeogeogr. Palaeoclim. Palaeoecol. 161, 337–359.

Dixit, Y., Hodell, D.A., Petrie, C.A., 2014. Abrupt weakening of the summer monsoon in northwest India ~4100 yr ago. Geology 42, 339–342.

Duan, F., Wang, Y., Liao, Z., Chen, S., Zhang, W., Shao, Q., 2018. Divergent influences of the Greenland and Antarctica climates on the Asian monsoon during a stadial to interstadial cycle. J. Asian Earth Sci. 159, 69–73.

Dutt, S., Gupta, A.K., Clemens, S.C., Cheng, H., Singh, R.K., Kathayat, G., et al., 2015. Abrupt changes in Indian summer monsoon strength during 33,800 to 5500 years BP. Geophys. Res. Lett. 42 (13), 5526–5532.

Dutt, S., Gupta, A.K., Wünnemann, B., Yan, D., 2018. A long arid interlude in the Indian summer monsoon during ~4,350 to 3,450 cal. yr BP contemporaneous to displacement of the Indus valley civilization. Q. Int. 482, 83–92.

Dutt, S., Gupta, A.K., Singh, M., Jaglan, S., Saravanan, P., Balachandran, P., et al., 2019. Climate variability and evolution of the Indus civilization. Q. Int. 507, 15–23.

Fairchild, I.J., Frisia, S., Borsato, A., Tooth, A.F., 2006. Speleothems. In: Nash, D.J., McLaren, S.J. (Eds.), Geochemical Sediments and Landscapes. Blackwells, Oxford.

Fanning, A.F., Weaver, A.J., 1997. Temporal-geographical meltwater influences on the North Atlantic conveyor: implications for the Younger Dryas. Paleoceanography 12 (2), 307–320.

Fleitmann, D., Burns, S.J., Mudelsee, M., Neff, U., Kramers, J., Mangini, A., et al., 2003. Holocene forcing of the Indian monsoon recorded in a stalagmite from southern Oman. Science 300 (5626), 1737–1739.

Gadgil, S., 2003. The Indian monsoon and its variability. Annu. Rev. Earth Planet. Sci. 31 (1), 429–467.

Govil, P., Naidu, P.D., 2011. Variations of Indian monsoon precipitation during the last 32 kyr reflected in the surface hydrography of the Western Bay of Bengal. Q. Sci. Rev. 30, 3871–3879.

Grousset, F.E., Pujol, C., Labeyrie, L., Auffret, G., Boelaert, A., 2000. Were the North Atlantic Heinrich events triggered by the behavior of the European ice sheets? Geology 28 (2), 123–126.

Gupta, A.K., Anderson, D.M., Overpeck, J.T., 2003. Abrupt changes in the Asian southwest monsoon during the Holocene and their links to the North Atlantic Ocean. Nature 421 (6921), 354.

Gupta, A.K., Das, M., Anderson, D.M., 2005. Solar influence on the Indian summer monsoon during the Holocene. Geophys. Res. Lett. 32, 17.

Gupta, A.K., Mohan, K., Das, M., Singh, R.K., 2013. Solar forcing of the Indian summer monsoon variability during the Ållerød period. Sci. Rep. 3, 2753.

Gupta, A.K., Yuvaraja, A., Prakasam, M., Clemens, S.C., Velu, A., 2015. Evolution of the South Asian monsoon wind system since the late Middle Miocene. Palaeogeogr. Palaeoclimatol. Palaeoecol. 438, 160–167.

Gupta, A.K., Dutt, S., Cheng, H., Singh, R.K., 2019. Abrupt changes in Indian summer monsoon strength during the last ~900 years and their linkages to socio-economic conditions in the Indian subcontinent. Palaeogeogr. Palaeoclim. Palaeoecol. 536, 109347.

Halley, E., 1686. An historical account of the trade winds and monsoons, observable in the seas between and near the tropics with an attempt to assign the physical cause of the said winds. Philos. Trans. R. Soc. Lond. 16, 153–168.

Heinrich, H., 1988. Origin and consequences of cyclic ice rafting in the northeast Atlantic Ocean during the past 130,000 years. Q. Res. 29 (2), 142–152.

Kathayat, G., Cheng, H., Sinha, A., Yi, L., Li, X., Zhang, H., et al., 2017. The Indian monsoon variability and civilization changes in the Indian subcontinent. Sci. Adv. 3, e1701296.

Kathayat, G., Cheng, H., Sinha, A., Berkelhammer, M., Zhang, H., Duan, P., et al., 2018. Evaluating the timing and structure of the 4.2 ka event in the Indian summer monsoon domain from an annually resolved speleothem record from Northeast India. Clim. Past 14, 1869–1879.

Kaushal, N., Breitenbach, S.F.M., Lechleitner, F.A., Sinha, A., Tewari, V.C., Ahmad, S. M., et al., 2018. The Indian summer monsoon from a speleothem $\delta^{18}O$ perspective—a review. Quaternary 1 (3), 29.

Kudrass, H.R., Hofmann, A., Doose, H., Emeis, K., Erlenkeuser, H., 2001. Modulation and amplification of climatic changes in the Northern Hemisphere by the Indian summer monsoon during the past 80 ky. Geology 29 (1), 63–66.

Kumar, K.K., Rajagopalan, B., Hoerling, M., Bates, G., Cane, M., 2006. Unravelling the mystery of Indian monsoon failure during El Nino. Science 314 (5696), 115–119. Available from: https://doi.org/10.1126/science.1131152.

Lechleitner, F.A., Breitenbach, S.F.M., Cheng, H., Plessen, B., Rehfeld, K., Goswami, B., et al., 2017. Climatic and in-cave influences on $\delta_{18}O$ and $\delta_{13}C$ in a stalagmite from northeastern India through the last deglaciation. Q. Res. 88, 458–471. Available from: https://doi.org/10.1017/qua.2017.72.

Leipe, C., Demske, D., Tarasov, P.E., HIMPAC project members, 2014. A Holocene pollen record from the northwestern Himalayan lake Tso Moriri: implications for palaeoclimatic and archaeological research. Q. Int. 348, 93–112.

Marzin, C., Kallel, N., Kageyama, M., Duplessy, J.-C., Braconnot, P., 2013. Glacial fluctuations of the Indian monsoon and their relationship with North Atlantic climate: new data and modelling experiments. Clim. Past 9 (5), 2135–2151.

Menzel, P., Gaye, B., Mishra, P.K., Anoop, A., Basavaiah, N., Marwan, N., et al., 2014. Linking Holocene drying trends from Lonar Lake in monsoonal central India to North Atlantic cooling events. Palaeogeogr. Palaeoclim. Palaeoecol. 410, 164–178.

Milliman, J.D., Syvitski, J.P., 1992. Geomorphic/tectonic control of sediment discharge to the ocean: the importance of small mountainous rivers. J. Geol. 100 (5), 525–544.

Miriyala, P., Sukumaran, N.P., Nath, B.N., Ramamurty, P.B., Sijinkumar, A.V., Vijayagopal, B., et al., 2017. Increased chemical weathering during the deglacial to mid-Holocene summer monsoon intensification. Sci. Rep. 7, 44310.

Mishra, P.K., Anoop, A., Schettler, G., Prasad, S., Jehangir, A., Menzel, P., et al., 2015. Reconstructed late quaternary hydrological changes from Lake Tso Moriri, NW Himalaya. Q. Int. 371, 76–86.

Naidu, P.D., Govil, P., 2017. Monsoon could trigger the global abrupt climate changes: new evidence from the Bay of Bengal. J. Ind. Geophys. Union 21 (4), 285–291.

Naik, S.S., Basak, C., Goldstein, S.L., Naidu, P.D., Naik, S.N., 2019. A 16-kyr record of ocean circulation and monsoon intensification from the Central Bay of Bengal. Geochem. Geophys. Geosyst. 20 (2), 872–882.

Nakamura, A., Yokoyama, Y., Maemoku, H., Yagi, H., Okamura, H., Matsouka, H., et al., 2016. Weak monsoon event at 4.2 ka recorded in sediment from Lake Rara, Himalayas. Q. Int. 397, 349–359.

Neff, U., Burns, S.J., Mangini, A., Mudelsee, M., Fleitmann, D., Matter, A., 2001. Strong coherence between solar variability and the monsoon in Oman between 9 and 6 kyr ago. Nature 411 (6835), 290.

Nilsson-Kerr, K., Anand, P., Sexton, P.F., 2019. Role of Asian summer monsoon subsystems in the inter-hemispheric progression of deglaciation. Nat. Geosci. 12 (4), 290.

Pandey, D.K., Clift, P.D., Kulhanek, D.K., and The Expedition 355 Scientists, 2016. Arabian sea monsoon. Proceedings of the International Ocean Discovery Program (International Ocean Discovery Program). https://doi.org/10.14379/iodp.proc.355.2016

Parthasarathy, B., Kumar, K.R., Munot, A.A., 1993. Homogeneous Indian monsoon rainfall variability and prediction. Proc. Indian. Acad. Sci.-Earth Planet. Sci. 102 (1), 121–155.
Partin, J.W., Quinn, T.M., Shen, C.-C., Okumura, Y., Cardenas, M.B., Siringan, F.P., et al., 2015. Gradual onset and recovery of the Younger Dryas abrupt climate event in the tropics. Nat. Commun. 6, 8061.
Prell, W.L., Murray, D.W., Clemens, S.C., Anderson, D.M., 1992. Evolution and variability of the Indian Ocean summer monsoon: evidence from the western Arabian Sea drilling program. Synth. Results Sci. Drill. Indian Ocean 70, 447–469.
Quade, J., Cater, M.L.J., Ojha, P.T., Adam, J., Harrison, T.M., 1995. Late Miocene environmental change in Nepal and the northern Indian subcontinent: stable isotopic evidence from paleosols. GSA Bull. 107, 1381–1397.
Rawat, S., Gupta, A.K., Sangode, S.J., Srivastava, P., Nainwal, H.C., 2015. Late Pleistocene–Holocene vegetation and Indian summer monsoon record from the Lahaul, Northwest Himalaya, India. Q. Sci. Rev. 114, 167–181.
Sandeep, K., Shankar, R., Warrier, A.K., 2017. A multi-proxy lake sediment record of Indian summer monsoon variability during the Holocene in southern India. Palaeogeogr. Palaeoclim. Palaeoecol. 476, 1–14.
Saraswat, R., Lea, D.W., Nigam, R., Mackensen, A., Naik, D.K., 2013. Deglaciation in the tropical Indian Ocean driven by interplay between the regional monsoon and global teleconnections. Earth Planet. Sci. Lett. 375, 166–175.
Satpathy, R.K., Steinke, S., Singh, A.D., 2019. Monsoon-induced changes in surface hydrography of the eastern Arabian Sea during the early Pleistocene. Geol. Mag. 1–11. Available from: https://doi.org/10.1017/S0016756819000098.
Schlolaut, G., Brauer, A., Nakagawa, T., Lamb, H.F., Tyler, J.J., Staff, R.A., et al., 2017. Evidence for a bi-partition of the Younger Dryas Stadial in East Asia associated with inversed climate characteristics compared to Europe. Sci. Rep. 7, 44983.
Shewale, M.P., Kumar, S., 2005. Climatological features of drought incidences in India. Meteorological Monograph (Climatology 21/2005). National Climate Centre, India Meteorological Department.
Sijinkumar, A.V., Clemens, S.C., Nath, B.N., Prell, W., Benshilla, R., Lengaigne, M., 2016. $\delta^{18}O$ and salinity variability from the last glacial maximum to recent in the Bay of Bengal and Andaman Sea. Q. Sci. Rev. 135, 79–91.
Sinha, A., Kathayat, G., Cheng, H., Breitenbach, S.F.M., Berkelhammer, M., Mudelsee, M., et al., 2015. Trends and oscillations in the Indian summer monsoon rainfall over the last two millennia. Nat. Commun. 6, 6309.
Srivastava, P., Agnihotri, R., Sharma, D., Meena, N., Sundriyal, Y.P., Saxena, A., et al., 2017. 8000-year monsoonal record from Himalaya revealing reinforcement of tropical and global climate systems since mid-Holocene. Sci. Rep. 7 (1), 14515.
Tada, R., Irino, T., Ikehara, K., Karasuda, A., Sugisaki, S., Xuan, C., et al., 2018. High-resolution and high-precision correlation of dark and light layers in the quaternary hemipelagic sediments of the Japan Sea recovered during IODP Expedition 346. Prog. Earth Planet. Sci. 5 (1), 19.
Tamsitt, V., Drake, H.F., Morrison, A.K., Talley, L.D., Dufour, C.O., Gray, A.R., et al., 2017. Spiralling pathways of global deep waters to the surface of the Southern Ocean. Nat. Commun. 8 (1), 172.
Tiwari, M., Ramesh, R., Bhushan, R., Sheshshayee, M.S., Somayajulu, B.L., Jull, A.T., et al., 2010. Did the Indo-Asian summer monsoon decrease during the Holocene following insolation? J. Q. Sci. 25 (7), 1179–1188.
Tripathy, G.R., Singh, S.K., Bhushan, R., Ramaswamy, V., 2011. Sr-Nd isotope composition of the Bay of Bengal sediments: impact of climate on erosion in the Himalaya. Geochem. J 45 (3), 175–186.

Tripathy, G.R., Singh, S.K., Ramaswamy, V., 2014. Major and trace element geochemistry of Bay of Bengal sediments: implications to provenances and their controlling factors. Palaeogeogr. Palaeoclim. Palaeoecol. 397, 20–30.
Wang, Y.J., Cheng, H., Edwards, R.L., An, Z.S., Wu, J.Y., Shen, C.-C., et al., 2001. A high-resolution absolute-dated late Pleistocene monsoon record from Hulu Cave, China. Science 294 (5550), 2345–2348.
Wang, P.X., Wang, B., Cheng, H., Fasullo, J., Guo, Z., Kiefer, T., et al., 2017. The global monsoon across time scales: mechanisms and outstanding issues. Earth Sci. Rev. 174, 84–121.
Wünnemann, B., Demske, D., Tarasov, P., Kotlia, B.S., Reinhardt, C., Bloemendal, J., et al., 2010. Hydrological evolution during the last 15 kyr in the Tso Kar lake basin (Ladakh, India), derived from geomorphological, sedimentological and palynological records. Q. Sci. Rev. 29, 1138–1155.

CHAPTER 7

Late Quaternary climatic imprints on clastic sediments of Broggerbreen peninsula, Ny-Ålesund region, Svalbard

Abhishek Verma, Deepak Y. Gajbhiye, Kato K. Ayemi, Pradeep Kumar, Amit Dharwadkar and Prakash Kumar Srivastava
Geological Survey of India, Faridabad, India

7.1 Introduction

Ny-Ålesund is situated on the Brogger peninsula of the Island of Spitsbergen in Svalbard. Svalbard is an archipelago in the Arctic Ocean and it is one of the northernmost landmass on the Earth lying between mainland Norway and North Pole, inside the Arctic Circle. The study area, confined to northern Ny-Ålesund, acquired its present landscape due to a set of Broggerbreen glaciers. The landscapes roll out to the north of study area, toward downstream of glacial Bayelva River flowing into the Kongsfjorden, eventually joining the Arctic Ocean. The area is highly sensitive toward climatic variations caused primarily by the complex interaction between the warm Spitsbergen Current and cold water mass from Barents Sea (Boitsov and Karsakov, 2012; Eriksen et al., 2015). The sensitivity of the climate is manifest in glacier variations in the area (Ahlmann, 1953). Nuth et al. (2010) have documented the glacier elevation changes in the area and contribution to sea-level rise.

7.2 Geomorphology and geological setup

Ny-Ålesund area is characterized by several geomorphic landforms such as sharp, residual hills, valley glaciers, outwash plains, meltwater streams, recessional/terminal, lateral, median, ground moraines, glacio-lakes, patterned ground, and fjords. The terrestrial landscape around Ny-Ålesund is designed by the surface processes governed during glacial and interglacial

Understanding Present and Past Arctic Environments
DOI: https://doi.org/10.1016/B978-0-12-822869-2.00002-5

period and modified by subsequent glacier movements (Overpeck et al., 1997; Singh et al., 2018). The highest point of the study area is Scheteligfjellet, which is 694 m above mean sea level.

Geologically, the area has a long history extending up to Pre-Cambrian. Svalbard region exposes rocks of all geologic ages besides witnessing multiple orogenic events giving rise to variety of geological structures (Hjelle, 1993). Ny-Ålesund area primarily exposes limestone and dolomite belonging to Carboniferous-Cretaceous sequence thrusted over the coal-bearing tertiary rocks. The basement rocks are constituted of phyllite and quartzite belonging to Pre-Cambrian to Silurian and Devonian. Tertiary rocks are represented by conglomerate and sandstone with subordinate shale and coal seams. The Ny-Ålesund area has been subjected to several orogenies from Grenvillian (~1 Ga) to Svalbardian (~450 Ma) affecting the subsequent evolution of the area (Harland et al., 1997; Dharwadkar et al., 2015).

7.3 Methodology

Available Landsat/Copernicus satellite images were used to identify the geomorphic features on mesoscale. These features were ground verified in the field and targeted for systematic terrestrial sediment sampling. Granulometry and statistical parameters (central tendencies of grain size distribution pattern, sorting, skewness, and kurtosis) have been calculated to decipher the distribution characteristics of sediments in the area. The samples were wet sieved for coarser fractions (sand) and gravitational settling method was used for separation of finer fractions (silt and clay) by standard pipette method. Obtained results were calculated and plotted to determine various statistical parameters (Folk, 1968, 1974; Krumbein and Pettijohn, 1938; Milner, 1962). Clay minerals were identified by X-ray diffraction (XRD) technique proposed by Jackson (1979). Clay mineralogy with XRD of the finest fraction of sediment samples has been used to understand the disintegration and decomposition of clastic sediments (Ehrmann et al., 1992). The granulometry of sediments helps in determining the depositional environment while the SEM describes the processes that have acted on the sediments. Quartz grains are very resistive toward weathering and able to preserve traces of sedimentary environmental processes (Krinsley and Doornkamp, 1973; Mahaney, 2002; Newsome and Ladd, 1999; Vos et al., 2014). Glacier and fluvial streams are responsible for the decomposition of the rocks and liberation of quartz grains (Schwamborn et al., 2012). For SEM, samples were prepared by method suggested by Krinsley and Doornkamp (1973). The processed quartz grains have been analyzed using high-resolution imaging LEO 440 SEM. The samples for optically simulated luminescence (OSL) dating have been systematically collected from the widely distributed morainic ridges of the study area in Ny-Ålesund.

7.4 Observations

7.4.1 Granulometry

Grain size parameters are used to get spatial and dynamic characteristic of terrestrial sediments. Almost all sediment samples have high percentage of sand and pebbles in comparison to silt and clay (Table 7.1 and Fig. 7.1). Finer comprising clay and silt vary from 0.24% to 4.26% and 1.92% to 9.51% in total weight, respectively. The pebble percentage varies between 9.49% and 53.10% and sand varies between 42.84% and 78.88% (Fig. 7.1). The mean grain size (Mz) varies between very coarse to coarse sand (-1.82Φ to 1.05Φ) having poorly to very poorly sorted sediments (1.90–2.68) characterized by platykurtic to very platykurtic (0.27–0.89) in range. Skewness of the sediment samples varies anonymously from strongly negative to strongly positive in the study area (Fig. 7.2). All samples are showing an asymmetrical and polymodal distribution pattern in nature (Fig. 7.3). It is also observed that most of the grains are angular in each size fraction. With increase in sand and silt percentages, mean grain size also increases. However, pebble percentage shows perfect negative correlation

Table 7.1 Granulometric statistical parameters.

Samples	Pebble	Sand	Silt	Clay	Mz	So	Sk	Kg
A1	27.59	64.75	3.41	4.26	−0.45	2.682	−0.028	0.825
A2	24.87	70.82	3.22	1.1	−0.6	1.897	0.584	0.345
A3	9.49	78.88	9.51	2.11	1.047	1.972	−0.453	0.855
A4	27.9	65.14	2.86	4.1	−0.167	2.494	−0.188	0.717
A5	18.59	77.77	3.39	0.24	0.133	2.222	−0.229	0.775
A6	18.17	77.06	3.5	1.27	0.05	2.173	−0.063	0.802
A7	53.1	42.84	2.33	1.73	−1.717	2.494	0.365	0.872
A8	20.19	69.17	6.96	3.68	0.343	2.379	−0.468	0.788
A9	44.39	49.65	2.91	3.04	−1.25	2.615	0.242	0.809
A10	41.55	53.96	1.92	2.57	−1.217	2.477	0.134	0.765
A11	49.26	47.51	2.09	1.14	−1.817	1.959	1.254	0.273
A12	16.67	76.28	3.53	3.53	0.267	2.239	−0.082	0.891

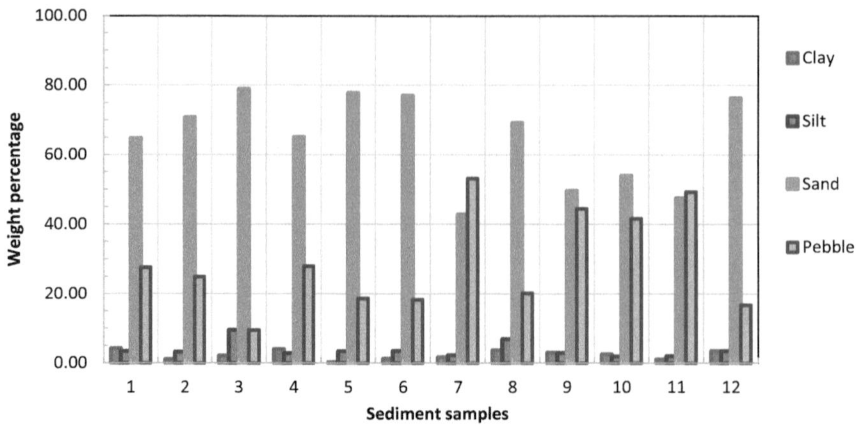

Figure 7.1 Weight percentages of different particle size in sediments of the study area.

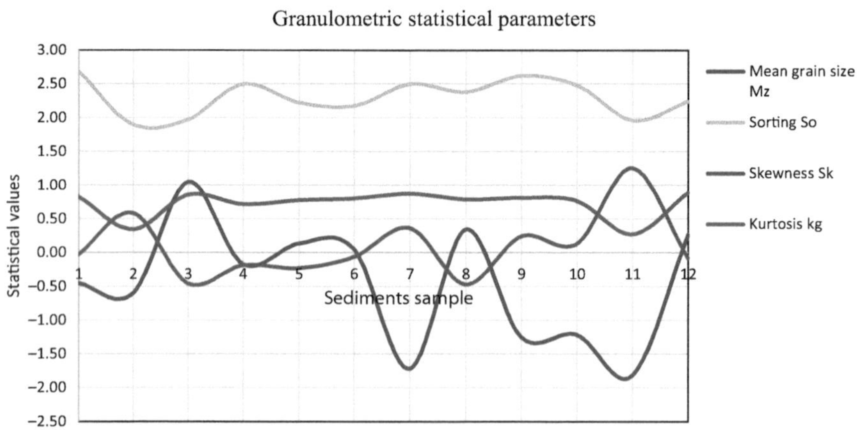

Figure 7.2 Granulometric statistical parameters of the sediments.

with the sand percentage and mean size of the sediments (Table 7.2). Sorting in diamicts shows a broad range and remained independent of grain size distribution parameters. Skewness is also irrelative with the mean grain size of each other samples with their geographic extent (Fig. 7.4). These type of correlations among granulometric parameters are typical character of glacial and glaciofluvial sediments in periglacial environment.

7.4.2 Quartz surface microtextures

Quartz is the most common rock-forming mineral and well resistant to weathering (Goldich, 1938). Varieties of weathering processes are preserved

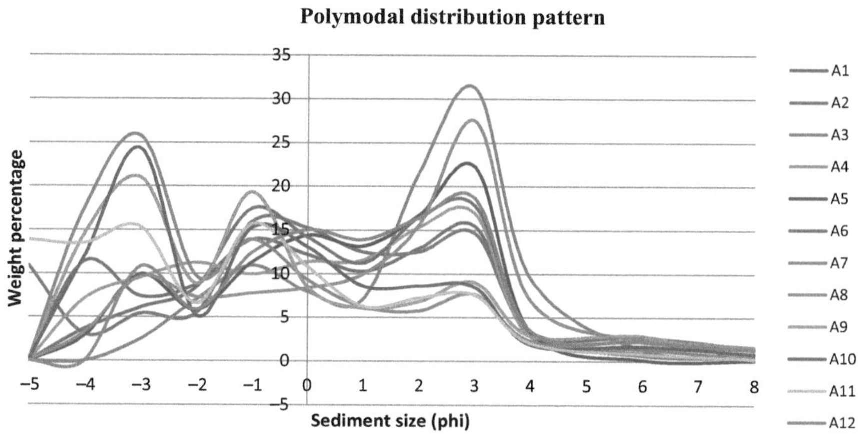

Figure 7.3 Grain size distribution pattern of the sediments collected from Ny-Ålesund, Svalbard.

Table 7.2 Correlation matrix of grain size parameters.

	Sand	Silt	Clay	Mz	So	Sk	Kg
Pebble	−0.97	−0.45	0	−0.95	0	0.5	0
Sand		0.31		0.86	−0.2	−0.42	0
Silt			0	0.59	−0.1	−0.38	0
Clay				0	0.43	0	0.2
Mz					0	−0.68	0.2
So						0	0.35
Sk							−0.6

on quartz grain surface in the form of microtextures and reflect sedimentation history (Margolis and Kennett, 1971; Krinsley and Doornkamp, 1973; Mahaney, 2002). Glacially originated quartz grains show characteristic surface features by the help of SEM (Krinsley and Funnell, 1965). Quartz surface microtextures are significantly used to ascribe the sequence of successive transitions from glacial to subglacial environments (Whalley and Krinsley, 1974). The characteristic mechanical features include conchoidal fractures, striations, imbricate breakage blocks, arcuate steps, curved or straight grooves, percussion fractures, evidences of abrasion, meandering ridges, collision pits, and many of the microscale traits. Angularity and relief of the grains are summed up in morphological textures. Chemical textures are identified as fresh overgrowth on grains, precipitation features, and adhered particles.

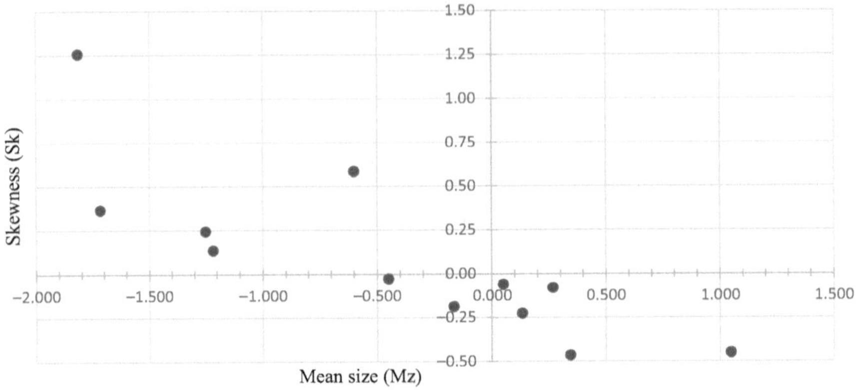

Figure 7.4 Mean grain size with nature of skewness of sediments.

The characteristic morphological, mechanical, and chemical textures of selected quartz grain from sediments of Ny-Ålesund area (Fig. 7.5) and frequency of occurrence (in %) of the textures are recorded (Table 7.3). Under SEM, quartz grains displayed broad variation in their surface microtextures. Edge rounding is the most common feature found on nearly 50% of the examined grains [Fig. 7.5 (A1−A10)]. The shapes vary between bladed [Fig. 7.5 (A2)], prolate [Fig. 7.6 (A1, A3, A6, and A10)], and subspherical [Fig. 7.5 (A4, A5, A7, A8, and A9)]. Mechanically upturned plates occur uniquely in grains [Fig. 7.5 (A6, A9, and A10)]. The grains typically exhibit high relief although many display medium relief [Fig. 7.5 (A1, A3, A4, A5, A6, A7, A8, and A9)]. Edge and surface abrasion vary between 20% and 60%. Other prominent microtextures are breakage blocks, randomly oriented V-shaped cracks, arcuate steps, conchoidal fractures, and subparallel linear fractures that occur on 20%−30% of the grains. Weathering surface varies between 5% and 80% with few precipitates, adhered particles, and new growths. Grooves, gouges, and troughs are relatively rare in all of the observed grains.

7.4.3 Clay mineral assemblage

Generally, clay minerals are a weathered product, originated from underlying rocks of the catchment. Few of the clay minerals reflect the prevailing climatic conditions from where they originate and later modified in the catchment (Srivastava et al., 2018). Clay minerals are mixed-layered minerals that are identified precisely by techniques such as XRD on the diffractogram by identification of peaks and its pattern. Nature of peaks and pattern of diffractogram are used to find the intensity of mechanical as

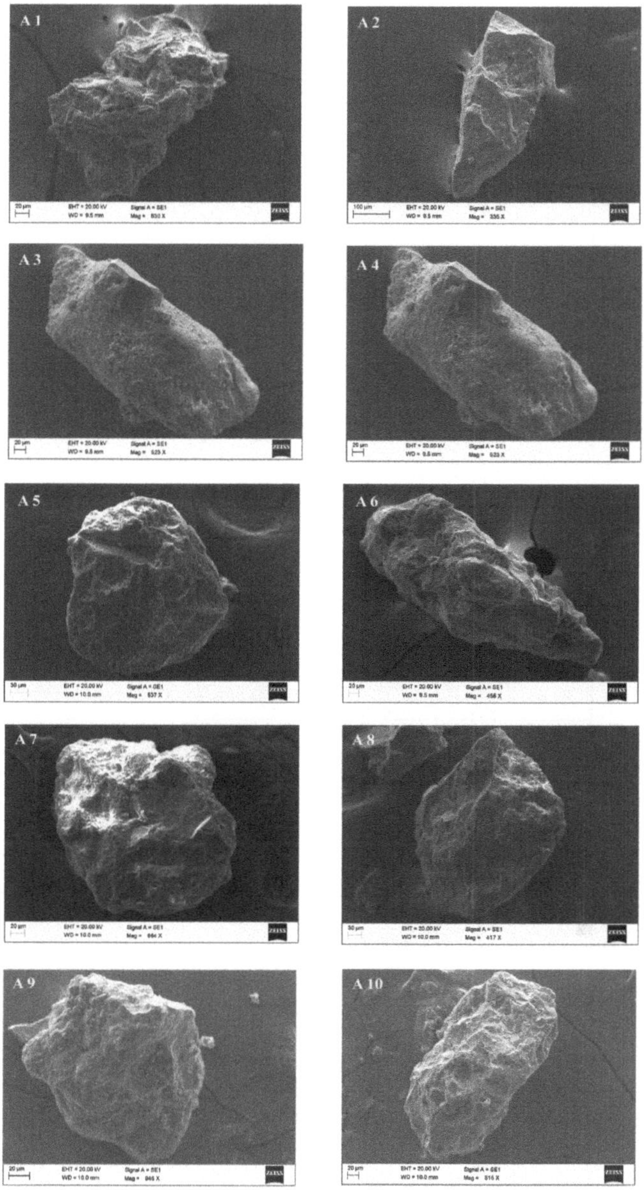

Figure 7.5 Scanning electron microscopy of quartz grain from sediments of Ny-Ålesund, Svalbard, Arctic region. A1: Prolate, medium, and small conchoidal fractures and arcuate steps. A2: Secondary precipitation of silica in a cavity and high relief. A3: Prolate, subangular, etched surfaces, and overgrowth. A4: Cavities indicating intense glacial action; overgrowth of silica. A5: Subspherical, imbricate blocks, and straight groove. A6: Prolate, medium, and small conchoidal fractures and arcuate steps. A7: Subspherical, medium pits, and small conchoidal fractures. A8: Etched surfaces and overgrowths. A9: Abraded and sinuous flow patterns, attributable to rotational effect during intense glacial grinding. A10: Prolate, small, and medium conchoidal fractures.

Table 7.3 SEM characteristics of quartz grains (in %) from sediment of eastern Broggerbreen peninsula, Svalbard, Arctic.

	Sample number									
	A1	A2	A3	A4	A5	A6	A7	A8	A9	A10
Mechanical textures										
Angular outline	70	90	50	20	10	80	20	20	30	80
Round outline	20	10	40	90	90	20	80	80	70	20
Low relief	10	30	25	10	15	15	20	20	20	35
Medium relief	70	50	55	75	75	60	70	60	60	45
High relief	20	30	20	15	10	25	10	20	20	20
Shape/sphericity	Prolate	Bladed	Prolate	Subspherical	Subspherical	Prolate	Subspherical	Subspherical	Subspherical	Prolate
Morphological texture										
Large conchoidal fractures	10	20			20	10	20	10	20	10
Small conchoidal fractures	20	5	5	10	5	20	25	5	5	20
Arcuate steps	15			10		15	10	5		15
Straight steps	10	5		5	5	10	20		5	10
Imbricated blocks	10	2			2	10			2	10
Large breakage blocks	5	30		5	40	5	10	15	30	5
Fractured plates		3		1	3		5	2	3	
Subparallel linear fractures	10	3		10	10	10		5	3	10
Curved grooves	5	1			1	5			1	5

Straight grooves/ striations		10			1		2		10	
V-shaped percussion cracks	10			10		10		5		10
Meandering ridges		5			2		15	2	2	
Edge abrasion	15	50	30	80	50	15	20	80	50	15
Surface abrasion	20	60		75	60	20	10	85	60	20
Deep troughs	5	10		2	10	5	5	10	10	5
Pitted surface/ collision pits	2	5			5	2	1		5	2
Chemical textures										
Preweathered surfaces	90	40	30	10	40	90	90	15	40	90
Weathered surfaces	10	60	70	90	60	10	10	85	60	10
Small precipitation features	15	2	15		2	15	15	5	2	15
Large precipitation features			10					5		
Adhering particles	3	10	10	1	10	3	2	20	10	3
New growth		15	1		15		2	2	15	

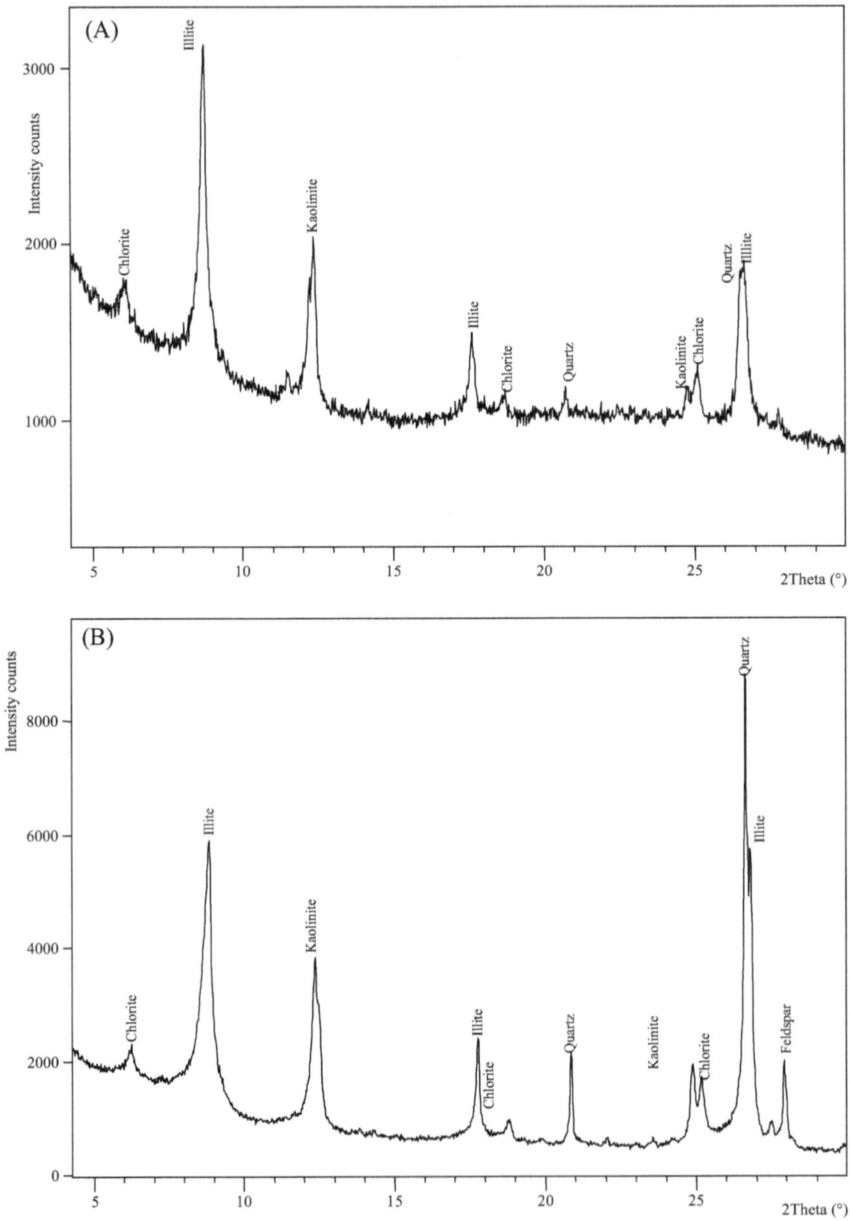

Figure 7.6 X-ray diffraction of clay minerals existing in the samples collected from the study area.

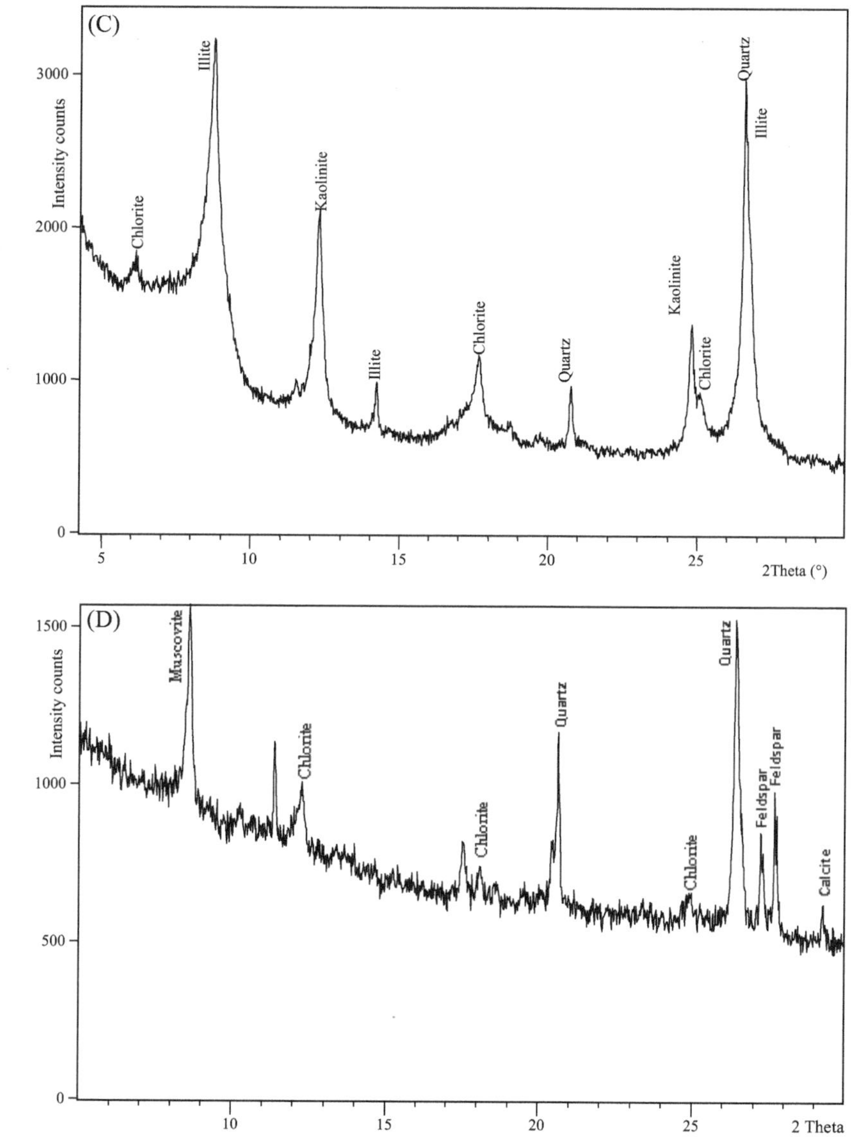

Figure 7.6 (continued)

well as chemical weathering of sediment. Moreover, these mineral diffractograms are a capable contrivance to get the paleoclimatic information (Sirocco and Lange, 1991; Dilli and Pant, 1994). The clay minerals identified by XRD of present samples include chlorite, illite, and kaolinite

(Fig. 7.6A–D). Along with clay minerals, unaltered or nontransformed minerals as feldspar, muscovite, calcite, and quartz are also detected. At the location A4, dominant clay minerals illite, kaolinite, and chlorite are observed along with unaltered quartz (Fig. 7.6A). Illite shows maximum relative intensity whereas kaolinite and chlorite show subsequently less intensity. The sharpness of the peaks indicates dominance of physical weathering. Location A3 having dominant minerals in clay fractions are illite, chlorite, and kaolinite with the presence of feldspar and quartz. Here also, quartz shows maximum relative intensity (Fig. 7.6B). Chlorite and feldspar have shown some effect of chemical weathering marking the presence of hydrolysis in the area. At the location of A4, illite and quartz show maximum intensity with the presence of identifiable amount of chlorite (Fig. 7.6C), indicating dominance of physical weathering with little chemical alteration during warm climatic conditions. Mineralogically, A5 shows similar characteristics to A4. Sample from location A2 consists of chlorite primarily, with muscovite, quartz, feldspar, and calcite (Fig. 7.6D). At locations A7 and A10 having closely similar assemblage, chlorite occurs with lower percentage of relative intensity than quartz, feldspar, and muscovite showing higher intensities. The absence of kaolinite as observed in diffractograms is notable. This reveals higher amount of chemical leaching and weathering was advanced under cold climatic condition.

There are two set of minerals characteristically found in different locations with nearly similar individualities throughout the study site. One set is dominated by illite, kaolinite, and chlorite, and the other by chlorite devoid of kaolinite and illite. Unaltered minerals such as quartz, feldspar, muscovite, and calcite are commonly existing all over the area.

7.4.4 Optically stimulated luminescence dating

Luminescence dating is the microdosimetric technique of estimating the time elapsed since the last day light exposure of the sediments. It uses quartz and feldspar as dosimeters. When the minerals are buried, they get concealed from daylight. The grains during burial are exposed to flux of radiation and begin to develop new signals due to exposure to flux of ionizing radiation. Sample preparation for OSL dating: Sample processing was carried out under subdued red light to get clean quartz or feldspar grains of desired size from natural samples by chemical and physical treatments. The samples were collected in specially designed opaque aluminum/steel tubes by horizontally inserting them in the sediment. Zones of contact, secondary calcification, and bioturbations were avoided. Risoe TL/OSL Reader has been used for OSL dating.

The first age of moraine sample is 19 ± 2 Ka that represents the late glacial span when the time between the Late Weichselian ice sheet started to retreat as the climate became warmer (Svendsen et al., 1992, 1996; Svendsen and Mangerud, 1997; Landvik et al., 1998). The second age of moraine sample is 32 ± 3 Ka that indicates earlier warm phase in an area.

7.5 Discussions

The sediments may have originated by variable degree of mechanical and chemical weathering, transported to the basin by different media and finally deposited under varying conditions. In a wider sense, the composition of sediments is mainly governed by geology of the source area, weathering processes, transporting agents, and climatic conditions. Sediment composition is also affected by distance of sediment transport and by diagenetic processes (Tucker, 1981). Coarser and angular sediment grains are generally produced from gneissic/granitic country rock and indicate less distance of transportation or from collision and abrasion (Pettijohn, 1984). These sediments are very susceptible to further grain size reduction as opposed to the sediments that have originated from a fine-grained country rock (Tulaczyk et al., 1998). However, coarse-grain glacial sediments will undergo interparticle collision that results in further reduction of grain size and production of a finer sediment (Boulton et al., 1974). The sediment samples studied during present work, show polymodal distribution (Fig. 7.3) and have more than one mode coarser and finer size than 2Φ that represents the sediments suffered by traction in basal zone of glacier (Boulton et al., 1974; Boulton, 1978). Almost sediments collected in the area are poorly sorted; higher in coarser weight percentage with angular grains indicates regular fluctuating energy conditions (Singh et al., 2018). Range of skewness also corroborates the variable energy condition of the area. Angular and unaltered sediment grains with the presence of ferromagnesian, feldspar, and other labile minerals are strong characteristic of immature glacial sediments (Reineck and Singh, 1980). The present observations suggest characterization of the unconsolidated sediments in the area as "till deposits." They may further be classified as ablation till due to the gradual recession observed in the area. The morainic ridges are made up of heterogeneous sediments with a range of particle sizes varying from boulders, pebbles, sand, and clay. Coarser grain percentages are high with pebbles also point toward existence of freshly derived and immature conditions, which are very common in an englacial environment. Till-covered areas and moraines as well as irregular surfaces signify accumulation of glacial debris during the retreat of glaciers in the region. The glacier acts like a conveyor of sediment load in supraglacial, englacial, and subglacial environments and ultimately in the periglacial precincts. After deposition, these diamicts got modified due to different geomorphic processes as a result of change in climatic conditions. It is also evident by sedimentary landforms that Ny-Ålesund is sculptured by surficial processes throughout the glacial–interglacial cycles. This observation is also substantiated by Singh and Ravindra (2011), who have also envisaged surging glaciers in the area.

The present study highlights the interplay among the glacial sediments along with climatic conditions, and the effectiveness of surface features in providing information on the sequence of events. The surface features of quartz grains were initially used by Whalley and Krinsley (1974) to attribute additional information about various glacial subenvironments. Studies of Krinsley and Doornkamp (1973), Margolis and Kennett (1971), Margolis and Krinsley (1974), Whalley and Krinsley (1974), Bull (1978), Helland and Diffendal (1993), and Mahaney and Kalm (1995) have been used to describe and interpret the surface microtextures of quartz grains. The significant morphological textures are angular outline, rounded outline, low relief, medium relief, and high relief. The mechanical textures were described as conchoidal fractures, arcuate steps, straight steps, imbricated blocks, large breakage blocks, subparallel linear fractures, straight grooves/striations, V-shaped percussion cracks, edge abrasion, surface abrasion, deep troughs, and pitted surface/collision pits. Chemical textures were noticed as precipitation features, adhering particles, and new growths. Numerous features are carved on quartz surface by glacial action, their microtextural frequency was recorded (Table 7.3).

The large number of grain characteristics with contrasting microtextural properties has been compiled by Mahaney et al. (2002), which facilitates comparison of data across regions. A potentially useful approach involves classifying mechanically induced microfractures into one of the three genetic groupings: (1) high-stress fractures, (2) percussion fractures, and (3) polygenetic fractures, or those created through a variety of other processes. It is observed that polygenetic nature and high-stress fractures dominantly reflected in majority of the quartz grains from Ny-Ålesund area are characteristic of till deposits (Mahaney et al., 1996). It is observed that the sediments of Ny-Ålesund experienced low degree of intergranular collision that resulted in V-shaped cracks and moderate edge rounding defining the percussion fractures. Moderate to high-stress microtextures require grain-to-grain contact under sufficient shear stress and have been reported from grains entrained in highly viscous debris flows and glacial environments (Mahaney, 2002). This study also suggests that majority of the quartz grains exhibit a polygenetic character with relatively high-stress micro-texture occurrence, which is ascribed to glacially influenced sediment. Accordingly, stress fractures are inferred to be the result of glacial transport. However, samples exhibiting polygenetic nature and low amount of percussion fracturing also indicate that grains were entrained fluvial, after initial glacial transport. This study effectively documents the degree of fluvial modification of glacially derived grains. Quartz grain microtexture and magnetic susceptibility studies indicate that the sediments suffered dominant glacial activity accompanied by meltwater action (Kar et al., 2018). Due to paucity of terrestrial records, the workers are relying on marine data sets to study the climatic variability of this region (Anderson and Brubaker, 1994). This erosional as well as depositional landform system along the margins of maritime-continental glacial environmental setup offers a great opportunity to understand paleoenvironmental fluctuations and processes. A sediment multiproxy approach has been undertaken to decipher the glacial and glaciofluvial history of Ny-Ålesund during Late Quaternary.

Clay minerals in terrestrial sediment mainly consist of illite, kaolinite, and chlorite observed in study area. Moreover, quartz, feldspar, calcite, and muscovite are also present in clay size fractions. Illite is usually an alteration product of feldspar, mica, and other aluminosilicate minerals under cold conditions of weathering and K^+ removal at low temperatures in a weak alkaline environment (Shi, 1995). In the present study area, illite is observed as a characteristic mineral of high latitudes, and a significant tool to interpret the weathering pattern of older sediments (Windom, 1976). The occurrence of illite pointed to the prevailing climatic conditions under which the sediments were deposited. It has been suggested that any metamorphic rock of low to high grade, from felsic to basic plutons and dykes can produce illite in glacial regime (Ehrmann et al., 2005). Broggerbreen peninsula is characterized by low-grade metamorphic rocks phyllite, mica schist, and gneiss, which contain quartz, feldspar, mica, and other aluminosilicate minerals, which may serve as source of illite under cold and dry weather conditions. Thus dominant illite-rich characteristic in the study area is a product of the physical weathering of rocks. Moreover, the natural environment characteristics of dry and cold climate and weak leaching action in the provenance area are conducive to the formation and conservation of illite. Kaolinite is a common mineral of the continent and has been widely reported as a weathering product under hot and humid condition (Biscaye, 1965; Singer, 1984). The kaolinite composition $Al_2SiO_5(OH)^+$ can be formed by any aluminosilicate by weathering, provided the K^+, Na^+, Ca^{++}, Mg^{++}, and Fe^{++} should easily leach from the parent rock. Kaolinite has widespread occurrence in the Arctic region. Since kaolinite cannot be formed under polar conditions, its occurrence in the sediment suggests warmer and wetter intervals (Birkenmajer, 1989; Elverhøi et al., 1989). The occurrence of kaolinite in the area could either be a result of chemical weathering in warmer periods or the mineral may be secondary in origin transported from distant source. In the study area, feldspar is one of the major constituents of the gneissic basement. The prevailing cold climate is not favorable for the formation of kaolinite, but the intervening warmer phase may have caused its formation. Therefore it is possible that temperature variations within the limits of the prevailing cold climate and humidity might have played a significant role in the formation of kaolinite through chemical weathering. Chlorite is considered to be a common mineral of detrital origin and a product of intense weathering under tropical conditions (Biscaye, 1965). Hillenbrand and Ehrmann (2003) suggested that chlorite of the northern polar area is mainly contributed by the physical weathering of basement metamorphic and igneous rocks and probably altered calc-alkaline volcanic rocks were the major source. The presence of detrital chlorite necessitates at least either a cool or arid weathering environment (Ghosh et al., 2019). Study area contains many chlorite minerals due to the low degree of metamorphism. Chlorite formed under physical weathering can be preserved in the glacial environment after formation and is subjected to glacial erosion.

During the Local Last Glacial Maximum (26–15 ka), Svalbard was completely covered by the late Weichselian ice sheet in the Barents Sea (Landvik et al., 1998). The last deglaciation began at ~15 ka, and the margin of late Weichselian ice sheet started to retreat as the climate became warmer (Svendsen et al., 1992, 1996; Svendsen and Mangerud 1997; Landvik et al., 1998). Between 13 and 12 ka, the ice-sheet margin retreated to the western coast of Spitsbergen, Svalbard (Lehman and Forman, 1992; Mangerud and Svendsen, 1992). The cirque glaciers retreated considerably or even disappeared from many catchments in western Svalbard only to reform during the late Holocene (Svendsen and Mangerud, 1997). Most glaciers in Spitsbergen advanced during the Little Ice Age to form extensive and well-preserved moraines, and reached their maximum (Svendsen et al., 1992; Werner, 1993). Since then, deglaciation has been continuous. At present, 60% of the land area of Svalbard is covered by glaciers, and the largest ice-free areas are in the central and western areas (Birks et al., 2004a,b). Western Svalbard is impacted by the North Atlantic Ocean warm current and has polar-maritime climate characteristics. In contrast, eastern Svalbard is influenced by cold currents from the north. Thus the obtained ages of the samples are 19 ± 2 and 32 ± 3 ka obtained from the study area, which represent time span of Last Glacial Maxima period (Mangerud et al., 1981) in Ny-Ålesund area.

7.6 Conclusions

The eastern part of Broggerbreen peninsula, Ny-Ålesund area shows influence of glacial and interglacial signatures during Late Quaternary time. The glacial signatures with effect of variable climate condition are highly evident in this region. The predominant mode of sediment transport is in basal environment with very little mixture of supraglacial and englacial sediment. Sediments are distinctly glacial in origin and were later modified by fluvioglacial fluvial action. During reworking major part of silt and clay-sized sediments was removed by the meltwater, which leads to enrichment of coarser sediments. The quartz surface microtextures show high-stress and polygenetic fractures, suggesting genetic similarities including sustained grain-to-grain shear stress in a glacial environment. Higher cryostatic pressure is also evident. Fluvial overprinting of glacially induced microtextures modifies the quartz grains due to change in climatic conditions. Existences of dominant glacial environment with intervals of warmer and wetter conditions have been evident from the clay characteristics. The area has undergone major interglaciation phases in pre- and post-LGM time in Late Quaternary. After that the glaciers have consistently receded since LGM resulting in the almost present-day disposition of the glaciers.

References

Ahlmann, H.W., 1953. Glacier variations and climatic fluctuations. Am. Geograph. Soc., Bowman Memorial Lectures, Ser 3 New York.

Anderson, P.M., Brubaker, L.B., 1994. Vegetation history of north central Alaska: a mapped summary of late Quaternary pollen data. Quat. Sci. Rev. 13 (1), 71–92.

Birkenmajer, K., 1989. A guide to tertiary geochronology of King George Island, West Antarctica. Pol. Polar Res. 10, 555–579.

Birks, H.J.B., Monteith, D.T., Rose, N.L., Jones, V.J., Peglar, S.M., 2004a. Recent environmental change and atmospheric contamination on Svalbard as recorded in lake sediments - modern limnology, vegetation and pollen deposition. J. Paleolimnol. 31.

Birks, H.J.B., Jones, V.J., Rose, N.L., 2004b. Recent environmental change and atmospheric contamination on Svalbard as recorded in lake sediments - an introduction. J. Paleolimnol. 31.

Biscaye, P.E., 1965. Mineralogy and sedimentation of recent deep-sea clay in the Atlantic Ocean and adjacent seas and oceans. Geol. Soc. Am. Bull. 76 (7), 803–831.

Boitsov, V.D., Karsakov, A.L., 2012. Atlantic water temperature and climate in the Barents Sea in 2000–2009. J. Mar. Sci. 69, 833–840.

Boulton, G.S., 1978. Boulder shapes and grain size distribution of debris as indicators of transport paths through a glacial and till genesis. Sedimentology 25, 773–799.

Boulton, G.S., Dent, D.L., Morris, E.M., 1974. Subglacial shearing and crushing, and the role of water pressures in tills from south- east Iceland. Geogr. Ann. A, Phys. Geogr. 56, 15–145.

Bull, P.A., 1978. A quantitative approach to scanning electron microscope analysis of cave sediments. In: Whally, W.B. (Ed.), Scanning Electron Microscopy in the Study of Sediments. Geo Abstracts, Norwich, pp. 201–228.

Dharwadkar, A., Roy, S.K., Swain, A.K., Mandal A., Kumar P. 2015. Study of Macro and Micro Structures in the Ductile Shear Zones of the West Spitsbergen Fold Thrust Belt and Implications on the Tectonic Evolution of the Svalbard Area of Arctic Region. Unpublished Report. Geological Survey of India, RP/CHQ/MIV/2013/098.

Dilli, K., Pant, R.K., 1994. Clay minerals as indicators of the provenance and paleoclimatic record of the Kashmir loess. J. Geol. Soc. India 44, 563–574.

Ehrmann, W.U., Melles, M., Kuhn, G., Grobe, H., 1992. Significance of clay mineral assemblages in the Antarctic Ocean. Mar. Geol. 107, 249–273.

Ehrmann, W., Setti, M., Marinoni, L., 2005. Clay minerals in Cenozoic sediments off Cape Roberts (McMurdo Sound, Antarctica) reveal palaeoclimatic history. Palaeogeogr. Palaeoclimatol. Palaeoecol. 229 (3), 187–211.

Elverhøi, A., Pfirman, S., Solheim, A., Larssen, B.B., 1989. Glaciomarine sedimentation and processes on High Arctic epi continental seas--exemplified by the northern Barents Sea. Mar. Geol. 85, 225–250.

Eriksen, E., Ingvaldsen, R.B., Nedreaas, K., Prozorkevich, D., 2015. The effect of recent warming on polar cod and beaked redfish juveniles in the Barents Sea. Reg. Stud. Mar. Sci. 2, 105–112.

Folk, R.L., 1968. Petrology of Sedimentary Rocks. Hemphill's Book Store, Austin, TX.

Folk, R.L., 1974. The Petrology of Sedimentary Rocks. Hemphill Publishing Co., Austin, TX, p. 182.

Ghosh, S., Mukhopadhyay, J., Chakraborty, A., 2019. Clay mineral and geochemical proxies for intense climate change in the Permian Gondwana Rock Record from Eastern India. AAAS Res. 2019, Article ID 8974075.

Goldich, S.S., 1938. A study in rock-weathering. J. Geol. 46, 17–58.

Harland, W.B., Anderson, L.M., Manasrah, D., Butterfield, N.J., 1997. The geology of Svalbard. Geol. Soc. Mem. 17.

Helland, P.E., Diffendal, R.F., 1993. Probable glacial climatic conditions in source areas during deposition of parts of the Ash Hollow Formation, Ogallala Group (late Tertiary), of western Nebraska. Am. J. Sci. 293 (8), 744–757.

Hillenbrand, C.D., Ehrmann, W., 2003. Late Neogene to Quaternary environmental changes in the Antarctic Peninsula region: evidence from drift sediments. Glob. Planet. Change 45, 165–191.

Hjelle, A., 1993. Simplified geological map and stratigraphical column for Svalbard. Geology of Svalbard. In POLARHÅNDBOK NO. 7, Norsk Polarinstitutt, Oslo.

Jackson, M.L., 1979. Soil Chemical Analysis-Advanced Course, second ed., 11th printing. University of Wisconsin, Madison, pp. 169–251.

Kar, A., Mazumdera, A., Mishraa, K., Patilb, S.K., Ravindra, R., Ranhotraa, P.S., et al., 2018. Climatic history of Ny-Alesund region, Svalbard, over the last 19,000 yr: Insights from quartz grain microtexture and magnetic susceptibility. Polar Sci. 18.

Krinsley, D.H., Doornkamp, J.C., 1973. Atlas of Quartz Sand Surface Textures, first ed. Cambridge University Press, Cambridge, 91 pp.

Krinsley, D.H., Funnell, B.M., 1965. Environmental history of quartz sand grains from the Lower and Middle Pleistocene of Norfolk, England. Quart. J. Geol. Soc. 121, 435–456. Available from: https://doi.org/10.1144/gsjgs.121.1.0435.

Krumbein, W.C., Pettijohn, F.J., 1938. Manual of Sedimentary Petrology. Appleton, Century, and Crofts, New York, 549 pp.

Landvik, J.Y., Bondevik, S., Elverhm, A., Fjeldskaar, W., Mangerud, J., Salvigsen, O., et al., 1998. The last glacial maximum of Svalbard and the Barents sea area: ice sheet extent and configuration. Quat. Sci. Rev. 17, 43–75.

Lehman, S.J., Forman, S.L., 1992. Late Weichselian glacier retreat in Kongsforden. West Spitsbergen. Svalbard. Quat. Res. 37 (2), 139–154. Available from: https://doi.org/10.1016/0033-5894(92)90078-W.

Mahaney, W.C., Kalm, V., 1995. Scanning electron microscopy of Pleistocene tills in Estonia. Boreas 24, 13–29.

Mahaney, W.C., Mahaney, W., 2002. Atlas of Sand Grain Surface, Textures and Applications. Oxford University Press, Oxford.

Mahaney, W.C., Claridge, G.C., Campbell, I., 1996. Microtextures on quartz grains in tills from Antarctica. Palaeogeogr. Palaeoclimatol. Palaeoecol. 121, 89–103.

Mangerud, J., Svendsen, J.I., 1992. The last interglacial-glacial period on Spitsbergen, Svalbard. Quat. Sci. Rev. l1, 633–664.

Mangerud, J., Sonstegaard, E., Sejrup, H.P., Haldorsen, S., 1981. A continuous Eemian-Early Weichselian sequence containing pollen and marine fossils at Fjsanger, western Norway. Boreas 10, 137–208.

Margolis, S.V., Kennett, J.P., 1971. Cenozoic paleoglacial history of Antarctica recorded in Subantarctic deep-sea cores. Am. J. Sci. 271, 1–36.

Margolis, S.V., Krinsley, D.H., 1974. Processes of formation and environmental occurrence of microfeatures on detrital quartz grains. Am. J. Sci. 274, 449–464.

Milner, H.B., 1962. Sedimentary Petrography. Macmillan Company, New York, p. 643.

Newsome, D., Ladd, P., 1999. The use of quartz grain microtextures in the study of the origin of sand terrains in Western Australia. Catena 35, 1–17.

Nuth, C., Moholdt, G., Kohler, J., Hagen, J.O., Kaab, A., 2010. Svalbard glacier elevation changes and contribution to sea level rise. J. Geophys. Res. 115 (1), F01008. Available from: https://doi.org/10.1029/2008JF001223.

Overpeck, J., Hughen, K., Hardy, D., Bradley, R., Case, R., Douglas, M., et al., 1997. Arctic Environmental change of the last four centuries. Science 278, 1251–1256.

Pettijohn, F.J., 1984. Sedimentary Rock, The Texture of Sediments, third Indian ed. CBS Publisher, New Delhi, pp. 31–50.

Reineck, H.E., Singh, I.B., 1980. Depositional Sedimentary Environment, Sediment Grain Parameters, second ed. Springer-Verlag, New York, pp. 132–148.

Schwamborn, G., Schirrmeister, L., Frütsch, F., Diekmann, B., 2012. Quartz weathering in freeze–thaw cycles: experiment and application to the El'gygytgyn Crater Lake

record for tracing Siberian permafrost history. Geografi ska Annaler: Series A, Phys. Geogr. 94, 481–499.
Shi, X., 1995. Progress and trends in marine clay minerals research. Mar. Geol. Lett. (Chin.) (1), 1–3.
Singer, A., 1984. The paleoclimatic interpretation of clay minerals in sediments -- a review. Earth-Sci. Rev. 21, 251–293.
Singh, D.S., Ravindra, R., 2011. Control of glacial and fluvial environments in the Ny-Alesund region. Arctic MAUSAM 62 (4), 641–646.
Singh, D.S., Dubey, C.A., Kumar, D., kumar, P., Ravindra, R., 2018. Climate events between 47.5 and 1 ka BP in glaciated terrain of the Ny-Alesund region, Arctic using geomorphology and sedimentology of diversified morphological zones. Polar Sci. (2018). Available from: https://doi.org/10.1016/j.polar.2018.05.003.
Sirocco, F., Lange, H., 1991. Clay mineral accumulation rates in the Arabian Sea during the late Quaternary. Mar. Geol. 97, 105–119.
Srivastava, H.B., Shrivastava, P.K., Roy, S.K., Beg, M.J., Asthana, R., Govil, P., et al., 2018. Transition in Late Quaternary Paleoclimate in Schirmacher Region, East Antarctica as Revealed from Lake Sediments. J. Geol. Soc. India 91, 1–14.
Svendsen, J.I., Mangerud, J., 1997. Holocene glacial and climatic variations on Spitsbergen, Svalbard. Holocene 7, 45–57.
Svendsen, J.L., Mangerud, J., Elverhm, A., Solheim, A., Schiillenhelm, R.T.E., 1992. The Late Weichselian glacial maximum on western Spitsbergen inferred from offshore sediment cores. Mar. Geol. 104, I–17.
Svendsen, J.I., Elverhøi, A., Mangerud, J., 1996. The retreat of the Barents Ice Sheet on the western Svalbard margin. Boreas 25, 244–256.
Tucker, M.E., 1981. Sedimentary Petrology: An Introduction. English Language Book Society/Blackwell Scientific Publications, pp. 10–24.
Tulaczyk, S., Kamb, B., Scherer, R.P., Engelhardt, H.F., 1998. Sedimentary processes at the base of a west Antarctic ice stream: constraints from textural and compositional properties of subglacial debris. J. Sediment. Res. 68 (3), 487–496. Available from: https://doi.org/10.2110/jsr.68.487.
Vos, K., Vandenberghe, N., Elsen, J., 2014. Surface textural analysis of quartz grains by scanning electron microscopy (SEM): from sample preparation to environmental interpretation. Earth-Sci. Rev. 128, 93–104.
Werner, A., 1993. Holocene moraine chronology, Spitsbergen, Svalbard: lichenometric evidence for multiple Neoglacial advances in the Arctic, Holocene, 3. pp. 128–137.
Whalley, W.B., Krinsley, D.H., 1974. A scanning electron microscope study of surface textures of quartz grains from glacial environments. Sedimentology 21, 87–105.
Windom, H.L., 1976. Lithogenous material in marine sediments. In: Riley, J.P., Chester, R. (Eds.), Chemical Oceanography, vol. 5. Academic Press, New York, pp. 103–135.

Further reading

Benn, D.I., Owen, L.A., 2002. Himalayan glacial sedimentary environments: a framework for reconstructing and dating the former extent of glaciers in high mountains. Quat. Inter. 97-98, 3–25.
Hagen, J.O., Liestol, O., 1990. Long- term glacier mass balance investigations in Svalbard. Ann. Glaciol. 14, 102–106.
Lamb, H.H., 1982. Climate: Present, Past and Future.
López-Moreno, J.I., Boike, J., Sanchez-Lorenzo, A., Pomeroy, A.W., 2016. Impact of climate warming on snow processes in Ny-Ålesund, a polar maritime site at Svalbard. Glob. Planet. Change. Available from: http://doi.org/10.1016/j.gloplacha.2016.09.006.

Singh, D.S., Ravindra, R., 2011a. Geomorphology of the MidreLoven Glacier, Ny-Alesund, Svalbard, Arctic. In: Singh, D.S., Chhabra, N.L. (Eds.), Geological Processes and Climate Change. Macmillan Publishers India Ltd., pp. 269–281.
Singh, D.S., Ravindra, R., 2011b. Control of glacial and fluvial environments 578 in the Ny-Alesund region. Arctic MAUSAM 62 (4), 641–646.
Van Meerbeeck, C.J., Renssen, H., Roche, D.M., 2009. How did Marine Isotope Stage 3 and Last Glacial Maximum climates differ? – Perspectives from equilibrium simulations. Clim. Past 5, 33–51.

CHAPTER 8

Reconstruction of Quaternary climate in Svalbard: CRN as proxy

K. Amrutha[1], Pankaj Kumar[2], Neloy Khare[3] and Jitendra Kumar Pattanaik[1]

[1]Department of Geology, Central University of Punjab, Bathinda, India
[2]AMS & Geochronology Group, Inter-University Accelerator Centre, New Delhi, India
[3]Ministry of Earth Sciences, Govt. of India, New Delhi, India

8.1 Introduction

Polar region experienced a large-scale melting of glaciers and sea ice retreat in the recent past. Drastic decrease of ice cover and its foreseen impact on ocean circulations, sea level, and ocean chemistry brought Arctic into the forefront of climate change studies (Streuff, 2013). High-latitude glaciers withstand minor climatic change. However, recent trend of global climatic conditions started influencing polar ice sheet and glacier to an alarming state (Benn and Evans, 2010; Dowdeswell et al., 2008; Kohler et al., 2007). Compared to lower latitudes, Arctic is significantly affected by minor changes in temperature and precipitation pattern, referred as "Arctic amplification" (Masson-Delmotte et al., 2013). Ice dynamics and impact of climate change on Antarctica ice sheet is different than the Arctic Circle as the surrounding environment and atmospheric circulations acting in these areas are diverse in nature. External forces like Milankovitch Cycles and solar irradiance; atmospheric forces like greenhouse gas concentrations, albedo, and cloud cover; and interaction among these forces are key factors responsible for prevailing climate and its changes in the Arctic Circle (Overpeck et al., 1997). Since few decades, Arctic Circle witnessed complex atmospheric, oceanic, and terrestrial changes (Syvitski, 2002). Location of Svalbard archipelago is very crucial in terms of the ocean circulation as it is lying between North Atlantic Ocean and Arctic Ocean. The climate here is largely governed by the North Atlantic Oscillation and thermohaline circulation. However, rise in annual average temperature of Arctic is linked to increase in CO_2 concentration and decrease in SO_4 gas in the atmosphere. Study suggests that role of SO_4 concentration in the atmosphere is more important than the

Understanding Present and Past Arctic Environments
DOI: https://doi.org/10.1016/B978-0-12-822869-2.00019-0

solar irradiation in the Arctic climate (Overpeck et al., 1997). Therefore it is essential to understand how climate has responded to external and atmospheric forces. Svalbard located in the North Atlantic sector of Arctic region has shown a high response to present climate changes due to multiple feedback mechanisms associated with oceanic, atmospheric, and cryosphere processes (Pithan and Mauritsen, 2014).

Multiproxy analysis of sediments and ice core provides ample scope to retrieve information about the paleoclimate. Some of the important methods are elemental and stable isotopic (O, H, and C) variations in the ice core, selected elemental ratio and microfossil analysis from sediment core, mapping of submerged glacial landforms/fjord systems, and borehole temperature profile. Physical, biogeochemical, magnetic, and isotopic analysis integrated with different geochronological techniques like cosmogenic radionuclide (CRN) dating and multivariate statistical analysis was applied to understand past changes in the thickness and retreat of ice sheets, sea ice extent, precipitation, and temperature variations in the Arctic. Over the past 65 Ma, Arctic has experienced the maximum change in the temperature (Bice et al., 2006) and since 4 Ma, temperature change is three to four times higher than the global average (Miller et al., 2010).

Climate change in Arctic region has an influence on the earth energy balance, which can affect the atmospheric circulation in the Northern Hemisphere and ocean circulation. These effects have global impact on sea level and greenhouse gas concentration (Miller et al., 2010). Rapid intrusion of Atlantic water into Arctic ocean leads to abrupt change in surface water circulation, ice drift, and atmospheric circulation (Maslanik et al., 1999) causing increase in cloudiness and surface air temperature (SAT; Serreze et al., 2000). These changes have resulted in decrease of snow cover and increase in glacier ablation and coastal erosion (Forbes and Syvitski, 1995; Lammers et al., 2001). However, the climate changes can be tracked by the immediate effect of the process. For example, warming of Arctic region is reflected in the sediment load carried by the rivers. Once the warming increases, Arctic rivers start to carry more sediment loads just as temperate rivers do. Stochastic sediment transport modelling in Arctic rivers show that for every 2°C warming, there is a 22% increase in sediment load and for every 20% increase in discharge; sediment load increase by 10% (Syvitski, 2002). To infer glacial activity from sediment deposits, it is required to understand different processes associated with the deposits such as paraglacial redeposition, mass wasting, lake stratification, and redox processes (Ballantyne, 2002; Vasskog et al., 2011; Lamoureux and Gilbert, 2004).

Reconstruction of Arctic glacial history becomes more reliable when it has input from maximum location of Arctic Circle. Results obtained from different parts of the Arctic need to be synthesized and evaluated for holistic view on impact of climate change on this region. Geomorphological mapping, multiple proxy analysis on sediment, and ice core samples combined with a suitable chronology for time constrain are essential to reconstruct the glacial history of the region (Bakke et al., 2013; Røthe et al., 2015; Wittmeier et al., 2015). Increase in the number of studies and the varieties of proxies applied for this study will make the process of reconstruction of glacial history of Arctic more reliable.

8.2 About Svalbard

Svalbard is located ~1100 km south of North pole, between 74–81°N and 10–35°E inside the Arctic Circle (Zwoliński et al., 2013). This location is along the dominant corridor of atmospheric moisture, between the Atlantic and the Arctic Basin (Drange et al., 2005). Svalbard archipelago covers an area of 61,020 km^2 and bordered by Arctic circumpolar basin from the north, Barents Sea from the east, and Greenland Sea from the west (National Ocean Service, 2009). At present, glaciers and ice caps cover roughly 57% of the archipelago (Nuth et al., 2013). The landscape of Svalbard is consisting of mountains with elevations up to 1700 m having partly alpine and partly plateau character; strandflats and fjord landscapes are common along the coast of Svalbard (Hjelle, 1993). Numerous islands like West Spitsbergen/Spitsbergen, North East land, Edge Island, Barents Island, and Charles land are parts of the Svalbard. Kongsfjorden–Krossfjorden is one of the important fjord systems in the Svalbard. Considering the Arctic Polar region, Svalbard is the warmest place at this latitude (Drange et al., 2005). North Atlantic Current, which supplies warm water to Svalbard from lower latitude across the Atlantic and along the coast of Norway and the associated large-scale atmospheric circulation, is responsible for this warm climatic condition (Mangerud and Svendsen, 2018). The climate of Svalbard is dependent on two circulation modes. One of the circulation modes brings mild air to the archipelago and results in high average temperature in winter and high precipitation. Other circulation type establishes a high-pressure zone over the Arctic Ocean and north Greenland (Hisdal, 1998).

The region has a sensitive climate due to its position at the northern extent of the North Atlantic Drift (West Spitsbergen Current; Fig. 8.1) and the southern border of the Arctic sea ice front (Aagaard et al., 1987; Rogers et al., 2005). Svalbard is categorized as having a dry High Arctic climate with periglacial conditions, extreme winter temperatures, and continuous permafrost (Christiansen et al., 2010). Svalbard has witnessed episode of fluctuations in glacier mass and ice front extent, which has changed the regional environmental conditions. Behavior of glaciers in Svalbard is strongly influenced by the oceanic currents from the surrounding basins (Barnhart et al., 2010). Svalbard margin provides an opportunity to understand the relationship between ice sheet variability, climate, and

Figure 8.1 Map of the Svalbard showing important locations and surrounding basins. *Modified after Birks, H.J.B., Jones, V.J., Rose, N.L., 2004. Recent environmental change and atmospheric contamination on Svalbard as recorded in lake sediments—an introduction. J. Paleolimnol. 31 (4), 403–410.*

oceanography. This margin is affected by both Atlantic and polar surface water and it is found that the advection of Atlantic water acted as a moisture source for the growth of ice sheets in this region. Advance and retreat of ice sheets in Arctic are closely linked to the flow of Atlantic water and polar waters from the Arctic Ocean.

Glacial processes and associated mechanisms or processes by which it affects the sedimentary record are important as the changes in climate can be studied from this sedimentation rate. Sediments deposited in this region are also an important inventory of paleoclimate. Fjords are formed as a result of glacial activity and represent a contact zone between marine and terrestrial environments (Howe et al., 2010). They play an important role in the reconstruction of glacial activity and related sedimentary processes. As the fjord systems are located in the higher latittude, they are receptive to climate change and very sensitive to prevailing oceanographic conditions. Hence, these systems are important for climate change research (Svendsen et al., 2002; Howe et al., 2003; Hald et al., 2004; Trusel et al., 2010). This summarizes the correlation between the glacial dynamics and the sedimentary processes associated with it and its role in climate changes (Trusel et al., 2010). In general, the correlation can be established by analyzing the changes in ice front. However, at some locations the correlation is poor as it is difficult to identify the ice front because of the debris cover glaciers (Rachlewicz et al., 2007).

8.2.1 Multiproxy analysis to understand the glacial history of Svalbard

Based on the location of the study area, nature, and availability of the samples, it is important to identify the suitable/viable proxies for analysis. Physical properties of sediment cores collected from lakes and sea provide useful information about the prevailing condition under which the sedimentation occurred. Contact zone of diamict and gravel till with sandy sills in a core sediment sample indicate the transformation of lake sediment from ice contact lake sediment to distal glacial-fed lake (Ballantyne, 2002). Diamict till is generally deposited when lake is covered with ice. Ice retreat caused progressive fining up of sediments, resulting into the silt sands, which is a characteristic of distal setting (Ballantyne, 2003). Core sediments characterized by stratification of silty and sandy beds indicate rapid deposition of reworked sediments during deglaciation (Müller, 1999). Presence of dewatering structures in the core can be related to the high sedimentation rate. However, high compactness of the core deposits may indicate the deformation mediated by stranded ice (Eyles et al., 1991). X-ray computed tomography of sediments can be used to visualize the samples and quantify the amount of sand-sized particles present (Røthe et al., 2018). Biotic and isotopic paleolimnological proxies are also used (Luoto et al., 2017) to understand glacial history. Biotic proxy such as fossil Chironomidae assemblages are sensitive to temperature changes and the stable oxygen isotopic signatures ($d^{18}O$) represent hydroclimatic condition (Luoto et al., 2017), which are useful in reconstructing past climate changes. Biological proxies such as pollen and organic matter integrated to dinoflagellate cysts can be used to determine climatic history (Singh et al., 2011).

8.3 Behavior of glaciers: causes and effects?

Landscape development in the Svalbard is the product of glacier advance and retreat. Since the termination of Little Ice Age (LIA), glaciers of Spitsbergen have been retreated and have contributed to the landscape evolution of that region (Hagen et al., 1993; Svendsen and Mangerud, 1997). Fluctuations in glacier mass balance due to glacier retreat and deglaciations impact the erosional pattern and the drainage system of the region, and hence have an influence on life by controlling the soil formation and local climate (Moreau et al., 2005). In the present interglacial period, the continental shelves and fjords have submerged and it has allowed the preservation of the geomorphology of the area (Burton, 2013).

Dynamics of glacial processes and associated sedimentary processes are closely related to glacial regime and reflect the climatic conditions of the area (Trusel et al., 2010). Surge-type glaciers are common in Svalbard as up to 43% of tidewater glaciers on Svalbard could be classified as surge type (Hagen, 1988; Dowdeswell et al., 1991; Liestøl, 1993; Hagen et al., 1993; Jiskoot et al., 2000; Błaszczyk et al., 2009). Surge-type glaciers have a cyclical occurrence; where during the quiescent/slow phase, ice accumulates in the elevated regions of the glacier and in the surge phase the ice flux is higher in the lower regions of the glaciers (Burton, 2013). Surging of glaciers has direct impacts on the geomorphology of the area.

Several authors have reconstructed the glacial history of the Svalbard region, using a variety of proxies (e.g., Birks et al., 1994; Elverhøi et al., 1995; Hald et al., 2004; Baeten et al., 2010; Dowdeswell et al., 2010; Jessen et al., 2010; Forwick and Vorren, 2011). However, a generalization remains difficult due to the manifold methods and locations used for the studies. Radiocarbon dating along Spitsbergen shelf shows that the large Barents Sea ice sheet has played an important role in the glacial history of Svalbard (Landvik et al., 1998). Marine geological studies on the Svalbard shelf during 1980s, through mapping and age estimates of glacial tills and subglacial flutes have also contributed in the documentation of glaciation phase of the region (Mangerud and Svendsen, 1992; Svendsen et al., 1992).

Glacier retreat rate is controlled by a number of factors. The rate varies according to the location of the glacier, type, size, thickness, thermal state and mass balance of the glacier, oscillations in equilibrium line altitude (ELA), dynamics of the debris content, and climatic condition (Rachlewicz et al., 2007). Climatic factors are the temperature of the region/climatic warming, duration of ablation season, proportion of winter and summer precipitation, etc. Information about the major ice sheet drainage patterns into the Arctic Oceans is not clear (Hormes et al., 2011). Field surveys, aerial image interpretation, CRN dating are the important proxies used in the reconstruction of glacial history of Svalbard region (Hormes et al., 2011). Reconstruction of glacial history based only on dating of glacial erratic boulders has limitations to fix the timing of deglaciation. Response of glaciers to climate change is represented by the fluctuation in the ice front position.

Data about glacial advance and retreat are important as the variation in retreat rate is related to the local differences in the size of glacial drainage basin, fjord depth, and the presence of bedrock outcrops (Burton, 2013). Study of landform assemblages in an area enables to understand the history of past sedimentary processes and from which past changes in glacial activity and climate can be assessed (Burton, 2013). These findings can be utilized to find the overall ice sheet history and to improve the understanding of the glacial dynamics within the area. As discussed in the previous section, climate changes are driven by different forcing factors and the response of climate to these factors vary spatially. It is important to select a proxy for climate reconstruction in such a way that the spatial variability is taken in consideration, it can be applied in variety of geographical area and a wide range of geological time can be covered (Miller et al., 2010).

8.4 CRN ^{10}Be as a proxy for climate reconstruction

Prior to discuss the use of ^{10}Be as a proxy for climate reconstruction in Polar region, it is important to know about the production and transport of ^{10}Be in archival systems. ^{10}Be is a radioactive isotope of beryllium, which decays to stable isotope Boron-10 at a very slow rate by emitting β particles with a decay constant of 4.99×10^{-7} year^{-1} ($T_{1/2}$ of ^{10}Be = 1.3987 Ma). ^{10}Be found in nature is the product of spallation reaction of secondary cosmic rays with stable isotopes of oxygen and nitrogen. The ^{10}Be thus produced in the upper atmosphere is termed as *garden variety* ^{10}*Be*, which finally comes down by dry and wet precipitation to become part of natural archive, for example, sediments, land, and ocean water. On the other hand, when high-energetic cosmic rays, defying the earth's magnetic field reach on the earth surface, interact with stable isotopes and form ^{10}Be in rock system and that is termed as *in-situ* ^{10}*Be*. In Arctic region, both *garden variety* and *in-situ* produced ^{10}Be are used as a proxy to infer the chronology and other information of corresponding archive. In spite of higher geomagnetic fields at poles, the production rate of *in-situ* ^{10}Be is higher at poles than equator because the earth's magnetic field lines are parallel to the cosmic ray flux and therefore deflections of cosmic rays are minimal at poles. Further, atmospheric produced ^{10}Be production rate increases as we go from equator toward poles but becomes constant beyond 60-degree Latitude. The concentration of CRNs in natural archives is very low (10^{-11} to 10^{-15}) and it is highly tedious and time-consuming to utilize their radioactivity nature in estimating their concentration. Therefore, instead of waiting for their decay, the CRNs are directly counted using an ultrasensitive technique called "accelerator mass spectrometry" (Kumar et al., 2011, 2015).

It is worth to mention that ^{14}C is an important time stamping tool, which is also produced mostly in the upper atmosphere by spallation of ^{14}N with secondary cosmic-ray neutrons. After its formation, it is attached with oxygen atoms to form $^{14}CO_2$, becomes part of global biogeochemical cycle along with stable carbon isotopes (^{12}C and^{13}C), and all living matters are labeled with ^{14}C. The ratio of ^{14}C with stable isotopes determines the age of the archive.

High Arctic region such as Svalbard where the biological productivity is very low and less chance of getting continuous sand sediment record make it difficult to apply radiocarbon dating and luminescence dating. (Alexanderson et al., 2014). However, the applicability of CRNs such as ^{10}Be in direct dating of glacial landforms and deposits makes geochronological studies possible (Gosse and Phillips, 2001).

Hormes et al. (2011) have studied the Svalbard archipelago using geomorphological mapping and CRN dating to develop an ice sheet configuration and the deglacial history of the region. They have studied eleven glacially transported erratic and eight bedrock samples from Nordaustlandet, eastern Svalbard. The concentration of CRNs accumulated in a surface is a measure of its surface history. ^{10}Be CRN is used to determine the exposure duration of boulders and bedrock surfaces. This study has yielded a new terrestrial data for the ice sheet configuration. CRN dates of boulders and bedrocks represent samples of a complex glacial and exposure history.

Henriksen et al. (2014) used ^{10}Be cosmogenic surface exposure dating on erratic boulders to reconstruct the deglaciation of the fjord, and identified the changing flow regime of the Kongsfjorden ice stream, as well as advances from local glaciers in the late Younger Dryas (YD) to Pre-Boreal. ^{10}Be surface exposure dating on isolated erratic and moraine boulders of Linne glacier, Linnerbreen in western Svalbard provides various information about the glacial history of the region and different climatic processes that might have happened during that period (Reusche et al., 2014). Linne is a regional glacier, which was separated from a larger ice sheet over Svalbard, and the above surface mentioned exposure dating interprets that the glacier retreat has happened between 13.6 ± 0.5 and 12.3 ± 0.5 Ka (Reusche et al., 2014). Even though it is questionable to have a glacier retreat during cold period, this glacial retreat could be climatically driven rather than a local change in glacier characteristics (Roe and O'Neal, 2009). North Atlantic Current has a significant role in determining the climate of Svalbard (Aagaard et al., 1987). West Spitsbergen Current, a branch of North Atlantic Current moderates the extent of sea ice around Svalbard (Onarheim et al., 2014).

Glacier retreat during the cold period is bit tricky; understanding this process with associated climate changes in the region showed that the retreat of Linnerbreen glacier occurred during the same time when the warm Atlantic waters penetrated west of Svalbard (Reusche et al., 2014). Since the glacial retreat in Linnerbreen is climatically driven, supposedly the observed retreat might have happened due to moisture starvation caused by increased extent of sea ice during the same time and/or due to elevated summer shortwave radiative forcing that might have overtook the effect of reduction in heat during Younger Dryas (Mangerud and Landvik, 2007). Hence, it can be said that the retreat of Linnerbreen glacier and other glaciers in Svalbard during Younger Dryas (YD) could be associated with a scarcity of moisture due to growth of sea ice and a warmer temperature causing reduction in glacier accumulation (Reusche et al., 2014).

The thermohaline circulation of the North Atlantic Ocean brings warm water to the coast of Svalbard causing glacial melting and related climate changes. The extent of sea ice cover in the Arctic Ocean also has an important role in determining the glacial extension and retreat in nearby areas. Increase in sea ice cover in Arctic Ocean with reduced temperature of polar waters results in reduced precipitation in the surrounding area causing reduction in moisture available for glacial growth and subsequently results in glacial retreat (Hebbeln et al., 1994; Gildor and Tziperman, 2001; Müller et al., 2009). However, for better understanding these phenomenon as a whole in Svalbard, additional CRN ages and associated studies during YD are needed, so that the general behavior and fluctuation in glacier mass in Svalbard can be understood.

Radiocarbon date helped to establish the deglaciation event between 13 and 12 Ka BP from the study area (Forman, 1990; Lehman and Forman, 1992; Svendsen et al., 1992; Elverhøi et al., 1995; Landvik et al., 1998; Forman et al., 2004). This is supported by CRN dating on islands in the northwest Svalbard, which revealed that there were some ice-free peaks present (Landvik et al., 2003). Therefore the occurrence of nunataks indicates thin ice cover in that region. The results of new CRN dating in the northwest Spitsbergen have revealed that cold-based ice was present during the Late Weichselian. The CRN dating (Gjermundsen et al., 2013) and multiproxy analysis of sediment cores (Jessen et al., 2010) provide a chronology of deglaciation earlier than previously thought (e.g., Elverhøi et al., 1993, 1995; Siegert and Dowdeswell, 1995; Andersen et al., 1996; Landvik et al., 1998; Mangerud et al., 1998; Dowdeswell and Elverhøi, 2002). However, looking at the age range and sequence of events, it demonstrates that there is a need for more data collection and modelling to understand the glacial history better.

Based on ^{10}Be surface exposure dating of glacial erratics, boulders and bedrock show time period of local glacial retreat at different regions of Svalbard. However, it is important to compile and study these glaciation–deglaciation processes to understand the regional scale glacial and deglacial history of Svalbard. Drastic change in ^{10}Be concentrations in a sediment core sample indicates a possibility of climate change and the average ^{10}Be concentration in each stage has an inverse correlation with the corresponding sedimentation rates (Eisenhauer et al., 1994). *Be* is an independent proxy that can be used for stratigraphic dating in Polar region.

Kumar et al. (2014) have studied the distribution of ^{10}Be, as well as the major element's concentrations and the grain size in surface and core sediment samples, collected from Krossfjorden and Kongsfjorden, Svalbard to understand the origin of ^{10}Be and its accumulation rate. The ^{10}Be concentration in sediments varies between 0.90×10^8 and 2.53×10^8 atoms per gram. Their results show that atmospherically produced ^{10}Be dominates over terrigenous (*in-situ*) sources of ^{10}Be in the studied samples. The calculated ^{10}Be accumulation rate varies from 3.1×10^6 to 8.6×10^6 atoms cm^{-2} $year^{-1}$, which is higher than the accumulation rate observed in the deep-sea sediment cores of the Arctic Ocean and the Norwegian Sea. Authors have attributed higher accumulation rate of ^{10}Be due to the higher influx of glacier meltwater into the fjord system. Findings indicate that glaciers in Svalbard are melting on faster rate in recent times and atmospherically produced ^{10}Be once trapped in the glaciers over a period of time is then released into the sediments due to higher outflow of meltwater from the Svalbard glacier to the basin.

8.5 Glacial history of Svalbard

Glacial history of the Svalbard is divided into initial growth phase between 2.3 and 1.6 Ma, expansion phase between 1.6 and 1.3 Ma and since 1.3 Ma, the Svalbard shelf has experienced episodes of glacial advances and retreats (Butt et al., 2000). Pleistocene glacial history studied from the sedimentary succession of the Svalbard marks the glaciation, deglaciation, and subsequent isostatic rebound happened during this period (Houmark-Nielsen and Funder, 1999). Glacial development in the Plio-Pleistocene time period in Svalbard is divided into three phases: initial growth phase, a transitional growth phase, and a final growth phase (Knies et al., 2009).

Holocene glacial history of Svalbard occurred in phases. Some studies indicate two-stage process, while majority of reports show Svalbard has a three-stage Holocene glacial history. These studies were conducted independently from different parts of the Svalbard using diverse methods. Spatial differences in the timing of glacial history occur due to individual catchment properties such as hypsometry and the moisture availability in the location (Røthe et al., 2018).

Uplift history of Broggerhalvoya peninsula, located on the southern part of the Kongsfjorden shows that the deglaciation of Svalbard might have occurred in two stages. This theory is based on the dating of palaeoshorelines, showing slow emergence of Broggerhalvoya peninsula prior to 10 Ka and uplift after 10 Ka (Barnhart et al., 2010). Isotopic data of late Holocene sediments collected from the west Svalbard region subdivide the millennial Surface Air Temperature (SAT) variation during winter into three phases. A cold stage with a progressive cooling of 9°C per century during 1800s, followed by the coldest period during the period of CE 1800–1900 and the third phase is a rapid warming commenced during 20th century (Divine et al., 2011). This rapid warming in the region was accompanied by a decline in sea ice extent in the study area. In a millennial perspective, the SAT shows a long-term cooling has occurred before the warming period. Since the end of 19th century, the temperature record of Svalbard has showed an increasing trend continued up to 1930s, followed by a cooler time period with the lowest temperature marked during 1960s and from 1960s a continuous warming has occurred till today (Nordli et al., 1996).

Three-stage Holocene climate history of Svalbard is controlled by different factors such as degree of melting of different ice sheets [e.g., Laurentide Ice Sheet (LIS)] resulting in fluctuation in freshwater contribution, variation in summer insolation and properties of Atlantic and Arctic waters (Van Der Bilt et al., 2015). Based on the physical, chemical, and magnetic analysis of glacial lake sediment core, a three-stage Holocene glacial history has been proposed. **Early Holocene period (11,270–6700 cal. yr BP)** was marked by active glaciers and modification of older glaciogenic sediments also continued during this period. However, a prolonged paraglacial activity was absent during this time. Around 9050 cal. yr BP, the glacial activity was at its peak reflecting Holocene glacier maximum (Van Der Bilt et al., 2015).

The Holocene glacial maximum is observed during 9500–9700 cal. yr BP and subsequently started to decrease due to an increase in inflow of Atlantic surface waters. All the increased glacial activities happened afterward are less pronounced (Van Der Bilt et al., 2015). Melting of LIS caused a climatic perturbation during 8.2 Ka BP resulting in both cold and dry condition where the glacier mass balance is determined by both temperature and precipitation (Kleiven et al., 2008; Rohling and Pälike, 2005; Oerlemans, 2005). During the time period, ~7400 and 6700 cal. yr BP catchment witnessed the disappearance of glaciers concurrent with an increase in temperature (Van Der Bilt et al., 2015). Studies carried out in the lake sediments in Varfluesjoen, Northern Spitsbergen show that in the early Holocene period the glaciers in the catchment were small or melted completely (Røthe et al., 2018). Warm and wet early and mid-Holocene was marked by three intervals (9500, 8500, and 7250 cal. yr BP) showing high snowmelt and a drier condition prevailed in mid-Holocene (Røthe et al., 2018). Glacial reconstruction from the west coast of Svalbard also shows that during early and mid-Holocene, many glaciers were small or completely melted (Svendsen and Mangerud, 1997; Røthe et al., 2015; Van Der Bilt et al., 2015; de Wet et al., 2018). Between ~5500 and 3800 cal. yr BP the climate was relatively warm corresponding to the presence of smaller glaciers compared to present day in Svalbard (Humlum et al., 2005) with Holocene climate optimum at ~8000–5000 cal. yr BP (Wanner et al., 2008).

Period (4250–4050 and 3380–3230 cal. yr BP) of **early neoglacial advances** is witnessed by centennial-scale cycles of glacier growth and decay. Long- and short-term regional forcing is responsible for this centennial-scale glacier advances (Van Der Bilt et al., 2015). Beginning of neoglacial period (3500 cal. yr BP) shows an increased glacial activity in the catchment of Varfluesjoen Lake (Røthe et al., 2018). The period 3230–700 cal. yr BP witnessed unfavorable conditions for glacial growth in Lake Hajeren. Hence, glaciers were absent between ~3230 and 1100 cal. yr BP (Van Der Bilt et al., 2015). This finding is little controversial as Reusche et al. (2014) and Røthe et al. (2015) propose a glacial maximum in western Svalbard during 1700 cal. yr BP. After 3000 cal. yr BP increase in sea ice cover and the influence of cold Arctic surface waters created an adverse condition for ice growth (Werner et al., 2013). Sea ice cover experienced a fluctuation after 2000 cal. yr BP that resulted in increase in moisture availability, the strengthening of warm Atlantic water inflow increased winter precipitation, and glacial growth occurred (Van Der Bilt et al., 2015). This time period experienced the coolest summers of the past 1800 years at around 1100 cal. yr BP.

In last 2000 years high glacial activity occurred between c.2000–900 and 750–350 cal. yr BP and in this duration, period with less glacial activity was recorded between 1900–1800, 1000–800, and 350–150 cal. yr BP (Røthe et al., 2018). This period ended with medieval climate anomaly (Bradley et al., 2003) and there was a gradual decrease in summer insolation with time leading toward a cold late Holocene. **Little Ice Age (LIA) (700 cal. yr BP – Present)** marks the period characterized by most favorable local conditions for glacial growth in 3800 years. After ~700 cal. yr BP, continuous presence of glaciers is observed (Van Der Bilt et al., 2015). This period experienced peak in glacial activity during 650 cal. yr BP and again at the end of 19th century. This late Holocene cooling ended at ~300–200 cal. yr BP with the end of LIA (Luoto et al., 2017). LIA is divided into three stages where the early stage was a cold and it enters into a temperate state during LIA maximum and returns to a cold stage during post-LIA (Lovell et al., 2018). During these stages, small glaciers in Svalbard also underwent a thermal cycling, which resulted in surge-type behavior of the glaciers (Sevestre and Benn, 2015).

Compared to LIA in continental northern Europe, LIA in Svalbard was very short-lived (Velle et al., 2011; D'Andrea et al., 2012; Zawiska et al., 2017). The observed glacier ELA records support this, where LIA is less represented or distinct in the Svalbard glacier record (Røthe et al., 2015). Climate has changed over the last 400 years from the LIA (17th to 19th century) to today's global warming (Overpeck et al., 1997); however, climate change has been identified mostly from the variation in sediment discharge. Understanding sediment discharge variations based on the landscape produced after a glacial history of a time period at certain location is still lacking, putting aside the argument that other factors also have an important role in determining the topography of a location.

8.6 Summary

Multiproxy analysis on ice core and sediment records helps to understand the climatic variation in regional scale to global level. However, surface exposure dating of glacial erratics, boulders, and bedrock using CRN ^{10}Be provides information about the local glacial retreat at different regions of Svalbard. Compilation of individual glaciation–deglaciation processes at different locations aids to comprehend the regional glacial history of Svalbard. Among CRN, ^{10}Be can be used as an independent proxy to understand the climatic change and to establish the chronological order of glacial events. Exploring the glaciers and their dynamics in Polar region and reconstruction of their behavior may help in understanding the future of the earth. Importance of the study of glacial dynamics can be seen from the study of glaciers from western Svalbard. Freshwater fluxes from these glaciers are reported to influence fjord sea ice, the hydrology of fjord and river and it has a role in determining the local biota also. In Svalbard, the paleoclimatic studies are not evenly distributed. The southern part of the archipelago is sparsely studied compared to other parts and there is a need for more studies from this region for the complete glacial history of Svalbard.

Acknowledgments

We are thankful to Vice Chancellor, Central University of Punjab, Bathinda and Director, Inter-University of Accelerator Centre, New Delhi for their support on the completion of this work. AK is thankful for the financial support towards PhD from INSPIRE, DST, New Delhi.

References

Aagaard, K., Foldvik, A., Hillman, S.R., 1987. The West Spitsbergen Current: disposition and water mass transformation. J. Geophys. Res.: Ocean. 92 (C4), 3778–3784.

Alexanderson, H., Backman, J., Cronin, T.M., Funder, S., Ingólfsson, Ó., Jakobsson, M., et al., 2014. An Arctic perspective on dating Mid-Late Pleistocene environmental history. Quat. Sci. Rev. 92, 9–31.

Andersen, E.S., Dokken, T.M., Elverhøi, A., Solheim, A., Fossen, I., 1996. Late Quaternary sedimentation and glacial history of the western Svalbard continental margin. Mar. Geol. 133 (3–4), 123–156.

Baeten, N.J., Forwick, M., Vogt, C., Vorren, T.O., 2010. Late Weichselian and Holocene sedimentary environments and glacial activity in Billefjorden, Svalbard. Geol. Soc., Lond., Special Publications 344 (1), 207–223.

Bakke, J., Bradley, R.S., Dahl, S., Balascio, N.L., 2013. Synchronous inter-hemispheric alpine glacier advances during the Antarctic Cold Reversal. AGUFM 2013, PP13E-07.

Ballantyne, C.K., 2002. Paraglacial geomorphology. Quat. Sci. Rev. 21 (18–19), 1935–2017.

Ballantyne, C., 2003. StacPollaidh-a late Devensian Nunatak? A comment on Bradwell and Krabbendam. Quat. Newsl. 34–38.

Barnhart, T.B., Carson, B., Bader, N., 2010. Glaciomarine Sediment Flux and Transportation Mechanisms. Kronebreen/Kongsvegan, Kongsfjorden, Svalbard.

Benn, D.I., Evans, D.J.A., 2010. Glaciers and Glaciation. Hodder Education, London, p. 802.

Bice, K.L., Birgel, D., Meyers, P.A., Dahl, K.A., Hinrichs, K.U., Norris, R.D., 2006. A multiple proxy and model study of Cretaceous upper ocean temperatures and atmospheric CO_2 concentrations. Paleoceanography 21 (2).

Birks, H.H., Paus, A., Svenndse, J.I., Alm, T., Mangerud, J., Landvik, J.Y., 1994. Late Weichselian environmental change in Norway, including Svalbard. J. Quat. Sci. 9 (2), 133–145.

Birks, H.J.B., Jones, V.J., Rose, N.L., 2004. Recent environmental change and atmospheric contamination on Svalbard as recorded in lake sediments—an introduction. J. Paleolimnol. 31 (4), 403–410.

Błaszczyk, M., Jania, J.A., Hagen, J.O., 2009. Tidewater glaciers of Svalbard: recent changes and estimates of calving fluxes. Pol. Polar Res. 85–142.

Bradley, R.S., Hughes, M.K., Diaz, H.F., 2003. Climate in medieval time. Science 302 (5644), 404–405.

Burton, D., 2013. Submarine Landforms and Past Ice Flow in the Krossfjorden System, Northwest Svalbard (Doctoral dissertation). University of Cambridge.

Butt, F.A., Elverhøi, A., Solheim, A., Forsberg, C.F., 2000. Deciphering Late Cenozoic development of the western Svalbard Margin from ODP Site 986 results. Mar. Geol. 169 (3–4), 373–390.

Christiansen, H.H., Etzelmüller, B., Isaksen, K., Juliussen, H., Farbrot, H., Humlum, O., et al., 2010. The thermal state of permafrost in the nordic area during the international polar year 2007–2009. Permafr. Periglac. Process. 21 (2), 156–181.

D'Andrea, W.J., Vaillencourt, D.A., Balascio, N.L., Werner, A., Roof, S.R., Retelle, M., et al., 2012. Mild Little Ice Age and unprecedented recent warmth in an 1800 year lake sediment record from Svalbard. Geology 40 (11), 1007–1010.

de Wet, G.A., Balascio, N.L., D'Andrea, W.J., Bakke, J., Bradley, R.S., Perren, B., 2018. Holocene glacier activity reconstructed from proglacial lake Gjøavatnet on Amsterdamøya, NW Svalbard. Quat. Sci. Rev. 183, 188–203.

Divine, D.V., Sjolte, J., Isaksson, E., Meijer, H.A.J., Van De Wal, R.S.W., Martma, T., et al., 2011. Modelling the regional climate and isotopic composition of Svalbard

precipitation using REMOiso: a comparison with available GNIP and ice core data. Hydrol. Process. 25 (24), 3748–3759.
Dowdeswell, J.A., Elverhøi, A., 2002. The timing of initiation of fast-flowing ice streams during a glacial cycle inferred from glacimarine sedimentation. Mar. Geol. 188 (1–2), 3–14.
Dowdeswell, J.A., Hamilton, G.S., Hagen, J.O., 1991. The duration of the active phase on surge-type glaciers: contrasts between Svalbard and other regions. J. Glaciol. 37 (127), 388–400.
Dowdeswell, J.A., Benham, T.J., Strozzi, T., Hagen, J.O., 2008. Iceberg calving flux and mass balance of the Austfonna ice cap on Nordaustlandet, Svalbard. J. Geophys. Res.: Earth Surf. 113 (F3).
Dowdeswell, J.A., Jakobsson, M., Hogan, K.A., O'Regan, M., Backman, J., Evans, J., et al., 2010. High-resolution geophysical observations of the Yermak Plateau and northern Svalbard margin: implications for ice-sheet grounding and deep-keeled icebergs. Quat. Sci. Rev. 29 (25–26), 3518–3531.
Drange, H., Dokken, T., Furevik, T., Gerdes, R., Berger, W. (Eds.), 2005. The Nordic Seas: An Integrated Perspective. American Geophysical Union (No. 158).
Eisenhauer, A., Spielhagen, R.F., Frank, M., Hentzschel, G., Mangini, A., Kubik, P.W., et al., 1994. 10Be records of sediment cores from high northern latitudes: implications for environmental and climatic changes. Earth Planet. Sci. Lett. 124 (1–4), 171–184.
Elverhøi, A., Fjeldskaar, W., Solheim, A., Nyland-Berg, M., Russwurm, L., 1993. The Barents Sea Ice Sheet—a model of its growth and decay during the last ice maximum. Quat. Sci. Rev. 12 (10), 863–873.
Elverhøi, A., Andersen, E.S., Dokken, T., Hebbeln, D., Spielhagen, R., Svendsen, J.I., et al., 1995. The growth and decay of the Late Weichselian ice sheet in western Svalbard and adjacent areas based on provenance studies of marine sediments. Quat. Res. 44 (3), 303–316.
Eyles, N., Mullins, H.T., Hine, A.C., 1991. The seismic stratigraphy of Okanagan Lake, British Columbia; a record of rapid deglaciation in a deep 'fiord-lake' basin. Sediment. Geol. 73 (1–2), 13–41.
Forbes, D., Syvitski, J.P.M., 1995. Paraglacial coasts. In: Woodruffe, C., Carter, R.W.G. (Eds.), Coastal Evolution. Cambridge University of Press, Cambridge, pp. 373–424.
Forman, S.L., 1990. Post-glacial relative sea-level history of northwestern Spitsbergen, Svalbard. Geol. Soc. Am. Bull. 102 (11), 1580–1590.
Forman, S.L., Lubinski, D.J., Ingólfsson, Ó., Zeeberg, J.J., Snyder, J.A., Siegert, M.J., et al., 2004. A review of postglacial emergence on Svalbard, Franz Josef Land and Novaya Zemlya, northern Eurasia. Quat. Sci. Rev. 23 (11–13), 1391–1434.
Forwick, M., Vorren, T., 2011. Submarine mass wasting in Isfjorden, Spitsbergen. Submarine Mass Movements and Their Consequences. Springer, pp. 711–722.
Gildor, H., Tziperman, E., 2001. A sea ice climate switch mechanism for the 100-kyr glacial cycles. J. Geophys. Res.: Ocean. 106 (C5), 9117–9133.
Gjermundsen, E.F., Briner, J.P., Akçar, N., Salvigsen, O., Kubik, P., Gantert, N., et al., 2013. Late Weichselian local ice dome configuration and chronology in Northwestern Svalbard: early thinning, late retreat. Quat. Sci. Rev. 72, 112–127.
Gosse, J.C., Phillips, F.M., 2001. Terrestrial *in-situ* cosmogenic nuclides: theory and application. Quat. Sci. Rev. 20 (14), 1475–1560.
Hagen, J.O., 1988. Glacier surge in Svalbard with examples from Usherbreen. Norsk Geografisk Tidsskrift-Norwegian J. Geogr. 42 (4), 203–213.
Hagen, J.O., Liestøl, O., Roland, E.R.I.K., Jørgensen, T., 1993. Glacier Atlas of Svalbard and Janmayen, vol. 129. Norskpolarinstitutt, Oslo, p. 141.
Hald, M., Ebbesen, H., Forwick, M., Godtliebsen, F., Khomenko, L., Korsun, S., et al., 2004. Holocene paleoceanography and glacial history of the West Spitsbergen area, Euro-Arctic margin. Quat. Sci. Rev. 23 (20–22), 2075–2088.

Hebbeln, D., Dokken, T., Andersen, E.S., Hald, M., Elverhøi, A., 1994. Moisture supply for northern ice-sheet growth during the Last Glacial Maximum. Nature 370 (6488), 357–360.
Henriksen, M., Alexanderson, H., Landvik, J.Y., Linge, H., Peterson, G., 2014. Dynamics and retreat of the Late Weichselian Kongsfjorden ice stream, NW Svalbard. Quat. Sci. Rev. 92, 235–245.
Hisdal, V., 1998. Svalbard Nature and History, 123. Norwegian Polar Institute, Polarhåndbok No. 12.
Hjelle, A., 1993. Geology of Svalbard. NorskPolarinstitutt, Oslo, p. 162.
Hormes, A., Akcar, N., Kubik, P.W., 2011. Cosmogenic radionuclide dating indicates ice-sheet configuration during MIS 2 on Nordaustlandet, Svalbard. Boreas 40 (4), 636–649.
Houmark-Nielsen, M., Funder, S., 1999. Pleistocene stratigraphy of Kongsfjordhallet, Spitsbergen, Svalbard. Polar Res. 18 (1), 39–50.
Howe, J.A., Moreton, S.G., Morri, C., Morris, P., 2003. Multibeam bathymetry and the depositional environments of Kongsfjorden and Krossfjorden, western Spitsbergen, Svalbard. Polar Res. 22 (2), 301–316.
Howe, J.A., Austin, W.E., Forwick, M., Paetzel, M., Harland, R., Cage, A.G., 2010. Fjord systems and archives: a review. Geol. Soc. 344 (1), 5–15. London, Special Publications.
Humlum, O., Elberling, B., Hormes, A., Fjordheim, K., Hansen, O.H., Heinemeier, J., 2005. Late-Holocene glacier growth in Svalbard, documented by subglacial relict vegetation and living soil microbes. Holocene 15 (3), 396–407.
Jessen, S.P., Rasmussen, T.L., Nielsen, T., Solheim, A., 2010. A new Late Weichselian and Holocene marine chronology for the western Svalbard slope 30,000–0 cal years BP. Quat. Sci. Rev. 29 (9–10), 1301–1312.
Jiskoot, H., Murray, T., Boyle, P., 2000. Controls on the distribution of surge-type glaciers in Svalbard. J. Glaciol. 46 (154), 412–422.
Kleiven, H.K.F., Kissel, C., Laj, C., Ninnemann, U.S., Richter, T.O., Cortijo, E., 2008. Reduced North Atlantic deep water coeval with the glacial Lake Agassiz freshwater outburst. Science 319 (5859), 60–64.
Knies, J., Matthiessen, J., Vogt, C., Laberg, J.S., Hjelstuen, B.O., Smelror, M., et al., 2009. The Plio-Pleistocene glaciation of the Barents Sea–Svalbard region: a new model based on revised chronostratigraphy. Quat. Sci. Rev. 28 (9–10), 812–829.
Kohler, J., James, T.D., Murray, T., Nuth, C., Brandt, O., Barrand, N.E., et al., 2007. Acceleration in thinning rate on western Svalbard glaciers. Geophys. Res. Lett. 34 (18).
Kumar, P., Pattanaik, J., Ojha, S., Gargari, S., Joshi, R., Roonwal, G., et al., 2011. ^{10}Be measurements at IUAC-AMS facility. J. Radioanal. Nucl. Chem. 290 (1), 179–182.
Kumar, P., Pattanaik, J.K., Khare, N., Chopra, S., Yadav, S., Balakrishnan, S., et al., 2014. Study of ^{10}Be in the sediments from the Krossfjorden and Kongsfjorden Fjord System, Svalbard. J. Radioanal. Nucl. Chem. 302 (2), 903–909.
Kumar, P., Chopra, S., Pattanaik, J.K., Ojha, S., Gargari, S., Joshi, R., et al., 2015. A new AMS facility at Inter University Accelerator Centre, New Delhi. Nucl. Instrum. Methods Phys. Res. Sect. B: Beam Interact. Mater. At. 361, 115–119.
Lammers, R.B., Shiklomanov, A.I., Vörösmarty, C.J., Fekete, B.M., Peterson, B.J., 2001. Assessment of contemporary Arctic river runoff based on observational discharge records. J. Geophys. Res.: Atmos. 106 (D4), 3321–3334.
Lamoureux, S.F., Gilbert, R., 2004. A 750-yr record of autumn snowfall and temperature variability and winter storminess recorded in the varved sediments of Bear Lake, Devon Island, Arctic Canada. Quat. Res. 61 (2), 134–147.
Landvik, J.Y., Bondebik, S., Elyerhoi, A., Fjeldskaar, W.I.L.L.Y., Mangerud, J.A.N., Siegert, S., et al., 1998. The last glacial maximum of Svalbard and the Barents Sea area: ice sheet extent and configuration. Quat. Sci. Rev. 17 (1), 43–76.

Landvik, J.Y., Brook, E.J., Gualtieri, L., Raisbeck, G., Salvigsen, O., Yiou, F., 2003. Northwest Svalbard during the last glaciation: ice-free areas existed. Geology 31 (10), 905–908.

Lehman, S.J., Forman, S.L., 1992. Late Weichselian glacier retreat in Kongsfjorden, west Spitsbergen, Svalbard. Quat. Res. 37 (2), 139–154.

Liestøl, O., 1993. Glaciers of Svalbard. Norway, Satellite Image Atlas of Glaciers of the World, USGS Professional Paper.

Lovell, H., Benn, D.I., Lukas, S., Ottesen, D., Luckman, A., Hardiman, M., et al., 2018. Multiple Late Holocene surges of a High-Arctic tidewater glacier system in Svalbard. Quat. Sci. Rev. 201, 162–185.

Luoto, T.P., Rantala, M.V., Tammelin, M.H., 2017. Tracking the limnoecological history of Lake Hiidenvesi (southern Finland) using the paleolimnological approach. Water Air Soil Pollut. 228 (12), 461.

Mangerud, J., Svendsen, J.I., 1992. The last interglacial-glacial period on Spitsbergen, Svalbard. Quat. Sci. Rev. 11 (6), 633–664.

Mangerud, J.A.N., Landvik, J.Y., 2007. Younger Dryas cirque glaciers in western Spitsbergen: smaller than during the Little Ice Age. Boreas 36 (3), 278–285.

Mangerud, J., Svendsen, J.I., 2018. The Holocene thermal maximum around Svalbard, Arctic North Atlantic; molluscs show early and exceptional warmth. Holocene 28 (1), 65–83.

Mangerud, J.A.N., Dokken, T., Hebbeln, D., Heggen, B., Ingolfsson, O., Landvik, J.Y., et al., 1998. Fluctuations of the Svalbard–Barents Sea Ice Sheet during the last 150 000 years. Quat. Sci. Rev. 17 (1–3), 11–42.

Maslanik, J. A., Serreze, M. C., Agnew, T. 1999. On the record reduction in 1998 western Arctic sea-ice cover. Geophys. Res. Lett., 26(13), 1905–1908.

Masson-Delmotte, V., Schulz, M., Abe-Ouchi, A., Beer, J., Ganopolski, A., Rouco, J.G., et al., 2013. Information from paleoclimate archives. In: Stocker, T.F., et al., (Eds.), Climate Change 2013: The Physical Science Basis Contribution of Working Group I to the Fifth Assessment Report of the Intergovernmental Panel on Climate Change. Cambridge, United Kingdom and New York, NY.

Miller, G.H., Brigham-Grette, J., Alley, R.B., Anderson, L., Bauch, H.A., Douglas, M.S. V., et al., 2010. Temperature and precipitation history of the Arctic. Quat. Sci. Rev. 29 (15–16), 1679–1715.

Moreau, M., Laffly, D., Joly, D., Brossard, T., 2005. Analysis of plant colonization on an arctic moraine since the end of the Little Ice Age using remotely sensed data and a Bayesian approach. Remote Sens. Environ. 99 (3), 244–253.

Müller, B.U., 1999. Paraglacial sedimentation and denudation processes in an Alpine valley of Switzerland. An approach to the quantification of sediment budgets. Geodinamica Acta 12 (5), 291–301.

Müller, J., Massé, G., Stein, R., Belt, S.T., 2009. Variability of sea-ice conditions in the Fram Strait over the past 30,000 years. Nat. Geosci. 2 (11), 772–776.

National Ocean Service, 2009. Annual Report.

Nordli, P.Ø., Hanssen-Bauer, I., Førland, E.J., 1996. Homogeneity Analyses of Temperature and Precipitation Series from Svalbard and Jan Mayen. Norske meteorologiskeinstitutt.

Nuth, C., Kohler, J., König, M., Deschwanden, A.V., Hagen, J.O.M., Kääb, A., et al., 2013. Decadal changes from a multi-temporal glacier inventory of Svalbard. Cryosphere 7 (5), 1603–1621.

Oerlemans, J., 2005. Extracting a climate signal from 169 glacier records. Science 308 (5722), 675–677.

Onarheim, I.H., Smedsrud, L.H., Ingvaldsen, R.B., Nilsen, F., 2014. Loss of sea ice during winter north of Svalbard. Tellus A: Dyn. Meteorol. Oceanogr. 66 (1), 23933.

Overpeck, J., Hughen, K., Hardy, D., Bradley, R., Case, R., Douglas, M., et al., 1997. Arctic environmental change of the last four centuries. Science 278 (5341), 1251–1256.
Pithan, F., Mauritsen, T., 2014. Arctic amplification dominated by temperature feedbacks in contemporary climate models. Nat. Geosci. 7 (3), 181–184.
Rachlewicz, G., Szczuciński, W., Ewertowski, M., 2007. Post-"Little Ice Age" retreat rates of glaciers around Billefjorden in central Spitsbergen, Svalbard. Pol. Polar Res. 159–186.
Reusche, M., Winsor, K., Carlson, A.E., Marcott, S.A., Rood, D.H., Novak, A., et al., 2014. 10Be surface exposure ages on the late-Pleistocene and Holocene history of Linnébreen on Svalbard. Quat. Sci. Rev. 89, 5–12.
Roe, G.H., O'Neal, M.A., 2009. The response of glaciers to intrinsic climate variability: observations and models of late-Holocene variations in the Pacific Northwest. J. Glaciol. 55 (193), 839–854.
Rogers, J.C., Yang, L., Li, L., 2005. The role of Fram Strait winter cyclones on sea ice flux and on Spitsbergen air temperatures. Geophys. Res. Lett. 32 (6).
Rohling, E.J., Pälike, H., 2005. Centennial-scale climate cooling with a sudden cold event around 8,200 years ago. Nature 434 (7036), 975–979.
Røthe, T.O., Bakke, J., Vasskog, K., Gjerde, M., D'Andrea, W.J., Bradley, R.S., 2015. Arctic Holocene glacier fluctuations reconstructed from lake sediments at Mitrahalvøya, Spitsbergen. Quat. Sci. Rev. 109, 111–125.
Røthe, T.O., Bakke, J., Støren, E.W., Bradley, R.S., 2018. Reconstructing Holocene glacier and climate fluctuations from lake sediments in Vårfluesjøen, northern Spitsbergen. Front. Earth Sci. 6, 91.
Serreze, M.C., Walsh, J.E., Chapin, F.S., Osterkamp, T., Dyurgerov, M., Romanovsky, V., et al., 2000. Observational evidence of recent change in the northern high-latitude environment. Clim. change 46 (1–2), 159–207.
Sevestre, H., Benn, D.I., 2015. Climatic and geometric controls on the global distribution of surge-type glaciers: implications for a unifying model of surging. J. Glaciol. 61 (228), 646–662.
Siegert, M.J., Dowdeswell, J.A., 1995. Numerical modeling of the late Weichselian Svalbard-Barents Sea ice sheet. Quat. Res. 43 (1), 1–13.
Singh, V., Farooqui, A., Mehrotra, N.C., Singh, D.S., Tewari, R., Jha, N., et al., 2011. Late pleistocene and early holocene climate of Ny-Alesund, Svalbard (Norway): a study based on biological proxies. J. Geol. Soc. India 78 (2), 109.
Streuff, K.T., 2013. Landform Assemblages in Inner Kongsfjorden, Svalbard: Evidence of Recent Glacial (Surge) Activity (Master's thesis). Universitetet i Tromsø.
Svendsen, J.I., Mangerud, J., 1997. Holocene glacial and climatic variations on Spitsbergen, Svalbard. Holocene 7 (1), 45–57.
Svendsen, J.I., Mangerud, J., Elverhøi, A., Solheim, A., Schüttenhelm, R.T., 1992. The Late Weichselian glacial maximum on western Spitsbergen inferred from offshore sediment cores. Mar. Geol. 104 (1–4), 1–17.
Svendsen, H., Beszczynska-Møller, A., Hagen, J.O., Lefauconnier, B., Tverberg, V., Gerland, S., et al., 2002. The physical environment of Kongsfjorden–Krossfjorden, an Arctic fjord system in Svalbard. Polar Res. 21 (1), 133–166.
Syvitski, J.P., 2002. Sediment discharge variability in Arctic rivers: implications for a warmer future. Polar Res. 21 (2), 323–330.
Trusel, L.D., Powell, R.D., Cumpston, R.M., Brigham-Grette, J., 2010. Modern glacimarine processes and potential future behaviour of Kronebreen and Kongsvegen polythermal tidewater glaciers, Kongsfjorden, Svalbard. Geol. Soc. 344 (1), 89–102. London, Special Publications.
Van Der Bilt, W.G., Bakke, J., Vasskog, K., D'Andrea, W.J., Bradley, R.S., Ólafsdóttir, S., 2015. Reconstruction of glacier variability from lake sediments reveals dynamic Holocene climate in Svalbard. Quat. Sci. Rev. 126, 201–218.

Vasskog, K., Nesje, A., Støren, E.N., Waldmann, N., Chapron, E., Ariztegui, D., 2011. A Holocene record of snow-avalanche and flood activity reconstructed from a lacustrine sedimentary sequence in Oldevatnet, western Norway. Holocene 21 (4), 597–614.

Velle, G., Kongshavn, K., Birks, H.J.B., 2011. Minimizing the edge-effect in environmental reconstructions by trimming the calibration set: chironomid-inferred temperatures from Spitsbergen. Holocene 21 (3), 417–430.

Wanner, H., Beer, J., Bütikofer, J., Crowley, T.J., Cubasch, U., Flückiger, J., et al., 2008. Mid-to Late Holocene climate change: an overview. Quat. Sci. Rev. 27 (19–20), 1791–1828.

Werner, K., Spielhagen, R.F., Bauch, D., Hass, H.C., Kandiano, E., 2013. Atlantic Water advection versus sea-ice advances in the eastern Fram Strait during the last 9 ka: multiproxy evidence for a two-phase Holocene. Paleoceanography 28 (2), 283–295.

Wittmeier, H.E., Bakke, J., Vasskog, K., Trachsel, M., 2015. Reconstructing Holocene glacier activity at Langfjordjøkelen, Arctic Norway, using multi-proxy fingerprinting of distal glacier-fed lake sediments. Quat. Sci. Rev. 114, 78–99.

Zawiska, I., Luoto, T.P., Nevalainen, L., Tylmann, W., Jensen, T.C., Obremska, M., et al., 2017. Climate variability and lake ecosystem responses in western Scandinavia (Norway) during the last Millennium. Palaeogeogr. Palaeoclimatol. Palaeoecol. 466, 231–239.

Zwoliński, Z., Giżejewski, J., Karczewski, A., Kasprzak, M., Lankauf, K.R., Migoń, P., et al., 2013. Geomorphological settings of Polish research areas on Spitsbergen. Landf. Anal. 22.

Further reading

French, H.M., 2007. The Periglacial Environment, third ed. John Wiley & Sons, Chichester, 458 p.

CHAPTER 9

Ultraviolet radiation—protecting pigments in plants growing at Arctic region

Jaswant Singh[1] and Rajni Khare[2]
[1]Department of Environmental Sciences, Dr. R.M.L. Avadh University, Ayodhya, India
[2]Delhi Public School, Greater Noida, India

9.1 Introduction

The Arctic can be defined as the region centered on the North Pole, where ice dominates the landscape, the soil is permanently frozen, and trees cannot grow. The Arctic is a frozen ocean, and it consists of the Arctic Ocean and parts of Russia, Sweden, Canada, Greenland, Norway, Alaska, Iceland, and Finland. The average temperature in winter is around −36°C and 0°C during summer. An essential aspect of the Arctic region is the stratospheric ozone depletion as in Antarctica so that this region experiences a high amount of ultraviolet (UV) radiation. However, ozone depletion is less severe in the Arctic than in Antarctica (Muller et al., 1997). Fig. 9.1 indicates the ozone-oxygen cycle in the ozone layer.

Arctic flora consists of mainly vascular plants, mosses, and algae. Lichens play a dominant role in Arctic ecosystems. During the Arctic spring and summer, ozone depletion occurs, resulting in a high level of UV radiation in the Arctic region (ACIA, 2004). UV-B radiation harms the Arctic plants' growth, development, morphology, physiology, and changes in the distribution pattern of flora (Callaghan et al., 2004a,b; Bjerke et al., 2005). Induction of UV-B absorbing pigments in the plants exposed to UV-B radiation in plants growing at Schirmacher Oasis of East Antarctica has also been reported (Singh et al., 2012; Singh and Singh, 2014). In the present communication, we have highlighted studies on the effects of UV-B radiation, its influence on the synthesis of plant pigments, and their protective roles in plants growing at the Arctic region.

Understanding Present and Past Arctic Environments
DOI: https://doi.org/10.1016/B978-0-12-822869-2.00012-8

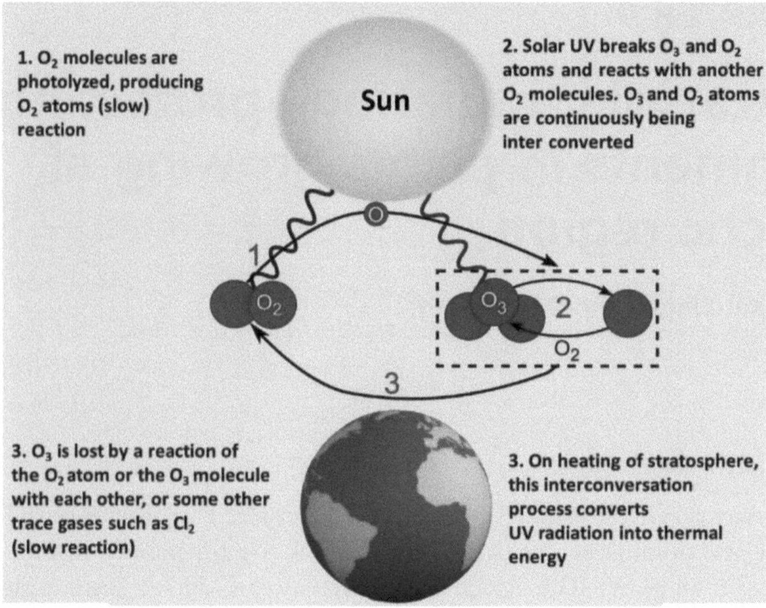

Figure 9.1 Ozone-oxygen cycle in the ozone layer (https://en.wikipedia.org/wiki/Ozone_layer).

9.2 Ozone layer and its importance

Ozone chemically consists of a molecule of three oxygen atoms. It is found mainly in the stratosphere, between 10 and 50 km from the Earth's surface. This area is in the upper atmosphere. By absorbing the sun's harmful UV radiations the ozone molecules eliminate a significant threat to life forms on Earth. UV rays can cause skin cancer and other diseases and deformities in plants and animals (Figs. 9.2 and 9.3).

9.3 Ozone depletion and ultraviolet radiation in the Arctic

The scenario of ozone depletion during the past two decades has been alarming. It was believed to have been reduced up to 50% in Antarctica's concentration during this time span. Due to the use of Chloroflorocarbons, a significant amount of harmful UV-B radiation is entering the atmosphere at a higher intensity.

The ozone depletion also occurs at the North Pole, but it is not as strong as in the South Pole. The increased power of UV-B radiation could be detrimental for the aquatic and terrestrial ecosystem. Ten percent reduction in the

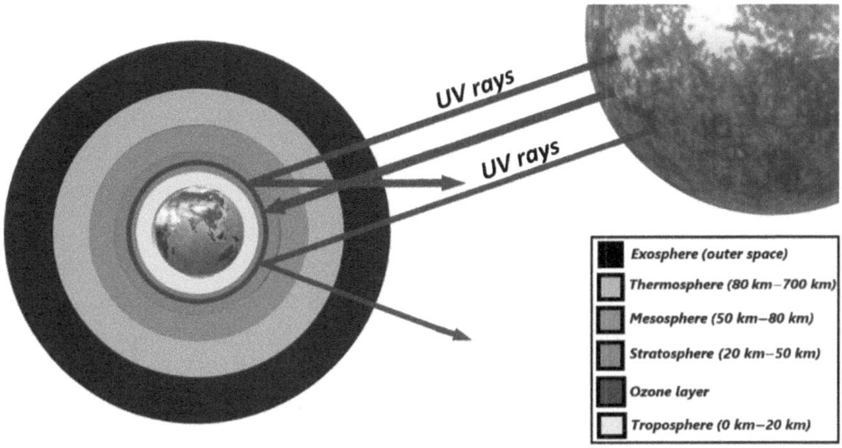

Figure 9.2 Ozone layer absorbs harmful UV radiations and protects life forms on Earth.

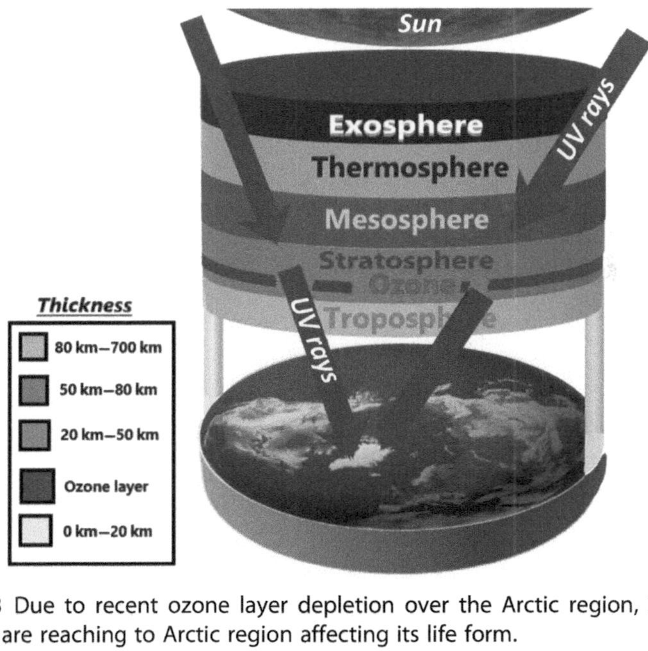

Figure 9.3 Due to recent ozone layer depletion over the Arctic region, harmful UV radiations are reaching to Arctic region affecting its life form.

stratospheric ozone level could result in a 15%–20% increase in UV radiation. Thus organisms that are not accustomed to this high UV radiation level will get adversely affected in terms of growth and development. If the primary producers are threatened, it won't be easy for the higher trophic level organisms at the poles to survive. Therefore increasing levels of UV-B radiation is now becoming a limiting factor for living organisms.

As per the documentation of ACIA (2004), from 1979 to 2000, the mean annual total ozone column pattern over the Arctic was about 3% per decade, while the tendency in mean spring total ozone column was about 5% per decade. Ozone in the stratosphere shields the Earth from much of the harmful UV radiation. Various workers have reported that the Arctic region is also facing adverse effects of high UV radiation. However, recent Arctic measurements show long-term decreases in total column ozone (Harris et al., 2002). Scientists have observed more frequent episodes of extremely low ozone, particularly during the springtime. Chubarova et al. (2002), Krzyscin (2000), and Kalliskota et al. (2000) reported an increase in UV radiation levels primarily in the spring when ozone depletion reaches the maximum.

Arctic ozone depletion is mainly affected by stratospheric temperatures, aerosols, solar zenith angle, surface albedo and altitude, and cloud formation over polar stratosphere. Radiation reflectance from snow can increase the biologically effective UV irradiance by more than 50%, due to a decrease in snow and ice cover on the surface of rivers, lakes, or oceans. The exposure of many organisms to damaging UV radiation increases during the ozone hole period (Goutail et al., 1999).

9.4 Arctic plants diversity

Vascular plants, bryophytes, are the main components of the Arctic terrestrial ecosystem, and algae are abundant in aquatic (fresh and marine) ecosystems (Afonina and Czernyadjeva, 1995; Andreev et al., 1996; Talbot et al., 1999; Konstantinova and Potemkin, 1996). The Arctic consists of about 3% species of the world flora, of which angiosperms are 0.7%, gymnosperms 1.6%, bryophytes 4%, and lichens 11% (Matveyeva and Chernov, 2000). The richness and composition of species vary in different parts of the Arctic. Elven (2007) reported that out of about 106 species including subspecies which are endemic to the Arctic, 69 species show a very restricted distribution. Further, Elven (2011) said that no species in the Arctic are considered invasive, but many species are at risk due to increasing human interference and climate change. However, most of the cryptogams have broad distributions in the Arctic region and have many adaptations to survive in changing climatic conditions (Callaghan et al., 2004c).

An algal component is an essential constituent of the food chain as they are the primary producers of fresh and marine food chains and a terrestrial ecosystem (Forest et al., 2011). They occur either in the upper water column (pelagic), related to sea ice (sympagic), or closely attached to bottom substrates (benthic). Haber (1995) and Legendre et al. (1992) reported that in the Arctic regions, marine diatoms have a high diversity and abundance in annual sea ice, pelagic waters, and benthic environments and contributing to more than 50% of total phytoplankton biomass and production (Armbrust, 2009).

9.5 Induction of UV-absorbing pigments in response to UV-B radiations

The most crucial protection type is the production of UV-absorbing pigments and their deposition near the exposed surfaces. The penetration of radiation to deeper cell layers can be prevented. Different organisms produce different kinds of UV-absorbing substances (Cockell and Knowland, 1999; Rozema et al., 2002). Many algae and some cyanobacteria produce mycosporine-like amino acids (MAAs). It includes mycosporine-glycine. Shick and Dunlap (2002) and Korbee et al. (2006) described the MAA's distribution, biosynthesis, induction by UV radiation, and other vital functions. Usnic acid in lichens is frequently increased by UV radiation exposure, although other factors are also important (Solhaug and Gauslaa, 2004). Gauslaa and Solhaug (2004) have shown the protective importance of the pigment parietin in *Xanthoria parietina* by increased UV protective pigment parietin synthesis. In bryophytes and pteridophytes, flavones and related compounds occur as glycosides. Webby et al. (1996) found a range of flavones and flavonol glycosides in a single moss species, *Bryum algens*. Feld et al. (2003) found 10 different flavones in a single liverwort species.

Wyatt et al. (1991) reported the presence of chrysoeriol and luteolin in *Lycopodium* species and some bryophytes. Algae and terrestrial plants recognize the UV radiation, adjust the production of UV-absorbing compounds accordingly and other enzymes that repair damage caused by UV radiation (Solhaug et al., 2003). To minimize the effects of high UV-B radiation, plants grow close to the ground and each other.

9.6 Ultraviolet radiations and adaptation in Arctic plants

Under the harsh climatic condition of Arctic region, cyanobacteria and algae develop adaptation mechanisms that protect and avoid or minimize UV radiation's adverse effects. The extracellular synthesis of protective compounds and the development of multilayered cell walls, mucilage sheets, and intracellular control of cell solute composition and viscosity. The symbiotic association among cyanobacteria and algae with fungi in lichens provides physical protection and synthesizes UV-absorbing pigments. Pigment synthesis in lichens of Arctic region occurs during excessive exposure to solar UV radiation, and pigments such as melanin, melanoids, and carotenoids protect against the UV radiation. The accumulation of parietin pigment in *X. parietina* is induced by UV-B radiation and provides protection (Gauslaa and Ustvedt, 2003). MacKenzie et al. (2002) reported that higher plants' carotenoids offer protection against excessive photosynthetically active light, and possibly also have a protective role against UV radiation.

Arctic plants have evolved several other adaptations that help the plants to survive under the harsh climatic and UV radiation conditions (Murray, 1995). The plants' smaller size makes survival possible, and the plants' shallow root system remains in contact with active soil (Convey, 2000). The arctic soil is low in nutrients because dead plants fail to decompose; they accumulate as peat. The Arctic plants are adapted to survive without much mineral nutrition. The cushion-shaped plants can compensate for the soil's infertility by intercepting windblown dust and small leaves minimizing the amount of water loss through leaf surface (Raven et al., 2005).

Many Arctic species can survive and grow under a layer of snow, and all polar plants can photosynthesize in freezing temperatures. The physiology of Arctic plants is well adapted to cope long winter and freezing weather conditions. During the summer, plants use prolonged sunlight to develop and produce flowers and seeds quickly. Dark-colored plants absorb more of the solar energy, and at the same time, the dark ground surface absorbs the heat of the spring sun, which warms the plant, and they can quickly resume active life after winter dormancy.

As per the Montreal Protocols, most countries have agreed to reduce production, use of CFCs, and other ozone-destroying substances. Recovery of the ozone hole is currently predicted by 2050. However, this is an area of intense research and validation. Therefore plants growing at the Arctic have to readapt according to changing environmental conditions. Poles are the only place on Earth where such sizeable predictable ozone loss occurs, and it provides a unique opportunity for conducting ozone-related research. Various studies conducted at Poles have provided valuable information on the UV-photoprotective pigments and several other processes (Singh et al., 2018).

9.7 Conclusion

A higher amount of UV radiation reaches the Arctic during spring, that's why plants synthesize more elevated amounts of UV-absorbing pigments and other protective compounds to cope with biologically active UV-B radiation effects. The synthesis of photoprotective pigments and different pigments and other adaptations indicates the survival of flora under such harsh climatic conditions. The synthesis of UV-absorbing pigments in response to UV-B radiation is of particular significance. The synthesis of protective pigments depends on the stress levels of UV radiation, which is based on plants' survival under the high UV-B radiation exposure at the Arctic and in other Polar regions of the globe.

References

ACIA, 2004. Arctic Climate Impact Assessment. Cambridge University Press, p. 140.

Afonina, O.M., Czernyadjeva, I.V., 1995. Mosses of the Russian Arctic: checklist and bibliography. Arctoa 5, 99–142.

Andreev, M., Kotlov, Y., Makarova, I., 1996. Checklist of lichens and lichenicolous fungi of the Russian Arctic. Bryologist 99, 137–169.

Armbrust, E.V., 2009. The life of diatoms in the world's oceans. Nature 459, 185–192.

Bjerke, J.W., Gwynn-Jones, D., Callaghan, T.V., 2005. Effects of enhanced UV-B radiation in the field on the concentration of phenolics and chlorophyll fluorescence in two boreal and arctic-alpine lichens. Environ. Exp. Bot. 53, 139–149.

Callaghan, T.V., Björn, L.O., Chernov, Y., Chapin III, F.S., Christensen, T.R., Huntley, B., et al., 2004a. Climate change and UV-B impacts on Arctic tundra and polar desert ecosystems: Rationale, concepts and approach to the assessment. Ambio 33, 393–397.

Callaghan, T.V., Björn, L.O., Chernov, Y., Chapin III, F.S., Christensen, T.R., Huntley, B., et al., 2004b. Climate change and UV-B impacts on Arctic tundra and polar desert ecosystems: Past changes in arctic terrestrial ecosystems, climate and UV-B radiation. Ambio 33, 385–479.

Callaghan, T.V., Bjorn, L.O., Chernov, Y., Chapin III, F.S., Christensen, T.R., Huntley, B., et al., 2004c. Climate change and UV-B impacts on Arctic tundra and polar desert ecosystems: responses to projected changes in climate and UV-B at the species level. Ambio 33, 386–392.

Chubarova, N.Y., Yurova, A.Y., Krotkov, N.A., Herman, J.R., Bhartia, P.K., 2002. Comparisons between ground-based measurements of broadband UV irradiance (300–380 nm) and TOMS UV estimates at Moscow for 1979–2000. Opt. Eng. 41 (12), 3070–3081.

Cockell, C.S., Knowland, J., 1999. Ultraviolet screening compounds. Biol. Rev. 74, 311–345.

Convey, P., 2000. How does cold constrain life cycles of terrestrial plants and animals? Cryo-Letters 21, 73–82.

Elven, R., 2007. Checklist of the Panarctic Flora (PAF) Vascular Plants. National Centre of Biosystematics, Natural History Museum, University of Oslo.

Elven, R., 2011. Annotated Checklist of the Panarctic Flora (PAF): Vascular Plants. National Centre of Biosystematics, Natural History Museum, University of Oslo.

Feld, H., Zapp, J., Becker, H., 2003. Secondary metabolites from the liverwort *Tylimanthus renifolius*. Phytochemistry (Oxf.) 64, 1335–1340.

Forest, A., Tremblay, J.-É., Gratton, Y., Martin, J., Gagnon, J., Darnis, G., et al., 2011. Biogenic carbon flows through the planktonic food web of the Amundsen Gulf (Arctic Ocean): a synthesis of field measurements and inverse modelling analyses. Prog. Oceanogr. 91, 410–436.

Gauslaa, Y., Ustvedt, E.M., 2003. Is parietin a UV-B or a blue-light screening pigment in the lichen *Xanthoria parietina*. Photochem. Photobiol. Sci. 2, 424–432.

Gauslaa, Y., Solhaug, K.A., 2004. Photoinhibition in lichens depend on cortical characteristics and hydration. Lichenologist 36, 133–143.

Goutail, F., Pommereau, J.P., Philips, C., 1999. Depletion of column ozone in the Arctic during the winters of 1993/94 and 1994/95. J. Atmos. Chem. 32, 1–34.

Haber, E., 1995. Benthic Marine Algae of the Canadian Arctic: Taxonomic Diversity, Rarity and Centres of Species Diversity. Report prepared for the Canadian Wildlife Service, Environment Canada, Hull.

Harris, N.R.P., Rex, M., Goutail, F., Knudsen, B.M., Manney, G.L., Muller, R., et al., 2002. Comparison of empirically derived ozone loss rates in the Arctic. J. Geophys. Res. 107 (D20), . Available from: https://doi.org/10.1029/2001JD000482In the Arctic vortex from 1988–89 to 1999–2000. Results from the Third European Stratospheric Experiment on Ozone (THESEO), Palermo, Italy, 2000.

Kalliskota, S., Kaurola, J., Taalas, P., Herman, J.R., Celarier, E.A., Krotkov, N., 2000. Comparison of daily UV doses estimated from Nimbus 7/TOMS measurements and ground-based spectroradiometric data. J. Geophys. Res. 105, 5059–5067.

Konstantinova, N.A., Potemkin, A.D., 1996. Liverworts of the Russian Arctic: an annotated checklist and bibliography. Arctoa 6, 125–150.

Korbee, N., Figueroa, F.L., Aquilera, J., 2006. Acumulación de aminoácidos tipo micosporina (MAAs): biosíntesis, fotocontrol y funciones ecofisiológicas. Rev. Chil. Historia Nat. 79, 119–132.

Krzyscin, J.W., 2000. Total ozone influence on the surface UV-B radiation in late spring-summer 1963–1997: an analysis of multiple timescales. J. Geophys. Res. 105 (D4), 4993–5000.

Legendre, L., Martineau, M.J., Therriault, J.C., Demers, S., 1992. Chlorophyll *a* biomass and growth of sea-ice microalgae along a salinity gradient (southeastern Hudson Bay, Canadian Arctic). Polar Biol. 12, 445–453.

MacKenzie, T.D.B., Krol, M., Huner, N.P.A., Campbell, D.A., 2002. Seasonal changes in chlorophyll fluorescence quenching and the induction and capacity of the photoprotective xanthophyll cycle in *Lobaria pulmonaria*. Can. J. Bot. 80, 255–261.

Matveyeva, N., Chernov, Y., 2000. Biodiversity of terrestrial ecosystems. In: Nuttall, M., Callaghan, T.V. (Eds.), The Arctic Environment, People, Policy. Harwood Academic Publishers, Amsterdam, pp. 233–274.

Muller, R., Crutzen, P.J., Gross, J.U., Bruhl, C., Russell, J.M., Gernant, H., McKenna, D.S., Tuck, A.F., 1997. Severe chemical loss in the Arctic during the winter of 1995–96. Nature 389, 709–712.

Murray, D.F., 1995. Causes of Arctic plant diversity: origin and evolution. In: Chapin III, F.S., Körner, C. (Eds.), the Arctic and Alpine Biodiversity Patterns, Causes and Ecosystem Consequences. Springer Verlag, Heidelberg, pp. 21–32.

Raven, P.H., Evert, R.F., Eichorn, S.E., 2005. Biology of Plants. Freeman and Company, New York.

Rozema, J., Bjorn, L.O., Bornman, J.F., Gaberscik, A., Hader, D.P., Trost, T., et al., 2002. The role of UV-B radiation in aquatic and terrestrial ecosystems – experimental and functional analysis of the evolution of UV-absorbing compounds. J. Photochem. Photobiol. B: Biol. 66, 2–12.

Shick, J.M., Dunlap, W.C., 2002. Mycosporine-like amino acids and related gaudosols: biosynthesis, accumulation, and UV protective functions in aquatic organisms. Ann. Rev. Physiol. 64, 223–262.

Singh, J., Singh, R.P., 2014. An adverse effect of UV-B radiation on plants growing at Schirmacher Oasis, East Antarctica. Toxicol. Int. 21 (1), 101–106.
Singh, J., Gautam, S., Pant, B.A., 2012. Effect of UV-B radiation on UV absorbing compounds and pigments of moss and lichen of Schirmacher Oasis region, East Antarctica. Cell. Mol. Biol. 58 (1), 80–84.
Singh, J., Singh, R.P., Khare, R., 2018. Influence of climate change on Antarctica flora. Polar Sci. 18, 94–101. ISSN 1873-9652.
Solhaug, K.A., Gauslaa, Y., 2004. Photosynthates stimulate the UVB induced fungal anthraquinone synthesis in the foliose lichen *Xanthoria parietina*. Plant Cell Environ. 27, 167–176.
Solhaug, K.A., Gauslaa, Y., Nybakken, L., Bilger, W., 2003. UV induction of sun-screening pigments in lichens. N. Phytol. 158, 91–100.
Talbot, S.S., Yurtsev, B.A., Murray, D.F., Argus, G.W., Bay, C., Elvebakk, A., 1999. Atlas of rare endemic vascular plants of the Arctic. In: Conservation of Arctic Flora and Fauna (CAFF) Technical Report No. 3. Territories. Sofia, Moscow.
Webby, R.F., Markham, K.R., Smith, R.I.L., 1996. Chemotypes of the Antarctic moss *Bryum algens* delineated by their flavonoid constituents. Biochem. Syst. Ecol. 24, 469–475.
Wyatt, B., Lane, D.M., Stoneburner, A., 1991. Chemosystematics of the Mniaceae. II. Flavonoids of Plagiomnium section rosulata. Bryologist 94, 443–448.

CHAPTER 10

Cyanobacteria in the polar regions: diversity, adaptation, and taxonomic problems

Prashant Singh, Naresh Kumar and Sagarika Pal
Laboratory of Cyanobacterial Systematics and Stress Biology, Department of Botany, Institute of Science, Banaras Hindu University, Varanasi, India

10.1 Introduction

The polar regions are characterized by a range of extreme and continuously harsh environmental conditions, which can eventually lead to a sustained pressure on all forms of life. Cyanobacteria are well suited and adapted to thrive and flourish in an extremely broad array of habitats having different extremes of temperatures. They can be found to be growing in hot arid desert soils, geothermal pools, tropical rain forests, polar tundra environments, ice shelves, and polar glacial surfaces (Fay, 1983). These astonishing range of habitats and niches in which the cyanobacteria survive can be attributed to the exceptional adaptability of these gram-negative photoautotrophs, which has been further enhanced and strengthened by their long-standing evolutionary history (Gaysina et al., 2019). Almost all the cyanobacteria possess the natural ability to support features such as respiration, oxygenic photosynthesis, buoyancy, and fixation of atmospheric nitrogen (Gaysina et al., 2019). These abilities further make them more special and provide them the ability to survive, grow, and even flourish in the polar regions in both regular and extreme stress periods. These abilities also further strengthen their position as the primary phototrophic contributors in all the natural polar microhabitats (Bryant, 2003). Cyanobacteria are exceptionally adapted to survive and thrive in the adverse environmental extremes of the polar regions. Some of these adaptations include changes in their physiology and biochemical aspects. Few examples of this include the ability to endure high levels of ultraviolet radiation and also withstand extreme phases of desiccation, which eventually allow them to survive the regularly occurring events of freezing

Understanding Present and Past Arctic Environments
DOI: https://doi.org/10.1016/B978-0-12-822869-2.00013-X

and thawing that happens in the polar regions. It is notable that cyanobacteria are the most dominant and persistent photoautotrophs of these regions and hence their contribution to the carbon and nitrogen cycling events is particularly more important and substantial. Irrespective of the microhabitats they thrive in, cyanobacteria are for sure most dominant and heavy contributors to the soil dynamics.

Apart from being one of the most vital primary producers, some dominant genera (*Nostoc*, *Phormidium*, *Leptolyngbya*, etc.) also contribute on a large magnitude by their capability to fix atmospheric nitrogen, and thus act as the keystone biotic sources of nitrogen for other biotic components (Jordan, 1978; Davey and Marchant, 1983; Lennihan et al., 1994; Solheim et al., 1996). It has been reported that nitrogen is usually one of the limiting factors for most of the polar microbial communities, thus also affecting adversely the primary production (Kvíderová, 2018). Hence, it has been found that most of the newly deglaciated areas have cyanobacteria as the first phototrophic colonizers, thus leading to the establishment and strengthening of primary succession.

The present review aims to briefly evaluate the cyanobacterial "way of life" in the polar regions by describing the components of some specific microhabitats along with discussing equally concisely some environmental conditions and the adaptive features that the cyanobacteria employ to survive and even flourish in these extremes. Also, some key concerns regarding the various issues that cyanobacteriologists (especially taxonomists) face while working with polar cyanobacteria have also been discussed as it was felt that although almost everyone faced these problems, but they were never discussed systematically.

10.2 Cyanobacterial diversity in different polar microhabitats

Different kinds of microhabitats exist in these extreme habitats, which also harbor different kinds of cyanobacterial components as a result of having different microenvironments (Figs. 10.1 and 10.2). In this section, we aim to discuss some of the microecosystems that exist in cold latitudes. The terrestrial habitats, on the basis of the dwellers, can be epilithic (organisms present on the rock surfaces and being visible), hypolithic (inhabitants living at the soil–rock interface, especially at places where pebbles appear translucent), and endolithic (organisms inhabit the interior surface of rocks and thus are usually not visible) (Cockell and Stokes, 2004; Omelon,

Figure 10.1 Algal growth across different habitats in the polar regions: (A) typical algal growth in stagnant water bodies; (B) floccose algal growth in water bodies; (C) algal growth in small periodic streams below glacial runoff; (D) algal growth in long-running streams having stones and pebbles; (E) ball-like algal growth on stone tops; and (F) reddish colored algal growth on stone tops in running streams.

Figure 10.2 Algal growth across different habitats in the polar regions: (A) ball-like algal growth in shady hidden crevices in running glacial streams; (B) luxuriant algal growth in running streams; (C) characteristic growth of *Nostoc commune* complex in meadows; (D) algal growth beneath ice in frozen streams; (E) cyanobacterial colonies (most probably *Gloeotrichia/Rivularia*) on stone tops in water bodies; and (F) a typical Cryoconite.

2008). Much of the information in this part is based on the review of Quesada and Vincent (2012).

10.2.1 The epilithic microhabitats

The epilithic microhabitats usually encounter high dose of ultraviolet radiation, appear as dark surface crusts or mats, and the dwellers generally appear in less severe climates such as those in maritime Antarctica. The typical dark coloration is usually due to the pigment scytonemin (Vincent, 2000). Notably, while these communities are usually tolerant to high radiation stress and desiccation, they exhibit a certain amount of sensitivity to salinity stress (Broady, 1986). Some of the common epilithic cyanobacterial genera are *Calothrix, Chamaesiphon, Chroococcidiopsis, Cyanothece, Homoeothrix, Myxosarcina, Nostoc, Phormidium, Pleurocapsa, Gloeocapsa, Scytonema, Stigonema*, and *Synechococcus* (Quesada and Vincent, 2012).

10.2.2 The hypolithic microhabitats

The hypolithic cyanobacteria are quite abundant in polar ecosystems (Thomas, 2005). They are known to be present under translucent rocks such as quartz (Cowan and Tow, 2004), but can also be found beneath opaque rocks such as dolomites (Cockell and Stokes, 2006). It is important to note that apart from being a good source of liquid water, these microhabitats also serve as a shelter to the cyanobacterial inhabitants, thus being much favored, as compared to open epilithic rock surfaces (Pointing et al., 2007). Additionally, they also help the dwellers to escape the ferocity of the extremely strong ultraviolet radiations (Cockell and Stokes, 2006). Another factor that makes the hypoliths good for cyanobacteria is the temperature buffer that provides both favorable temperature and some amount of shade (Wynn-Williams, 2000). Algal growth in these microhabitats rests typically between the lithic and the edaphic transient zones. The genus *Chroococcidiopsis* is usually the dominant genus (but can be absent if the habitat has high salinity), while *Gloeocapsa*-like cyanobacteria is also known to be frequent though doubts have been cast on the relatively simpler morphological descriptions of members of this microhabitat as DNA-based studies have shown that certain non-Chroococcales could adopt similar shapes in hypolithic ecosystems (Pointing et al., 2007). Other cyanobacterial genera that could be found in hypolithic microhabitats are *Aphanothece, Calothrix, Lyngbya, Nodularia, Nostoc, Phormidium,*

Plectonema, Pleurocapsa, Scytonema, Tolypothrix, and *Synechococcus* (Quesada and Vincent, 2012).

10.2.3 The endolithic microhabitats

The endolithic microhabitats can further be euendolithic (living forms that can bore inside the rocks and live inside the holes) (Cockell and Herrera, 2007), cryptoendolithic (living forms that dwell in the spaces between the porous rocks) (Omelon et al., 2007), and lastly chasmoendolithic (living forms that dwell inside rock cracks and fissures) (Büdel et al., 2008).

Endolithic microenvironments could be among the favorite habitats for cyanobacterial communities because of sufficient quantity of water being present, comparatively better and constant temperatures, and finally significant protection from high ultraviolet radiation (Wynn-Williams, 2000). Also, they have quite reduced habitat loss rates that could otherwise occur by wind erosion and rock surface detachment (Cockell and Herrera, 2007). Prominent cyanobacterial genera that have been found to be occurring in endolithic microenvironments are *Aphanocapsa, Aphanothece, Chroococcidiopsis, Anabaena, Calothrix, Cyanothece, Eucapsis, Hormathonema, Homoeothrix, Leptolyngbya, Lyngbya, Microchaete, Microcoleus, Nodularia, Gloeocapsa, Oscillatoria, Nostoc, Phormidium, Plectonema, Pleurocapsa, Schizothrix, Stigonema, Tolypothrix,* and *Synechococcus* (Quesada and Vincent, 2012).

10.2.4 Soil and associated crusts

The biological soil crusts (BSCs) are important microecosystems showing omnipresence in both the polar regions. In fact, they constitute the main primary producers in the extremely dry polar deserts of the Arctic and Antarctic regions (Yoshitake et al., 2010; Colesie et al., 2014; Williams et al., 2017; Borchhardt et al., 2017; Rippin et al., 2018). Common cyanobacterial genera that are typically found to be occurring in BSCs are *Chroococcus, Gloeocapsa, Leptolyngbya, Microcoleus, Nostoc, Phormidium,* and *Scytonema* (Hoffmann, 1989; Zidarova, 2008; Büdel et al., 2016; Pushkareva et al., 2016). In spite of being extremely hardy and resistant microhabitats, the resilience of the BSCs is regularly tested by the continuous effects of climate change. Hence, these BSCs can also serve to be good proxies for understanding global climate change as the composition and abundance of these BSCs are reported to be changing (Evans and Lange, 2001; Johnson et al., 2012; Zelikova et al., 2012). Continuously increasing temperatures and high irradiance along with irregular and unpredictable precipitation patterns all across the Arctic and Antarctic regions have now led to a probable change in the species diversity and community structure of BSCs due to invasion by other foreign living forms (Evans and Lange, 2001; Chown et al., 2012; Pushkareva et al., 2016; Lee et al., 2017).

Soil as a habitat in the polar regions is exposed to intense bright solar radiation, low humidity, fluctuating temperatures in short intervals, osmotic and salinity stress, soil erosion due to winds, and inconsistent snowfall, thus in total making them usually not the most favorable habitat for growth of cyanobacterial communities (Vincent, 1988). In spite of all these limitations and tough conditions, cyanobacterial communities still dominate in the soil habitats (Fernández-Carazo et al., 2011; Michaud et al., 2012; Vincent, 1988). There have been several reports of edaphic cyanobacteria all around the polar regions (Vincent, 1988). It has been documented that mainly the cyanobacteria belonging to the orders Chroococcales, Oscillatoriales, and Nostocales are present in the Antarctic soils (Cavacini, 2001; Wood et al., 2008). Along with the different constituents of the soil, humidity is also an equally important factor in deciding the diversity of soil cyanobacteria (Wood et al., 2008). Another important factor is the presence of the nutritional source near the communities as it has been seen that if the natural reservoir of nutrients is nearer, then the communities get better support for growth and survival (Wood et al., 2008). The most common members of the order Oscillatoriales found in most of the studied soils belong to the genera *Leptolyngbya* and *Phormidium* (Cavacini, 2001; Mataloni et al., 2000). The amount and duration of water availability are known to be important parameters in deciding the distribution of cyanobacteria in the soil habitats (Vincent, 1988).

In mineral soils that have periodic presence/absence of water, it is usually the members of the order Nostocales that tend to dominate, but in moist or wet soils, the Oscillatoriales generally dominate. Lastly, the members of the order Chroococcales are found to be more in association with other taxa in moist but unflushed habitats. In the High Arctic regions, around 18 different cyanobacterial species were enumerated from barren soils, belonging to 7 genera including the highly diverse and frequently occurring *Leptolyngbya* and *Phormidium* (Kastovska et al., 2005). The study of trapped polar cyanobacteria and fossils is one of the most interesting prospects of working in the field of polar microbiology. These fossils serve as a treasure of information for studying the formation of earth and evolution of life. Extremely ancient communities of edaphic cyanobacteria have been preserved in the polar permafrost, which serve as valuable fossil remnants of the past ages along with sometimes also being shown to have revival capability even after being trapped for millions of years (Erokhina et al., 2000; Vishnivetskaya et al., 2002, 2003; Vishnivetskaya, 2009). The Arctic permafrost regions have been a source of many members of the orders Oscillatoriales and Nostocales (Vishnivetskaya et al., 2003; Vishnivetskaya, 2009) with many showing close phylogenetic affiliations with the genera *Leptolyngbya, Microcoleus, Phormidium, Nostoc,* and *Anabaena.*

10.2.5 Cryoconites

Cryoconites are dark incoherent-granular sediments that are seen usually on the ablation surface of glaciers. These hole-like entities are formed when sediment is blown onto the ice and is heated by solar energy, thus causing it to melt into the glacier's surface. As these dark cryoconite holes can be easily visually identified on a glacier's shining surface, and because they are distinct, clear, and self-contained, isolated pockets of life occurring in extremely tough terrains, they have been studied more seriously by microbial ecologists who see them as ideal systems for natural microcosm experiments (Ambrosini et al., 2017; Sommers et al., 2018). They have also been referred as islands within a sea of glacial ice with each cryoconite hole harboring its own specific microbial community (Stanish et al., 2013). Cyanobacteria have been found to be interesting and frequent components of many cryoconites (Takeuchi and Li, 2008). They have been reported from polar glaciers (Mueller et al., 2001; Stibal et al., 2006) as well as other mountainous glaciers around the world (Xiang et al., 2009). In an Arctic glacier, the cyanobacterial communities found in supraglacial sediments have been found to be dominated by the genera *Leptolyngbya, Nostoc,* and *Phormidium* (Stibal et al., 2006), while in high mountain glaciers the cyanobacterial communities are usually represented by members of the order Oscillatoriales (Xiang et al., 2009). Cryoconite holes explored in Antarctica have given strong evidence of various cyanobacteria being the dominant colonizers with their photosynthates being the primary source of nutrients for the rest of the associated heterotrophic communities along with an incredible array of microinvertebrates like nematodes, rotifers, and even tardigrades (Mueller et al., 2001). The commonly occurring cryoconite-dwelling cyanobacteria all over the world include the genera *Leptolyngbya, Crinalium, Phormidium, Aphanothece, Aphanocapsa, Gloeocapsa*, and *Synechococcus*.

10.3 Diverse conditions in the polar regions

The extreme conditions of the polar regions force the cyanobacteria to keep adapting and keep modifying their survival strategies. Some of the common forms of extremities may include extreme low temperatures (made more complicated by freeze–thaw processes), desiccation, salinity, and wide irradiance fluctuations (including both photosynthetically active radiation and ultraviolet light) (Convey et al., 2014; Kvíderová, 2018). All these factors usually are connected and thus act in a more or less synergistic fashion, which ultimately leads to reduced biological activity. Cyanobacteria, being the photoautotrophic components of these habitats, are the critical factors that ensurek continued microbial dynamics and sustained productivity, which can thereafter be used by the other heterotrophic populations comprising of Actinobacteria, Proteobacteria, Firmicutes, and Bacteroidetes along with several other smaller numbers of organisms in higher trophic levels (Aislabie et al., 2006; de Los Ríos et al., 2004; Stomeo et al., 2012). We will now discuss briefly three major environmental stresses and the mechanisms that help the cyanobacteria to counter these conditions (Fig. 10.3).

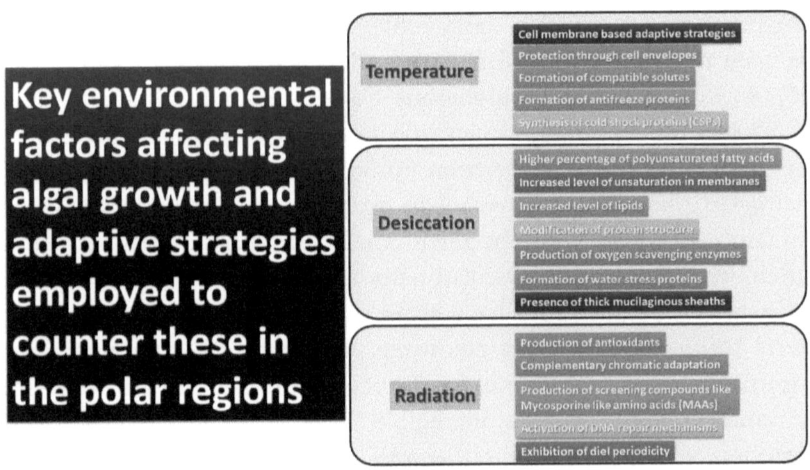

Figure 10.3 Key environmental factors affecting algal growth and adaptive strategies employed to counter these in the polar regions.

10.3.1 Temperature-related stress (low temperature and freeze–thaw)

It is a well-established fact now that majority of the polar cyanobacteria are psychrotolerant while very few are psychrophilic (Vincent, 2000; Quesada and Vincent, 2012). This means that they can grow well at temperatures near to 10°C–15°C and may also survive at a bit higher temperature (this higher temperature though, may vary). Along with these optimal temperatures, they can also survive at subzero temperatures (Kvíderová, 2018). In spite of being very resilient and persistent (Friedmann et al., 1993; Sun and Friedmann, 1999; Tashyreva and Elster, 2016), they usually have very slow growth rates (Vincent, 2007), which is in fact a complication when attempting the isolation and purification of the collected samples. For countering the extreme low temperatures, cyanobacteria are known to adopt different kinds of strategies in which the cell membrane usually plays a pivotal role. There is generally an elevated level of unsaturated fatty acids, number of cis-double bonds, methylated branches, and shortened fatty acid chains (Gounot and Russell, 1999) when the cyanobacteria encounter low temperatures. In fact, the unsaturation of the membrane lipids is a consistent feature, which also indicates the ability of the cyanobacteria to counter low temperatures. Another mechanism to counter cold stress is the ability to synthesize cold shock proteins that allow the cyanobacteria to continue growing at lower temperatures (Gounot and Russell, 1999; Raymond-Bouchard and Whyte, 2017). The basic aim of employing cold-tolerance methods is to preserve the basic structure and ensure the critical functioning of the proteins. Low temperature exposure increases the strength of the polar interactions on the protein surface while the nonpolar forces become weaker (Kvíderová, 2018). This also affects the enzymatic machineries and hence the overall aim of protein-centered strategies is to maintain the optimal folding, keep up the basic functioning, and prevent the dissociation of the individual subunits (Walles et al., 1999). Some studies have also demonstrated the presence of a sensory mechanism that initially gets activated by lower membrane fluidity resulting in the activation of the destaurase enzymes. The increased expression of these destaurases leads to an enhanced accumulation of destaurated fatty acids and continued maintenance of the membrane fluidity (Murata and Wada, 1995; Nishida and Murata, 1996).

The stress that the cyanobacteria encounter during low temperature is probably less complicated if considering the freeze–thaw stress, which can be infrequent and less predictable, thus being more dangerous.

Interestingly, this becomes a greater threat during the beginning and at the end of the summer season as compared to the actual polar summers (Elster et al., 2012). The response of the cyanobacterial cells to freezing and thawing stress depends upon their growth stage, the amount and the rate of freezing and thawing, and also on the surrounding environment (MacLeod and Calcott, 1976; Smith, 1995). It has been reported that in general sense, the polar cyanobacteria are usually more resistant to freezing as compared to polar algae (Davey, 1989; Šabacká and Elster, 2006) with the survival even being seen at temperatures as low as −100°C in certain conditions (Šabacká and Elster, 2006). Interestingly, the freeze−thaw rate was measured and it was found to be quite slow around approximately several degrees per hour (Elster et al., 2012). This process of freezing and thawing in itself causes a variety of stresses that keeps the entire cyanobacterial population under a kind of unstable environment, thus not allowing the physiology of the organism to be at any level near to the optima. The frequent change in temperature, the continuous shift among the various phases of water, the infrequent and abrupt changes in pH, salt concentration, and the formation of ice crystals are all the phenomena that are associated with freeze−thaw stress. All these happenings are also interconnected to each other and so may not operate actually in isolation. Hence, this combined effect poses serious survival threats to the cyanobacteria. Ice crystals have been known to penetrate intracellular structures and cause problems in the osmotic gradient and by reactive oxygen species (ROS) production too (Tanghe et al., 2003). Cyanobacteria are known to employ different strategies to counter this freeze−thaw stress and one of them is the tendency to form antifreeze proteins (AFPs). These AFPs are usually present in the cells in low concentrations and act by decreasing the freezing point, thus inhibiting ice crystal formation (Chao et al., 1996). They have been detected in cyanobacterial mats containing the genera *Nostoc* and *Phormidium* (Raymond and Fritsen, 2000), though they are still to be detected in axenic cultures, which is a matter of interest. Notably, AFPs have also been detected in bacteria in few studies (Gilbert et al., 2004; Kawahara, 2017). Another way that the cyanobacteria could employ to counter freeze−thaw−related problems is the formation of compatible solutes like dimethylsulfoxide (DMSO), which are known to provide protection during freezing. But the production and accumulation of the compatible solutes are still to be understood properly in cyanobacteria (Klähn and Hagemann, 2011). Another protective feature known in cyanobacteria is the cell envelopes composed of the extracellular polysaccharides. These are known to provide protection and buffer the cells inside from many kinds of abiotic stress factors. The extracellular polysaccharides are known to slow down the flow of liquid water during the freeze—thaw events and thus help the cells in avoiding sudden cell volume changes (Vincent, 2000).

10.3.2 Desiccation-related stress

Desiccation is a common and also an important type of stress in polar habitats especially in the hydroterrestrial and terrestrial habitats during the summer season (Elster, 2002; Elster and Benson, 2004). An enormous difference in water content is usually visible in various smaller microhabitats, which could be all very close to each other. One of the most classical and frequently seen example of this is visible in the meadows, which are usually dry at the top and fully saturated with water at the bottom (Elster et al., 2012). One of the most common dwellers of these meadows is the cyanobacteria belonging to the genus *Nostoc*, especially the *Nostoc commune* complex. Astonishingly, the strong survival ability of these *N. commune* members was proved to be around 55 (Shirkey et al., 2003), 87 (Lipman, 1941), or even more than 100 years (Cameron, 1962). The various types of problems that desiccation-related stress can cause in cyanobacteria need to be understood first before we dwell on the ways to counter this stress. It has been noted that the entire physiological machinery of the organism is brought down with important processes such as nitrogen fixation, photosynthesis, and respiration being affected adversely. The order of events that eventually breakdown the physiological machinery of the cyanobacteria are diminished nitrogen fixation, followed by lowered level of photosynthesis, and ultimately a cease in the respiratory events. While the downregulation of the regulatory machineries is quite fast, the upregulation upon rehydration and return of favorable conditions is slow with the order of improvement being first respiration followed by photosynthesis and a very slow revival of the process of nitrogen fixation (Scherer et al., 1984; Davey, 1989; Scherer and Zhong, 1991; Hawes et al., 1992; Potts, 1999, 2000; Qiu and Gao, 2001). While it is understandable that the stress factors influence the physiology of the cyanobacteria very badly, but there do exist efficient mechanisms that help these organisms to survive these assortments of physiological complexities. Osmotic and mechanical stresses are common and eventually lead to an imbalance in water relations and solute potentials. Extremely low levels of water potentials that can cause immediate destruction are known to be tolerated by cryptoendolithic cyanobacteria belonging to the genera *Chroococcus* and *Chroococcidiopsis* (Potts and Friedmann, 1981; Palmer and Friedmann, 1990). *Nostoc* is another immensely resilient cyanobacterial genus that can withstand strong physiological variations and still survive in the polar habitats (Kvíderová et al., 2011).

Desiccation response has been known to be associated with an elevated level of lipids in *Calothrix* sp. (Singh, 2001) and higher percentage of polyunsaturated membrane lipids in *Nostoc* (Olie and Potts, 1986). Some other physiological reactions include the production of oxygen scavenging enzymes in response to ROS formation, thus ultimately minimizing membrane damage, modification of protein structure (like phosphorylation or disulfide linkage of coenzyme A), and increased level of unsaturation in membranes, genome multiplication and activation of DNA repair mechanisms like photoreactivation, excision repair, and postreplication repair (Potts, 1994; Kvíderová, 2018). An interesting mode of adaptation and desiccation tolerance is the ability of cyanobacteria to form water stress proteins (WSPs), late embryogenesis abundant (LEA) proteins, and dehydrins (Potts, 1994). The LEA proteins help stabilize other proteins and membranes during drying events, especially in the presence of sugars such as trehalose (Hand et al., 2011) while the dehydrins can inhibit coagulation of macromolecules, thus ensuring the optimal functioning of various physiological processes (Close, 1997). The presence of both the previously mentioned proteins was detected in cyanobacteria (Close and Lammers, 1993) while some specific WSPs were also detected in *N. commune* colonies that were postulated to play some role in the modification of extracellular envelopes (Potts, 1999). A common feature associated with all the desiccation-resistant cyanobacterial taxa is the presence of thick mucilaginous sheaths or envelopes (Caiola et al., 1996; Potts, 1994, 1999, 2000; Tamaru et al., 2005) usually composed of mainly mucopolysaccharides, which retain water and delay desiccation (Caiola et al., 1996; Pereira et al., 2009). Comparatively larger cellular envelopes have been reported in *Chroococcidiopsis* sp. during long-term phases of desiccation (Caiola et al., 1996).

10.3.3 Radiation-related stress

Along with being the main source of energy, light is one of the main factors that affects the physiology and thus the stress adaptive capacity of cyanobacteria in the polar habitats. Also, it is notable that the capacity to tolerate light and survive varies from species to species with those being even in the same locality could be showing differential response (Aigner et al., 2017). The process of photosynthesis has been shown to be occurring even in an extremely wide range of temperature from almost −7°C to +75°C (Castenholz, 1969; Davey, 1989). In general, the rate of photosynthesis increases continuously to an optimum, after which it could fall steeply (Davison, 1991). Also, it has been reported that cyanobacteria dwelling in low-temperature environments can reach higher photosynthetic rates at lower temperatures; their optimum and upper temperature limits of photosynthesis thus being lower than the algal counterparts of the warmer regions (Kuebler et al., 1991). Generally, at lower temperatures, the rate of individual steps in electron transport (Kanazawa et al., 1992; Renger and Hanssum, 1992; Kramer and Crofts, 1993), CO_2 diffusion (Beardall and Giordano, 2002), enzymatic activity (Stitt et al., 1987; Sukenik et al., 1987), and turnover of membrane proteins (Melis, 1999) are slower and thus could limit photosynthesis in any photosynthetic organism. Due to this, the saturation irradiance is lower and photoinhibition occurs at lower irradiance levels (Powels et al., 1983; Bradbury and Baker, 1984; Greer et al., 1986; Smillie and Hetherington, 1988) and this phenomenon called "photoinhibition of low temperatures" must be thus considered in polar cyanobacteria (Torzillo et al., 1998). The phenomenon of complementary chromatic adaptation is another interesting feature of cyanobacterial cells that comes into play especially in the polar regions. Cyanobacterial mats typically are important in context of this phenomenon as the spectral composition of light varies in different layers of the mats (Quesada et al., 1999). As compared to the top layers, the inner deeper layers have been found to have more chlorophyll-a and thus more pigments absorbing in blue and green regions (Bonilla et al., 2005; Hawes and Schwarz, 1999, 2001). Thus, on these types of layered mats, the topmost surface communities have more of protective functions that can thus facilitate the photosynthesis happening in the deeper layers (Sasaki and Mizuno, 2017). The process of this chromatic adaptation has been well explored in *Calothrix* sp. and *Synechococcus* sp. (Grossman et al., 1994; Six et al., 2004). The mechanism of this phenomenon has been elucidated in detail when studying the cyanobacterium *Calothrix* sp. PCC 7601 that synthesizes more phycoerythrin (PE) when grown under green light while more phycocyanin (PC) was synthesized on being grown under red light (Grossman et al., 1994). Before this study, Tandeau de Marsac (1977) had divided cyanobacteria into three groups I, II, and III according to their reaction to the spectral composition. It was postulated that group I taxa changed the size and number of phycobilisomes (PBS) in sync with the photosystem arrangement though there were no changes in the PBS absorption characteristics. The group II taxa changed the level of PE in the PBS while the group III could change the level of PE and PC both. It is also reported that if PBS-absorbed light is dominant then PSI:PSII ratio is high while this ratio is lower if chlorophyll-a absorbed light is more abundant (Fujita et al., 1994).

Light can be a limiting factor for all polar cyanobacteria and thus, they are generally adapted to grow in low light conditions in microhabitats like crevices of soil or rock, bottom sediments of permanently frozen lakes, or in the lower layers of microbial mats (Vincent, 2000). Small-sized colonies (1−2 μm wide) are reported to decrease the degree of self-shading and thus help the cyanobacteria to efficiently utilize dim light (Vincent, 2000). Well-adapted taxa in low light intensities are usually known to have high amount of Chl*a* and increased number of thylakoids (Berner and Sukenik, 1998). In that case, the PS I:PS II ratio is also greater than 1 (Kawamura et al., 1979). The opposite type of scenario is encountered on the exposed surfaces such as rock, soil crusts, or glaciers where the irradiance is high. These exposed surfaces also encounter an extremely high dose of UV radiation. This high amount of UV radiation usually affects negatively all the primary producers, particularly the cyanobacteria (Jokela et al., 1993; Döhler, 1994; Caldwell et al., 1998; Häder et al., 1998, 2015; Rozema et al., 2002; Callaghan et al., 2004a,b; Wulff et al., 2008). Cyanobacteria are in general capable of tolerating high doses of UV radiation (Seckbach and Oren, 2007). Also, it is prudent to generalize that usually the surface dwelling, highly exposed species are more tolerant to UV radiation. These resistant species seems to have bigger cells, and can grow in large colonies, coenobia, or mats, thus providing protection to internal cells or filaments (Pattanaik et al., 2007). The ability and strategy to tolerate the high dose of UV radiation seem to be a species-specific quality. For example, there is usually a high number of actively moving filaments in the upper layers of *Scytonema mats* while there is a low pigment concentration in surface cells on Oscillatorean mats (Quesada et al., 1999). Generally, cyanobacteria display four strategies of survival against UV radiation and also acclimation in high light intensities: (1) selection of favorable and conducive habitats; (2) production of screening compounds like mycosporine-like amino acids; (3) production of antioxidants; and (4) DNA repair mechanisms (Vincent, 2000). Another interesting mechanism of terrestrial cyanobacteria survival in light-related stress is the exhibition of diel periodicity during the process of photosynthesis. It has been reported that during the early developmental stages, the rate of photosynthesis is high in the early morning while, during the late development stages, the photosynthetic activity almost ended in late morning or around midday (Pushkareva et al., 2017). It is notable that the short days and long phases of darkness during the winter season severely limit the growth and physiology of all Arctic photoautotrophs (Cockell et al., 1999). It has been reported that the cyanobacterial mats endure the extreme winters as typical freeze-tolerant dry crusts (Vincent and Howard-Williams, 1986; Hawes et al., 1992; Tashyreva and Elster, 2016).

10.4 Some key concerns during culture-based studies on polar cyanobacteria

Based on previous experiences, some key points that need to be emphasized while working on polar cyanobacteria are as follows:

1. Culture-based studies on polar cyanobacteria are usually tough, challenging, and become complicated in long-term perspectives.
2. One of the main problems is the adjustment and transition of the culture from being present in a low-temperature ecosystem to its ability to survive in a low-temperature incubator having light. It is important to note that the mere usage of low-temperature systems may not be able to actually mimic the low-temperature natural ecosystem. Unfortunately, this problem is still typically unsolved and accounts for the majority of the loss of samples that happens during the initial acclimation process in the laboratories.
3. Another key problem faced after the initial acclimation process is the slow growth and a tendency of the samples to slowly decay even before any discrete colony appears.
4. Inoculation of samples in BG11 supplemented plates and liquid medium at the time of collection itself is an interesting strategy that could give the sample a bit of time to get adjusted in the medium. But, proper maintenance in the extreme field conditions, logistic limitations, and transport facilities are a big concern that could make this strategy more cumbersome and time-consuming.
5. Collection of samples/mats followed by drying and subsequent transportation is another strategy that is known to work. But, growing the dried samples in the laboratory is usually a complicated process as it is time-consuming and is frequently accompanied by loss of samples or simply their inability to grow again.

6. The changed appearance and morphology of the laboratory-grown culture as compared to the natural samples are again big concerns especially if taxonomic studies are planned. There should be a general congruence between the natural sample and the purified laboratory-grown culture in taxonomic perspectives; unfortunately, this usually does not happen with polar cyanobacteria.

10.5 Concluding notes

Polar cyanobacteria are interesting but challenging samples due to many reasons. The diverse array of microhabitats they dwell in and the successful adaptive strategies that they employ to thrive in these tough environments are remarkable. The astounding success of these organisms is a result of their ability to adapt, evolve, and replenish themselves. Thus working with the polar cyanobacteria samples is indeed an attractive field with the added perks of being in a pristine environment. But the challenges faced during sampling and postsampling handling of cyanobacterial samples is a topic that needs attention and proper treatment. Loss of samples in case of studies based on culturomics is a routine process and has severely masked the proper exploration of cyanobacterial cultures from these extreme habitats. The dearth of new cyanobacterial taxa being discovered from the polar regions can be simply understood by the fact that one of the most prominent and dominant cyanobacterial genus *Nostoc* has had number of morphotypic genera described from various parts of the world (Řeháková et al., 2007; Hrouzek et al., 2013; Genuário et al., 2015; Bagchi et al., 2017; Hentschke et al., 2017; Saraf et al., 2019) but not even a single new taxa has been described and discovered from the polar habitats. This strange occurrence could although be because of the supposed dominance of one single taxa in these extreme regions but apart from this, it is also because of the challenges faced in isolation, purification, and culturing of cyanobacteria from these regions. In the future, it is envisaged that better strategies that could mimic the polar conditions holistically must be employed to gather better insights about the diversity and taxonomic complexity of polar cyanobacteria.

Acknowledgments

We thank the Head, Department of Botany, BHU for providing the necessary facilities and encouragement. Large part of this work comes from the experiences of the field access support provided to PS through NCPOR (MoES) as a part of Indian Scientific Expedition to the Arctic (2017, 2019).

References

Aigner, S., Herburger, K., Holzinger, A., Karsten, U., 2017. Epilithic *Chamaesiphon* (Synechococcales, Cyanobacteria) species in mountain streams of the Alps—interspecific differences in photo-physiological traits. J. Appl. Phycol. 30, 1125–1134.

Aislabie, J.M., Chhour, K.L., Saul, D.J., Miyauchi, S., Ayton, J., Paetzold, R.F., et al., 2006. Dominant bacteria in soils of Marble point and Wright valley, Victoria land, Antarctica. Soil Biol. Biochem. 38 (10), 3041–3056.

Ambrosini, R., Musitelli, F., Navarra, F., Tagliaferri, I., Gandolfi, I., Bestetti, G., et al., 2017. Diversity and assembling processes of bacterial communities in cryoconite holes of a Karakoram glacier. Microb. Ecol. 73 (4), 827–837.

Bagchi, S.N., Dubey, N., Singh, P., 2017. Phylogenetically distant clade of *Nostoc*-like taxa with the description of *Aliinostoc* gen. nov. and *Aliinostoc morphoplasticum* sp. nov. Int. J. Syst. Evol. Microbiol. 67, 3329–3338.

Beardall, J., Giordano, M., 2002. Ecological implications of microalgal and cyanobacterial CO_2 concentrating mechanisms, and their regulation. Funct. Plant Biol. 29 (3), 335–347.

Berner, T., Sukenik, A., 1998. Photoacclimation in photosynthetic microorganisms: an ultrastructural response. Isr. J. Plant Sci. 46 (2), 141–146.

Bonilla, S., Villeneuve, V., Vincent, W.F., 2005. Benthic and planktonic algal communities in a high arctic lake: pigment structure and contrasting responses to nutrient enrichment. J. Phycol. 41, 1120–1130.

Borchhardt, N., Schiefelbein, U., Abarca, N., Boy, J., Mikhailyuk, T., Sipman, H.J., et al., 2017. Diversity of algae and lichens in biological soil crusts of Ardley and King George Island, Antarctica. Antarctic Sci. 29, 229–237.

Bradbury, M., Baker, N.R., 1984. A quantitative determination of photochemical and non-photochemical quenching during the slow phase of the chlorophyll fluorescence induction curve of bean leaves. Biochim. Biophys. Acta (BBA)-Bioenergetics 765 (3), 275–281.

Broady, P.A., 1986. Ecology and taxonomy of the Vestfold Hills. In: Pickard, J. (Ed.), Antarctic Oasis: Terrestrial Environments and History of the Vestfold Hills. Academic, Sydney, pp. 165–202.

Bryant, D.A., 2003. The beauty in small things revealed. Proc. Natl. Acad. Sci. 100 (17), 9647–9649.

Büdel, B., Bendix, J., Bicker, F.R., Allan Green, T.G., 2008. Dewfall as a water source frequently activates the endolithic cyanobacterial communities in the granites of Taylor valley, Antarctica. J. Phycol. 44 (6), 1415–1424.

Büdel, B., Dulić, T., Darienko, T., Rybalka, N., Friedl, T., 2016. Cyanobacteria and algae of biological soil crusts. In: Weber, B., Büdel, B., Belnap, J. (Eds.), Biological Soil Crusts: An Organizing Principle in Drylands. Springer, Switzerland, pp. 55–80.

Caiola, M.G., Billi, D., Friedmann, E.I., 1996. Effect of desiccation on envelopes of the cyanobacterium *Chroococcidiopsis* sp. (Chroococcales). Eur. J. Phycol. 31 (1), 97–105.

Caldwell, M.M., Björn, L.O., Bornman, J.F., Flint, S.D., Kulandaivelu, G., Teramura, A. H., et al., 1998. Effects of increased solar ultraviolet radiation on terrestrial ecosystems. J. Photochem. Photobiol. B: Biol. 46 (1–3), 40–52.

Callaghan, T.V., Björn, L.O., Chernov, Y., Chapin, T., Christensen, T.R., Huntley, B., et al., 2004a. Biodiversity, distributions and adaptations of Arctic species in the context of environmental change. AMBIO: J. Hum. Environ. 33 (7), 404–417.

Callaghan, T.V., Björn, L.O., Chernov, Y., Chapin, T., Christensen, T.R., Huntley, B., et al., 2004b. Responses to projected changes in climate and UV-B at the species level. AMBIO: J. Hum. Environ. 33 (7), 418–435.

Cameron, R.E., 1962. Species of *Nostoc vaucher* occurring in the Sonoran Desert in Arizona. Trans. Am. Microsc. Soc. 81 (4), 379–384.

Castenholz, R.W., 1969. Thermophilic blue-green algae and the thermal environment. Bacteriol. Rev. 33, 476–504.
Cavacini, P., 2001. Soil algae from northern Victoria Land (Antarctica). Polar Biosci. 14, 45–60.
Chao, H., Davies, P.L., Carpenter, J.F., 1996. Effects of antifreeze proteins on red blood cell survival during cryopreservation. J. Exp. Biol. 199 (9), 2071–2076.
Chown, S., Huiskes, A.H., Gremmen, N.J., Lee, J.E., Terauds, A., Crosbie, K., et al., 2012. Continent-wide risk assessment for the establishment of nonindigenous species in Antarctica. Proc. Natl. Acad. Sci. 109, 4938–4943.
Close, T.J., 1997. Dehydrins: a commonalty in the response of plants to dehydration and low temperature. Plant Physiol. 100 (2), 291–296.
Close, T.J., Lammers, P.J., 1993. An osmotic stress protein of cyanobacteria is immunologically related to plant dehydrins. Plant Physiol. 101 (3), 773–779.
Cockell, C.S., Stokes, M.D., 2004. Widespread colonization by polar hypoliths. Nature 431 (7007), 414.
Cockell, C.S., Stokes, M.D., 2006. Hypolithic colonization of opaque rocks in the Arctic and Antarctic polar desert. Arct. Antarctic Alp. Res. 38 (3), 335–342.
Cockell, C.S., Herrera, A., 2007. Why are some microorganisms boring? Trends Microbiol. 16, 101–106.
Cockell, C.S., Stokes, M.D., Korsmeyer, K.E., 1999. Overwintering strategies of Antarctic organisms. Environ. Rev. 8 (1), 1–19.
Colesie, C., Gommeaux, M., Green, T.A., Büdel, B., 2014. Biological soil crusts in continental Antarctica: Garwood Valley, southern Victoria Land, and Diamond Hill, Darwin Mountains region. Antarctic Sci. 26, 115–123.
Convey, P., Chown, S.L., Clarke, A., Barnes, D.K., Bokhorst, S., Cummings, V., et al., 2014. The spatial structure of Antarctic biodiversity. Ecol. Monogr. 84 (2), 203–244.
Cowan, D.A., Tow, A.L., 2004. Endangered antarctic environments. Annu. Rev. Microbiol. 58, 649–690.
Davey, M.C., 1989. The effects of freezing and desiccation on photosynthesis and survival of terrestrial Antarctic algae and cyanobacteria. Polar Biol. 10 (1), 29–36.
Davey, A., Marchant, H.J., 1983. Seasonal variation in nitrogen fixation by *Nostoc commune* Vaucher at the Vestfold Hills, Antarctica. Phycologia 22 (4), 377–385.
Davison, I.R., 1991. Environmental effects on algal photosynthesis: temperature. J. Phycol. 27, 2–8.
de Los Ríos, A., Ascaso, C., Wierzchos, J., Fernández-Valiente, E., Quesada, A., 2004. Microstructural characterization of cyanobacterial mats from the McMurdo Ice Shelf, Antarctica. Appl. Environ. Microbiol. 70 (1), 569–580.
Döhler, G., 1994. UV-effects on the nitrogen metabolism of marine phytoplankton and adaptation to UV radiation. Stratospheric Ozone Depletion/UV-B Radiation in the Biosphere. Springer, pp. 163–174.
Elster, J., 2002. Ecological classification of terrestrial algal communities in polar environments. Geoecology of Antarctic Ice-Free Coastal Landscapes. Springer-Verlag, Berlin/Heidelberg, pp. 303–326.
Elster, J., Benson, E.E., 2004. Life in the polar terrestrial environment with a focus on algae and cyanobacteria. Life in the Frozen State. CRC Press, pp. 137–176.
Elster, J., Kvíderová, J., Hájek, T., Láska, K., Šimek, M., 2012. Impact of warming on *Nostoc* colonies (Cyanobacteria) in a wet hummock meadow, Spitsbergen. Pol. Polar Res. 33 (4), 395–420.
Erokhina, L.G., Spirina, E.V., Shatilovich, A.V., Gilichinskiĭ, D.A., 2000. Chromatic adaptation of viable ancient cyanobacteria from arctic permafrost. Microbiology 69, 855–856.
Evans, R.D., Lange, O.L., 2001. Biological Soil Crusts and ecosystem nitro-gen and carbon dynamics. In: Belnap, J., Lange, O.L. (Eds.), Biological Soil Crusts: Structure, Function, and Management. Springer, Heidelberg, pp. 263–279.

Fay, P., 1983. The Blue-Greens (Cyanophyta-Cyanobacteria). The Institute of Biology's Studies in Biology no. 160. Edward Ernold, Baltimore, MD, 88 pp.
Fernández-Carazo, R., Hodgson, D.A., Convey, P., Wilmotte, A., 2011. Low cyanobacterial diversity in biotopes of the Transantarctic Mountains and Shackleton Range (80-82oS), Antarctica. FEMS Microbiol. Ecol. 77, 503−517.
Friedmann, E.I., Kappen, L., Meyer, M.A., Nienow, J.A., 1993. Long-term productivity in the cryptoendolithic microbial community of the Ross Desert, Antarctica. Microb. Ecol. 25 (1), 51−69.
Fujita, Y., Murakami, A., Aizawa, K., Ohki, K., 1994. Short-term and long-term adaptation of the photosynthetic apparatus: homeostatic properties of thylakoids. The Molecular Biology of Cyanobacteria. Springer, Dordrecht, pp. 677−692.
Gaysina, L.A., Saraf, A., Singh, P., 2019. Cyanobacteria in diverse habitats. In: Mishra, A. K., Tiwari, D.N., Rai, A.N. (Eds.), Cyanobacteria: From Basic Science to Applications, vol. 1. Elsevier Academic Press, pp. 1−28.
Genuário, D.B., Vaz, M.G.M.V., Hentschke, G.S., Sant'Anna, C.L., Fiore, M.F., 2015. *Halotia* gen. nov., a phylogenetically and physiologically coherent cyanobacterial genus isolated from marine coastal environments. Int. J. Syst. Evol. Microbiol. 65, 663−675.
Gilbert, J.A., Hill, P.J., Dodd, C.E., Laybourn-Parry, J., 2004. Demonstration of antifreeze protein activity in Antarctic lake bacteria. Microbiology 150 (1), 171−180.
Gounot, A.M., Russell, N.J., 1999. Physiology of cold-adapted microorganisms. Cold-Adapted Organisms. Springer, Berlin, Heidelberg, pp. 33−55.
Greer, D.H., Berry, J.A., Björkman, O., 1986. Photoinhibition of photosynthesis in intact bean leaves: role of light and temperature, and requirement for chloroplast-protein synthesis during recover. Planta 168, 253−260.
Grossman, A.R., Schaefer, M.R., Chiang, G.G., Collier, J.L., 1994. The responses of cyanobacteria to environmental conditions: light and nutrients. The Molecular Biology of Cyanobacteria. Springer, Dordrecht, pp. 641−675.
Häder, D.P., Kumar, H.D., Smith, R.C., Worrest, R.C., 1998. Effects on aquatic ecosystems. J. Photochem. Photobiol. B: Biol. 46 (1−3), 53−68.
Häder, D.P., Williamson, C.E., Wängberg, S.Å., Rautio, M., Rose, K.C., Gao, K., et al., 2015. Effects of UV radiation on aquatic ecosystems and interactions with other environmental factors. Photochem. Photobiol. Sci. 14 (1), 108−126.
Hand, S.C., Menze, M.A., Toner, M., Boswell, L., Moore, D., 2011. LEA proteins during water stress: not just for plants anymore. Annu. Rev. Physiol. 73, 115−134.
Hawes, I., Schwarz, A.M., 1999. Photosynthesis in an extreme shade environment: benthic microbial mats from Lake Hoare, a permanently ice-covered Antarctic lake. J. Phycol. 35 (3), 448−459.
Hawes, I., Schwarz, A.M., 2001. Absorption and utilization of irradiance by cyanobacterial mats in two ice-covered Antarctic lakes with contrasting light climates. J. Phycol. 37 (1), 5−15.
Hawes, I., Howard-Williams, C., Vincent, W.F., 1992. Desiccation and recovery of Antarctic cyanobacterial mats. Polar Biol. 12 (6−7), 587−594.
Hentschke, G.S., Johansen, J.R., Pietrasiak, N., Rigonato, J., Fiore, M.F., Sant'Anna, C. L., 2017. *Komarekiella atlantica* gen. et sp. nov. (*Nostocaceae, Cyanobacteria*): a new subaerial taxon from the Atlantic Rainforest and Kauai, Hawaii. Fottea 17, 178−190.
Hoffmann, L., 1989. Algae of terrestrial habitats. Bot. Rev. 55, 77−105.
Hrouzek, P., Lukešová, A., Mareš, J., Ventura, S., 2013. Description of the cyanobacterial genus *Desmonostoc* gen. nov. including *D. muscorum* comb. nov. as a distinct, phylogenetically coherent taxon related to the genus *Nostoc*. Fottea 13, 201−213.
Johnson, S.L., Kuske, C.R., Carney, T.D., Housman, D.C., Gallegos-Graves, L.V., Belnap, J., 2012. Increased temperature and altered summer precipitation have differential effects on biological soil crusts in a dryland ecosystem. Glob. Change Biol. 18, 2583−2593.

Jokela, K., Leszczynski, K., Visuri, R., 1993. Effects of Arctic ozone depletion and snow on UV exposure in Finland. Photochem. Photobiol. 58 (4), 559–566.
Jordan, R.P., 1978. The Snowmelt Hydrology of a Small Alpine Watershed (Doctoral dissertation). British, University of Columbia.
Kanazawa, A., Kramer, D., Crofts, A., 1992. Temperature dependence of PS2 electron transfer reactions measured by flash-induced fluorescence changes. In: Murata, N. (Ed.), Research in Photosynthesis, vol. II. Kluwer Academic Publishers, Dordrecht.
Kastovska, K., Elster, J., Stibal, M., Šantrůčková, H., 2005. Microbial assemblages in soil microbial succession after glacial retreat in Svalbard (high Arctic). Microb. Ecol. 50, 396–407.
Kawahara, H., 2017. Cryoprotectants and ice-binding proteins. Psychrophiles: From Biodiversity to Biotechnology. Springer, Cham, pp. 237–257.
Kawamura, M., Mimuro, M., Fujita, Y., 1979. Quantitative relationship between two reaction centers in the photosynthetic system of blue-green algae. Plant Cell Physiol. 20 (4), 697–705.
Klähn, S., Hagemann, M., 2011. Compatible solute biosynthesis in cyanobacteria. Environ. Microbiol. 13 (3), 551–562.
Kramer, D.M., Crofts, A.R., 1993. The controlled reduction of high and low potential chains of the bf complex by plastoquinol. Biochim. Biophys. Acta (BBA)-Bioenergetics 1184, 193–201.
Kuebler, J.E., Davison, I.R., Yarish, C., 1991. Photosynthetic temperature adaptation in the red algae *Lomentaria baileyana* and *Lomentaria orchadensis*. Br. Phycol. J. 26, 9–19.
Kvíderová, J., 2018. Internal structure and photosynthetic performance of *Nostoc* sp. colonies in the high Arctic. Acta Soc. Bot. Pol. 87 (4).
Kvíderová, J., Elster, J., Šimek, M., 2011. In situ response of *Nostoc commune* sl colonies to desiccation in Central Svalbard, Norwegian High Arctic. Fottea 11 (1), 87–97.
Lee, J.R., Raymond, B., Bracegirdle, T.J., Chades, I., Fuller, R.A., Shaw, J.D., et al., 2017. Climate change drives expansion of Antarctic ice-free habitat. Nature 547, 49–54.
Lennihan, R., Chapin, D.M., Dickson, L.G., 1994. Nitrogen fixation and photosynthesis in high arctic forms of *Nostoc commune*. Can. J. Bot. 72 (7), 940–945.
Lipman, C.B., 1941. The successful revival of *Nostoc commune* from a herbarium specimen eighty-seven years old. Bull. Torrey Bot. Club 664–666.
MacLeod, R.A., Calcott, P.H., 1976. Cold shock and freezing damage to microbes. In: Gray, T.R.G., Postagen, R. (Eds.), The Survival of Vegetative Microbes. Cambridge University Press, Cambridge, pp. 81–109.
Mataloni, G., Tell, G., Wynn-Williams, D.D., 2000. Structure and diversity of soil algal communities from Cierva Point (Antarctic Peninsula). Polar Biol. 23, 205–211.
Melis, A., 1999. Photosystem-II damage and repair cycle in chloroplasts: what modulates the rate of photodamage in vivo? Trends Plant Sci. 4 (4), 130–135.
Michaud, A.B., Šabacká, M, Priscu, J.C., 2012. Cyanobacterial diversity across landscape units in a polar desert: Taylor Valley, Antarctica. FEMS Microbiol. Ecol. 82 (2), 268–278.
Mueller, D.R., Vincent, W.F., Pollard, W.H., Fritsen, C.H., 2001. Glacial cryoconite ecosystems: a bipolar comparison of algal communities and habitats. Nova Hedwig 123, 173–198.
Murata, N., Wada, H., 1995. Acyl-lipid desaturases and their importance in the tolerance and acclimatization to cold of cyanobacteria. Biochem. J. 308 (1), 1.
Nishida, I., Murata, N., 1996. Chilling sensitivity in plants and cyanobacteria: the crucial contribution of membrane lipids. Annu. Rev. Plant Biol. 47 (1), 541–568.
Olie, J.J., Potts, M., 1986. Purification and biochemical analysis of cytoplasmic membrane from desiccation-tolerant cyanobacterium *Nostoc commune* UTEX 584. Appl. Environ. Microbiol. 52, 706–710.

Omelon, C.R., 2008. Endolithic microbial communities in polar desert habitats. Geomicrobiology 25, 404–414.

Omelon, C.R., Pollard, W.H., Ferris, F.G., 2007. Inorganic species distribution and microbial diversity within high arctic cryptoendolithic habitats. Microb. Ecol. 54 (4), 740–752.

Palmer, R.J., Friedmann, E.I., 1990. Water relations and photosynthesis in the cryptoendolithic microbial habitat of hot and cold deserts. Microb. Ecol. 19, 111–118.

Pattanaik, B., Schumann, R., Karsten, U., 2007. Effects of ultraviolet radiation on cyanobacteria and their protective mechanisms. Algae and Cyanobacteria in Extreme Environments. Springer, Dordrecht, pp. 29–45.

Pereira, S., Zille, A., Micheletti, E., Moradas-Ferreira, P., De Philippis, R., Tamagnini, P., 2009. Complexity of cyanobacterial exopolysaccharides: composition, structures, inducing factors and putative genes involved in their biosynthesis and assembly. FEMS Microbiol. Rev. 33 (5), 917–941.

Pointing, S.B., Warren-Rhodes, K.A., Lacap, D.C., Rhodes, K.L., McKay, C.P., 2007. Hypolithic community shifts occur as a result of liquid water availability along environmental gradients in China's hot and cold hyper arid deserts. Environ. Microbiol. 9, 414–424.

Potts, M., 1994. Desiccation tolerance of prokaryotes. Microbiol. Rev. 58 (4), 755–805.

Potts, M., 1999. Mechanisms of desiccation tolerance in cyanobacteria. Eur. J. Phycol. 34, 319–328.

Potts, M., 2000. Nostoc. In: Whitton, B.A., Potts, M. (Eds.), The Ecology of Cyanobacteria. Kluwer Academic Publishers, Dordrecht, pp. 465–504.

Potts, M., Friedmann, E.I., 1981. Effects of water stress on cryptoendolithic cyanobacteria from hot desert rocks. Arch. Microbiol. 130 (4), 267–271.

Powels, S.B., Berry, J.A., Björkman, O., 1983. Interaction between light and chilling temperatures on the inhibition of photosynthesis in chilling-sensitive plants. Plant Cell Environ. 6, 117–123.

Pushkareva, E., Johansen, J.R., Elster, J., 2016. A review of the ecology, ecophysiology and biodiversity of microalgae in Arctic soil crusts. Polar Biol. 39, 2227–2240.

Pushkareva, E., Kvíderová, J., Šimek, M., Elster, J., 2017. Nitrogen fixation and diurnal changes of photosynthetic activity in Arctic soil crusts at different development stage. Eur. J. Soil Biol. 79, 21–30.

Qiu, B., Gao, K., 2001. Photosynthetic characteristics of the terrestrial blue-green alga, *Nostoc flagelliforme*. Eur. J. Phycol. 36 (02), 147–156.

Quesada, A., Vincent, W.F., 2012. Cyanobacteria in the cryosphere: snow, ice and extreme cold. In: Whitton, B.A. (Ed.), Ecology of Cyanobacteria II: Their Diversity in Space and Time. Springer, New York, pp. 387–399.

Quesada, A., Vincent, W.F., Lean, D.R., 1999. Community and pigment structure of Arctic cyanobacterial assemblages: the occurrence and distribution of UV-absorbing compounds. FEMS Microbiol. Ecol. 28 (4), 315–323.

Raymond, J.A., Fritsen, C.H., 2000. Ice-active substances associated with Antarctic freshwater and terrestrial photosynthetic organisms. Antarctic Sci. 12 (4), 418–424.

Raymond-Bouchard, I., Whyte, L.G., 2017. From transcriptomes to metatranscriptomes: cold adaptation and active metabolisms of psychrophiles from cold environments. In: Margesin, R. (Ed.), Psychrophiles: From Biodiversity to Biotechnology. Springer, Cham, pp. 437–457.

Řeháková, K., Johansen, J.R., Casamatta, D.A., Xuesong, L., Vincent, J., 2007. Morphological and molecular characterization of selected desert soil cyanobacteria: three species new to science including *Mojavia pulchra* gen. et sp. nov. Phycologia 46, 481–502.

Renger, G., Hanssum, B., 1992. Studies on the reaction coordinates of the water oxidase in PSII membrane fragments from spinach. FEBS Lett. 299, 28–32.

Rippin, M., Borchhardt, N., Williams, L., Colesie, C., Jung, P., Büdel, B., et al., 2018. Genus richness of microalgae and Cyanobacteria in biological soil crusts from Svalbard and Livingston Island: morphological versus molecular approaches. Polar Biol. 41 (5), 909–923.

Rozema, J., Björn, L.O., Bornman, J.F., Gaberščik, A., Häder, D.P., Trošt, T., et al., 2002. The role of UV-B radiation in aquatic and terrestrial ecosystems—an experimental and functional analysis of the evolution of UV-absorbing compounds. J. Photochem. Photobiol. B: Biol. 66 (1), 2–12.

Šabacká, M., Elster, J., 2006. Response of cyanobacteria and algae from Antarctic wetland habitats to freezing and desiccation stress. Polar Biol. 30 (1), 31–37.

Saraf, A.G., Dawda, H.G., Singh, P., 2019. *Desikacharya* gen. nov., a phylogenetically distinct genus of *cyanobacteria* along with the description of two new species, *Desikacharya nostocoides* sp. nov. and *Desikacharya soli* sp. nov., and reclassification of *Nostoc thermotolerans* to *Desikachary athermotolerans* comb. nov. Int. J. Syst. Evol. Microbiol. 69, 307–315.

Sasaki, A., Mizuno, A.N., 2017. Partitioning light spectra: adaptive stratification of phytobenthic communities in Antarctic lakes. J. Theor. Biol. 424, 1–10.

Scherer, S., Zhong, Z.P., 1991. Desiccation independence of terrestrial *Nostoc commune* ecotypes (Cyanobacteria). Microb. Ecol. 22 (3), 271–283.

Scherer, S., Ernst, A., Chen, T.W., Böger, P., 1984. Rewetting of drought-resistant blue-green algae: time course of water uptake and reappearance of respiration, photosynthesis, and nitrogen fixation. Oecologia 62 (3), 418–423.

Seckbach, J., Oren, A., 2007. Oxygenic photosynthetic microorganisms in extreme environments. Algae and Cyanobacteria in Extreme Environments. Springer, Dordrecht, pp. 3–25.

Shirkey, B., McMaster, N.J., Smith, S.C., Wright, D.J., Rodriguez, H., Jaruga, P., et al., 2003. Genomic DNA of *Nostoc commune* (Cyanobacteria) becomes covalently modified during long-term (decades) desiccation but is protected from oxidative damage and degradation. Nucleic Acids Res. 31 (12), 2995–3005.

Singh, A.P., 2001. Study on Lipids of Some Desiccation Tolerant Cyanobacteria. Banaras Hindu University, Varanasi.

Six, C., Thomas, J.C., Brahamsha, B., Lemoine, Y., Partensky, F., 2004. Photophysiology of the marine cyanobacterium *Synechococcus* sp. WH8102, a new model organism. Aquat. Microb. Ecol. 35 (1), 17–29.

Smillie, R.M., Hetherington, S.E., 1988. Stress tolerance and stress-induced injury in crop plants measured by chlorophyll fluorescence in vivo: chilling, freezing, ice cover, heat and high light. Plant Physiol. 72, 1043–1050.

Smith, M.G., 1995. Survival of *E. coli* and *Salmonella* after chilling and freezing in liquid media. J. Food Sci. 60, 509–512.

Solheim, B., Endal, A., Vigstad, H., 1996. Nitrogen fixation in Arctic vegetation and soils from Svalbard, Norway. Polar Biol. 16 (1), 35–40.

Sommers, P., Darcy, J.L., Gendron, E.M., Stanish, L.F., Bagshaw, E.A., Porazinska, D.L., et al., 2018. Diversity patterns of microbial eukaryotes mirror those of bacteria in Antarctic cryoconite holes. FEMS Microbiol. Ecol. 94 (1), 167.

Stanish, L.F., O'Neill, S.P., Gonzalez, A., Legg, T.M., Knelman, J., McKnight, D.M., et al., 2013. Bacteria and diatom co-occurrence patterns in microbial mats from polar desert streams. Environ. Microbiol. 15 (4), 1115–1131.

Stibal, M., Šabacká, M., Kaštovská, K., 2006. Microbial communities on glacier surfaces in Svalbard: impact of physical and chemical properties on abundance and structure of cyanobacteria and algae. Microb. Ecol. 52, 644–654.

Stitt, M., Huber, S., Kerr, P., 1987. Control of photosynthetic sucrose formation. In: Hatch, M.D., Boardman, N.K. (Eds.), Photosynthesis. Academic Press, New York, NY, pp. 220–275.

Stomeo, F., Makhalanyane, T.P., Valverde, A., Pointing, S.B., Stevens, M.I., Cary, C.S., et al., 2012. Abiotic factors influence microbial diversity in permanently cold soil horizons of a maritime-associated Antarctic Dry Valley. FEMS Microbiol. Ecol. 82 (2), 326–340.

Sukenik, A., Bennett, J., Falkowski, P., 1987. Light-saturated photosynthesis—limitation of electron transport or carbon fixation? Biochim. Biophys. Acta (BBA)-Bioenergetics 891, 205–215.

Sun, H.J., Friedmann, E.I., 1999. Growth on geological time scales in the Antarctic cryptoendolithic microbial community. Geomicrobiol. J. 16 (2), 193–202.

Takeuchi, N., Li, Z.Q., 2008. Characteristics of surface dust on Urumqi Glacier No.1 in the Tien Shan Mountains, China. Arctic Antarctic Alp. Res. 40, 744–750.

Tamaru, Y., Takani, Y., Yoshida, T., Sakamoto, T., 2005. Crucial role of extracellular polysaccharides in desiccation and freezing tolerance in the terrestrial cyanobacterium *Nostoc commune*. Appl. Environ. Microbiol. 71 (11), 7327–7333.

Tandeau de Marsac, N., 1977. Occurrence and nature of chromatic adaptation in cyanobacteria. J. Bacteriol. 130 (1), 82–91.

Tanghe, A., Van Dijck, P., Thevelein, J.M., 2003. Determinants of freeze tolerance in microorganisms, physiological importance, and biotechnological applications. Adv. Appl. Microbiol. 53, 129–176.

Tashyreva, D., Elster, J., 2016. Annual cycles of two cyanobacterial mat communities in hydro-terrestrial habitats of the High Arctic. Microb. Ecol. 71 (4), 887–900.

Thomas, D.N., 2005. Photosynthetic microbes in freezing deserts. Trends Microbiol. 13, 87–88.

Torzillo, G., Bernardini, P., Masojídek, J., 1998. On-line monitoring of chlorophyll fluorescence to assess the extent of photoinhibition of photosynthesis induced by high oxygen concentration and low temperature and its effect on the productivity of outdoor cultures of Spirulina platensis (Cyanobacteria). J. Phycol. 34 (3), 504–510.

Vincent, W.F., 1988. Microbial Ecosystems of Antarctica. Cambridge University Press, Cambridge, p. 304.

Vincent, W.F., 2000. Cyanobacterial dominance in the Polar regions. In: Whitton, B.A., Potts, M. (Eds.), The Ecology of Cyanobacteria. Their Diversity in Time and Space. Kluwer Academic Publishers, Dordrecht, 321–340, 669.

Vincent, W.F., 2007. Cold tolerance in cyanobacteria and life in the cryosphere. Algae and Cyanobacteria in Extreme Environments. Springer, Dordrecht, pp. 287–301.

Vincent, W.F., Howard-Williams, C.L.I.V.E., 1986. Antarctic stream ecosystems: physiological ecology of a blue-green algal epilithon. Freshw. Biol. 16 (2), 219–233.

Vishnivetskaya, T.A., 2009. Viable cyanobacteria and green algae from permafrost darkness. In: Margesin, R. (Ed.), Permafrost Soils, Soil Biology, 16. Springer, Heidelberg, pp. 73–84. 348.

Vishnivetskaya, T.A., Vorobyova, E.A., Gilichinsky, D.A., 2002. Viable green algae and cyanobacteria within terrestrial permafrost. In: Sawaya-Lacoste, H. (Ed.), Proceedings of the Second European Workshop on Exo-astrobiology, 295–298. ESA Publications Division, Noordwijk, p. 625.

Vishnivetskaya, T.A., Spirina, E.V., Shatilovich, A.V., Erokhina, L.G., Vorobyova, E.A., Gilichinsky, D.A., 2003. The resistance of viable permafrost algae to simulated environmental stresses: implications for astrobiology. Int. J. Astrobiol. 2, 171–177.

Walles, S., Srahna, M., Devos, N., Gerday, C., Loppes, R., Matagne, R.F., 1999. Protein adaptation to low temperatures: a comparative study of a-tubulin sequences in mesophilic and psychrophilic algae. Extremophiles 3, 221–226.

Williams, L., Borchhardt, N., Colesie, C., Baum, C., Komsic-Buchmann, K., Rippin, M., et al., 2017. Biological soil crusts of Arctic Svalbard and of Livingston Island, Antarctica. Polar Biol. 40, 399–411.

Wood, S.A., Rueckert, A., Cowan, D.A., Cary, S.C., 2008. Sources of edaphic cyanobacterial diversity in the dry valleys of eastern Antarctica. ISME J. 2, 308–320.
Wulff, A., Zacher, K., Hanelt, D., Al-Handal, A., Wiencke, C., 2008. UV radiation-a threat to Antarctic benthic marine diatoms? Antarctic Sci. 20 (1), 13–20.
Wynn-Williams, D.D., 2000. Cyanobacteria in deserts-life at the limit? In: Whitton, B.A., Potts, M. (Eds.), The Ecology of Cyanobacteria. Their Diversity in Time and Space, 341–366. Kluwer Academic Publishers, Dordrecht, p. 669.
Xiang, S.R., Shang, T.C., Chen, Y., Yao, T.D., 2009. Deposition and postdeposition mechanisms as possible drivers of microbial population variability in glacier ice. FEMS Microbiol. Ecol. 70, 165–176.
Yoshitake, S., Uchida, M., Koizumi, H., Kanda, H., Nakatsubo, T., 2010. Production of biological soil crusts in the early stage of primary succession on a high Arctic glacier foreland. N. Phytol. 186, 451–460.
Zelikova, T.J., Housman, D.C., Grote, E.E., Neher, D.A., Belnap, J., 2012. Warming and increased precipitation frequency on the Colorado Plateau: implications for biological soil crusts and soil processes. Plant Soil 355, 265–282.
Zidarova, R., 2008. Algae from Livingston Island (S Shetland Islands): a checklist. Phytol. Balcanica 14, 19–35.

CHAPTER 11

Arctic biodiversity amidst looming climate apocalypse: current status and way forward

Alka Rani[1], Kriti Gupta[2], Khem Chand Saini[1], Sahil Narwal[1] and Felix Bast[1]
[1]Department of Botany, Central University of Punjab, Ghudda-VPO, Bathinda, India
[2]Department of Botany, DAV College, Bathinda, India

11.1 Climate change in the Arctic

The Arctic region is a relatively small fraction of the globe's surface, undergoing remarkable environmental changes (Winton, 2008; Overland, 2016). With the changing climatic conditions, many people have their eyes on the Arctic region. Climate change is reshaping the Arctic at a much faster rate. Alterations to temperature regimes, surface currents, sea-ice cover, and other vital processes are the critical factors for climate change. According to IPCC (2014a,b), the rising concentration of greenhouse gases (GHGs) is the primary indicator of global climate change in arctic ecosystems. The warming of the Arctic is more than twice the global average (Serreze and Barry, 2011) and is called as Arctic Amplification (AA), and it is thought to be primarily driven by sea-ice loss (Pistone et al., 2014; Screen and Simmonds, 2010). AA requires immediate attention and ways to mitigate the effects of climate change there than elsewhere in the world.

11.1.1 Why does Arctic biodiversity matter?

The terrestrial ecosystems of the Arctic region are impacted heavily by climate change, with the significant changes in earth-surface phenomena, which decline in glacier ice and snow, melting of permafrost, increase in vegetation cover and shrub encroachment, these alterations mediated changes in trophic interactions (Sturm et al., 2001; Ims et al., 2013; Cooper, 2014). The ecosystems and biodiversity are extremely fragile that even minor environmental perturbations will have an

Understanding Present and Past Arctic Environments
DOI: https://doi.org/10.1016/B978-0-12-822869-2.00003-7

immediate and long-lasting effect in the Arctic region (Kevan et al., 1995). The biodiversity is heavily affected by several anthropogenic activities not from within the Arctic but also from outside. The growing demand for shipping and large-scale exploitation of Arctic oil gas, other mineral resources together have adverse effects on the ecosystems of the Arctic continent. Climate change influences marine communities by altering species composition and relative abundances, potentially changing the food web structure considerably (Kortsch et al., 2015). That's why the biodiversity area is the focal area of the Arctic Council, which continuously works on the Conservation of Arctic Fauna and Flora (CAFF). Scientists recorded a high rate of snowmelt in Greenland during summer. Canada's last ice shelf located in the territory of Nunavut, melted into the sea in early August. High temperatures were recorded in the parts of Svalbard and Siberia, a group of Norwegian islands in the Arctic Ocean, in late summers.

Recently the Arctic is considered a treeless plain with only sparse tundra vegetation comprising sedges, grasses, and a few flowering plants compared to the Russian sediment cores which mainly contained pollen grains from the trees such as fir, hemlock, larch, and spur. It shows that boreal woodlands, which today end several miles farther south and west in Russia and at the Arctic Circle in Alaska, arrived today at all the route to the Arctic Ocean across Arctic Russia and North America.

The climate change in Arctic region led to the northward distribution of boreal species such as phytoplanktons (Hegseth and Sundfjord, 2008), zooplankton (Weydmann et al., 2014), zoobenthos (Bluhm et al., 2009), fish (Mueter and Litzow, 2008), seabirds (Piatt and Kitaysky, 2002; CAFF, 2010) and marine mammals (Moore, 2008). This changing pattern of distribution of species replaced the long-lived and slow-growing Arctic organisms with the smaller and short-lived boreal counterparts (Berge et al., 2005; Wesławski et al., 2010). In recent decades, the Arctic Ocean and its adjacent marginal seas have registered unprecedented warming (Hoegh-Guldberg and Bruno, 2010; Overland and Wang, 2016) and concomitant ecosystem changes (Feng et al., 2016; Fossheim et al., 2015; Grebmeier et al., 2006b). For example, the Bering and Chukchi Seas have seen mean sea surface temperatures (SSTs) anomalies of 2°C–8°C since 2015 in August relative to 1982–2010 (Richter-Menge et al., 2019). Most of the Bering Sea has remained ice-free for most of the winter since 2017 while the

Chukchi Sea experienced one of the lowest September ice extents in the satellite record in 2019 (Richter-Menge et al., 2019). These changes drive the expansion of temperate species into Arctic waters (Fossheim et al., 2015; Mueter and Litzow, 2008; Poloczanska et al., 2013). In the Pacific sector of the Arctic, prolonged alternating episodes of warming and cooling have significantly reshuffled spatial distributions and species abundances of marine benthic communities (Alabia et al., 2018; Duffy-Anderson et al., 2017; Grebmeier, 2012; Stevenson and Lauth, 2019). The Svalbard archipelago showed an increasing warming trend as the air temperature in this region increases in recent decades (Nordli et al., 2014). Rain-on-snow' events (hereafter ROS events) occurred in Svalbard due to the strong oceanic influence on the weather systems inside the archipelago (Svendsen et al., 2002), which leads to an increasing frequency of winter warming (Førland et al., 2012). These events significantly affected the entire terrestrial ecosystem by changing snowpack properties and subsnowpack soil temperatures (Putkonen and Roe, 2003; Rennert et al., 2009) and in turn affected Svalbard reindeer behavior in terms of habitat use and diet (Hansen et al., 2010; Stien et al., 2010) and population sex ratios (Peeters, 2014). These ROS events likely represent one of the most important facets of ongoing climate change for Arctic terrestrial ecosystems (Thompson et al., 2013). Climate change and anthropogenic activities enhance the rate and extent of biological invasions (Hellmann et al., 2008). These changes affect native communities at an alarming rate by decreasing productivity, altering food web dynamics, and modifying habitat complexity (Cheung et al., 2009; Hoegh-Guldberg and Bruno, 2010). The Arctic region is now under unprecedented threat of biological invasions due to climate warming and increased human activity (Matishov et al., 2011; Miller and Ruiz, 2014; Ricciardi et al., 2017). The reduction in sea-ice content will provide suitable conditions for more southern species and negatively affect native biota. Mollusks and fishes, for example, may spread from the Pacific across the Arctic to the Atlantic Ocean under warmer climate as happened in the mid-Pliocene (Vermeij and Roopnarine, 2008; Wisz et al., 2015). Shorebirds are the most numerous and species-rich taxa of Arctic waterbirds, but some flyway populations, especially those using the East Asian–Australasian Flyway, are declining (Clemens et al., 2016; Piersma et al., 2016) due to shifts in the patterns of seasons.

11.1.2 Why does the Arctic suffer more than Antarctica?

Climate change directly affect the Arctic and Antarctic ecosystems (Post et al., 2009; Turner et al., 2009). The Arctic, a land of both larger and smaller continents, covers approximately 12 million km^2 area out of which roughly 2.5 million km^2 covered with ice and the remaining area is considered as either barren (<50% vegetation cover) or tundra, which covers about 3–5 million km^2 of the area, respectively (Millennium Ecosystem Assessment, 2005). Arctic ecosystems are highly sensitive to changes in climate variables and the duration of specific weather conditions. The main climatic changes in Arctic ecosystems are temperature, precipitation, water availability, nutrient sources, and availability. Increasing temperature leads to decreasing permafrost, which leads to increasing C-degradation and GHG emissions. That's why there is always a change in vegetation type from forest to shrub and tundra, which constitutes the primary ecosystem type in this region. This transition coincides with average June temperatures of around 10°C, which varies latitudinally between the continents due to differences in geography and local weather patterns. Due to local warming in the Arctic, there is an enhanced plant growth that caused a migration of the tree line toward the north (Serreze et al., 2000; Callaghan et al., 2011). According to satellite measurements that began in the late 1970s, the Arctic sea ice has declined by about 12% per decade. Permafrost is one of the most important potential source for carbon loss, due to which the northern circumpolar alone stored about 1672 pg C or roughly twice the amount of C found in the atmosphere (Tarnocai et al., 2009) leading to an increase in temperature. This increase in temperature prompts the more efflux of C from Arctic soils altogether, which would additionally quicken environmental change locally and internationally (Schuur et al., 2009).

The idyllic icy picture of the Arctic northernmost snow caps turned into a land filled with greenery that gets transformed due to the drastic change in the climate. NASA conducted various studies and concluded that the Arctic tundra is turning green in its uppermost regions of Alaska, the Northern coast of Canada, Siberia, and Quebec and Labrador. Such intriguing changes attributed to higher temperatures, warm air along with higher soil moisture and temperature. This sort of landscape change has hazardous climatic ramifications in the long run, and they severely disrupt the regional water cycle and the carbon cycle and energy. Arctic climate change responses thus have far detrimental impacts.

By contrast, continent Antarctica has one of the harshest environments on the Earth. This continent is the driest, coldest, and windiest on planet Earth (Convey, 1996). Only 0.3% of the Antarctic area is ice-free during the summer period, and life in Antarctica is mainly confined to these ice-free areas of coastal outcrops, offshore islands, nunataks, mountain ranges, and oases. In addition, the Antarctic environment also has extreme UV irradiation, owing to a large ozone hole above the continent, long and dark polar winters that last for 6 months at the South Pole. Owing to these factors, the biodiversity of Antarctica remains the least among other continents. The main agents for dispersal of flora and fauna, including potential invasive species are scientists and tourists, to and within Antarctica (Chown et al., 2012). Antarctic biodiversity is more constrained due to climatic conditions, dispersal limitations, and habitable land surface area. Antarctica's maritime provinces have well-developed plant communities in terms of ground cover, where primary productivity is very low. These areas are limited to two species of vascular plants, approximately 111 mosses and 273 lichens (Ochyra et al., 2008; Øvstedal and Smith, 2009).

The Antarctic Treaty System prohibits mining and related activities, direct human-induced impacts in Antarctica. Maritime stations such as Faraday/Vernadsky Station showed an increase in temperature by 0.56°C per decade (Turner et al., 2005; Comiso, 2006). In Greenland the temperature increased by 1.19°C per decade in between 1982 and 2005. It appears that all of the Arctic is warming (Millennium Ecosystem Assessment, 2005) despite some regions showing a cooling trend in the early part of the last century. There is little evidence for any significant long-term increase in temperature over continental Antarctica (Steig et al. 2009), although short-term (decadal) cooling trends were recorded in areas such as the McMurdo Dry Valleys (Doran et al., 2002; Turner et al., 2005). Precipitation and general patterns are not well documented in Antarctica (Turner et al., 2009).

Climate models predict a more directional warming, and increased precipitation, throughout the Arctic region(Alabia et al., 2018; Arctic Council, 1996). In the Arctic, it is noticed that the mean annual temperature increased by 4°C–7°C and precipitation by 5%–10% on average (but up to 35% in some high Arctic regions; ACIA, 2005). All these climatic changes affected the Arctic region more than Antarctica.

11.1.3 What are the scientific evidences of changing climate impacting Arctic biodiversity?

According to IPCC (2014a,b), rising concentrations of GHGs drive widespread changes in the global physical climate and its ecosystems. Ice cover in the Arctic sea shrinks to a low point in mid-September every year. In 2020, it measured just 1.44 million square miles (3.74 million square kilometers)—the second-most lowest value in a long time since satellites started taking estimations. There are several Arctic climate indicators for Arctic climate change. These include:

1. Air temperature
2. Permafrost
3. Carbon cycling
4. Arctic humidification
5. Arctic greening
6. Fires
7. Decrease in terrestrial snow cover
8. Arctic Ocean sea ice
9. Land ice

11.1.3.1 Air temperature

Arctic air temperature changes from 1971 to 2017 indicated warming by 2.7°C at the annual scale; 3.1°C in the cold season (October–May); and 1.8°C in the warm season (June–September). A large number of processes contribute to the amplified Arctic temperature variations (Pithan and Mauritsen, 2014). Thus AA is higher in the winter and smallest in the summer (June–September). Moisture advection (Zhang et al., 2013; Neff et al., 2014) and freezing of sea ice in the Arctic (Markus et al., 2009) are the main contributors for increasing cold season air temperatures, which produce maximum Arctic warming in autumn and winter. Miller et al. (2010) from, palaeo data, concluded that AA was higher between 3 and 4.

11.1.3.2 Permafrost

Permafrost temperatures have been increasing by 2°C–2.5°C for Arctic regions from the past 47 years. The permafrost thaw leads to increased organic sediments into the freshwater ecosystem by hill slope slumping, increasing or decreasing the biodiversity. According to Lund et al. (2014) and Parazoo et al. (2018), reduced permafrost area contributes to amplified warming due to ground latent heat sink reduction. Thawing lead to permafrost collapse severely affected the hydrology, vegetation composition, and biogeochemical cycling of the area (Johansson et al., 2006). Continuous permafrost in upland tundra areas with increasing active layer depths leads to soil drying and limits the vegetation growth (Liljedahl et al., 2016).

11.1.3.3 Carbon cycling

Carbon cycling is influenced by terrestrial carbon exchange and its coupling between the ocean and land. The substantial organic matter stocks of carbon in the Arctic contained in permafrost and presented within the methane hydrates, which exist beneath both subterranean and subsea permafrost of the Arctic affect carbon cycling. During the growing season, CO_2 uptake increased in the tundra ecosystem. Further increased temperature affects tundra CO_2 and CH_4 emissions. Jeong et al. (2018) found accelerating rates of carbon cycling of atmospheric CO_2 measurements from Barrow, Alaska (71.29N, 156.79W) and concluded that temperature dependencies of respiration and carbon uptake increased in the cold season. Hugelius et al. (2014) estimated that Arctic soils contain ~50% of the world's global soil carbon, and hence the potential release is enormous.

11.1.3.4 Arctic humidification

Increasing humidity in the land and coastal regions of the Arctic surface increases downward longwave radiation (Zhang et al., 2001) leading to AA (Pithan and Mauritsen, 2014). Changing precipitation patterns (e.g., shifts from snow to rain), plant composition, and productivity impacted distribution of animals and demographics both directly and indirectly (Schmidt et al., 2015; Kankaanpaa et al., 2018).

11.1.3.5 Arctic greening

It is an overall increase in vegetation cover in the Arctic, which has been observed across tundra ecosystems over the last 30 years (Bhatt et al., 2017). Greening occurred at warm sampling sites because of increased summer temperatures of air and soil, soil moisture. Satellite-derived vegetation indices (often termed the Arctic greening) (Raynolds et al., 2012) reported signs of in situ increase in vegetation height, biomass, cover, and abundance (Bhatt et al., 2017; Guay et al., 2014) associated with warming (Street et al., 2007). Tundra greenness increased (greening) at ~37.3% at sampling sites from 1985 to 2016. As a result of extensive greening, the grassy tundra is getting converted into shrublands, and shrublands are further becoming denser and thicker.

11.1.3.6 Arctic fires

The Zombie Fires in the Arctic region are becoming frequent. It is a fire initiated from a previous growing season. These fires spread slowly under the ground that is made of carbon-rich peat. The fire reignites when the weather warms. These fires in the Arctic are spreading to areas that were formerly fire-resistant. As the north region of the Arctic Circle (the Tundra) is drying up, the lower vegetation in this region catches fire. These wildfires are most common in the permafrost regions of Siberia due to rising temperatures (CAFF, Arctic Biodiversity and the Post, 2020; Duffy Anderson et al., 2017; CBD, 2010). Half of these regions are burnt on ancient carbon-rich peat soils. These fires have the potential of turning a carbon sink into a carbon source, which in turn increases global warming.

11.1.3.7 Decrease in terrestrial snow cover

Snow is a major driver for Arctic ecosystem functioning, affecting the permafrost thaw, surface energy balance, hydrology, plant phenology, and GHG exchange. The timing of snowmelt is key for both increasing CO_2 and CH_4 emissions (Parmentier et al., 2011; Lund et al., 2012) and growing season (Mastepanov et al., 2013; Pirk et al., 2016). Longer snow-free seasons extended the growth period of the plant, enhanced CO_2 uptake, and at the same time increased respiration rate too. Changing snow cover largely impacted ecosystems outside of the growing season. Warm spells in winter remove this protective cover and cause plant damage (Phoenix and Bjerke, 2016). ROS events also lead to complete melting of snow cover and exposed vegetation to cold conditions again. The damage caused by these affects carbon cycling and vegetation growth in the following growing season (Parmentier et al., 2018), which is linked with mass caribou mortality (Tyler, 2010).

11.1.3.8 Arctic Ocean sea ice

Sea-ice extent and volume are continuing their downward trends. The past decade had low-record in summer 2012, and it is the lowest decade ever in the satellite era beginning in the 1970s. These unprecedented changes in Arctic sea ice change in extent, area, thickness, and spatial distribution. Along with Arctic sea ice decline, there is emerging evidence for a loss of biodiversity in sea-ice habitats.

11.1.3.9 Arctic land ice

The Arctic was the largest global source of sea-level rise contributor from the last many years. Increasing temperature is the main cause of decreasing land ice.

11.2 Current status of Arctic biodiversity

According to the literature, around 21,000 plants, animals, and fungal species are reported from the Arctic. A considerable percentage of these species are either endemic to the Arctic or familiar with the boreal zone. But climate-driven range dynamics have left little room for specializing local conditions and speciation on regional spatial scales. Consequently, there are few species with minimal distributions. More than 20 cycles of Pleistocene glaciation forced species to migrate, adapt, or go extinct (Feng and Poulsen, 2016; Kevan et al., 1995; Steig et al., 2019). Many terrestrial species occupied southern refugia during glaciations and recolonized northern areas during interglacials. Ice-free refugia persisted within the Arctic proper; species occupying these refugia diverged in isolation, promoting Arctic diversification. The most significant Arctic refugium was Beringia and adjacent parts of Siberia. The Arctic region is declared a global "hot spot" continuously changing and warming faster than other parts of the Earth (Descamps et al., 2017).

11.2.1 Arctic vegetation

Arctic vegetation is very diverse, including bryophytes, lichens, and vascular plants. Vascular plants in the Arctic are relatively deprived. There are 21 floristic provinces and 5 subzones divisions in the Arctic territory based on species richness and composition. Talbot (1999) reported 2180 species from the southernmost low Arctic subzone E and 102 species from the northernmost high Arctic subzone A. About 5% of the vascular plant flora is endemic to the Arctic. Apart from different species of grasses and forbs, the endemic species of Arctic include the various genera of *Puccinellia* (salt marsh and goosegrass), *Papaver* (poppy), *Draba* (whitlow-grass), *Oxytropis*, and *Potentilla*, excluding woody species. The density of Arctic endemism increases from north to south. Northern Iceland, North Alaska—Yukon Territory, and Western Alaska are supposed to inhabit some invasive species. *Prunus padus* is a naturalized non native species found only in North Alaska–Yukon Territory (Wasowicz et al., 2020). The Arctic has 341 non native vascular plants that belong to 39 families and 180 genera viz. Asteraceae (48 taxa), Poaceae (51 taxa), Brassicaceae (45), *Rumex* (12), *Ranunculus* (7), *Trifolium* (7), *Poa* (8), and *Vicia* (7) (Taylor et al., 2020; Wasowicz et al., 2020). Arctic woody plant species have less extended shoot above the snow surface and creep alongside the ground or form dwarf shrubs (Vowles and Björk, 2019). Dwarf shrubs are reported, especially in the Sub- and Low-Arctic, whereas forbs and graminoids are the dominant vascular plant groups in the High-Arctic (Virtanen et al., 2016). Bryophytes, mosses, and lichens represent one of the most important groups that are less dependent on soil resources than vascular plants because they lack true roots or tissues dedicated to water and nutrients transport (Iversen et al., 2015). Saniewski et al. (2020) reported that *Racomitrium lanuginosum* (Hedw.) Brid. (bryophyte) and *Cetrariella delisei* (Bory ex Schaer.) (lichen) are the best bioindicators for changes due to radioactivity (^{137}Cs) contamination in the Arctic region. Different species of lichens and bryophytes collected by Saniewski et al. (2020) during 1985 and 2017 are listed in Table 11.1. Wietrzyk-Pełka et al. (2020) established 270 plots in a grid in eight glacier forelands, and the mature tundra surrounding them to investigate the mosses and liverworts and 105

Table 11.1 A detailed list of cryptogamic species collected between 1985 and 2017 (Saniewski et al., 2020).

Species	Group	Region
Cetrariella delisei	Lichens	Kaffioyra
Flavocetraria nivalis	Lichens	Bellsund
F. nivalis	Lichens	Bellsund, E Sorkapp Land, Petuniabukta, N-W Sorkapp Land
C. delisei	Lichens	Bellsund, Ny-Ålesund, Petuniabukta
Cetraria islandica	Lichens	Bellsund, N-W Sorkapp Land
Sphaerophorus globosus	Lichens	Bellsund, N-W Sorkapp Land
Sanionia uncinata	Bryophytes	Bellsund, Ny-Ålesund
Ptilidium ciliare	Bryophytes	Bellsund
Racomitrium lanuginosum	Bryophytes	Bellsund, Adventdalen, Kaffioyra, N-W Sorkapp Land
S. uncinata	Bryophytes	N-W Sorkapp Land, N-W Sorkapp Land
P. ciliare	Bryophytes	Kaffioyra, N-W Sorkapp Land

species were recorded, 85 species of mosses and 20 species of liverworts. *Sanionia uncinata* (Hedw.) Loeske and species of the family Polytrichaceae, as well as the genera *Bryum* and *Drepanocladus*, were reported in the Kongsfjorden area (Wietrzyk-Pełka et al., 2020).

Werner et al. (2007), during the Winter Arctic Polynya Study (WARPS) aboard RV (ARK XIX/1 of RV) "Polarstern" expedition from March–April 2003, determined the richness of sea-ice algae and heterotrophic protists during late winter. The team reported a total of 40 different taxa and many unidentified algae within pack ice during late winter. *Bleakeleya* cf. *notata, Haslea wawrikae, Lioloma* sp., *Prorocentrum* cf. *compressum*, and *Korschikoviella/Keratococcus* sp. are reported for the first time from Arctic sea ice. Apart from the utmost diverse Algae diatoms (Bacillariophyceae) microalgal with 24 species, Chrysophytes (*Chrysolykos* sp., *Dinobryon faculiferum*, unidentified) also existed. Kinetoplastida, Choanoflagellate, Ciliata, and heterotrophic nanoflagellates were heterotrophic protists found in the Arctic.

Arctic marine waters inhabitant algae are small, solitary photosynthetic cells and are the primary producers that contain different types of pigments, and reproduce by spore and gametes formation (Payer et al., 2013). They comprise numerous heterogeneous and evolutionarily diverse clusters, including both microalgae and macroalgae, along with cyanobacteria (Adl et al., 2012). Two varieties viz. ice algae and phytoplankton of Arctic microalgae are established. Ice algae are associated with ice crystals present in the interstitial water among crystals or remain attached to the undersurface of the ice crystals (Horner et al., 1988). In contrast, phytoplankton remains suspended in the upper layer of the water column. Poulin et al. (2011) reported 1874 phytoplankton and 1027 ice algae belonging to 2016 taxa of Arctic water. Most temperate or cold-water species inhabit the Arctic but few of them are warm water species, distributed globally (Hasle et al., 1996). There exist differences in the species composition along with different areas of Arctic due to the advection of algal cells by the fluxes in the Arctic (Lizotte, 2001). The composition of the phytoplankton varies seasonally; ice communities are widespread throughout the Arctic (Sukhanova et al., 2009). Solitary diatoms are also found in interstitial communities, whereas the widely held colonial algae are most commonly reported in year-old ice from sub ice communities or from the offshore regions with the exception of *Melosira arctica* (von Quillfeldt et al., 2003). Continuous discharge of river water alters the composition and results in the warming of seawater, which consequently affects the phytoplankton composition, which is adapted to cold waters (Kraberg et al., 2013). Due to climatic warming, the North Pacific planktonic diatom *Neodenticula seminae* reappears in lower Arctic zones (Poulin et al., 2011). Seaweed diversity is maximum in temperate regions of the Arctic, which extend in the Northern Hemisphere from the 9°C to 10°C summer SST isotherm in the north to the 20°C to 23°C winter SST isotherm in the south (Lüning, 1990; Van den Hoek, 1975). Coralline algae (Corallinales, Rhodophyta) are important constituents of calcifying benthic marine groups. *Lithothamnion glaciale*, coralline algae, was collected from northwest Svalbard at 40–50 m depth near Cape Rubin (Büdenbender et al., 2011). Athanasiadis (2008) identified two new species *Leptophytum jenneborgii* and *Leptophytum arcticum*, of the genus *Leptophytum* Adey, which are endemic to NW Spitsbergen. Adey et al. (2008) reanalyzed the origin of Arctic endemic algal flora and evolutionary pattern among the Arctic, Atlantic Sub-Arctic, and North Pacific, through the thermogeographic model and evolutionary trees–based molecular data and proved that 91% of the marine algae occur in the North Pacific, 6% befall in the Boreal Atlantic Ocean, and single species is constrained to the Sub-Arctic. It was concluded that the Great Trans-Arctic Biotic Interchange, which occurs at 3.5 mya, with the opening of the Bering Strait inoculated the Arctic with a North Pacific marine fauna, which was known to be the "center of origin" (Adey et al., 2008). In the Arctic >4350 fungal species exist, comprising 2600 fungi and 1750 lichens. A total of 15% decline in lichen species was reported from low to high Arctic (Daniëls, 1982). Boreal zone hot spot act as a home for- 700 lichens species and 100 lichenicolous fungal species (Elvebakk and Bjerke, 2006), 709 lichens, 94 lichenicolous fungi were reported from Northern Norway (Spribille et al., 2010) and 668 lichens, 98 lichenicolous fungi from Alaska. A total of 143 endemic lichen species are recorded from the Arctic.

11.2.2 Zooplankton

Zooplankton is either microscopic, such as, jellyfish, comb jellies, and medusa or multicellular, such as, copepods, also known as ocean's grass and grazers. Multiple projects from 1975 to 2007 reported around 174 zooplanktonic species from the central Arctic region. It was reported that many species from the North Atlantic and Pacific are introduced to the Arctic due to the influx of river water from the Atlantic and Pacific.

Larvaceans (pelagic tunicates), pteropod mollusks (pelagic snails), and chaetognaths (arrow worms) are important taxonomic groups of Arctic, and crustaceans like copepods and amphipods are endemic species of the Arctic (Gradinger, 1999). From the data collected by in situ observations using remotely operated vehicles and net collections, it was reported that gelatinous megaplankton viz. "true" jellyfishes (medusae) were significant in the central Arctic. Approximately 6% of Arctic zooplankton species were categorized as ice-associated (cryopelagic) by Kosobokova et al. (2011), whereas 39 invertebrate species were recorded as ice-associated by Gardiner and Dick (2010). However, the exact borderline between cryopelagic and cryobenthic is still unclear (Gardiner and Dick, 2010; Kosobokova et al., 2011). Acidification of Arctic water effects a variety of calcifying species, either by interfering with the formation of exoskeletons (e.g., crustaceans and mollusks) or by disturbing the formation of sensory organs (e.g., statoliths of cephalopods and otoliths of fish larvae). *Limacina helicina* is a pteropod species that comprises the aragonitic shell that is highly vulnerable to dissolution in an acidic environment (Hunt et al., 2010).

11.2.3 Benthos

In 2007–2008 Ocean Diversity (ArcOD) project (http://www.arcodiv.org; Bluhm et al., 2011) was conducted by the Census of Marine Life (http://www.coml.org; Snelgrove, 2010) and the International Polar, which contributed to revealing the biodiversity of the benthic flora in the deepsea, coastal region, and shelf of Arctic marine ecosystem (Snelgrove, 2010). From 14 out of the 19 marine Arctic shelf regions, 2636 species, including 668 Annelida, 847 Arthropoda, 228 Echinodermata, 392 Mollusca, and 501 species of other phyla were reported. From numerous parts of the central Arctic (deeper than 500 m), 1125 taxa of benthic species were compiled by Bodil et al. (2011). This inventory was specifically dominated by Arthropoda (366), Annelida (194), Nematoda (140), and Foraminifera (197).

11.2.4 Nekton

There are two groups of Arctic fishes; one group includes marine species, whereas the other group includes anadromous species that specifically drift from saltwater to spawn in freshwater, for example, salmonids and coregonids. Anadromous species occur both along the coast and freshwater. In the Arctic region, 242–633 marine fishes are belonging to 106 families and approximately 18–49 freshwater fishes were recorded (Christiansen et al., 2014; Mecklenburg et al., 2011). A total of 92% of Arctic fish species are bony fishes, whereas only 8% are cartilaginous fishes (Mecklenburg et al., 2013). The decline in fish diversity was recorded extending from Arctic gateways toward the Atlantic (Norwegian and Barents Seas) and Pacific Oceans (Bering and Chukchi Seas) (Christiansen et al., 2013; Coad and Reist, 2004). Cryopelagic species, residing in sea ice, including the circumpolar polar cod *Boreogadus* and the ice cod *Arctogadus*, act as a significant prey species for marine mammals and larger fish. The demersal fauna of the Arctic pseudoabyss represents endemic species (Chernova, 2011). Invertebrate nekton diversity is lesser as compared to fishes in the Arctic. Nesis (2001) reported seven cephalopods, including squid, octopus, and cuttlefish species in the Arctic. Benthic octopods from the Bering and Chukchi Seas are the only poorly known Arctic endemic species (Renaud et al., 2015). Continuous ocean warming may result in the drifting of native species and appearances of warm water species; such shifting can cause drastic changes to the fish community (Hollowed et al., 2013). Mesopelagic fish, for example, the Myctophid (Lanternfish) *Benthosema glaciale* and the sternoptychid (family Sternoptychidae) *Maurolicus muelleri*, is found in southern Newfoundland and Norwegian fjords (Folkvord et al., 2016). These cold-water fishes are abundant in the high seas and likely to move from the sub-Arctic region to the full Arctic.

11.2.5 Arctic birds

Arctic waters act as a host to millions of marine birds during summer, which comes to nest. The maximum breeding density of seabirds is reported from the Northern Hemisphere in Arctic water (Cairns et al., 2008). Ganter et al. (2013) listed 44 species of seabirds, among which 1 marine goose (brant) and 10 sea ducks (eiders and scoters) breed in the Arctic, whereas 23 seabirds species, 7 sea ducks, and the brant, which are endemic to the high Arctic, breed there. A maximum number of Arctic birds (approximately 34 species including ivory gull, little auk, Ross's gull, and Sabine's gull) belong to the order Charadriiformes (Ganter et al., 2013). Smith et al. (2020) summarized the status and trends of Arctic terrestrial birds comprising 88 species, 228 subspecies, or distinct flyway populations, which were recorded across different regions, collected from raw data, derived from published sources or from expert opinion. Arctic terrestrial birds are ruled by insectivores, encompassing 41 of the 88 species (47%) (Smith et al., 2020). Lapland longspur (*Calcarius lapponicus*) is the most abundant Arctic terrestrial bird with a global population size of approximately 130 M individuals (Flight, 2017). Arctic tundra is dominated by ptarmigan and tundra swans (*Cygnus columbianus*), a herbivorous bird. Willow and rock ptarmigan (*Lagopus lagopus*, *Lagopus mutus*) are also reported from the Arctic (Cotter, 1999). Arctic land birds including the snowy owl (*Bubo scandiacus*), the raven (*Corvus corax*), the gyrfalcon (*Falco rusticolus*), and the Arctic redpoll (*Acanthis hornemanni*) along with many Arctic marine bird including the black guillemot (*Cepphus grylle*), ivory gull (*Pagophila eburnean*), thick-billed murre (*Uria lomvia*), and Ross's gull (*Rhodostethia rosea*) sustain over extensive Arctic land throughout the winter (Smith et al., 2020), whereas Arctic shorebirds (93% of species) migrate in winter to boreal peripheral, temperate regions (waterfowl, birds of prey, owls, auks, passerines, and gulls), tropics (Sabine's gull *Xema sabini*), Antarctic or sub-Antarctic regions (skuas *Stercorarius* spp.) (Barry et al., 2013a,b). Therefore migratory birds connect the Arctic to all other parts of the world and are affected, either positively or negatively, by events happening exterior to the Arctic, also known as extra-Arctic effects. Hence the conservation of Arctic birds, especially species facing extinction (Siberian crane *Leucogeranus leucogeranus* or the spoon-billed sandpiper *Eurynorhynchus pygmeus*) requires international cooperation. There are approximately 200 breeding bird species in the Arctic, which constitute 2% of the worldwide avian biodiversity. The majority of Arctic avian biodiversity is contributed by Charadriiformes (76 spp.) (shorebirds, gulls, and auks) and Anseriformes (waterfowl) (32 spp.) (Anderson and Wong, 1992). The ongoing scenario of global climatic changes disproportionately affects the avian population at high latitudes (Change, 2007). A decline in the community of Arctic breeding red knot *Calidris canutus* ssp. *rufa* occurs due to changes in feeding environments in Delaware Bay, United States (List and Concern, 2014). Eskimo curlew *Numenius borealis* becomes extinct due to excessive hunting on its migration ranges in Eastern North America (Butchart et al., 2006).

11.2.6 Mammals

Mammalian diversity is very low in the Arctic, approximately 2% of global mammalian diversity, which includes 35 species of marine and 67 species of terrestrial mammals, was reported from this biome (Assessment, 2013). Arctic waters are the inhabitant of 35 species of marine mammals, out of which 7 species, including beluga, narwhal, ringed seal, bowhead whale, bearded seal, polar bear, and walrus, inhabit the Arctic whole year-round and use sea ice for a few months. Hooded seal, spotted seal, harp seal, and ribbon seal reside in open waters of the Arctic and sub-Arctic year-round but use sea ice for pupping during the winter and spring (Jørgensen et al., 2016). Twenty-four species of marine mammal, including nineteen species of cetaceans, four species of pinnipeds, the sea otter, and a carnivore, are reported from the low Arctic waters, which migrate seasonally to the Arctic for feed. Tannerfeldt and Angerbjörn (1998) reported that in Fennoscandia, the historic Arctic fox population was approximately 10,000–20,000 during the late 1800s and early 1900s. The population suffers a drastic decline due to overharvesting for its costly fur (Hamel et al., 2013; Tannerfeldt and Angerbjörn, 1998). At the beginning of the early 20th century, the Scandinavia Arctic fox (*Alopex lagopus*) is categorized as critically endangered due to a drastic decline in population size, which is hypothesized to be the consequence of increased interspecific competition or variations in prey subtleties. *A. lagopus* has failed to recover naturally, even after the protection of more than 65 years (Dalén et al., 2006).

11.2.7 Conservation status

Climatic change is a major threat to Arctic biodiversity. Arctic temperature is increasing at an alarming rate over the last two decades, instigating widespread deviations in its wildlife. Therefore to ensure the conservation of Arctic Biodiversity, the Arctic Council (AC) was established in 1996 as a product of the Ottawa Declaration. AC is an intergovernmental forum stimulating interaction, collaboration, and harmonization among the Arctic States, Arctic indigenous communities, and other Arctic inhabitants (Kuersten, 2016). It comprises eight member states viz. the Kingdom of Denmark (including Greenland and the Faroe Islands), Canada, Iceland, Finland, Norway, Sweden, Russia, and the United States and six indigenous organizations recognized as Permanent Participants (Arctic Athabaskan Council, Aleut International Association, Inuit Circumpolar Council, Gwich'in Council International, Saami Council (Council, 1996). Majority of the Council's work is commenced by six working groups.

1. The Arctic Contaminants Action Program (ACAP)
2. Arctic Monitoring and Assessment Programme (AMAP)

3. Conservation of Arctic Flora and Fauna (CAFF)
4. Emergency Prevention, Preparedness and Response (EPPR)
5. Protection of the Arctic Marine Environment (PAME) and
6. The Sustainable Development Working Group (SDWG)

CAFF is a subsidiary body of the AC designed to mandate the conservation of Arctic biodiversity and to share its verdicts with the governments and tenants of the Arctic, thereby assisting to encourage practices that safeguard the sustainability of the Arctic's biodiversity (Barry et al., 2020). The Circumpolar Biodiversity Monitoring Program (CBMP) endeavors for the conservation of Arctic's biological diversity, significantly reduces its loss, and aids in the sustainable use of the Arctic's living stocks by the indigenous people of the Arctic and stakeholders inside and outside the Arctic (Logan et al., 2018).

The Arctic is exceptional in biological, physical, and chemical properties with extreme conditions for survival and maximum genetic variations among its species. The impact of environmental changes, explicitly warming, is more significant and earlier than other parts of the globe. Consequently, Arctic biodiversity faces vulnerabilities (Grebmeier et al., 2010; Overland, 2009; Stevenson and Lauth, 2019). For example, Arctic seabirds nest on land, but they depend on the ocean, lakes, and rivers for feed. Many Arctic fishes are anadromous, viz., Arctic Char, Salmon, etc. Seals inhabit in and on the sea ice but hunt in the ocean. Arctic migratory birds migrate to every part of the world, excluding the inland of Antarctica. Therefore the conservation of Arctic breeding birds all the way through their migratory arrays is a global challenge as it covers almost all of the world's major marine and terrestrial ecosystems, requiring extraordinary international cooperation, which can be accomplished through the CBMP. Apart from migrating birds, many other species belonging to the land and marine ecosystem migrate to the Arctic, hunting for food resources. Numerous species of whales survive on the Arctic's nutrient-rich cold water. There exist major knowledge gaps and fragmented data regarding the status and trends of Arctic biodiversity, which contribute to the complex food webs of the Arctic. Many species (taxa) lack significant data preventing passable classification. Fragmentary data specify a circumpolar decline in certain migratory species, for example, some seabirds and shorebirds, whereas certain species are recovering, for example, whales and some geese from previous over-exploitation intensities. The number of globally threatened Arctic species or taxa has increased in recent decades; presently, the list comprises 73 vascular plant species, 43 mammals, 12 fishes, and 16 birds. Invasion of alien species is another major concern, as they might influence indigenous flora and fauna. Hence as a consequence, Arctic biodiversity monitoring is crucial for maintaining the virgin state of the Arctic and to certify a sustainable environment for the residents of the Arctic (Poloczanska et al., 2013; Smith et al., 2018; Svendsen, 2002). Monitoring is the solitary approach to assess the continuously ongoing natural and anthropogenic changes in the Arctic and diminish the adverse impacts (Vongraven et al., 2009). In 2000 the Arctic Climate Impact Assessment (ACIA) was adopted at the Ministerial Conference of the AC at Point Barrow, Alaska, reporting Arctic warming, its consequences, the shift in vegetation zones, species' diversity, distributions, and reducing sea ice (Petersen et al., 2004). Biodiversity indicators are chosen to monitor progress toward enhanced conservation and reduce substantial losses of biodiversity (Barry et al., 2013a,b). Arctic biodiversity indicators are adopted from the CBD, CBMP (Gill, 2008):

- Arctic Species Trend Index (ASTI)
- Arctic Red List Index
- Arctic Land Cover Change Index
- Arctic Habitat Fragmentation Index and
- Arctic Human Well-Being Index

11.3 Threats to the Arctic biodiversity

The Arctic faces some of the very harsh environmental conditions such as extreme cold, drought, strong winds, short growing season, and extended darkness. Due to these extreme conditions, the Arctic has highly specialized organisms, some of them are endemic that evolved to adapt and survive the extreme conditions. Even though the Arctic region is remotely placed, its ecosystem and species are facing threats from within and outside of northern latitudes, like anthropogenic disturbances, climate change, overexploitation of endemic and migratory species, landscape alteration, habitat loss and fragmentation, resource extraction, and the shifting distributions of prey and pathogens (Christensen et al., 2020).

11.3.1 Climate change

The most severe threat is climate change that is affecting not just the Arctic but the whole world. The effects of climate change are visible all over the world, but they are more pronounced in high latitudes (ACIA 2005; IPCC 2018). It is predicted that the Arctic may face warming twice that of the world average (Collins et al., 2013; Smith et al., 2020). Already the climate change has caused serious changes to the Arctic biodiversity. As soon as the breeding season comes to an end, the Arctic breeding birds fly from boreal to tropic habitats, some of them even fly to the Antarctic. During their journey, the birds face many challenges such as anthropogenic pressures and loss of habitat (Studds et al., 2017; Smith et al., 2020).

11.3.2 Effect on freshwater ecosystems

The Arctic region also supports a wide range of freshwater ecosystems harboring diverse organisms. These ecosystems are also facing serious threats due to climate change. According to a report, these freshwater ecosystems are undergoing some very serious changes (Lento et al., 2019a, b; Heino et al., 2020) and due to these changes, the opportunities to gather information about these ecosystems are rapidly disappearing. In the Arctic, these environmental changes are mainly anthropogenic such as increase in global warming (Heino et al., 2009, 2020), long-range transport of pollutants, eutrophication, brownification, and land-use alterations, and these activities are increasing in the Arctic (Hayden et al., 2019; Heino et al., 2020). In the past few years, the Arctic temperature has been increasing (https://www.ncdc.noaa.gov/sotc/), and in 2019, record-breaking summer was observed across the Arctic, Greenland, Canada, Alaska, and some parts of Siberia. This increase in temperature led to numerous wildfires and unprecedented melting of the ice sheets (http://nsidc.org/greenland-today/). Another problem emerging in the Arctic is the permafrost thaw; this occurs from increasing temperature and precipitation (Kokelj et al., 2015; Heino et al., 2020). This will have many consequences that include jeopardizing infrastructures across the Arctic (Hjort et al., 2018; Heino et al., 2020) as well as higher loads of sediment and solutes affecting water quality. Global warming severely impacted the abundance and distribution of species present in freshwater (Heino et al., 2009, 2020) and living conditions of indigenous peoples (Ford and Pearce, 2010; Heino et al., 2020). Due to permafrost thaw drainage (Smith et al., 2005; Heino et al., 2020), evaporation has increased causing desiccation of ponds (Smol and Douglas, 2007a; Heino et al., 2020), which led to the disappearance of aquatic ecosystems.

11.3.3 Permafrost thaw

The permafrost thaw leads to increased organic sediments into the freshwater ecosystem by hill slope slumping and thus can increase or decrease the biodiversity. For example, thaw slumps could introduce substantial amounts of sediments to the river causing a significant decline in the benthic macro invertebrate abundance and periphyton biomass. The chemistry and biota of lakes can also be affected by increasing the clarity of water and productivity in these systems. The increasing concentration of nutrients in freshwater ecosystems due to thawing permafrost or anthropogenic activities poses an ecosystem-level threat in many regions of the Arctic. Decreased ice cover and increased nutrient enrichment resulted in the increased benthic primary productivity in rivers and shifting from benthic to pelagic primary production in lakes.

11.3.4 Anthropogenic activities

The Arctic is also facing increased land-use pressures from various anthropogenic activities such as sand extraction from river banks, increased road construction, hydropower generation, mineral extraction, and oil and gas exploration. With the increase in global warming, these activities will also increase in the Arctic due to the accessibility of remote locations. Global warming is caused by the amount of carbon dioxide (CO_2) in the atmosphere while at the same time, the dissolution of CO_2 in the upper ocean causes ocean acidification (IPCC, 2014c; Hansel et al., 2020). Along with overfishing, the ocean acidification has been known to act as a potential major stressor to the subsidiary fisheries and marine fish (Fossheim et al., 2015; Robinson et al., 2015; Kleisner et al., 2017; Hansel et al., 2020). Global warming also causes ocean warming, which also affects the distribution of species and their vital rates such as growth and mortality (Bjornsson et al., 2007; Hansen et al., 2010). Global warming interacts with land-use alterations like an expansion of agriculture in less remote regions of the Arctic, and more and more shrubs and trees in the tundra landscape. The terrestrial and aquatic ecosystems are closely linked with each other, so any change in the land use and land cover will seriously affect the freshwater environment.

The Arctic freshwater ecosystem is more productive due to greater macrophyte cover and increased cyanobacterial and algal blooms. The increased number of primary producers has a concomitant effect on the consumers and predators, which range from zooplankton and benthic macroinvertebrates to fish and waterfowl. With these changes, the ecosystem will also reorganize, thereby reassembling the biological communities in the Arctic freshwaters. Areas other than the Arctic are also having high primary production over the past few decades, due to extended growing seasons and warmer conditions.

11.3.5 Brownification

Another major environmental change that is occurring in the Arctic freshwaters is the brownification. Brownification is caused by increased concentrations of dissolved organic matter. It is partly a result of the combined effects of landscape alterations and global warming in high latitudes. Due to brownification, the water transparency decreases, ecosystem processes and food webs are altered, and ultimately affecting fish production(Richter-Menge et al., 2019; Steig et al., 2009). This is a serious concern as the high Arctic freshwaters have clear water.

11.3.6 Diversity and distribution of species

Temperatures in the Arctic have increased by more than double the global average over 20 years (IPCC, 2019; Taylor et al., 2020). Due to this increase in temperature, the diversity and distribution of animals and plants are expected to shift and ultimately the interaction among the species (Callaghan et al., 2004; Meltofte, 2013; Taylor et al., 2020). A total of 7 out of the 14 muscid fly species declined in the Arctic (Loboda et al., 2018; Taylor et al., 2020), and some of these species are important pollinators (Tiusanen et al., 2016; Taylor et al., 2020), and their larvae may have some role in the freshwater systems (Michelsen, 2015; Taylor et al., 2020). Some major decomposer invertebrates also declined in some habitats between 2007 and 2017 in Greenland (Callaghan et al., 2004; Taylor et al., 2020). There was a significant decrease in the abundance of spider species from 1996 to 2014 in Zackenberg. The decline of these species can be attributed to the climate change in the Arctic (Bowden et al., 2018; Gillespie et al., 2020; Taylor et al., 2020). The invertebrate groups that were thought to form the diet of birds and mammals also declined in Zackenberg from 1996 to 2016 (Koltz et al., 2018; Taylor et al., 2020).

11.3.7 Sea ice

Another threat is the loss of sea ice, which has some serious ecological implications (Post et al., 2013; Stroeve et al., 2007; Peeters et al., 2019).

Sea ice acts as a corridor for the dispersal across water bodies in polar bears (Laidre et al., 2018), Arctic fox (Geffen et al., 2007; Norén et al., 2011), caribou (Jenkins et al., 2016; Leblond et al., 2016; Miller et al., 2005), and vascular plants also (Alsos et al., 2016). Sea-ice loss is attributed to increased temperature and AA (Serreze et al., 2009) so the increase in isolation of Arctic island populations is foreseen (Frankham, 1997, 1998). The island population lacks genetic diversity and shows high inbreeding, and due to inbreeding, they are prone to loss of genetic diversity either through genetic drift or extinction risk following environmental perturbations (Frankham, 1998; Lande, 1998; Saccheri et al., 1998). Therefore dispersal and recolonization by sea ice are crucial for the Arctic island populations (Hanski, 1998; Post et al., 2013; Vuilleumier et al., 2010).

11.3.8 Oil and gas exploration

Due to increased economic and political interest to exploit oil and gas reserves, the Arctic is considered as the next frontier for oil and gas exploration (AMAP, 2010; Borgerson, 2013; Trump et al., 2018; Wegeberg et al., 2019). The climate change has receded the sea-ice extent and prolonged the ice-free period, thus opening new routes for maritime traffic in the Arctic (AMAP, 2010; Smith and Stephenson, 2013; Sulistiyono et al., 2015). But the harsh Arctic conditions and increased traffic also increase the threat of shipping accidents and oil spills into the marine environment (Fingas and Hollebone, 2003; Arctic Council, 2009). The aquatic species that are endemic to the marine ecosystem may decline due to the oil spills.

11.4 Arctic and India—historical background

An eminent Indian scholar, Tilak, wrote a book titled *The Arctic Home in the Vedas* in 1903, where he mentioned India's ancestral links with the Arctic. According to him, the Aryans originated from the Arctic region, traveled southwards, and later divided into two branches. One branch went to mainland Europe while the other branch came to India (Sinha and Gupta, 2014). On February 9, 1920, under the British Empire, India became part of the Spitsbergen Treaty, commonly called the Svalbard Treaty that described Norway's supremacy over the Arctic archipelago of Svalbard (Sakhuja, 2013; Pronina et al., 2020). After accession to the independence and development of national capitalism, India proceeded its activity in the region. In 1981 the Ministry of Ocean organized the first expedition to the Arctic. The Svalbard archipelago constitutes a total landmass of 61,020 km^2, including numerous islands between mainland Norway and the North Pole. Only Spitsbergen, the largest among all islands, was always inhabited. Around 60% of Svalbard is covered by ice year-round, and less than 10% supports vegetation (Grydehøj et al., 2012; Thomas, 2018).

11.4.1 Current scenario

India is the most recent Asian country to participate in Arctic research but has more than 30 years experience with Antarctica. Now, India is actively taking an interest in development of the Arctic region. Although the country has not articulated an official "Arctic policy," the Ministry of External Affairs released a short document in June 2013 outlining India's interests. India's Arctic research is mainly focused on atmospheric, marine, and glaciological science, and there is a growing domestic interest in governance issues (Stensdal, 2012). The Arctic approach of India is based on a product of three factors—environmental protection, economy, and policy (Lunev, 2012; Sinha, 2019). The country's interest in the Arctic is also directed toward the research on flora and fauna of various forms of life in the Arctic region. India launched its Arctic research program and mounted its first scientific expedition to the Arctic Ocean in 2007 with a prime focus on climate change (Lackenbauer, 2013). India inaugurated its first scientific research station, "Himadri" on July 1, 2008 at the International Research Base in Ny-Ålesund in Svalbard to conduct work in glaciology, atmospheric sciences, biochemistry, geological mapping, and paleoclimatology. This functions under the guidance of NCAOR (National Centre for the Antarctic and Ocean Research, Goa), under the Ministry of Earth Sciences (MoEs) (Stensdal, 2012; Chaturvedi, 2013; Sinha, 2019). The ministry was established in 2006, when the Ministry of Ocean Development was merged with IMD (India Meteorological Department), the NCMRWF (National Centre for Medium-Range Weather Forecasting), IITM (Indian Institute of Tropical Meteorology), and the EREC (Earth Risk Evaluation Centre) (Gewelt, 2016). It focuses on Earth System Sciences for better management on weather forecasting, monsoon, climate change, including disasters such as cyclones, earthquakes, and tsunamis, research on polar regions, seas around India, and develop technology for exploration and sustainable utilization of ocean resources (Nayak, 2008; Pronina et al., 2020). In July 2018 National Centre for Antarctic and Ocean Research (NCAOR) was renamed as the National Centre for Polar and Oceanic Research (NCPOR) (Nanda and Council, 2019).

The AC, a high-level multilateral forum was established in 1996 through the Ottawa Declaration, offers a valuable platform for discussions on Arctic affairs. Eight countries signed it: Norway, the United States, Sweden, Russia, Denmark, Iceland, Canada, and Finland, were called the "member states" in the AC. In May 2013 at the Kiruna Ministerial Meeting of the AC, India was given observer status in the AC and four other Asian countries: China, Japan, Singapore, and South Korea. India is also a part of the International Arctic Science Committee (IASC), a non-governmental organization that facilitates and advocates cooperative research between all countries active in Arctic research (Gewelt, 2016). India made an agreement with the Norwegian Polar Research Institute for scientific cooperation and a Norwegian state-owned company for logistical support and maintenance at the research base. The country also joined the Council of the IASC in 2012 and has committed more than $12 million (US) to Arctic research (Lackenbauer, 2017). The main objectives behind India's interest in the Arctic are the following:

1. Environmental protection is a key factor behind the involvement of India in the Arctic (Nanda and Council, 2019). One of the prime objectives is to study the dynamics of Arctic glaciers with the focus on its effect on sea-level changes, then to compare the rate of changes in Arctic ice with that of Himalayan glaciers (Shaumyan and Zhuravel, 2016). The Indian scientists use the information on Arctic ice melting to monitor the biogeochemical aspects of the Himalayan ice, that is, its impact on the hydrology, ecology, and climate. Arctic ice melting affects the condition of the ice at the Tibetan Plateau, where major rivers originate in India. Any changes in ice conditions on top of the Himalayan Ridge can cause severe floods and cold result in disastrous consequences for the entire Indo-Ganges valley. The increase in sea level due to the melting of glaciers can lead to flooding of a considerable part of India with more than 100 million people. Also, the glaciers of the Himalaya provide the source for the major Asian rivers, including the Mekong and the Brahmaputra, and the vast population is dependent on them for agricultural activities. Therefore monitoring Arctic ice melting is one of the priorities for Indian scientists working in the Arctic.

2. To study the teleconnections between the Arctic climate and the Indian monsoon by evaluating the ice core and sediment records from the Arctic glaciers and the Arctic Ocean. The teleconnection between the tropical Indian Ocean and the Arctic is critical for India's agricultural sector. Agriculture accounts for about 15% of Indian Gross Domestic Product and thus reduction in agricultural productivity due to monsoon precipitation directly influences the Indian economy (Lackenbauer, 2017; Sinha, 2019; Pronina et al., 2020).
3. By using satellite data to characterize sea ice in the Arctic and estimate the effect of global warming in the northern polar region. The melting of Arctic ice and increasing of freshwater in the region prevent heat from escaping, resulting in warmer waters to the Indian Ocean, which alters Indian monsoons. This directly or indirectly affects agriculture as well as the population (Nanda and Council, 2019).
4. To carry out a comprehensive assessment of flora and fauna of the Arctic vis-à-vis their response to anthropogenic activities. Besides, it is proposed to undertake a comparative study of the life forms from both the Polar regions. The Arctic is believed to hold 13% of the world's undiscovered oil and up to 30% of the world's undiscovered natural gas supplies. Studies indicate that the Arctic could hold the last remaining undiscovered hydrocarbon resources on Earth. India has also evinced interest in the Arctic fossil reserve, and the largest Indian oil company, the OVL (Oil and Natural Gas, Videsh Limited), has approached Rosneft, a primary state-run Russian energy company, for exploration in the Yamal Peninsula in the Russian portion of the Arctic (Bhagwat, 2020; Gadihoke, 2012).

11.4.2 Svalbard Global Seed Vault

The Svalbard Global Seed Vault (SGSV) is commonly known as "Global Seed Vault" or "Doomsday Vault." The story of SGSV was started in 1984 when the Nordic gene bank put a backup of Nordic seed collection in a coal mine in Svalbard. The Seed Vault was opened on February 26, 2008 to provide the safety net for the international conservation system of plant genetic resources (PGR), managed in partnership by the Government of Norway, the Nordic Genetic Resources Centre (NordGen), and Global Crop Diversity Trust (GCDT) (Reddy, 2017; Steffensen et al., 2019). It has deep permafrost, securing seeds at very low temperatures (Asdal and Guarino, 2018). The Seed Vault also provides insurance against the loss of seeds in gene banks, as well as a safe haven for seeds during extreme regional or global crises. The Seed Vault is owned by the Norwegian government and managed through an agreement between the Norwegian Ministry for Agriculture and Food, the Global Crop Diversity Trust, and the Nordic Genetic Resource Centre (Gruber, 2017; Reddy, 2017). According to Food and Agriculture Organization of the United Nations (FAO), the four largest gene banks at the national level are—(1) National Centre for Genetic Resources Preservation (NCGRP) in the United States of America, (2) Institute of Crop Germplasm Resources, Chinese Academy of Agricultural Sciences (ICGR-CAAS) in China, (3) Indian Council of Agricultural Research, National Bureau of Plant Genetic Resources (ICAR–NBPGR) in India, and (4) N.I. Vavilov All Russian Scientific Research Institute of Plant Industry (VIR) in the Russian Federation (Jacob et al., 2015). Currently, in the SGSV database (https://seedvault.nordgen.org/), a total of 1,060,073 accessions were deposited by 88 depositors across the world.

The Seed Vault depositor agreement was signed between NBPGR and the Norwegian Ministry of Agriculture and Food in 2008. The first time 25 accessions of pigeon pea (*Cajanus cajan*) were deposited in SGSV on April 9, 2014, and later some deposits in 2017 by ICAR–NBPGR, India. To date, from ICAR–NBPGR, 225 seed samples belonging to three taxa were deposited in SGSV, which includes pigeon pea (*C. cajan*), rice (*Oryza sativa*), and sorghum (*Sorghum bicolor*) (Steffensen et al., 2019). India plans to deposit duplicates of 26,000 ICAR–NBPGR accessions registered in the Multilateral System (MLS) of the Plant Treaty in the Seed Vault shortly (Steffensen et al., 2019). The primary aim of SGSV is to conserve copies of all unique plant genetic material from different gene banks. Furthermore, the deposited accessions are guaranteed by the Government of Norway.

11.4.3 Ways to save Arctic biodiversity

Climate change is the most dangerous threat to the biodiversity of the Arctic and exacerbates all other threats. Changing patterns of high temperatures, precipitation, and winds are likely to give rise to very different climates in the Arctic. Arctic summer sea-ice cover and the amount of multiyear ice are diminishing at an accelerating rate. Especially in colder Arctic waters, the concentrations of carbon dioxide increased in the atmosphere, leading to ocean acidification worldwide. Warming also resulted in the loss of permafrost and glaciers, affecting hydrology, vegetation, erosion patterns, and other features of terrestrial ecosystems. Anthropogenic activities increased in the Arctic due to improved access, rising global demand for resources. Oil spills increased risks from pollution as the development of the Arctic proceeds. More invasive species reached the Arctic as more ships traveled north and more roads built. These activities lead to habitat degradation. And more activity may mean more people, who may increase fishing and hunting pressures. Individually, each of these challenges places stress on Arctic biodiversity. Together, they create a web of stresses on Arctic biodiversity. So, biodiversity must be conserved holistically in the Arctic region.

11.4.4 Role of Arctic Council

To address the conservation challenges that face Arctic biodiversity now and in the decades to come, international cooperation is completely needed. In addition to AC initiatives underway, there are other conventions such as Aichi Biodiversity that is under the United Nations Convention on Biological Diversity (CBD) also acted for the conservation of biodiversity (Smith et al., 2015). This convention included many regulatory and nonregulatory measures in place or under development to provide consistent standards and approaches for developments in the Arctic.

CAFF working group of the AC—The CBMP was formed to address the pan-Arctic need for relevant, integrated biodiversity monitoring information. The CBMP is an international network of scientists, managers, conservation organizations, government agencies, and Arctic community experts and leaders that collaborate to develop and implement long-term and comprehensive plans for monitoring the integrity of Arctic biomes: marine (Gill et al., 2011), freshwater (Culp et al., 2012), terrestrial (Christiansen et al., 2013), and coastal (Jones et al., 2019). The AC provides a framework for collecting of data, harmonizing this region from time to time to determine the trends and status of Arctic biodiversity. It supports ecosystems and processes in order to identify the priorities and knowledge gaps for monitoring and provide timely information to managers, decision-makers, and communities within the Arctic and globally. The Council was established as a forum for sustainable development and environmental protection and addressed several issues: transboundary pollution, environmental degradation, Arctic flora and fauna, and climate change. The main outputs of the AC included scientific monitoring and assessments, technical recommendations and guidelines via its working groups. CAFF issued a marine biodiversity status report in 2017 and a freshwater report in 2019 (Lento et al., 2019a,b). Furthermore, the culmination of CAFF as a forum for scientific cooperation and knowledge generation came with the release of the ABA at the May 2013 Arctic Council Ministerial Meeting in Kiruna, Sweden (CAFF, 2013). It is an ambitious exercise covering all life forms in the Arctic—from microorganisms, plants and insects to birds and mammals, at sea and on land. It provides a comprehensive description of the status and trends of Arctic biodiversity and describes stressors, knowledge gaps, and conservation and research priorities. In addition to mapping out the status and trends of species and ecosystem, ABA analyzes the direct causes of biodiversity loss.

Many NGOs that have influenced the preparation and follow-up of the CBD and other biodiversity-related conventions, two organizations stand out: The International Union for the Conservation of Nature (IUCN) and the World Wildlife Fund (WWF). These are the two leading organizations in relation to the AC and Arctic biodiversity(Grebmeier et al., 2006b). The IUCN and CAFF have established cooperation with the CAFF flora expert group and served as an IUCN Arctic Plants specialist group. Moreover, the IUCN plays a special role in the "Agreement on the Conservation of Polar Bears and Their Habitat" signed in 1973, by the five range countries: Canada, Denmark (Greenland), Norway (Svalbard), the United States (Alaska), and the USSR. The IUCN Polar Bear Specialist Group, part of the IUCN Species Survival Commission, acts as a de facto scientific and technical body for the agreement. This group was established in 1968, following the First International Scientific Meeting on the Polar Bear in 1965 (Durner et al., 2018). The IUCN has identified globally significant marine sites in the Arctic that warrant protection and might qualify for World Heritage status (Speer et al., 2017). It has also been involved in a project assessing the number of microplastics and their impact on the Arctic Ocean (IUCN; Microplastics: A global disaster in the Arctic Ocean). On the other hand, WWF has had a Global Arctic Programme since 1992 and has offices in all Arctic countries except Iceland. It provides the scientific input to the AC. The WWF has had a more policy-oriented Arctic profile than IUCN and provides recommendations, suggestions for AC chairmanship programs to strengthen conservation and sustainable development aspects (WWF–Global Arctic Programme, 2016). It also produces *The Circle*—an Arctic magazine published four times a year each covering a theme relevant to the AC agenda.

The Circle no. 4, 2018, is especially devoted to Arctic biodiversity. WWF assessed the six areas of the AC's work important for the state of conservation in the Arctic: Conservation areas, Biodiversity, Shipping, Cooperation on Oil Spill Prevention, Preparedness and Response, Black Carbon and Adaptation, and Ecosystem-Based Management. More than the IUCN, the WWF exercised policy advocacy, mainly through its Arctic Council Conservation Scorecard on the performance of Arctic states and AC governance arrangements.

11.4.5 Role of the UN in response to changing climate

The United Nations 13th goal is on climate action. This goal states: take urgent action to combat climate change and its impact. As countries move toward rebuilding their economies after Covid-19 recovery, recovery plans can shape the 21st-century economy in ways that are green, clean, safe, healthy, and more resilient. Keeping in this view, the UN Secretary-General has proposed six climate-positive actions for governments to take:

1. Green transition: Investments must accelerate the decarbonization of all aspects of our economy.
2. Green jobs and sustainable and inclusive growth.
3. Green economy: Making societies and people more resilient through a transition that is fair to all and leaves no one behind.
4. Invest in sustainable solutions: Fossil fuel subsidies must end, and polluters must pay for their pollution.
5. Confront all climate risks.
6. Cooperation—No country can succeed alone.

Acknowledgments

KCS gratefully acknowledges Council of Scientific and Industrial Research (CSIR), New Delhi, India for the financial support toward PhD. AR acknowledges the Indian Council of Medical Research (ICMR), New Delhi, India for the SRF. FB acknowledges grant-in-aid from SERB-Core research Grant (CRG/2019/005499).

References

ACIA, 2005. Arctic climate impact assessment report. Cambridge University Press, Cambridge Arctic. Nat. Clim. Change 5, 673–677.

Adey, W.H., Lindstrom, S.C., Hommersand, M.H., Müller, K.M., 2008. The Biogeographic Origin of Arctic Endemic Seaweeds: a Thermogeographic View 1. J. Phycol. 44 (6), 1384–1394.

Adl, S.M., Simpson, A.G., Lane, C.E., Lukeš, J., Bass, D., Bowser, S.S., et al., 2012. The revised classification of eukaryotes. J. Eukaryot. Microbiol. 59 (5), 429–514.
Alabia, I.D., Molinos, J.G., Saitoh, S.I., Hirawake, T, Mueter, F.J., 2018. Distribution shifts of marine taxa in the Pacific Arctic under contemporary climate changes. Divers. Distrib. 24, 1583–1597.
Alsos, I.G., Ehrich, D., Seidenkrantz, M.-S., Bennike, O., Kirchhefer, A.J., Geirsdottir, A., 2016. The role of sea ice for vascular plant dispersal in the Arctic. Biol. Lett. 12 (9), 20160264. Available from: https://doi.org/10.1098/rsbl.2016.0264.
AMAP, 2010. Assessment 2007: Oil and Gas Activities in the Arctic – Effects and Potential Effects 1 Arctic Monitoring and Assessment Programme (AMAP), Oslo, vii + 423 pp.
Anderson, R., Wong, P., 1992. Western Palaearctic and Ethiopian species of Skrjabinoclava (Nematoda: Acuarioidea) in Icelandic shorebirds (Aves: Charadriiformes) en route to breed in the New World and Greenland. Can. J. Zool. 70 (10), 1861–1877.
Arctic Council, 1996. Ottawa Declaration, Arctic Council Secretariat. Available online: http://hdl.handle.net/11374/85 (accessed 01.05.20.).
Arctic Council, 2009. Arctic Marine Shipping Assessment 2009 Report. Second Printing.
Asdal, Å., Guarino, L., 2018. The Svalbard Global Seed Vault: 10 years—1 million samples. Biopreserv. Biobank. 16 (5), 391–392.
Assessment, C.A.B., 2013. Status and Trends in Arctic Biodiversity. Conservation of Arctic Flora and Fauna, Akureyri.
Athanasiadis, A., 2008. Two endemic coralline red algae on the northern limits of the latest glaciation. Taxon 57 (1), 223–230.
Barry, T., Berteaux, D., Bültmann, H., Christiansen, J.S., Cook, J.A., Dahlberg, A., et al., 2013a. Arctic Biodiversity Assessment 2013, Conservation of Arctic Flora and Fauna (CAFF).
Barry, T., Christensen, T., Payne, J., Gill, M., 2013b. Circumpolar Biodiversity Monitoring Program Strategic Plan, 2013-2017: Phase II Implementation of the CBMP. CAFF Monitoring Series Report Nr. 8.
Barry, T., Daviðsdóttir, B., Einarsson, N., Young, O.R., 2020. How does the Arctic Council support conservation of Arctic biodiversity? Sustainability 12 (12), 5042.
Berge, J., Johnsen, G., Nilsen, F., et al., 2005. Ocean temperature oscillations enable reappearance of blue mussel *Mytilus edulis* in Svalbard after 1000 years of absence. Mar. Ecol. Prog. Ser. 303, 167–175.
Bhagwat, D., 2020. Russia and India in the Arctic: the need for greater synergy. Arctic North 38, 73–90.
Bhatt, U.S., et al., 2017. Changing seasonality of panarctic tundra vegetation in relationship to climatic variables. Environ. Res. Lett. 12, 055003.
Bjornsson, B., Steinarsson, A., Árnason, T., 2007. Growth model for Atlantic cod (*Gadus morhua*): effects of temperature and body weight on growth rate. Aquaculture 271, 216–226.
Bluhm, B.A., Iken, S.L., Mincks, B.I., et al., 2009. Community structure of epibenthic megafauna in the Chukchi Sea. Aquat. Biol. 7, 269–293.
Bluhm, B.A., Gradinger, R., Hopcroft, R.R., 2011. Arctic ocean diversity: synthesis. Mar. Biodivers. 41 (1), 1–4.
Bodil, B.A., Ambrose, W.G., Bergmann, M., Clough, L.M., Gebruk, A.V., Hasemann, C., et al., 2011. Diversity of the arctic deep-sea benthos. Mar. Biodivers. 41 (1), 87–107.
Borgerson, S., 2013. The coming Arctic boom: as the ice melts, the region heats up. Foreign Aff. 92, 76.

Bowden, J.J., Hansen, O.L.P., Olsen, K., Schmidt, N.M., Høye, T.T., 2018. Drivers of inter-annual variation and long-term change in high-Arctic spider species abundances. Polar Biol. 41, 1635–1649.

Büdenbender, J., Riebesell, U., Form, A., 2011. Calcification of the Arctic coralline red algae *Lithothamnion glaciale* in response to elevated CO_2. Mar. Ecol. Prog. Ser. 441, 79–87.

Butchart, S.H., Stattersfield, A.J., Brooks, T.M., 2006. Going or gone: defining 'Possibly Extinct' species to give a truer picture of recent extinctions. Bull.-Br. Ornithol. Club 126, 7.

CAFF, 2010. Arctic Biodiversity Trends 2010: Selected Indicators of Change. CAFF International Secretariat, Akureyri.

CAFF, 2013. Arctic Biodiversity Assessment: Report for Policy Makers. CAFF, Akureyri.

CAFF, 2017. State of the Arctic Marine Biodiversity Report. Conservation of Arctic Flora and Fauna International Secretariat, AkureyriRetrieved from. Available from: https://www.arcticbiodiversity.is/marine.

CAFF. Arctic Biodiversity and the Post, 2020. Framework, Report Submitted to the Chairs of the Open-Ended Working Group on the Post-2020 Global Biodiversity Framework on Its Second Meeting. CAFF International Secretariat, Akureyri, pp. 1–13.

Cairns, D.K., Gaston, A.J., Huettmann, F., 2008. Endothermy, ectothermy and the global structure of marine vertebrate communities. Mar. Ecol. Prog. Ser. 356, 239–250.

Callaghan, T.V., Björn, L.O., Chernov, Y., Chapin, T., Christensen, T.R., Huntley, B., et al., 2004. Biodiversity, distributions and adaptations of arctic species in the context of environmental change. AMBIO 33, 404–417.

Callaghan, T.V., Tweedie, C.E., Akerman, J., Andrews, C., Bergstedt, J., Butler, M.G., et al., 2011. Multidecadal changes in tundra environments and ecosystems: synthesis of the International Polar Year-Back to the Future Project (IPY-BTF). AMBIO 40, 705–716.

CBD. UN Convention on Biological Diversity (CBD): Decision X/13 New and Emerging Issues, 2010. Available online: https://www.cbd.int/decision/cop/?id = 12279 (accessed 11.01.20.).

Change, C., 2007. Synthesis Report. Contribution of Working Groups I, II and III to the Fourth Assessment Report of the Intergovernmental Panel on Climate Change. IPCC, Geneva.

Chaturvedi, S., 2013. China and India in the 'receding' Arctic: rhetoric, routes and resources. Jadavpur J. Int. Relat. 17 (1), 41–68.

Chernova, N., 2011. Distribution patterns and chorological analysis of fish fauna of the Arctic region. J. Ichthyol. 51 (10), 825–924.

Cheung, W.W.L., Lam, V.W.Y., Sarmiento, J.L., Kearney, K., Watson, R., Pauly, D., 2009. Projecting global marine biodiversity impacts under climate change scenarios. Fish. Fish. 10, 235–251. Available from: https://doi.org/10.1111/j.1467-2979.2008.00315.x.

Chown, S.L., Huiskes, A.H.L., Gremmen, N.J.M., Lee, J.E., Terauds, A., Crosbie, K., et al., 2012. Continent-wide risk assessment for the establishment of non-indigenous species in Antarctica. Proc. Natl. Acad. Sci. U.S.A. 109, 4938–4943.

Christensen, T., Barry, T., Taylor, J.J., et al., 2020. Developing a circumpolar programme for the monitoring of Arctic terrestrial biodiversity. AMBIO 49, 655–665.

Christiansen, J., Reist, J., Brown, R., Brykov, V., Christensen, G., Christoffersen, K., et al., 2013. Chapter 6—Fishes. In: Melthofte, H. (Ed.), Arctic Biodiversity Assessment. Status and Trends in Arctic Biodiversity. Conservation of Arctic Flora and Fauna, Akureyi.

Christiansen, J.S., Mecklenburg, C.W., Karamushko, O.V., 2014. Arctic marine fishes and their fisheries in light of global change. Glob. Change Biol. 20 (2), 352–359.
Clemens, R.S., Rogers, D., Hansen, B.D., et al., 2016. Continental-scale decreases in shorebird populations in Australia. Emu 116, 119–135.
Coad, B.W., Reist, J.D., 2004. Annotated List of the Arctic Marine Fishes of Canada. Fisheries and Oceans Canada Winnipeg, Canada.
Collins, M., Knutti, R., Arblaster, J., Dufresne, J.-L., Fichefet, T., Friedlingstein, P., et al., 2013. Long-term climate change: projections, commitments and irreversibility. In: Stocker, T.F., Qin, D., Plattner, G.-K., Tignor, M., Allen, S.K., Boschung, J., Nauels, A., Xia, Y. (Eds.), Climate Change 2013: The Physical Science Basis. Contribution of Working Group I to the Fifth Assessment Report of the Intergovernmental Panel on Climate Change. Cambridge University Press, Cambridge.
Comiso, J.C., 2006. Arctic warming signals from satellite observations. Weather 61, 70–76.
Convey, P., 1996. Overwintering strategies of terrestrial invertebrates from Antarctica – the significance of flexibility in extremely seasonal environments. Eur. J. Entomol. 93, 489–505.
Cooper, E.J., 2014. Warmer shorter winters disrupt Arctic terrestrial ecosystems. Annu. Rev. Ecol. Evol. Syst. 45, 271–295. Arctic Tundra. In: Fram Centre Report Series. Tromsø, Fram Centre.
Cotter, R.C., 1999. The reproductive biology of Rock Ptarmigan (*Lagopus mutus*) in the central Canadian arctic. Arctic 23–32.
Council, A., 1996. Declaration on the Establishment of the Arctic Council 1996-Ottawa Declaration.
Culp, J.M., Goedkoop, W., Lento, J., Christoffersen, K.S., Frenzel, S., Gudbergsson, G., et al., 2012. The Arctic Freshwater Biodiversity Monitoring Plan. CAFF International Secretariat, CAFF Monitoring Series Report no. 7. CAFF International Secretariat, Akureyri. https://caff.is/freshwater/freshwater-monitoring-publications/196-arctic-freshwater-biodiversity-monitoring-plan. ISBN 978-9935-431- 19-6.
Dalén, L., Kvaløy, K., Linnell, J., Elmhagen, B., Strand, O., Tannerfeldt, M., et al., 2006. Population structure in a critically endangered arctic fox population: does genetics matter? Mol. Ecol. 15 (10), 2809–2819.
Daniëls, F.J.A, 1982. Vegetation of the Angmagssalik District, Southeast Greenland. IV: Shrub, Dwarf Shrub and Terricolous Lichens. Meddelelser Om Groenland. Bioscience. ISSN 0106-1054
Descamps, S., Aars, J., Fuglei, E., Kovacs, K.M., Lydersen, C., Pavlova, O., et al., 2017. Climate change impacts on wildlife in a High Arctic archipelago—Svalbard, Norway. Glob. Change Biol. 23 (2), 490–502.
Doran, P.T., Priscu, J.C., Lyons, W.B., Walsh, J.E., Fountain, A.G., McKnight, D.M., et al., 2002. Antarctic climate cooling and terrestrial ecosystem response. Nature 415, 517–520.
Durner, G.M., Laidre, K.L., York, G.S., 2018. Polar Bears: Proceedings of the 18th Working Meeting of the IUCN/SSC Polar Bear Specialist Group, 7–11 June 2016, Anchorage, Alaska. IUCN, Gland, Switzerland and Cambridge. Retrieved from: https://portals.iucn.org/library/sites/library/files/documents/SSC-OP-063-En.pdf
Duffy-Anderson, J.T., Stabeno, P.J., Siddon, E.C., Andrews, A.G., Cooper, D.W., Eisner, L.B., et al., 2017. Return of warm conditions in the Southeastern Bering Sea: Phytoplankton to Fish. PLoS One 6.
Elvebakk, A., Bjerke, J.W., 2006. The Skibotn area in North Norway—an example of very high lichen species richness far to the north: a supplement with an annotated list of species. Mycotaxon 96, 141.

Feng, R, Poulsen, C.J., 2016. Refinement of Eocene lapse rates, fossil leaf altimetry, and North American Cordilleran surface elevation estimates. Earth Planet Sci. Lett. 436, 130–141.

Fingas, M.F., Hollebone, B.P., 2003. Review of behaviour of oil in freezing environments. Mar. Pollut. Bull. 47, 333–340.

Flight, P.I., 2017. Avian Conservation Assessment Database, Version 2017.

Folkvord, A., Gundersen, G., Albretsen, J., Asplin, L., Kaartvedt, S., Giske, J., 2016. Impact of hatch date on early life growth and survival of Mueller's pearlside (*Maurolicus muelleri*) larvae and life-history consequences. Can. J. Fish. Aquat. Sci. 73 (2), 163–176.

Ford, J.D., Pearce, T., 2010. What we know, do not know, and need to know about climate change vulnerability in the western Canadian Arctic: a systematic literature review. Environ. Res. Lett. 5, 014008. Available from: https://doi.org/10.1088/1748-9326/5/1/014008.

Førland, E.J., Benestad, R., Hanssen-Bauer, I., Haugen, J.E., Skaugen, T.E., 2012. Temperature and precipitation development at Svalbard 1900–2100. Adv. Meteorol. 2012, 1–14.

Fossheim, M., Primicerio, R., Johannesen, E., Ingvaldsen, R.B., Aschan, M.M., Dolgov, A.V., 2015. Recent warming leads to a rapid borealization of fish communities in the Arctic. Nat. Clim. Change 5 (7), 673–677. Available from: https://doi.org/10.1038/nclimate2647.

Frankham, R., 1997. Do island populations have less genetic variation than mainland populations? Heredity 78, 311–327. Available from: https://doi.org/10.1038/hdy.1997.46.

Frankham, R., 1998. Inbreeding and extinction: island populations. Conserv. Biol. 12 (3), 665–675. Available from: https://doi.org/10.1046/j.1523-1739.1998.96456.x.

Gadihoke, N., 2012. Arctic melt: the outlook for India. Maritime Affairs. J. Natl. Marit. Found. India 8 (1), 1–12.

Ganter, B., Gaston, A.J., Anker-Nilssen, T., Blancher, P., Boertmann, D., 2013. Chapter 4: Birds. Arctic Biodiversity Assessment: Status and Trends in Arctic Biodiversity. Conservation of Arctic Flora and Fauna, Akureyri, pp. 143–180.

Gardiner, K., Dick, T.A., 2010. Arctic cephalopod distributions and their associated predators. Polar Res. 29 (2), 209–227.

Geffen, E., Waidyaratne, S., Dalén, L., Angerbjörn, A., Vila, C., Hersteinsson, P., et al., 2007. Sea ice occurrence predicts genetic isolation in the Arctic fox. Mol. Ecol. 16 (20), 4241–4255. Available from: https://doi.org/10.1111/j.1365-294X.2007.03507.x.

Gewelt, A., 2016. India in the Arctic: Science, Geopolitics and Soft Power-Perspectives on Contemporary Indian Foreign Policy (Master's thesis).

Gill, M., 2008. CAFF CBMP Report No. 12-A Strategy for Developing Indices and Indicators to Track Status and Trends in Arctic Biodiversity. Conservation of Arctic Flora and Fauna Working Group (CAFF).

Gill, M.J., Crane, K., Hindrum, R., Arneberg, P., Bysveen, I., Denisenko, N.V., et al., 2011. Arctic Marine Biodiversity Monitoring Plan. CAFF Monitoring Series Report no. 3. CAFF International Secretariat, Akureyri. https://caff.is/monitoring-series/view_document/3-arctic-marinebiodiversity-monitoring-plan. ISBN 1.978-9979-9778-7-2.

Gillespie, M.A.K., Alfredsson, M., Barrio, I.C., Bowden, J., Convey, P., Coulson, S.J., et al., 2020. Circumpolar terrestrial arthropod monitoring: a review of ongoing activities, opportunities and challenges, with a focus on spiders. In: Schmidt, N.M., Jóhannesdóttir, H. (Eds.), Terrestrial Biodiversity in a Rapidly Changing Arctic, vol. 49. AMBIOSpecial Issue. Available from: https://doi.org/10.1007/s13280-019-01185-y.

Gradinger, R., 1999. Vertical fine structure of the biomass and composition of algal communities in Arctic pack ice. Mar. Biol. 133 (4), 745–754.
Grebmeier, J.M., 2012. Shifting patterns of life in the Pacific Arctic and sub-arctic seas. Ann. Rev. Mar. Sci. 133 (4), 63–78.
Grebmeier, J.M., Moore, S.E., Overland, J.E., Frey, K.E., Gradinger, R, 2010. Biological response to recent pacific Arctic sea ice retreats. EOS 91, 161–163.
Grebmeier, J.M., Overland, J.E., Moore, S.E., Farley, K.E., et al., 2006b. A major ecosystem shift in the northern Bering Sea. Science 311, 1461–1464.
Gruber, K., 2017. Agrobiodiversity: the living library. Nature 544 (7651), S8–S10.
Grydehøj, A., Grydehøj, A., Ackrén, M., 2012. The globalization of the Arctic: negotiating sovereignty and building communities in Svalbard, Norway. Isl. Stud. J. 7 (1), 99–118.
Guay, K.C., et al., 2014. Vegetation productivity patterns at high northern latitudes: a multi-sensor satellite data assessment. Glob. Change Biol. 20, 3147–3158.
Hamel, S., Killengreen, S., Henden, J.-A., Yoccoz, N., Ims, R., 2013. Disentangling the importance of interspecific competition, food availability, and habitat in species occupancy: recolonization of the endangered Fennoscandian arctic fox. Biol. Conserv. 160, 114–120.
Hansel, M.C., Schmidt, J.O., Stiasny, M.H., Stöven, M.T., Voss, R., Quaas, M.F., 2020. Ocean warming and acidification may drag down the commercial Arctic cod fishery by 2100. PLoS One 15 (4), e0231589. Available from: https://doi.org/10.1371/journal.pone.0231589.
Hansen, B.B., Aanes, R., Saether, B.E., 2010. Partial seasonal migration in high-arctic Svalbard reindeer (*Rangifer tarandus platyrhynchus*). Can. J. Zool.-Rev. Can. Zool. 88, 1202–1209.
Hanski, I., 1998. Metapopulation dynamics. Nature 396, 41–49. Available from: https://doi.org/10.1038/23876.
Hasle, G.R., Syvertsen, E.E., Steidinger, K.A., Tangen, K., Tomas, C.R., 1996. Identifying Marine Diatoms and Dinoflagellates. Elsevier.
Hayden, B., Harrod, C., Thomas, S.M., Eloranta, A.P., Myllykangas, J., Siwertsson, A., et al., 2019. From clear lakes to murky waters – tracing the functional response of high-latitude lake communities to concurrent 'greening' and 'browning'. Ecol. Lett. 22, 807–816. Available from: https://doi.org/10.1111/ele.13238.
Hegseth, E.N., Sundfjord, A., 2008. Intrusion and blooming of Atlantic phytoplankton species in the high Arctic. J. Mar. Syst. 74, 108–119.
Heino, J., Virkkala, R., Toivonen, H., 2009. Climate change and freshwater biodiversity: detected patterns, future trends and adaptations in northern regions. Biol. Rev. 84, 39–54. Available from: https://doi.org/10.1111/j.1469-185X.2008.00060.x.
Heino, J., Culp, J.M., Erkinaro, J., Goedkoop, W., Lento, J., Ruhland, K.M., et al., 2020. Abruptly and irreversibly changing Arctic freshwaters urgently require standardized monitoring. Br. Ecol. Soc. 57 (7), 1192–1198. Available from: https://doi.org/10.1111/1365-2664.13645.
Hellmann, J.J., Byers, J.E., Bierwagen, B.G., Dukes, J.S., 2008. Five potential consequences of climate change for invasive species. Conserv. Biol. 22, 534–543. Available from: https://doi.org/10.1111/j.1523-1739.2008.00951.x.
Hjort, J., Karjalainen, O., Aalto, J., Westermann, S., Romanovsky, V.E., Nelson, F.E., et al., 2018. Degrading permafrost puts Arctic infrastructure at risk by mid-century. Nat. Commun. 9, 5147. Available from: https://doi.org/10.1038/s41467-018-07557-4.
Hoegh-Guldberg, O., Bruno, J.F., 2010. The impact of climate change on the world's marine ecosystems. Science 328, 1523–1528. Available from: https://doi.org/10.1126/science.1189930.
Hollowed, A.B., Planque, B., Loeng, H., 2013. Potential movement of fish and shellfish stocks from the sub-Arctic to the Arctic Ocean. Fish Oceanogr. 22 (5), 355–370.

Horner, R.A., Syvertsen, E.E., Thomas, D.P., Lange, C., 1988. Proposed terminology and reporting units for sea ice algal assemblages. Polar Biol. 8 (4), 249–253.
Hugelius, G., et al., 2014. Estimated stocks of circumpolar permafrost carbon with quantified uncertainty ranges and identified data gaps. Biogeosciences 11, 6573–6593.
Hunt, B., Strugnell, J., Bednarsek, N., Linse, K., Nelson, R.J., Pakhomov, E., et al., 2010. Poles apart: the "bipolar" pteropod species *Limacina helicina* is genetically distinct between the Arctic and Antarctic oceans. PLoS One 5 (3), e9835.
Ims, R.A., Jepsen, J.U., Stien, A., Yoccoz, N.G., 2013. COAT – Climate-Ecological Observatory.
IPCC, 2014a. In: Field, C.B. (Ed.), Summary for Policymakers Climate Change 2014: Impacts, Adaptation, and Vulnerability. Part A: Global and Sectoral Aspects. Contribution of Working Group II to the Fifth Assessment Report of the Intergovernmental Panel on Climate Change. Cambridge University Press, Cambridge and New York, pp. 1–32.
IPCC, 2014b. Climate Change 2013: The Physical Science Basis. Contribution of Working Group I to the Fifth Assessment Report of the Intergovernmental Panel on Climate Change. Cambridge University Press, Cambridge.
IPCC, 2014c. In: Pachauri, R.K., Meyer, L.A. (Eds.), Climate Change 2014: Synthesis Report. Contribution of Working Groups I, II and III to the Fifth Assessment Report of the Intergovernmental Panel on Climate Change. Cambridge University Press.
IPCC, 2018. Global Warming of 1.5C: An IPCC Special Report on the Impacts of Global Warming of 1.5C Above Pre-Industrial Levels and Related Global Greenhouse Gas Emission Pathways, in the Context of Strengthening the Global Response to the Threat of Climate Change, Sustainable Development, and Efforts to Eradicate Poverty. Masson-Delmotte, V., Zhai, P., Portner, H.O., Roberts D., Skea, J., Shukla, P.R., et al. (Eds.), in press. https://www.ipcc.ch/site/assets/uploads/sites/2/2018/07/SR15_SPM_High_Res.pdf
IPCC, 2019. IPCC Special Report on the Ocean and Cryosphere in a Changing Climate. Portner, H.-O., Roberts, D.C., Masson-Delmotte, V., Zhai, P., Tignor, M., Poloczanska, E., et al. (Eds.) https://www.ipcc.ch/srocc/download-report/.
IUCN. Microplastics: A Global Disaster in the Arctic Ocean. Retrieved 2 July 2020 from: https://www.iucn.org/content/microplastics-global-disasterarctic-Ocean
Iversen, C.M., Sloan, V.L., Sullivan, P.F., Euskirchen, E.S., McGuire, A.D., Norby, R.J., et al., 2015. The unseen iceberg: plant roots in arctic tundra. N. Phytol. 205 (1), 34–58.
Jacob, S.R., Tyagi, V., Agrawal, A., Chakrabarty, S.K., Tyagi, R.K., 2015. Indian plant germplasm on the global platter: an analysis. PLoS One 10 (5), e0126634.
Jenkins, D.A., Lecomte, N., Schaefer, J.A., Olsen, S.M., Swingedouw, D., Côté, S.D., et al., 2016. Loss of connectivity among island-dwelling Peary caribou following sea ice decline. Biol. Lett. 12 (9), 20160235. Available from: https://doi.org/10.1098/rsbl.2016.0235.
Jeong, S.-J., et al., 2018. Accelerating rates of Arctic carbon cycling revealed by long-term atmospheric CO_2 measurements. Sci. Adv. 4, eaao1167.
Johansson, M., Christensen, T.R., Akerman, H.J., Callaghan, T.V., 2006. What determines the current presence or absence of permafrost in the Tornetrask region, a subarctic landscape in Northern Sweden? AMBIO 35, 190–197.
Jones, T., McLennan, D., Behe, C., Arvnes, M., Wesseberg, S., Sergienko, L., et al., 2019. Arctic Coastal Biodiversity Monitoring Plan. Conservation of Arctic Flora and Fauna International Secretariat, Akureyri. https://caff.is/coastal/coastal-monitoring-publications/473-arctic-coastal-biodiversity-monitoring-plan. ISBN 978-9935-431-76-9.

Jørgensen, L., Archambault, P., Armstrong, C., Dolgov, A., Edinger, E., Gaston, T., et al., 2016. Chapter 36G: Arctic Ocean. The First Global Integrated Marine Assessment: World Ocean Assessment I. Oceans and Law of the Sea, United Nations.

Kankaanpaa, T., Skov, K., Abrego, N., Lund, M., Schmidt, N.M., Roslin, T., 2018. Spatiotemporal snowmelt patterns within a high Arctic landscape, with implications for flora and fauna. Arct. Antarct. Alp. Res. 50.

Kevan, P.G., Forbes, B.C., Kevan, S.M., Behan-Pelletier, V, 1995. Vehicle tracks on high Arctic tundra: their effects on the soils, vegetation and soil arthropods. J. Appl. Ecol. 32, 655−667.

Kleisner, K.M., Fogarty, M.J., McGee, S., Hare, J.A., Moret, S., Perretti, C.T., et al., 2017. Marine species distribution shifts on the U.S. Northeast Continental Shelf under continued ocean warming. Prog. Oceanogr. 153, 24−36. Available from: https://doi.org/10.1016/j.pocean.2017.04.001.

Kokelj, S.V., Tunnicliffe, J., Lacelle, D., Lantz, T.C., Chin, K.S., Fraser, R., 2015. Increased precipitation drives mega slump development and destabilization of ice-rich permafrost terrain, northwestern Canada. Glob. Planet. Change 129, 56−68. Available from: https://doi.org/10.1016/j.gloplacha.2015.02.008.

Koltz, A.M., Schmidt, N.M., Høye, T.T., 2018. Differential arthropod responses to warming are altering the structure of Arctic communities. R. Soc. Open. Sci. Available from: https://doi.org/10.1098/rsos.171503.

Kortsch, S., et al., 2015. Proc. R. Soc. B282. Available from: https://doi.org/10.1098/rspb.2015.1546.

Kosobokova, K.N., Hopcroft, R.R., Hirche, H.-J., 2011. Patterns of zooplankton diversity through the depths of the Arctic's central basins. Mar. Biodivers. 41 (1), 29−50.

Kraberg, A.C., Druzhkova, E., Heim, B., Loeder, M., Wiltshire, K.H., 2013. Phytoplankton community structure in the Lena Delta (Siberia, Russia) in relation to hydrography. Biogeosciences 10 (11), 7263.

Kuersten, A., 2016. The Arctic Five versus the Arctic Council. Arct. Yearb. 2016, 389−395.

Lackenbauer, P.W., 2013. India's Arctic engagement: emerging perspectives. Arct. Yearb. 1−19.

Lackenbauer, P.W., 2017. India and the arctic: revisionist aspirations, arctic realities. Jindal Glob. Law Rev. 8 (1), 23−54.

Laidre, K.L., Born, E.W., Atkinson, S.N., Wiig, Ø., Andersen, L.W., Lunn, N.J., et al., 2018. Range contraction and increasing isolation of a polar bear subpopulation in an era of sea-ice loss. Ecol. Evol. 8 (4), 2062−2075. Available from: https://doi.org/10.1002/ece3.3809.

Lande, R., 1998. Anthropogenic, ecological and genetic factors in extinction and conservation. Res. Popul. Ecol. 40 (3), 259−269. Available from: https://doi.org/10.1007/bf02763457.

Leblond, M., St-Laurent, M.-H., Côté, S.D., 2016. Caribou, water, and ice − fine-scale movements of a migratory arctic ungulate in the context of climate change. Mov. Ecol. 4 (1), 14. Available from: https://doi.org/10.1186/s40462-016-0079-4.

Lento, J., Goedkoop, W., Culp, J., Christoffersen, K.S., Lárusson, K.F., Fefilova, E., et al., 2019a. State of the Arctic Freshwater Biodiversity Monitoring Report. Retrieved from https://caff.is/freshwater/freshwatermonitoringpublications/stateofthearcticfreshwaterbiodiversitymonitoringreport.

Lento, J., Goedkoop, W., Culp, L., Christoffersen, K., Fefilova, E., Guðbergsson, G., et al., 2019b. State of the Arctic Freshwater Biodiversity. Conservation of Arctic Flora and Fauna International Secretariat, Akureyri.

Liljedahl, A.K., et al., 2016. Pan-Arctic ice-wedge degradation in warming permafrost and its influence on tundra hydrology. Nat. Geosci. 9, 312 − + .

List, I.R., Concern, L., 2014. Proposal for adding five subspecies of red knot (*Calidris canutus*) to the CMS Cooperative Action List during the 2014–2017 Triennium. In: Paper Presented at the Convention on Migratory Species.

Lizotte, M.P., 2001. The contributions of sea ice algae to Antarctic marine primary production. Am. Zool. 41 (1), 57–73.

Loboda, S., Savage, J., Buddle, C.M., Schmidt, N.M., Høye, T.T., 2018. Declining diversity and abundance of High Arctic fly assemblages over two decades of rapid climate warming. Ecography 41, 265–277. Available from: https://doi.org/10.1111/ecog.02747.

Logan, S., Christensen, T., Barry, T., Price, C., Lárussoon, K., 2018. Circumpolar Biodiversity Monitoring Program Strategic Plan: 2018-2021.

Lund, M., Falk, J.M., Friborg, T., Mbufong, H.N., Sigsgaard, C., Soegaard, H., et al., 2012. Trends in CO_2 exchange in a high Arctic tundra heath, 2000–2010. J. Geophys. Res. 117.

Lund, M., Hansen, B.U., Pedersen, S.H., Stiegler, C., Tamstorf, M.P., 2014. Characteristics of summer-time energy exchange in a high Arctic tundra heath 2000–2010. Tellus B 66.

Lunev, S., 2012. India Goes to the Arctic. Russian International Affairs Council.

Lüning, K., 1990. Seaweeds: Their Environment, Biogeography, and Ecophysiology. John Wiley & Sons.

Markus, T., Stroeve, J.C., Miller, J., 2009. Recent changes in Arctic sea ice melt onset, freeze up, and melt season length. J. Geophys. Res.-Oceans 114.

Mastepanov, M., Sigsgaard, C., Strom, L., Tamstorf, M.P., Lund, M., Christensen, T.R., 2013. Revisiting factors controlling methane emissions from high-Arctic tundra. Biogeosciences 10, 5139–5158.

Matishov, G.G., Makarevich, P.R., Ishkulov, D.G., 2011. "Invaders" and "non-invaders": causes and consequences of their appearance. Izvestia Samara Sci. Cent. Russ. Acad. Sci. 13, 1357–1366 (in Russian).

Mecklenburg, C.W., Møller, P.R., Steinke, D., 2011. Biodiversity of arctic marine fishes: taxonomy and zoogeography. Mar. Biodivers. 41 (1), 109–140.

Mecklenburg, C., Byrkjedal, I., Christiansen, J., Karamushko, O., Lynghammar, A., Møller, P., 2013. List of Marine Fishes of the Arctic Region Annotated with Common Names and Zoogeographic Characterizations. Conservation of Arctic Flora and Fauna.

Meltofte, H. (Ed.), 2013. Arctic Biodiversity Assessment. Status and Trends in Arctic Biodiversity. Conservation of Arctic Flora and Fauna, Akureyri.

Michelsen, V., 2015. Anthomyiidae, Fannidae, Muscidae and Scathophagidae (the Muscidae family group). In: Bocher, J., Kristensen, K., Pape, T., Vilhelmsen, L. (Eds.), The Greenland Entomofauna: An Identification Manual of Insects, Spiders and Their Allies. Koninklijke Brill, Leiden.

Millennium Ecosystem Assessment, 2005. Ecosystems and Human Well-being: Current State and Trends. Findings of the Condition and Trends Working Group, vol. 1. Island Press, Washington, DC.

Miller, A.W., Ruiz, G.M., 2014. Arctic shipping and marine invaders. Nat. Clim. Change 4, 413–416. Available from: https://doi.org/10.1038/nclimate2244.

Miller, F.L., Barry, S.J., Calvert, W.A., 2005. Sea-ice crossings by caribou in the south-central Canadian Arctic Archipelago and their ecological importance. Rangifer 25 (4), 77–88. Available from: https://doi.org/10.7557/2.25.4.1773.

Miller, G.H., Alley, R.B., Brigham-Grette, J., Fitzpatrick, J.J., Polyak, L., Serreze, M.C., et al., 2010. Arctic amplification: can the past constrain the future? Quat. Sci. Rev. 29, 1779–1790.

Moore, S.E., 2008. Marine mammals as ecosystem sentinels. J. Mammal. 89, 534–540.

Mueter, F.J., Litzow, M.A., 2008. Sea ice retreat alters the biogeography of the Bering Sea continental shelf. Ecol. Appl. 18, 309–320.

Nanda, D., Council, A., 2019. India's Arctic potential. Occasional Pap. 186, 1–30.

Nayak, S., 2008. Polar research In India. Indian J. Mar. Sci. 37 (4), 352–357.

Neff, W., Compo, G.P., Ralph, F.M., Shupe, M.D., 2014. Continental heat anomalies and the extreme melting of the Greenland ice surface in 2012 and 1889. J. Geophys. Res.-Atmos. 119, 6520–6536.

Nesis, K.N., 2001. West-Arctic and East-Arctic distributional ranges of cephalopods. Sarsia 86 (1), 1–11.

Nordli, O., Przybylak, R., Ogilvie, A.E.J., Isaksen, K., 2014. Long-term temperature trends and variability on Spitsbergen: the extended Svalbard Airport temperature series, 1898–2012. Polar Res. 33, 21349.

Norén, K., Carmichael, L., Fuglei, E., Eide, N.E., Hersteinsson, P., Angerbjörn, A., 2011. Pulses of movement across the sea ice: population connectivity and temporal genetic structure in the arctic fox. Oecologia 166 (4), 973–984. Available from: https://doi.org/10.1007/s00442-011-1939-7.

Ochyra, R., Bednarek-Ochyra, H., Smith, R.I.L., 2008. New and rare moss species from the Antarctic. Nova Hedwig 87, 457–477.

Overland, J.E., 2009. The case of global warming in the Arctic. Influence of climate change on the changing Arctic and sub- Arctic conditions. Springer, pp. 13–23.

Overland, J.E., Wang, M.Y., 2016. Recent extreme arctic temperatures are due to a split polar vortex. J. Clim. 29, 5609–5616.

Øvstedal, D.O., Smith, R.I.L., 2009. Further additions to the lichen flora of Antarctica and South Georgia. Nova Hedwig 88, 157–168.

Parazoo, N.C., Koven, C.D., Lawrence, D.M., Romanovsky, V., Miller, C.E., 2018. Detecting the permafrost carbon feedback: talik formation and increased cold-season respiration as precursors to sink-to-source transitions. Cryosphere 12, 123–144.

Parmentier, F.J.W., van der Molen, M.K., van Huissteden, J., Karsanaev, S.A., Kononov, A.V., Suzdalov, D.A., et al., 2011. Longer growing seasons do not increase net carbon uptake in the northeastern Siberian tundra. J. Geophys. Res. 116.

Parmentier, F.J.W., Rasse, D.P., Lund, M., Bjerke, J.W., Drake, B.G., Weldon, S., et al., 2018. Vulnerability and resilience of the carbon exchange of a subarctic peatland to an extreme winter event. Environ. Res. Lett. 13.

Payer, D., Barry, T., Berteaux, D., Bültmann, H., Cristiansen, J., Cook, J., et al., 2013. Arctic Biodiversity Assessment: Status and Trends in Arctic biodiversity. Narayana Press, p. 674.

Peeters, B., 2014. Effects of Climate and Animal Density on Adult Sex-Ratio Variation in a Large Herbivore (Unpublished M.Sc. thesis). Norwegian University of Science and Technology, Trondheim.

Peeters, B., Moullec, M.L., Raeymaekers, J.A.M., Marquez, J.F., Røed, K.H., Pedersen, Å.Ø., et al., 2019. Sea ice loss increases genetic isolation in a high Arctic ungulate metapopulation. Wiley 26 (4). Available from: https://doi.org/10.1111/gcb.14965.

Petersen, A., Zockler, C., Gunnarsdottir, M.V., 2004. Circumpolar Biodiversity Monitoring Program Framework Document-CBMP Report No. 1.

Phoenix, G.K., Bjerke, J.W., 2016. Arctic browning: extreme events and trends reversing arctic greening. Glob. Change Biol. 22, 2960–2962.

Piatt, J.F., Kitaysky, A.S., 2002. Tufted puffin: *Fratercula cirrhata*. In: Poole, A. (Ed.), The Birds of North America Online. Cornell Laboratory of Ornithology, Ithaca, NY. Available online at:. Available from: http://bna.birds.cornell.edu/bna/species/708.

Piersma, T., Lok, T., Chen, Y., et al., 2016. Simultaneous declines in summer survival of three shorebird species signals a flyway at risk. J. Appl. Ecol. 53, 479–490.

Pirk, N., Mastepanov, M., Parmentier, F.J.W., Lund, M., Crill, P., Christensen, T.R., 2016. Calculations of automatic chamber flux measurements of methane and carbon dioxide using short time series of concentrations. Biogeosciences 13, 903–912.

Pistone, K., Eisenman, I., Ramanathan, V., 2014. Observational determination of albedo decrease caused by vanishing Arctic Sea ice. Proc. Natl. Acad. Sci. U.S.A. 111, 3322–3326. Available from: https://doi.org/10.1073/pnas.1318201111.

Pithan, F., Mauritsen, T., 2014. Arctic amplification dominated by temperature feedbacks in contemporary climate models. Nat. Geosci. 7, 181–184.

Post, E., Forchhammer, M.C., Bret-Harte, M.S., Callaghan, T.V., Christensen, T.R., Elberling, B., et al., 2009. Ecological dynamics across the Arctic associated with recent climate change. Science 325, 1355–1358.

Poloczanska, E.S., Brown, C.J., Sydeman, W.J., Kiessling, W., et al., 2013. Global imprint of climate change on marine life. Nat. Clim. Chang. 3, 919–925.

Post, E., Bhatt, U.S., Bitz, C.M., Brodie, J.F., Fulton, T.L., Hebblewhite, M., et al., 2013. Ecological consequences of sea-ice decline. Science 341 (6145), 519–524. Available from: https://doi.org/10.1126/science.1235225.

Poulin, M., Daugbjerg, N., Gradinger, R., Ilyash, L., Ratkova, T., von Quillfeldt, C., 2011. The pan-Arctic biodiversity of marine pelagic and sea-ice unicellular eukaryotes: a first-attempt assessment. Mar. Biodivers. 41 (1), 13–28.

Pronina, V., Eidemiller, K.Y., Khazov, V., Rubtsova, A., 2020. The Arctic policy of India. In: Paper Presented at the IOP Conference Series: Earth and Environmental Science.

Putkonen, J., Roe, G., 2003. Rain-on-snow events impact soil temperatures and affect ungulate survival. Geophys. Res. Lett. 30, 1188.

Raynolds, M.K., Walker, D.A., Epstein, H.E., Pinzon, J.E., Tucker, C.J., 2012. A new estimate of tundra-biome phytomass from trans-Arctic field data and AVHRR NDVI. Remote Sens. Lett. 3, 403–411.

Reddy, D.E., 2017. Emerging trends in seed banking for food and agriculture: an international perspective. J. Agric. Food Inf. 18 (2), 145–160.

Renaud, P.E., Sejr, M.K., Bluhm, B.A., Sirenko, B., Ellingsen, I.H., 2015. The future of Arctic benthos: expansion, invasion, and biodiversity. Prog. Oceanogr. 139, 244–257.

Rennert, K.J., Roe, G., Putkonen, J., Bitz, C.M., 2009. Soil thermal and ecological impacts of rain on snow events in the circumpolar Arctic. J. Clim. 22, 2302–2315.

Ricciardi, A., Blackburn, T.M., Carlton, J.T., Dick, J.T.A., Hulme, P.E., Iacarella, J.C., et al., 2017. Invasion science: a horizon scan of emerging challenges and opportunities. Trends Ecol. Evol. 32, 464–474. Available from: https://doi.org/10.1016/j.tree.2017.03.007.

Richter-Menge, J, Druckenmiller, M.L., Jeffries, M, 2019. Arctic Report Card. NOAA. Available from: www.arctic.noaa.gov/Report-Card.

Robinson, L.M., Gledhill, D.C., Moltschaniwskyj, N.A., Hobday, A.J., Frusher, S., Barrett, N., et al., 2015. Rapid assessment of an ocean warming hotspot reveals "high" confidence in potential species' range extensions. Glob. Environ. Change 31, 28–37. Available from: https://doi.org/10.1016/j.gloenvcha.2014.12.003.

Saccheri, I., Kuussaari, M., Kankare, M., Vikman, P., Fortelius, W., Hanski, I., 1998. Inbreeding and extinction in a butterfly metapopulation. Nature 392 (6675), 491–494. Available from: https://doi.org/10.1038/33136.

Sakhuja, V., 2013. India and the Melting Arctic. The Institute of Peace and Conflict Studies, New Delhi, p. 3804.

Saniewski, M., Wietrzyk-Pełka, P., Zalewska, T., Olech, M., Węgrzyn, M.H., 2020. Bryophytes and lichens as fallout originated radionuclide indicators in the Svalbard archipelago (high arctic). Polar Sci. 100536.

Schmidt, N.M., Pedersen, S.H., Mosbacher, J.B., Hansen, L.H., 2015. Long-term patterns of muskox (*Ovibos moschatus*) demographics in high arctic Greenland. Polar Biol. 38, 1667–1675.
Schuur, E.A.G., Vogel, J.G., Crummer, K.G., Lee, H., Sickman, J.O., Osterkamp, T.E., 2009. The effect of permafrost thaw on old carbon release and net carbon exchange from tundra. Nature 459, 556–559.
Screen, J.A., Simmonds, I., 2010. The central role of diminishing sea ice in recent Arctic temperature amplification. Nature 464, 1334–1337. Available from: https://doi.org/10.1038/nature09051.
Serreze, M.C., Barry, R.G., 2011. Processes and impacts of Arctic amplification: a research synthesis. Glob. Planet. Change 77, 85–96.
Serreze, M.C., Walsh, J.E., Chapin III, F.S., Osterkamp, T., Dyurgerov, M., Romanovsky, V., et al., 2000. Observational evidence of recent changes in the northern high-latitude environment. Clim. Change 46, 159–207.
Serreze, M.C., Barrett, A.P., Stroeve, J.C., Kindig, D.N., Holland, M.M., 2009. The emergence of surface-based Arctic amplification. Cryosphere 3 (1), 11–19. Available from: https://doi.org/10.5194/tc-3-11-2009.
Shaumyan, T.L., Zhuravel, V.P., 2016. India and the Arctic: environment, economy and politics. Arctic 24, 153.
Sinha, U.K., 2019. India in the Arctic: a multidimensional approach. Vestnik of Saint Petersburg University. Int. Relat. 12 (1), 113–126.
Sinha, U.K., Gupta, A., 2014. The Arctic and India: strategic awareness and scientific engagement. Strategic Anal. 38 (6), 872–885.
Smith, L.C., Stephenson, S.R., 2013. New Trans-Arctic shipping routes navigable by midcentury. Proc. Natl. Acad. Sci. 110, E1191–E1195.
Smith, L.C., Sheng, Y., MacDonald, G.M., Hinzman, L.D., 2005. Disappearing Arctic lakes. Science 308, 1429. Available from: https://doi.org/10.1126/science.1108142.
Smith, R., Barry, T., Kateras, F., 2015. Thoughts on the Arctic Biodiversity Congress. Biodiversity 16, 47–50.
Smith, S.L., Chartrand, J, Ducheshe, C, 2018. Report on 2017 field activities and collection of ground-thermal and active-layer data in the Mackenzie corridor, North West territories. Geological Survey of Canada Open file 8492, 109pp.
Smith, P.A., McKinnon, L., Meltofte, H., et al., 2020. Status and trends of tundra birds across the circumpolar Arctic. AMBIO 49, 732–748. Available from: https://doi.org/10.1007/s13280-019-01308-5.
Smol, J.P., Douglas, M.S.V., 2007. Crossing the final ecological threshold in high Arctic ponds. Proc. Natl. Acad. Sci. U S A 104 (30), 12395–12397.
Snelgrove, P.V., 2010. Discoveries of the Census of Marine Life. Cambridge University Press, Cambridge.
Speer, L., Nelson, R., Casier, R., Gavrilo, M., von Quillfeldt, C., Cleary, J., et al., 2017. Natural Marine World Heritage in the Arctic Ocean. Report of an Expert Workshop and Review Process. IUCN, GlandRetrieved from. Available from: https://portals.iucn.org/library/sites/library/files/documents/2017-006.pdf.
Spribille, T., Pérez-Ortega, S., Tønsberg, T., Schirokauer, D., 2010. Lichens and lichenicolous fungi of the Klondike Gold Rush National Historic Park, Alaska, in a global biodiversity context. Bryologist 113 (3), 439–515.
Steffensen, L.L., Axelsson, J., Asdal, Å., 2019. Global Seed Vault: a pledge to mankind. Indian J. Plant. Genet. Resour. 32 (3), 303–306.
Steig, E.J., Schneider, D.P., Rutherford, S.D., Mann, M.E., Comiso, J.C., Shindell, D.T., 2009. Warming of the Antarctic ice-sheet surface since the 1957 international geophysical year. Nature 457, 459–463.

Steig, E.J., Schneider, D.P., Rutherford, S.D., Mann, M.E., Comiso, J.C., Stevenson, D. E., et al., 2019. Bottom trawl surveys in the northern Bering Sea indicate recent shifts in the distribution of marine species. Polar Biol. 42, 407–421.

Stensdal, I., 2012. Asian Arctic research 2005–2012: harder, better, faster, stronger. Polar Res. 1–39.

Stevenson, D.E., Lauth, R.R., 2019. Bottom trawl surveys in the Northern Bering Sea indicate recent shifts in the distribution of marine species. Polar Biology 42, 407–421.

Stien, A., Loe, L.E., Mysterud, A., Severinsen, T., Kohler, J., Langvatn, R., 2010. Icing events trigger range displacement in a high-arctic ungulate. Ecology 91, 915–920.

Street, L., Shaver, G., Williams, M., Van Wijk, M., 2007. What is the relationship between changes in canopy leaf area and changes in photosynthetic CO_2 flux in arctic ecosystems? J. Ecol. 95, 139–150.

Stroeve, J., Holland, M.M., Meier, W., Scambos, T., Serreze, M., 2007. Arctic sea ice decline: faster than forecast. Geophys. Res. Lett. 34 (9), L09501. Available from: https://doi.org/10.1029/2007gl029703.

Studds, C.E., Kendall, B.E., Murray, N.J., Wilson, H.B., Rogers, D.I., Clemens, R.S., et al., 2017. Rapid population decline in migratory shorebirds relying on Yellow Sea tidal mudflats as stopover sites. Nat. Commun. 8, 14895. Available from: https://doi.org/10.1038/ncomms14895.

Sturm, M., Racine, C., Tape, K., 2001. Climate change - increasing shrub abundance in the Arctic. Nature 411, 546–547.

Sukhanova, I.N., Flint, M.V., Pautova, L.A., Stockwell, D.A., Grebmeier, J.M., Sergeeva, V.M., 2009. Phytoplankton of the western Arctic in the spring and summer of 2002: structure and seasonal changes. Deep Sea Res. Part II: Topical Stud. Oceanogr. 56 (17), 1223–1236.

Sulistiyono, H., Khan, F., Lye, L., Yang, M., 2015. A risk-based approach to developing design temperatures for vessels operating in low temperature environments. Ocean Eng. 108, 813–819.

Svendsen, H., Beszczynska-Møller, A., Hagen, J.O., Lefauconnier, B., Tverberg, V., Gerland, S., et al., 2002. The physical environment of Kongsfjorden Krossfjorden, an Arctic fjord system in Svalbard. Polar Res. 21 (1), 133–166.

Talbot, S.S., 1999. Atlas of Rare Endemic Vascular Plants of the Arctic. US Fish and Wildlife Service.

Tannerfeldt, M., Angerbjörn, A., 1998. Fluctuating resources and the evolution of litter size in the arctic fox. Oikos 545–559.

Tarnocai, C., Canadell, J.G., Schuur, E.A.G., Kuhry, P., Mazhitova, G., Zimov, S., 2009. Soil organic carbon pools in the northern circumpolar region. Glob. Biogeochem. Cycles 23, GB2023. Available from: https://doi.org/10.1029/2008GB03327.

Taylor, J.J., Lawler, J.P., Aronsson, M., Barry, T., Bjorkman, A.D., Christensen, T., et al., 2020. Arctic terrestrial biodiversity status and trends: a synopsis of science supporting the CBMP State of Arctic Terrestrial Biodiversity Report. AMBIO 49 (3), 833–847.

Thomas, J., 2018. Svalbard's the Place (Doctoral dissertation).

Thompson, R.M., Beardall, J., Beringer, J., Grace, M., Sardina, P., 2013. Means and extremes: building variability into community-level climate change experiments. Ecol. Lett. 16, 799–806.

Tiusanen, M., Hebert, P.D.N., Schmidt, N.M., Roslin, T., 2016. One fly to rule them all-muscid flies are the key pollinators in the Arctic. Proc. R. Soc. B-Biol. Sci. 283, 8. Available from: https://doi.org/10.1098/rspb.2016.1271.

Trump, B.D., Kadenic, M., Linkov, I., 2018. A sustainable Arctic: making hard decisions. Arct. Antarct. Alp. Res. 50 (1), e1438345-2–e143845-10.

Turner, J., Colwell, S.R., Marshall, G.J., Lachlan-Cope, T.A., Carleton, A.M., Jones, P. D., et al., 2005. Antarctic climate change during the last 50 years. Int. J. Climatol. 25, 279–294.

Turner, J., Bindschadler, R., Convey, P., Di Prisco, G., Fahrbach, E., Gutt, J., et al., 2009. Antarctic Climate Change and the Environment. Scientific Committee for Antarctic Research, Cambridge, p. 554.

Tyler, N.J.C., 2010. Climate, snow, ice, crashes, and declines in populations of reindeer and caribou (*Rangifer tarandus* L.). Ecol. Monogr. 80, 197–219.

Van den Hoek, C., 1975. Phytogeographic provinces along the coasts of the northern Atlantic Ocean. Phycologia 14 (4), 317–330.

Vermeij, G.J., Roopnarine, P.D., 2008. The coming Arctic invasions. Science 321, 780–781. Available from: https://doi.org/10.1126/science.1160852.

Virtanen, R., Oksanen, L., Oksanen, T., Cohen, J., Forbes, B.C., Johansen, B., et al., 2016. Where do the treeless tundra areas of northern highlands fit in the global biome system: toward an ecologically natural subdivision of the tundra biome. Ecol. Evol. 6 (1), 143–158.

von Quillfeldt, C.H., Ambrose, W.G., Clough, L.M., 2003. High number of diatom species in first-year ice from the Chukchi Sea. Polar Biol. 26 (12), 806–818.

Vongraven, D., Arneberg, P., Bysveen, I., Crane, K., Denisenko, N., Gill, M., et al., 2009. CAFF CBMP Report No. 19-Circumpolar Biodiversity Marine Monitoring Plan-Background Paper.

Vowles, T., Björk, R.G., 2019. Implications of evergreen shrub expansion in the Arctic. J. Ecol. 107 (2), 650–655.

Vuilleumier, S., Bolker, B.M., Leveque, O., 2010. Effects of colonization asymmetries on metapopulation persistence. Theor. Popul. Biol. 78 (3), 225–238. Available from: https://doi.org/10.1016/j.tpb.2010.06.007.

Wasowicz, P., Sennikov, A.N., Westergaard, K.B., Spellman, K., Carlson, M., Gillespie, L.J., et al., 2020. Non-native vascular flora of the Arctic: taxonomic richness, distribution and pathways. AMBIO 49 (3), 693–703.

Wegeberg, S., Hansson, S.V., Beest, F.M.V., Rasmussen, J.F., Gustavson, K., 2019. Smooth or smothering? The self-cleaning potential and photosynthetic effects of oil spill on arctic macro-algae *Fucus distichus*. Mar. Pollut. Bull. 150. Available from: https://doi.org/10.1016/j.marpolbul.2019.110604.

Werner, I., Ikävalko, J., Schünemann, H., 2007. Sea-ice algae in Arctic pack ice during late winter. Polar Biol. 30 (11), 1493–1504.

Węsławski, J.M., Wiktor Jr, J., Kotwicki, L., 2010. Increase in biodiversity in the arctic rocky littoral, Sorkappland, Svalbard, after 20 years of climate warming. Mar. Biodivers. 40, 123–130.

Weydmann, A., Carstensen, J., Goszczko, I., et al., 2014. Shift towards the dominance of boreal species in the Arctic: inter-annual and spatial zooplankton variability in the West Spitsbergen current. Mar. Ecol. Prog. Ser. 501, 41–52.

Wietrzyk-Pełka, P., Cykowska-Marzencka, B., Maruo, F., Szymański, W., Węgrzyn, M.H., 2020. Mosses and liverworts in the glacier forelands and mature tundra of Svalbard (High Arctic): diversity, ecology, and community composition. Pol. Polar Res. 41 (2).

Winton, M., 2008. Sea ice – albedo feedback and nonlinear Arctic climate change. In: Et, D.W., Bitz, C.M., Tremblay, L.-B. (Eds.), Arctic Sea Ice Decline: Observations, Projections, Mechanisms, and Implications. American Geophysical Union, Washington, DC, pp. 111–132.

Wisz, M.S., Broennimann, O., Grønkjær, P., Møller, P.R., Olsen, S.M., Swingedouw, D., et al., 2015. Arctic warming will promote Atlantic-Pacific fish interchange. Nat. Clim. Change 5, 261–265. Available from: https://doi.org/10.1038/nclimate2500.

WWF—Global Arctic Programme, 2016. WWF Observer Report to the Arctic Council. Retrieved from: https://oaarchive.arctic-council.org/bitstream/handle/11374/1882/EDOCS-3643-v1A-2016-06-01_WWF_Observer_Review-Report.

Zhang, T., Stamnes, K., Bowling, S.A., 2001. Impact of the atmospheric thickness on the atmospheric downwelling longwave radiation and snowmelt under clear-sky conditions in the Arctic and subarctic. J. Clim. 14, 920—939.

Zhang, X.D., He, J.X., Zhang, J., Polyakov, I., Gerdes, R., Inoue, J., et al., 2013. Enhanced poleward moisture transport and amplified northern high-latitude wetting trend. Nat. Clim. Change 3, 47—51.

CHAPTER 12

Mass balance and spatio-temporal change in the area of Vestre Broggerbreen glacier, Ny-Ålesund, Svalbard, Arctic, between 1993 and 2018

Mohd Soheb and Alagappan Ramanathan
School of Environmental Sciences, Jawaharlal Nehru University, New Delhi, India

12.1 Introduction

Ice cover evolution is one of the best indicators of climate change conditions as the impact of global warming is first and the fastest on the ice cover. However, the processes behind the evolution are difficult to understand due to their complex interaction among atmosphere, land, and internal mechanisms (Østby et al., 2017). Almost 10% of the total land area on Earth is under ice cover in the form of ice caps, ice sheets, glaciers, etc. Ice cover outside the Antarctica and Greenland ice sheets account for only ~4% of the area and would yield a global sea-level rise of ~0.5 m (Dyurgerov and Meier, 1997). Small glaciers and ice caps comprise only ~3% of the total glacierized area but are very sensitive to climate changes and have contributed to ~20% of the sea-level rise during the last century (Cooper et al., 2002). As per the recent IPCC report on Polar regions, Svalbard and Jan Mayen had the lowest change observed between 2006 and 2015 of -9 ± 5 Gt year^{-1}. The biggest change in the Arctic region was found in Greenland followed by Arctic Canada north and south. Overall, the Arctic region lost mass at a rate of -213 ± 29 Gt year^{-1} contributing to a sea-level rise of 0.59 ± 0.08 mm year^{-1} (Meredith et al., 2019).

The Arctic region is changing swiftly in response to recent global warming. Understanding these changes, their causes, and predicting their future trajectories are a few of the many critical challenges in the present time. Observations from the recent changes over the last few decades are of key importance to assess the longer trend, rate, and magnitude of the impact (Cofaigh et al., 2018).

Understanding Present and Past Arctic Environments
DOI: https://doi.org/10.1016/B978-0-12-822869-2.00023-2

Svalbard Arctic possesses unique climatic condition due to the presence of diverse cryospheric elements such as sea ice, ice caps, permafrost, and glaciers with irregularly distributed in space and more significantly located in very different climatic regimes. The warm western Spitsbergen Current keeps the western coast ice free and having distinct characteristics from the rest of Svalbard Archipelago. Studies at Ny-Ålesund show two distinct atmospheric conditions in the annual cycle. The southerly winter flow bring warm and moist air to the Svalbard region, while colder and dry air mass during summer is brought from the north-east. General Circulation Model predictions suggest that the ongoing global and regional warming will affect the Arctic first. The huge ice masses stored in the Arctic have a significant role in governing the global environment including a significant contribution to sea-level rise; it needs our concern to develop our knowledge in the field of Arctic glaciology.

The glacier network provides a highly useful tool for monitoring spatial and temporal changes in climate so that it is extensively used for reconstructing and modeling past and future climatic scenarios. India expanded its polar cryospheric studies after the inclusion in the Arctic council in 2013 by adopting Vestre Broggerbreen and Feiringbreen glaciers near Ny-Ålesund, to carry out mass balance studies. These glaciers are influenced by sea ice, ocean, and atmospheric circulation and have unique climatic conditions of the polar environment that provide ample opportunity to understand glacier behavior and their interrelationship with climate. Therefore the present study focuses on change in glacier area and mass balance. Glacier area change and mass balance are important parameters to understand the health of a glacier under the changing climate as it gives information on both horizontal and vertical changes. Here, we have analyzed the aerial change of Vestre Broggerbreen glacier using the available Landsat imagery of 1993, 1998, 2010, and 2018 and also reported the in situ mass balance of the same glacier for the available study period 2011–17. Precipitation and temperature from Ny-Ålesund were analyzed to see the impact on area change and mass balance of the glacier.

12.2 Study area

Svalbard archipelago located on the north of mainland Norway is divided into few smaller islands. The largest island of Spitsbergen, with the main settlement of Longyearbyen and Ny-Ålesund, consists of mostly glaciers covering ~60% of its land area. Vestre Broggerbreen glacier lies at ~3 km south-west from the nearest settlement of Ny-Ålesund. It is a valley-type glacier with an area and length of ~3.58 km^2 and ~3.83 km, respectively

(Fig. 12.1). Vestre Broggerbreen is a debris-free north-east-oriented glacier extending from an elevation of ~100 to ~700 m asl. The region has one of the longest meteorological records in the Arctic, which indicates a relatively mild climate because of its surrounding ocean currents and the general atmospheric circulations (Nordli et al., 2014). The mean annual average temperature and precipitation during 1993–2018 was −4.05°C and 486 mm at Ny-Ålesund. The winter temperature is expected to increase by 10°C with an increase in annual precipitation by 2100 (Førland et al., 2011). Table 12.1 and Fig. 12.1 provide all the study area–related information.

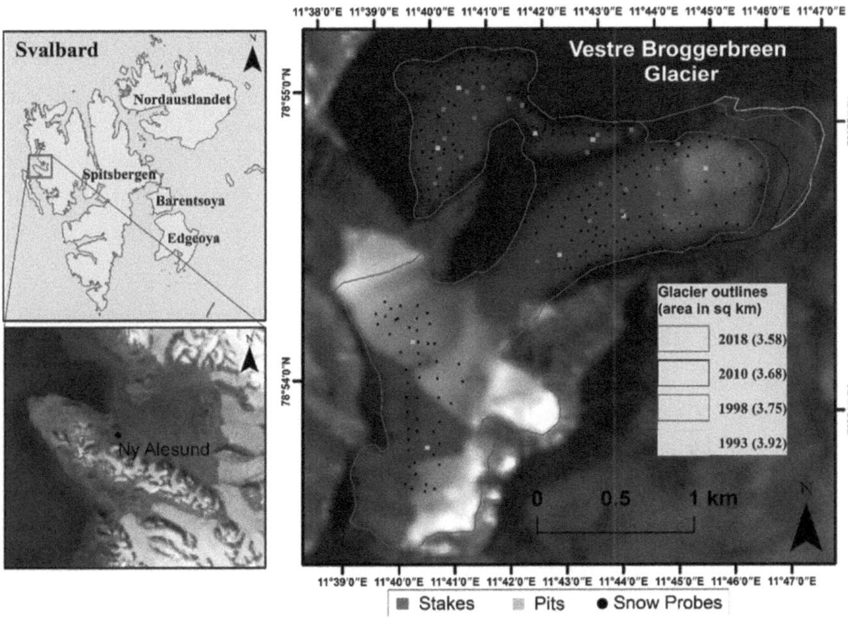

Figure 12.1 Location map of Vestre Broggerbreen glacier, ablation stakes, accumulation pits, and snow probes, Ny-Ålesund, Svalbard, Arctic.

Table 12.1 Details of the study area.

Details of Vestre Broggerbreen glacier	
Coordinates	78°54′43.83″N, 11°44′1.32″E
Region	Ny-Ålesund, Svalbard, Arctic
Area (2018)	~3.58 km^2
Length (2018)	~3.83 km
Orientation	North-east
Max and min elevation	~100–700 m asl
Mean annual temperature	~ −4.05°C (Ny-Ålesund, 1993–2018)
Mean annual precipitation	~486 mm (Ny-Ålesund, 1993–2018)

12.3 Methods

12.3.1 Glacier area and hypsometry

To detect and analyze the changes in Vestre Broggerbreen glacier, a multitemporal remote sensing technique was applied. The study utilizes freely available orthorectified and georeferenced images from 1993, 1998, 2010, and 2018 of Landsat series (TM, ETM + , and OLI) to derive glacier boundaries using a standardized semiautomatic approach based on red and shortwave infrared band ratio (Racoviteanu et al., 2009; Paul, 2001). This method is robust and time-efficient in delineating clean ice, but a manual correction of glacier boundaries is always required to correct the misclassified pixels due to snow patches, shadowed areas, cloud patches, etc. The threshold required for the image segmentation varies for different regions, and for the present study area, a threshold of 1.4 was used. Unfortunately, no suitable Landsat imagery is available before 1993 either due to cloud or snow cover. Glacier hypsometry was prepared using a digital elevation model (DEM) of the Norwegian Polar Institute (2014). Details of the imagery used in the present study are given in Table 12.2.

12.3.2 Mass balance

Annual glacier-wide mass balance measurements of Vestre Broggerbreen glacier started in 2011 using the direct glaciological method (Cuffey and Paterson, 2010). Total accumulation and ablation measurements were performed each year at the end of winter and summer season. Accumulation of solid precipitation was measured through snow pits and snow probes. During accumulation measurements, ~9 snow pits were dug to obtain the snow water equivalent from the density measurements. The density measurements were performed at every 10–20 cm profile depending upon the type of snow and refreeze layers. Additional snow probing (> 200 points) was also carried out throughout the glacier to measure the snow depth at various altitudes/locations (Fig. 12.1). Ablation measurement was carried out using a network of ~30 ablation stakes inserted 10–12 m deep in the ice with Heucke steam drill (Heucke, 1999) at different elevations. Ice density was assumed to be 900 kg m^{-3}, and in the presence of snow, the density was measured in the field.

The sum of accumulation and ablation was then integrated over the entire glacier surface area, and the mass balance was obtained using the following equation:

$$b_n = \sum b_i \frac{S_i}{S} \tag{12.1}$$

Table 12.2 Details of the satellite imagery used in the study.

ID	Sensor	Acquisition date	Source
LC08_L1TP_023241_20180817_20180829_01_T2	Landsat OLI	August 17, 2018	https://Earthexplorer.usgs.gov/
LE07_L1GT_220003_20100823_20161213_01_T2	Landsat ETM +	August 23, 2010	
LT05_L1GS_218003_19980715_20161223_01_T2	Landsat TM	July 15, 1998	
LT05_L1GS_220003_19930917_20180227_01_T2	Landsat TM	September 17, 1993	
Digital terrain model (DTM), Norwegian Polar Institute (2014)			https://data.npolar.no/

where b_i is the mass balance of an altitudinal area (s_i) obtained from the corresponding ablation and accumulation measurements, and S is the total surface area of the glacier (Cuffey and Paterson, 2010; Wagnon et al., 2007). The glacier area was delineated using a semiautomatic approach. Additional visual interpretation of the Landsat imagery was also made with the aid of several field surveys and photographs. The hypsometry of the glacier was extracted using DEM of the Norwegian Polar Institute (2014) (Table 12.2).

12.4 Results and discussion

12.4.1 Area change

The glacier areas for 1993, 1998, 2010, and 2018 have been calculated using a semiautomatic approach involving a band ratio algorithm and manual corrections on Landsat TM, ETM + , and OLI images. Details of changes in the area of Vestre Broggerbreen glacier are presented in Figs. 12.1–12.4. It was found that the area has decreased with varying rates during different periods. The glacier area of Vestre Broggerbreen was found to be 3.96, 3.90, 3.66, and 3.58 km^2 in 1993, 1998, 2010, and 2018, respectively (Figs. 12.1 and 12.2). The glacier witnessed the highest retreat of 0.24 km^2 (6.39%) during 1998–2010 at a rate of 0.02 $km^2\ a^{-1}$ followed by 2010–18 where the retreat was found to be 0.09 km^2 (2.44%) at 0.011 $km^2\ a^{-1}$. The period 1993–98 witnessed the lowest retreat during the whole study period with a retreat of 0.06 km^2 (1.53%) at 0.012 $km^2\ a^{-1}$. Overall, Vestre Broggerbreen glacier lost a total area of 0.39 km^2, equivalent to ~10% of the area, at a rate of 0.016 $km^2\ a^{-1}$ in 25 years (Fig. 12.3).

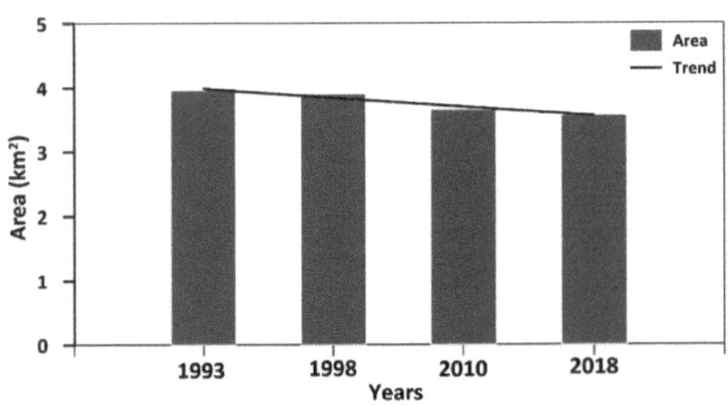

Figure 12.2 Area (km^2) of Vestre Broggerbreen in 1993, 1998, 2010, and 2018.

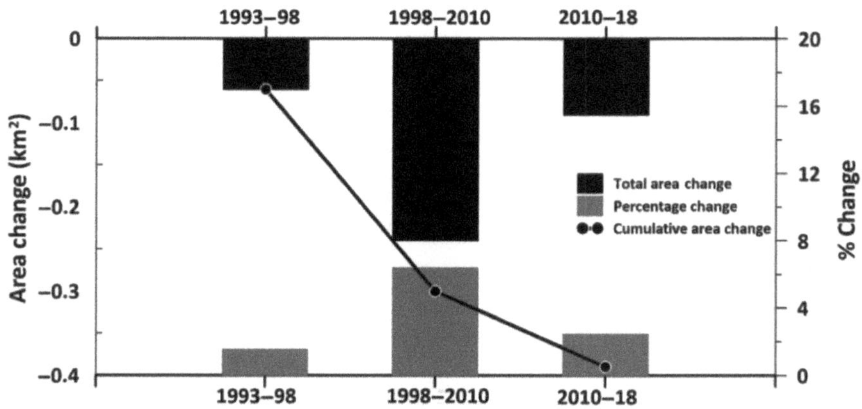

Figure 12.3 Area, percentage area, and cumulative area change of Vestre Broggerbreen glacier for 1993–98, 1998–2010, and 2010–18 periods.

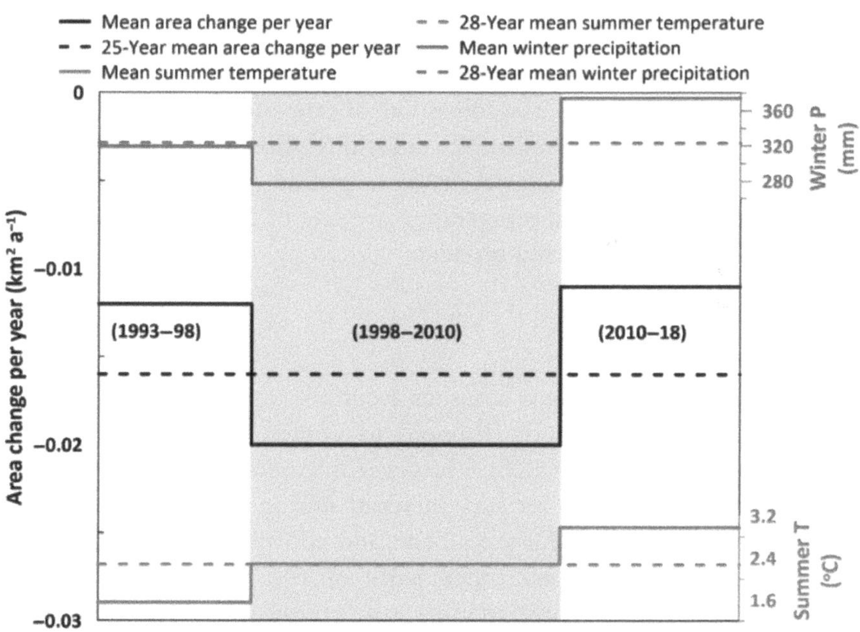

Figure 12.4 Periodical change in the area per year versus mean winter precipitation and summer temperature. The dashed lines represent the 25-year means.

The relationship among area change, temperature, and precipitation has also been analyzed. Since the area change information is only available for three periods, that is, 1993–98, 1998–2010, and 2010–18, the information was compared against mean summer/winter and 25-year mean summer/winter temperature and precipitation. It was found that the highest retreat rate of 1998–2010 (0.02 $km^2 a^{-1}$) was due to ~47 mm lower mean winter precipitation against the 25-year averages. However, the mean summer temperature during 1998–2010 was similar to the 25-year average, thus indicating winter precipitation control over the change in the area. The lowest retreat rate of the whole study period was found to be during 2010–18 (0.011 $km^2 a^{-1}$). During this period, both summer temperature and winter precipitation means were higher than the 25-year averages, thus resulting in compensating one another's impact on the glacier retreat. The mean annual temperature and precipitation were found to be 0.7°C and 50 mm higher than the 25-year averages. The period 1993–98 witnessed a similar retreat rate as in 2010–18 with different impacts of summer temperature and winter precipitation over the retreat. During this time mean summer temperature was 0.75°C lower with similar precipitation as of 25-year averages resulting in a retreat similar to the period 2010–18. This indicates that the summer temperature was most probably controlling the retreat during 1993–98. The study of glacier dynamics is complicated as it involves lots of processes other than temperature and precipitation, for example, radiations (short and longwave), surface temperature, supra glacier meltwater, ground heat flux, and albedo. However, precipitation and temperature are one of the major drivers for a retreat of Vestre Broggerbreen glacier.

12.4.2 Mass balance

Vestre Broggerbreen glacier is a winter accumulation type valley glacier as the majority of accumulation happens in winter, whereas in summer, ablation is dominant (Fig. 12.5). However, in theory, solid precipitation (depending upon the amount and duration) during summer can alter the ablation process by changing the albedo and contribute to accumulation. It has been reported from other parts of the world, for example, Himalaya, that the solid precipitation hampers the ablation during summer, but for this glacier, there is not enough data to prove the same.

Annual and cumulative MB of Vestre Broggerbreen glacier for six hydrological years (2011–17) with respective summer, winter, and annual precipitation and temperature are shown in Fig. 12.6. Mass balance during the entire study period was negatively ranging between −0.08 and −1.22 m w.e. The glacier witnessed a higher loss in 2015/16 (−1.22 m w.e.) followed by 2012/13 (−0.94 m w.e.) and 2014/15 (−0.79 m w.e.),

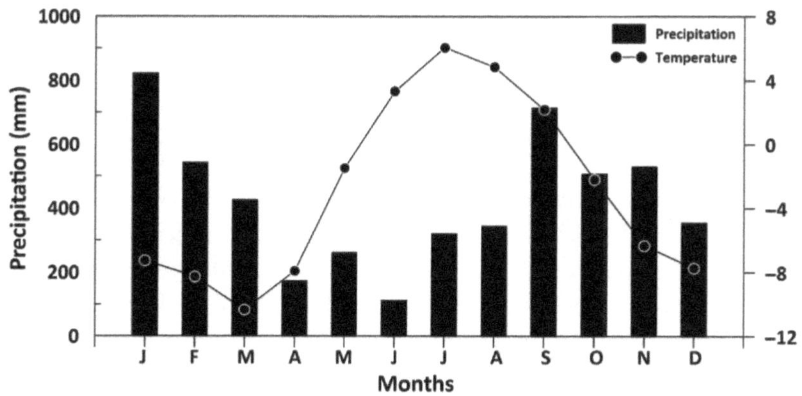

Figure 12.5 Monthly mean precipitation and temperature from 2011 to 2017 at Ny-Ålesund, Svalbard, Arctic.

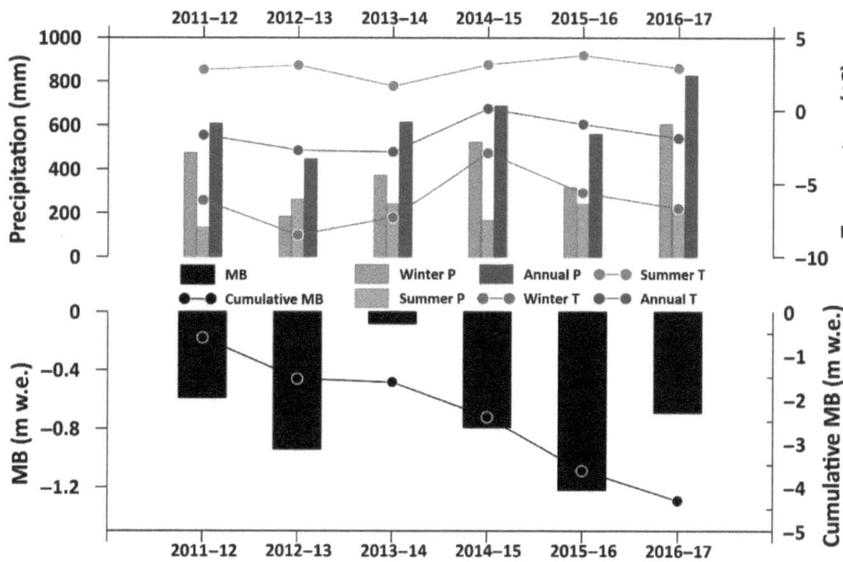

Figure 12.6 Annual and cumulative mass balance, winter, summer, and annual temperature and precipitation from 2011 to 2017.

while the loss was comparatively lower during 2011/12 (−0.59 m w.e.) and 2016/17 (−0.69 m w.e.). During 2013/14, the glacier was in a nearly balanced condition of −0.08 m w.e. (Patel et al., 2019).

It has been observed that during high negative mass balance years (2015–16 and 2012–13), winter precipitation was comparatively lower together with a high summer temperature. Whereas during the years with lower negative mass balances (2011–12, 2014–15, and 2016–17), winter precipitation was found to be comparatively higher with higher summer temperature, thus compensating the impact of summer temperature to some extent. The lowest mass balance year (2013–14) was due to the lower summer temperature and moderate winter precipitation. A strong relationship between summer temperature and mass balance was observed with $R^2 = 0.97$ at $P < .05$. However, no good relationship between mass balance and other parameters (summer, winter and annual precipitation, winter and annual temperature) was found. Throughout the entire study period, the hydrological year 2015/16 witnessed the highest mean summer temperature (3.78°C) leading to the highest mass loss, whereas the year 2013/14 witnessed the lowest mean summer temperature resulting in a least negative mass balance. This suggests that summer temperature is the major driver for mass balance.

12.5 Conclusion

The study reports changes in the area from 1993 to 2018 and mass balance from 2011 to 2017. The area changes were calculated using a combination of band ratio technique and manual corrections on the available Landsat images of summertime. It was found that the glaciated area of Vestre Broggerbreen glacier has decreased from 3.96 km^2 in 1993 to 3.57 km^2 in 2018, at a varying rate ranging between 0.011 and 0.02 km^2, resulting in a total area loss of 0.39 km^2 (~10% at 0.016 $km^2\,a^{-1}$). A comparatively rapid decrease in the glaciated area was found during 1998–2010 (0.02 $km^2\,a^{-1}$), whereas a similar retreat rate as in 1993–98 (0.011 $km^2\,a^{-1}$) was found in 2010–18 (0.012 $km^2\,a^{-1}$). One of the major factors of influencing these retreats was winter precipitation and summer temperature.

Mass balance was calculated through direct glaciological method using ablation and accumulation information. Ablation measurements were carried out by measuring a network of ~30 stakes installed on the glacier at the end of every summer, whereas accumulation measurements were performed through several snow pits together with a network of 200+ snow depth measurements each year at the end of every winter. Mass balance of Vestre Broggerbreen glacier was found to be negative throughout the entire study period (2011–2018). Mass balance ranged between −0.08 (2013–14) and −1.22 m w.e. (2015–16) with a cumulative mass balance of −4.31 m w.e. There is a strong relationship between mass balance and summer temperature with $R^2 = 0.97$ at $P < .05$; however, no good correlation was found with others studied parameters (winter, summer, annual precipitation and winter, annual temperature).

Acknowledgments

The authors acknowledge the Jawaharlal Nehru University for the laboratory and library facilities. We also thank NCPOR, GSI Govt. of India for the help during fieldwork and expeditions. We thank all the members who directly or indirectly contributed to the study. The authors also thank the USGS for the Landsat images and the Norwegian Meteorological Institute for the Ny-Ålesund temperature and precipitation data.

References

Cofaigh, C.Ó., Kirchner, N., Fedorov, G., Noormets, R., de Vernal, A., 2018. Arctic environmental change beyond instrumental records: introduction and overview. Arktos 4 (1), 27. Available from: https://doi.org/10.1007/s41063-018-0061-z.

Cooper, R.N., Houghton, J.T., McCarthy, J.J., Metz, B., 2002. Climate change 2001: the scientific basis. Foreign Aff. 81 (1), 208. Available from: https://doi.org/10.2307/20033020.

Cuffey, K., Paterson, W.S.B., 2010. The Physics of Glaciers, fourth ed. Butterworth-Heinemann/Elsevier.

Dyurgerov, M.B., Meier, M.F., 1997. Mass balance of mountain and subpolar glaciers: a new global assessment for 1961-1990. Arct. Alp. Res. 29 (4), 379. Available from: https://doi.org/10.2307/1551986.

Førland, E.J., Benestad, R., Hanssen-Bauer, I., Haugen, J.E., Skaugen, T.E., 2011. Temperature and precipitation development at Svalbard 1900–2100. Adv. Meteorol. 2011, 1–14. Available from: https://doi.org/10.1155/2011/893790.

Heucke, E., 1999. A light portable steam-driven ice drill suitable for drilling holes in ice and firn. Geografiska Annaler 7.

Meredith, M., Sommerkorn, M., Cassotta, S., Derksen, C., Ekaykin, A., Hollowed, A., et al., 2019. Polar Regions. In: Pörtner, H.-O., Roberts, D.C., Masson-Delmotte, V., Zhai, P., Tignor, M., Poloczanska, E., Mintenbeck, K., Alegría, A., Nicolai, M., Okem, A., Petzold, J., Rama, B., Weyer, N.M. (Eds.), IPCC Special Report on the Ocean and Cryosphere in a Changing Climate. In press.

Nordli, Ø., Przybylak, R., Ogilvie, A.E.J., Isaksen, K., 2014. Long-term temperature trends and variability on Spitsbergen: the extended Svalbard Airport temperature series, 1898–2012. Polar Res. 33 (1), 21349. Available from: https://doi.org/10.3402/polar.v33.21349.

Norwegian Polar Institute, 2014. Terrengmodell Svalbard (S0 Terrengmodell) [DTM]. Norwegian Polar Institute. https://doi.org/10.21334/npolar.2014.dce53a47.

Østby, T.I., Schuler, T.V., Hagen, J.O., Hock, R., Kohler, J., Reijmer, C.H., 2017. Diagnosing the decline in climatic mass balance of glaciers in Svalbard over 1957–2014. Cryosphere 11 (1), 191–215. Available from: https://doi.org/10.5194/tc-11-191-2017.

Patel, L., Sharma, P., Thamban, M., 2019. Spatio-temporal variability of snow water equivalent over the Vestre Broggerbreen and Feiringbreen glaciers, Ny-Ålesund, Svalbard. J. Earth Syst. Sci. 128 (7), 183. Available from: https://doi.org/10.1007/s12040-019-1224-4.

Paul, F. 2001. Evaluation of different methods for glacier mapping using Landsat TM. Workshop on Land Ice and Snow, Dresden/FRG, 16-17 June, 2000. EARSeL eProceedings 1, 239-45.

Racoviteanu, A.E., Paul, F., Raup, B., Khalsa, S.J.S., Armstrong, R., 2009. Challenges and recommendations in mapping of glacier parameters from space: results of the 2008 Global Land Ice Measurements from Space (GLIMS) workshop, Boulder, Colorado, USA. Ann. Glaciol. 50 (53), 53–69. Available from: https://doi.org/10.3189/172756410790595804.

Wagnon, P., Linda, A., Arnaud, Y., Kumar, R., Sharma, P., Vincent, C., et al., 2007. Four years of mass balance on Chhota Shigri Glacier, Himachal Pradesh, India, a new benchmark glacier in the western Himalaya. J. Glaciol. 53 (183), 603–611. Available from: https://doi.org/10.3189/002214307784409306.

CHAPTER 13

Spatiotemporal variability of snowmelt onset across Svalbard inferred from scatterometer data (2000–2017)

Alvarinho J. Luis[1], Kirti K. Mahanta[2] and Shridhar D. Jawak[3]
[1]Polar Remote Sensing Section, National Centre for Polar and Ocean Research, Ministry of Earth Sciences, Vasco da Gama, India
[2]Odisha Space Applications Centre (ORSAC), Department of Science & Technology, Government of Odisha, Bhubaneswar, India
[3]Svalbard Integrated Arctic Earth Observing System (SIOS), SIOS Knowledge Centre, Svalbard Science Centre, Longyearbyen, Norway

13.1 Introduction

The Arctic region has undergone a significant change with a decline in the sea ice extent (Fetterer et al., 2017), increases in the air temperature (Box et al., 2019), enhancement in the glacier retreat (Kaser et al., 2006), and accelerated thawing of permafrost leading to more carbon release (Turetsky et al., 2019), a positive trend in the Arctic precipitation and river discharge (Vihma et al., 2016). Melting Arctic land ice is the largest global source of sea-level rise, accounting for 48% of the contribution during 2003–10 (AMAP, 2017) and 30% of the total sea-level rise since 1992 (Box et al., 2018). In the global climate system, the cryosphere–climate feedback mechanisms operate over a wide range of spatial and temporal scales. Snow and ice have a remarkable effect on the climate, as they modulate the energy exchanges between the surface and the atmosphere because of their physical properties. One of the most important properties of the cryosphere is albedo, which has a positive feedback on climate, with significant change from 80%–90% for dry snow to about 50% for the snow melt. Therefore monitoring the spatiotemporal variability of the snow melt is therefore essential to account for the glacier surface energy and the mass balance and melt-water runoff (Wiscombe and Warren, 1980; Winther, 1993).

Understanding Present and Past Arctic Environments
DOI: https://doi.org/10.1016/B978-0-12-822869-2.00009-8

Before the advent of remote sensing, observations of the geophysical condition of the ice sheets were possible only through a limited number of in situ measurements. Thanks to the remoteness of Svalbard and the harsh climate, the remote sensing data are the only reasonable means to obtain daily year-round observations of the entire islands' snow cover properties. Microwave remote sensing has an obvious advantage in detecting the ice sheet snow melt independent of the weather, night condition, and clouds. The microwave scatterometers measure backscatter (σ^0), calculated as the ratio of emitted energy to the energy received.

Several researchers have attempted to detect the snow melt over the Arctic sea ice using passive and active microwave sensors (Anderson, 1987; Garrity, 1992). Using microwave with frequencies ranging from 0.3 to 300 GHz and different polarization (*HH* or *VV*), remote sensing can be accomplished either through passive microwave observations by measuring the emitted radiation with a radiometer or by using an active microwave sensor to measure the intensity of the echo or the return signal sent by a radar. The changes in the snow surface are observed by a scatterometer as changes in the microwave signature, therefore by closely monitoring the microwave signature and its variation over time, the temporal variability of the physical properties, such as melt and refreeze, can be estimated.

Microwave remote sensing provides numerous advantages which are lacking in other satellite sensors, which include observations on the snow cover area, the snow water equivalent, and the detection of liquid water in the snowpack at the onset of melt (Kunzi et al., 1982). The physical characteristics of the snowpack determine its microwave properties, such as the volume scattering by the snow grains within the snow layer, resulting in microwave emission at the top of the snow surface being less than the ground emission. The properties affecting the microwave response from a snowpack include the depth and water equivalent, the liquid water content, density, grain size and shape, temperature and stratification, as well as the snow state and land cover. The high sensitivity of the active microwave radiation (Tedesco, 2007) to snow melt and higher spatial resolutions (Long, 2010; Ulaby et al., 1982) make it an ideal tool to monitor the state of snow cover, compared to the passive microwave sensors.

The microwave measurements of T_b and σ^0 are appropriate variables for estimating the melt duration (MD) and extent. The occurrence of even small amounts of liquid water in the snowpack radically changes the dielectric properties of the snow at microwave frequencies and polarizations, causing significant changes in the microwave measurements, enabling effective melt detection (Ashcraft and Long, 2006). Scatterometers have been used for monitoring melt processes over the ice sheets of Antarctica and Greenland (Abdalati, 2007; Abdalati and Steffen, 2001; Li et al., 2017; Steffen et al., 2004; Wang et al., 2007; Rotschky et al., 2011; Wang et al., 2005) and the low-latitude mountainous regions (Panday et al., 2011). In contrast, relatively few studies have used backscatter data to detect and monitor the timing of melt onset on the Arctic ice caps (Smith et al., 2003), the melt extent and duration on the Greenland ice sheet (Joshi et al., 2001), and the melt onset of Arctic sea ice (cf. Forster et al., 2001). The monitoring of mass balance is often carried out on fairly small glaciers that have been selected partly for logistical reasons rather than the optimum scientific priorities (Hagen et al., 2003). Moreover, the remoteness of many Arctic glaciers combined with the harsh weather conditions complicates the incessant acquisition of reliable data.

Using a sequence of daily σ^0 images from the NASA scatterometer, Forster et al. (2001) found a spatially continuous advancement of decreasing σ^0 due to snowmelt onset on sea ice across the Arctic Ocean in spring 1997. Their efforts led to a snowmelt map with the melt onset dates for the Arctic Ocean using a snowmelt-onset detection algorithm based on the time derivative of σ^0. Using the *Ku*-band (13.4 GHz) *V*-polarized σ^0 from Seawinds/QuikSCAT, Kunz and Long (2006) determined periods of surface freeze and melt cycles on the Antarctic ice shelves. Using Seawinds/QuikSCAT scatterometer-based σ^0, innovative algorithms have been developed to monitor the ice layer extent, snowmelt, and the snow accumulation on Greenland with validations using in situ data from the Greenland Climate Network, and Antarctica (Nghiem et al., 2007). Their results revealed a significant increase in the snow accumulation rate on the Greenland ice sheet, including the west flank in January–March 2005. These studies demonstrate the effective use of σ^0 from microwave scatterometers to map melt–freeze cycles in the rugged and inaccessible terrain.

Over the last three decades, the Svalbard Archipelago has experienced Arctic amplification, which is an increase in the air temperature nearly twice that of the global average (cf. Cohen et al., 2014). What is its impact on the reduction in snow coverage due to enhanced melting in many regions? To gain an insight into this question, we address the spatiotemporal variability of two parameters for the summer period: total days of melt or MD and melt onset date (SMO), by employing a suite of microwave scatterometers to detect the surface melt and MD across

Svalbard, based on the high sensitivity of radar backscatter to liquid water in snow. We compare the σ^0 using the air temperature data from the closest weather stations to Svalbard. Extending the work of Rotschky et al. (2011), we computed SMO and MD using Ocean *Scatterometer* (OSCAT), advanced scatterometer (ASCAT), and OSCAT-2 for the post-2009 period for the glaciated areas of Svalbard.

The chapter is organized as follows. The study area is introduced in Section 13.2, and the data and methodology used are outlined in Section 13.3. The results are presented in Section 13.4 and the inferences are discussed in Section 13.5. The conclusions of this work are summarized in Section 13.6.

13.2 Study area and data

13.2.1 Study area

The archipelago of Svalbard (74°–81°N, 10°–35°E, see Fig. 13.1) is located in the North Atlantic sector of the Arctic Ocean. Of the ~57% glacierized regions (Hagen et al., 2003) there is a mix of cirque and valley glaciers, ice fields, and ice caps (Nuth et al., 2013). The in situ mass balance inventory is available for about 30% of Svalbard's glaciers (Nuth et al., 2013), implying a need for remote sensing data to better understand the regional variations in their dynamics. The Svalbard glaciers show a declining mass–balance trend, when the first glaciological measurements began in 1967 (Østby et al., 2017). There are main islands that include the largest island Spitsbergen, and also Nordaustlandet and Edgeøya, which cover a total area of about 61,022 km^2. The archipelago is situated to the north-western sector of the Barents Sea.

The Gulf Stream transports warm Atlantic water northwards along the coast of Norway, where it bifurcates into two main branches. One branch is the West Spitsbergen Current that flows along the west coast of Spitsbergen. The cold East Spitsbergen Current flows along the east coast of the archipelago from north of Nordaustlandet to the south of Edgeøya, and the southern parts of Spitsbergen, before it turns north on the west coast of Spitsbergen. The climate of Svalbard is highly influenced by the variability and trends in the wintertime sea ice extent (Divine and Dick, 2006; Day et al., 2012), as well as changes in the atmospheric circulation patterns (Hanssen-Bauer and Førland, 1998).

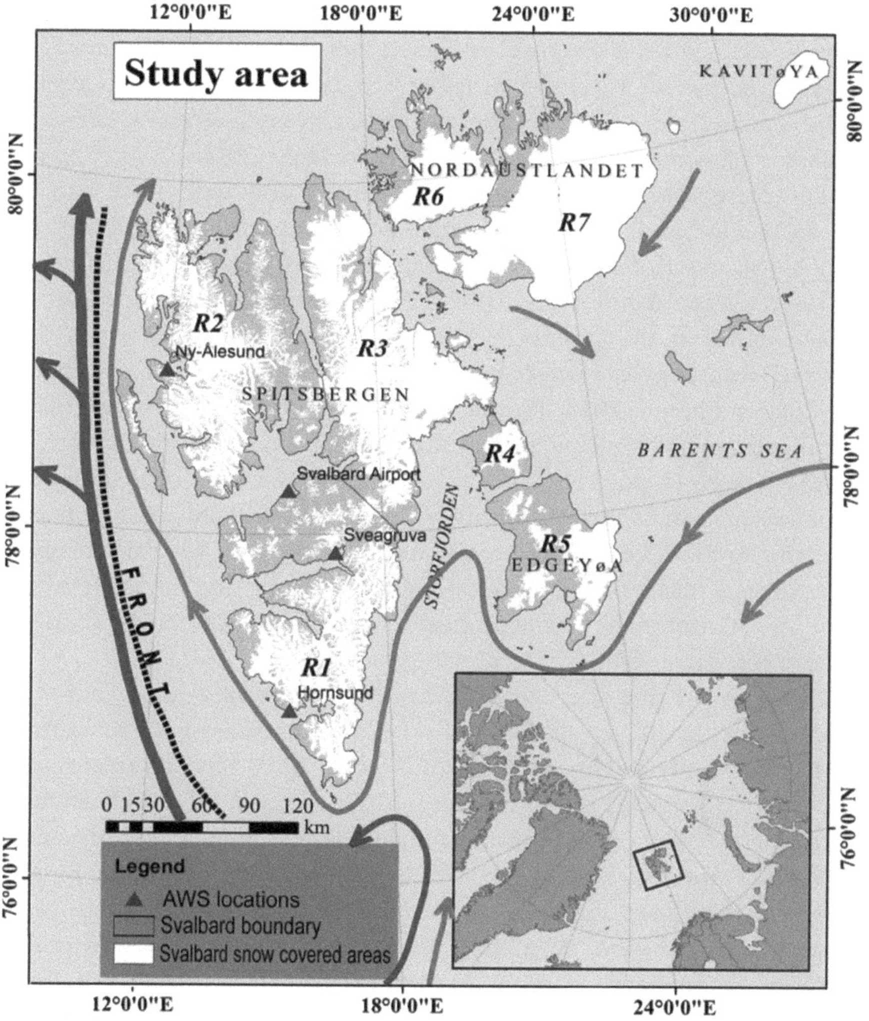

Figure 13.1 Study area indicating the different regions (R1–R7) referred to in this study. The glaciated regions on Svalbard are shown in white. Shown are northward-flowing warm West Spitsbergen Current (thick line), the cold East Spitsbergen Current (thin line), and the frontal area between the two types of current. Meteorological data from weather stations used in this study are indicated by the symbol ▲. *Adapted from Svendsen, H., Beszczynska-Møller, A., Hagen, J.O., Lefauconnier, B., Tverberg, V., Gerland, S., et al., 2002. The physical environment of Kongsfjorden–Krossfjorden, an Arctic fjord system in Svalbard. Polar Res. 21 (1), 133–166. https://doi.org/10.3402/polar.v21i1.6479L.*

Using a coupled surface energy balance—snow/firn/soil model with 3-hourly temporal and 1 km × 1 km spatial resolution, van Pelt et al. (2019) inferred an area-averaged positive climatic mass balance of 0.09 m w.e. a^{-1} (water equivalent per annum) and a significant negative long-term trend of −0.06 m w.e./a/decade for the entirety of Svalbard for the period 1957—2018.

The meteorological conditions are spatially and temporally variable due to the location at the confluence zone between the cold and dry polar air masses from the north and more warm and humid air masses from the warm Atlantic currents to the southwest. A strong seasonal difference exists in the observed temperature trends, with a more pronounced warming in winter/spring (Førland et al., 2011a,b,c; Bintanja and van der Linden, 2013; Nordli et al., 2014). The region is dominated by a highly variable precipitation in the form of rain and snow throughout the year due to its position at the polar front, formed as a consequence of a convergence of warm north Atlantic current and cold ocean currents from the Arctic Ocean, Nordic seas, and Barents Sea. Overall, Svalbard experiences a relatively wet and mild climate, due to the influence of the maritime environment and the heat provided by the North Atlantic Current (Hisdal, 1985).

Over the last three decades, Svalbard has experienced its greatest temperature rise (Nordli et al., 2014; Isaksen et al., 2016), with a warming trend of 2.6°C per century over the 1898—2012 period for the air temperature (T_a) data recorded at Svalbard airport (Nordli et al., 2014). The atmospheric circulation over Svalbard is strongly influenced by cold northeasterly polar winds from the Barents Sea, which are frequently interrupted by passages of low-pressure disturbances from the southwest, transporting moist, maritime air northward. As a consequence, there are both west—east and south—north gradients of snow accumulation and T_a in Svalbard (Sand et al., 2003), onto which large interannual variations are overlaid (Førland and Hanssen-Bauer, 2003). The long-term trends in the annual precipitation tend to be positive over Svalbard, but the precipitation level has been relatively constant since 1990 (Førland et al., 2011a,b,c), despite of the fact that temperatures have continued to rise. This may have a bearing on the regime shift in the large-scale atmospheric circulation in the area (Overland and Wang, 2010). The archipelago is frequently influenced by extratropical cyclones associated with the North Atlantic cyclone track, resulting in large precipitation events in winter. The precipitation at Ny Ålesund is distributed fairly evenly over much of the year, with annual total precipitation amounting to ~423 mm (Serreze et al., 2015).

13.2.2 Data

13.2.2.1 Scatterometer backscatter data

In this study, the σ^0 data have been used to monitor Svalbard's melt and refreeze cycles over the 2000–17 period. For each year we registered SMO and MD using the σ^0 observations from a suite of scatterometers spanning 2000–17 (Table 13.1). Since the melt in Svalbard can occur at any time of the year, even in winter, this study is restricted to the summer period from April to September months, where an appreciably high melt is observed. Compared to other satellite radars, the microwave scatterometers obtain data at a high temporal resolution, but at a coarser spatial resolution. They have been used elsewhere to map the annual melt extent and duration across major Arctic ice caps (Smith et al., 2003; Wang et al., 2005) and the Greenland Ice Sheet (e.g., Wismann, 2000; Nghiem et al., 2001; Steffen et al., 2004). The data from *Ku*-band (13.4 GHz) SeaWinds/QuikSCAT satellite spanned June 1999 to November 2009. The *H*-polarized, slice-based scatterometer image reconstruction (SIR) products (Hicks and Long, 2011; Long et al., 2001) incidence angle normalized σ^0 at 40° incidence were used. The SIR algorithm produces significantly enhanced spatial resolution of σ^0 images by taking advantage of the measurement overlap from multiple passes of the scatterometer over the target site, rendering a resolution better than the main lobe aperture resolution of the sensor (Early and Long, 2001). As demonstrated by Sharp and Wang (2009), the spatial resolution enhancement using the SIR algorithm (Early and Long, 2001; Long and Hicks, 2005) enables melt studies to be carried out at finer spatial scales. Furthermore, the temporal resolution of SIR data is adequate for investigating the seasonal melt–freeze dynamics (Mortin et al., 2014).

The OSCAT, which was designed and built by the Indian Space Research Organization, operates at 13.515 GHz with a pencil beam system that works in both *HH* and *VV* polarizations. The data spanned November 2009 to February 2014, consisting of daily images with *HH* polarization. The ASCAT is a microwave scatterometer launched by the European Space Agency and operated by the European Organization for the Exploitation of Meteorological Satellite. The satellite carries a *C*-band operated at 5.26 GHz and uses *VV* polarization to measure the returned backscatter. Since ASCAT is unable to cover the entire Arctic region daily, we used 2-day aggregate ASCAT SIR data from March 2014 to September 2016 normalized to a 40° incidence angle.

Table 13.1 Characteristics of scatterometers used in the study.

Satellite	QuikSCAT	Oceansat-2	Meteorological Operational Satellite-A	Scatsat-1
Owner	NASA	ISRO	ESA	ISRO
Sensor/ Instillment	SeaWinds	OSCAT	ASCAT	OSCAT-2
Band and frequency	Ku (13.4 GHz)	Ku (13.515 GHz)	C: (5.255 GHz)	Ku (13.515 GHz)
Polarization	HH	HH	VV	HH
Swatli (kin)	1400 swath-HH (inner beam) 1800 swath for W (outer beam)	1400 (IIII and W) 1400–1800 (VV)	Two swaths of 550 on either side of satellite pass	1800 (VV) 1400 (HH)
Type	24-hr composite			
Native resolution	Egg: 25 × 35 km Slice: 6 × 25 km	Egg: 30 × 68 km Slice: 6 × 30 km	SZO: 25 × 50 km SZF: 4 × 20 km	Egg: 30 × 68 km Slice: 6 × 30 km
Temporal resolution	Daily			
Spatial resolution	2.25 km		4.45 km	2.25 km
Incidence angle (range)	46° and 54.4°	49° and 57°	Variable: 25° – 65°	57°
Products	σ^0			
Dataset date	July 01, 2000 to November 23, 2009	November 05, 2009 to February 28, 2014	July 01, 2013 to June 30, 2017	October 20, 2016 to February 28, 2018
Data source	Scatterometer Climate Record Pathfinder, BYU http://www.scp.byu.edu			MOSDAC https://www.mosdac.gov.in

OSCAT-2 is a *Ku*-band scanning scatterometer microwave radar with an incidence angle of 57° and is operated at 13.515 GHz. It was launched onboard Scatsat-1 satellite in October 2016 by the Indian Space Research Organisation. We used data for the year 2017. The missing backscatter data for more than 100 days in the summer of 2010 were excluded for the analysis.

The analysis of σ^0 from *Ku*-band scatterometers is preferred because these measurements are made using dual polarization, which can provide additional information when discriminating surface characteristics. On the other hand, European Remote Sensing scatterometers (ERS-1 and -2) were operated in *C*-band (5.4 GHz or 5.4 cm) at single polarization (*VV*), with an incidence angle ranging from 18° to 57°, illuminating a 500 km wide swath on the right hand side of the satellite. At a lower incidence angle, the σ^0 from snow is stronger at *H* polarization than at *V* polarization (Wang et al., 2007). Using ERS wind scatterometers, the melt onset and temporally integrated backscatter reduction were estimated for 14 small Arctic ice caps from 1992 to 2000, which showed an interannual and regional variability in the timing of the melt onset that were comparable, averaging 33 ± 9 and 38 ± 13 days, respectively.

13.2.2.2 Weather station data

The weather data from Svalbard (▲ in Fig. 13.1) were used to calibrate and validate the melt detection method. The details of the AWS locations and sensor accuracy are given in Table 13.2. The weather stations at Svalbard are maintained by the Norwegian Meteorological Institute and provide a continuous T_a record and precipitation (accumulated over 24 hours) over a long period (Førland et al., 2011a,b,c), but restricted to southern and western Spitsbergen. Precipitation is reported only once every day, with 24 hours accumulated precipitation in mm. Standard observation times for weather stations are 7:00, 13:00, and 19:00 Norwegian standard time. Since 1989, the annual T_a at the different stations over the Svalbard region has increased by 1.0°C–1.2°C decade^{-1}, with a dramatic wintertime warming of 2°C–3°C decade^{-1} (Førland et al., 2011a,b,c). The Svalbard region has experienced an increase in annual precipitation (Førland et al., 2011a,b,c).

13.3 Methods

13.3.1 Melt onset detection criteria

The major detection methods of snow melting and freezing from scatterometer data are classified into the following four categories. The first is

Table 13.2 Details of sensors operated at the weather stations in Svalbard.

	Geographic location	Region	WMO number	In-operation since	Sensor	Data resolution
Ny-Alesund	11.93°E, 78.92°N	R2 (western Spitsbergen)	01007	August 1974	Temp. accuracy: ± 2°C Wind accuracy: 0.4 m s^{-1} (for wind up to 60 m s^{-1}) Make: Vaisala	Daily mean
Svalbard Airport	15.50°E, 78.24°N	R1 (eastern Spitsbergen)	01008	August 1975	Temp. accuracy: ± 2°C Wind accuracy: 0.4 m s^{-1} (for wind up to 60 m s^{-1}) Make: Vaisala	Daily mean
Sveagruva	16.72°E, 77.89°N	R1 (eastern Spitsbergen)	01005	May 1978	Temp. accuracy: ± 2°C Wind accuracy: 0.4 m s^{-1} (for wind up to 60 m s^{-1}) Make: Vaisala	Daily mean
Hornsund	15.53°E, 77.00°N	R1 (eastern Spitsbergen)	01003	June 1985	Temp. accuracy: ± 2°C Wind accuracy: 0.4 m s^{-1} (for wind up to 60 m s^{-1}) Make: Vaisala	Daily mean

by using a single-threshold method (Nghiem et al., 2001; Rotschky et al., 2011). The second method is by using a double-threshold method (Brown et al., 2007; Wang et al., 2008, 2005). These two methods are used to determine the threshold value for classifying melting and non-melting snow. The third method is based on the physical models, wherein the threshold used for discriminating melting and nonmelting snow is derived from the physical characteristics of snow, such as snow depth and snow liquid moisture content (Ashcraft and Long, 2006; Hicks and Long, 2011). The fourth method uses algorithms to detect melting over the ice sheets, for example, the double-threshold snowmelt detection method which is based on the backscattering difference between ascending orbit pass and descending orbit pass (Steffen et al., 2004), the classification method which combines the polarization channel difference, the maximum likelihood model (Kunz and Long, 2006), and so on.

For melt detection, there is a need to employ an empirical threshold method that is commonly applied elsewhere (Steffen et al., 2004; Sharp and Wang, 2009). The microwaves easily penetrate a dry snowpack, thus the backscatter contributions from surface scattering at the air–snow interface can be neglected. From field observations (Rott et al., 1993) it has been found that the penetration depth is around 8 m at 10.3 GHz (X-band) for a dry winter snowpack. The intensive backscattering in winter is thus largely a function of volume scattering from snow grains and internal layers of the snowpack (Stiles and Ulaby, 1980). Regional differences as well as smaller temporal fluctuations in σ^0_{mean} arise from the topography and from the variations in snowpack properties, that is, snow density, dielectric properties, internal layering, effective grain size, and surface roughness (Legrésy et al., 1998; Long and Drinkwater, 1994; Rott et al., 1993) due to fresh snowfall, snow compaction, grain growth, formation of sun or wind crusts, or changes in snow temperature. For wet snow, which is highly absorptive for microwaves (Ulaby et al., 1982), volume scattering is absent due to superficial penetration into the snowpack (Mätzler and Hüppi, 1989).

Using the single threshold method, this study reports SMO and MD for each pixel for the period 2000–17. This approach is based on the dielectric properties of wet snow and dry snow. The high dielectric constant for wet snow significantly reduces volume scattering while increasing microwave absorption, resulting in a reduced backscatter coefficient with the introduction of a small quantity of liquid water (Stiles and Ulaby, 1980). Since the near-surface ice or the land area can affect the backscatter, we masked the nonglaciated areas over Svalbard. To detect the spatial variations in backscatter resulting from differences in snowpack morphology, a dynamic σ^0 threshold ($\sigma^0{}_{\mathrm{Tr}}$) was used. Thus, the primary algorithm specifies for every day in each pixel, a melting pixel (MP) is defined as:

$$\mathrm{MP} = 1, \quad \text{if} \ \ \sigma_i^0 < \left(c \times \overline{\sigma_w^0}\right) \tag{13.1}$$

$$\mathrm{MP} = 1, \quad \text{if} \ \ \sigma_i^0 \geq \left(c \times \overline{\sigma_w^0}\right), \tag{13.2}$$

where σ_i^0 is the daily backscatter value for a pixel in dB, $\overline{\sigma_w^0}$ is the mean winter σ^0 for each image pixel spanning January to March, and c is a constant assigned a value of 1.35, which is empirically calculated using T_{a} from the weather station and σ^0 data, following Rotschky et al. (2011).

MD was calculated over the glaciated areas of Svalbard (Fig. 13.1) and then spatially mapped for each year spanning 2000–17. The glaciated areas were identified using the map hosted by the Norwegian Polar Institute (https://geodata.npolar.no/). MD is cumulative of total melt days during the summer. The summer melt onset (SMO) is the day on which the backscatter coefficient remained below σ_{Tr}^0 for 10 or more consecutive days. We set days ≥ 10 as the threshold to distinguish SMO from earlier short-term melt events that cause considerable σ^0 variation. Rotschky et al. (2011) used the threshold as 7 days. As the snow starts melting at the beginning of summer, a sharp dip in σ^0 was encountered. Likewise, in late summer/autumn, it increases gradually. In the late summer, when σ^0 recovers slowly, the choice of c has a slight impact on MD because of the slope and signal noise at the end of the melt season. Since no sharp change on σ^0 to the end of melting was observed, we modified the threshold by $\pm 10\%$ which yielded MD within ± 10 days.

13.4 Results

First, we compare σ^0 derived from *Ku*-b and QuikSCAT and OSCAT and *C*-band ASCAT to check the incidence angle behavior of the backscatter signal. We plotted σ^0 over the overlapping period 5–16 November, 2009, averaged over R7 in Fig. 13.2. The shorter wavelength of QuikSCAT ($\sim$3 cm) restricts deeper penetration into the snowpack, so lower σ^0 values are observed compared to *C*-band ASCAT. The root mean square error in σ^0 between QuikSCAT and OSCAT (ASCAT)

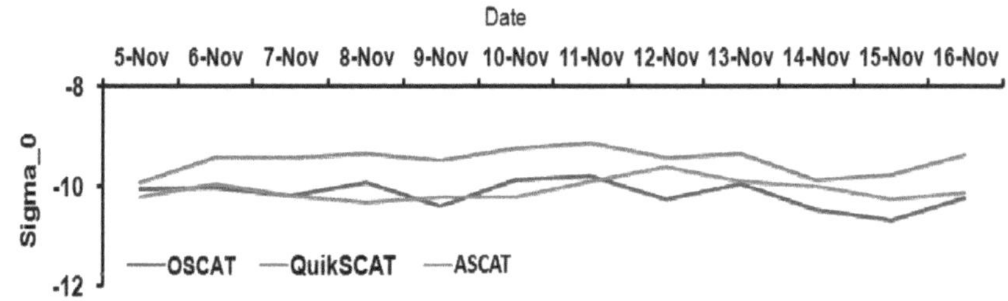

Figure 13.2 Time series of backscatter for QuikSCAT, OSCAT, and ASCAT for an overlapping period averaged for R7. *OSCAT*, Ocean scatterometer; *ASCAT*, advanced scatterometer.

amounts to 0.49 dB (0.42 dB), and that between the OSCAT and ASCAT was found to be 0.099 dB.

13.4.1 Comparison of backscatter with air temperature record

The time series of σ^0 and T_a for sample regions, R1 and R2, are shown in Fig. 13.3A and B. We have provided a zoomed time series in Fig. 13.3B and C for 2003 for Sveagruva and for 2013 at Svalbard airport. The time series graph shows a sharp reduction of σ^0 during June, which is the beginning of summer in Svalbard. It is observed that a sharp increase in σ^0 for different scatterometers coincides with 0°C crossing in air temperature. There was a sharp reduction in σ^0 when T_a drops to 0°C. However, σ^0 exhibits low fluctuations throughout the year when the temperature is below 0°C. The higher T_a affects the snow or ice surface and increases the surface melting of the snowpack.

Snow is an inhomogeneous medium, which consists of ice particles, air, and liquid water during the melting condition. The liquid water in wet snow increases the permittivity of the snow layer (Ulaby et al., 1982). The backscatter of wet snow is highly dependent on the snow layer wetness and roughness of the snow cover (Fung, 1994). In the case of wet snow, surface scattering dominates over volume scattering because of a dramatic increase in the dielectric loss factor of the layer, which increases the absorption coefficient and reduces the penetration depth. During the summer period, the penetration depth is confined to the uppermost 3–4 cm (Liu et al., 2006), which leads to a drop in σ^0. Ashcraft and Long (2006) suggested that a wet snow layer of 3.8 cm thickness and 1% liquid moisture content resulted in a decrease in σ^0 of the order of 2.7 dB. This study observed a 7 dB reduction of σ^0 in June 2000 (Fig. 13.3) due to water content present in the snow/ice, and also due to melt water over the surface in the form of melt streams and

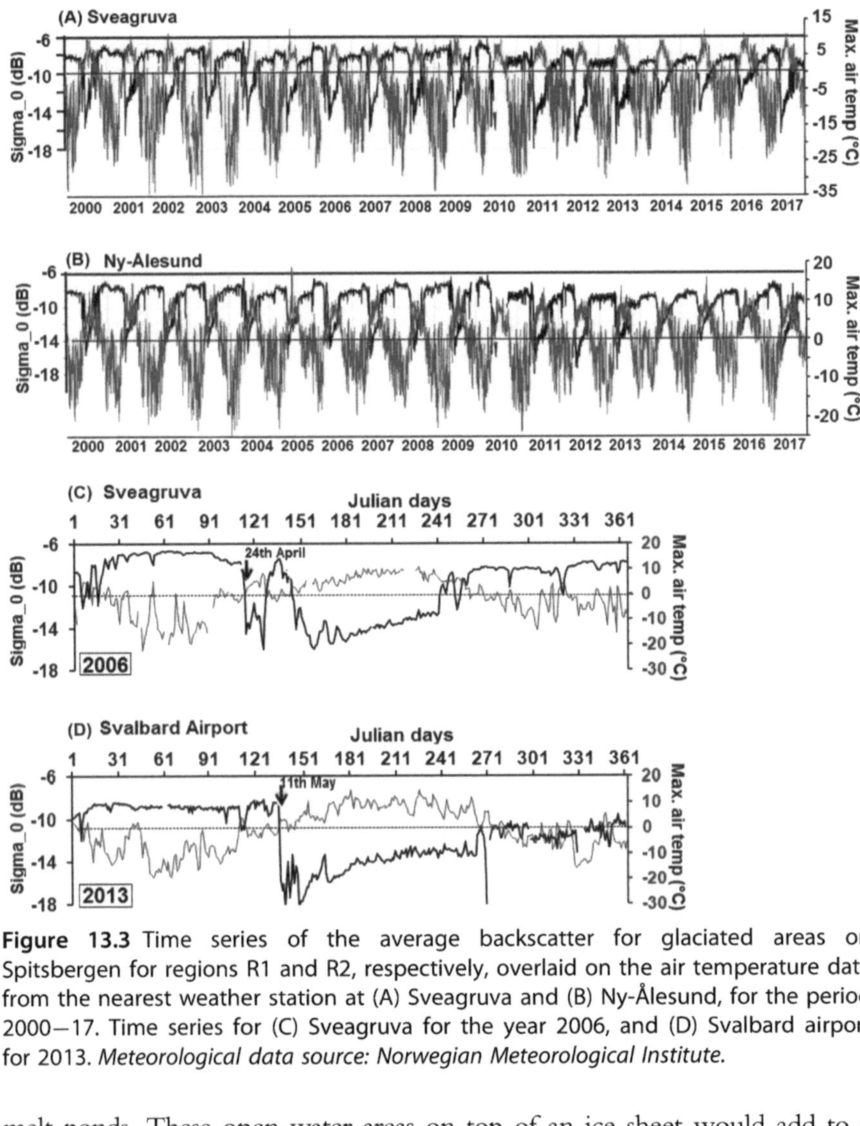

Figure 13.3 Time series of the average backscatter for glaciated areas on Spitsbergen for regions R1 and R2, respectively, overlaid on the air temperature data from the nearest weather station at (A) Sveagruva and (B) Ny-Ålesund, for the period 2000–17. Time series for (C) Sveagruva for the year 2006, and (D) Svalbard airport for 2013. *Meteorological data source: Norwegian Meteorological Institute.*

melt ponds. These open-water areas on top of an ice sheet would add to a further decrease in backscatter.

13.4.2 Spatial patterns of melt onset/duration

The 18-year mean and standard deviation (SD) for MD and SMO are portrayed in Fig. 13.4. In general, we find higher MD (mean ± SD: 66 ± 13 days) and SMO (146 ± 15 day, SMO: 26th May) in the southernmost and western

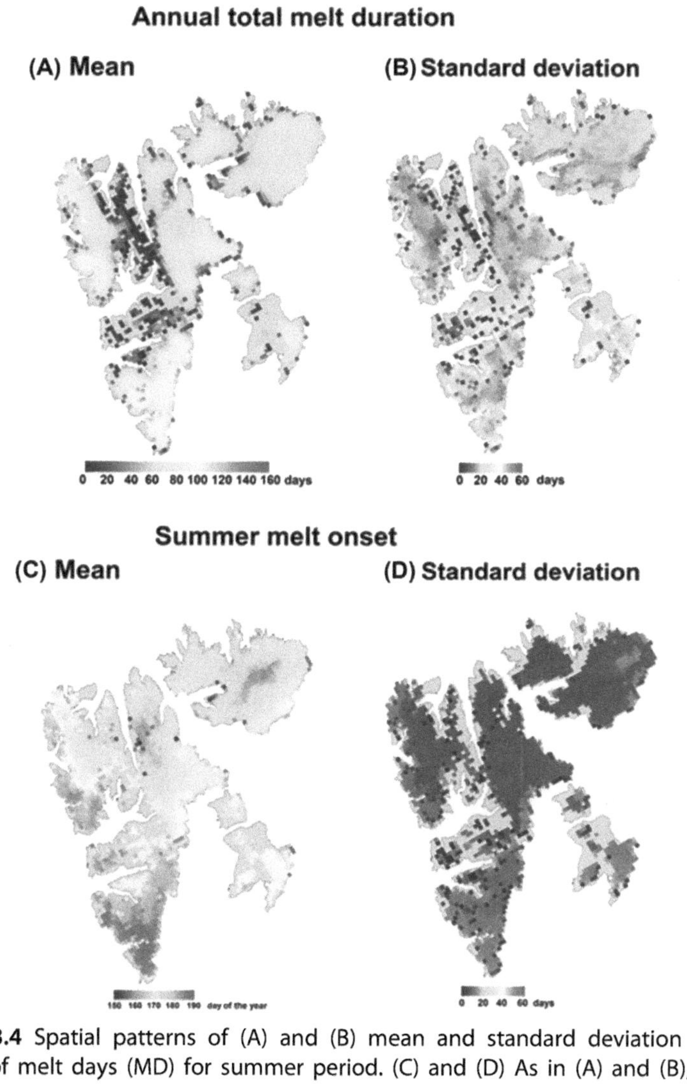

Figure 13.4 Spatial patterns of (A) and (B) mean and standard deviation of total number of melt days (MD) for summer period. (C) and (D) As in (A) and (B), but for summer-onset melt date, for the 2000–17 period.

regions of Svalbard (R1 and R2), compared to the northern and eastern regions (mean MD: 63 ± 16 days; SMO: 162 ± 14 day, 11th June). This may be attributed to the influence of the warm Spitsbergen Current along the western Svalbard coast (Fig. 13.1). The SMO, which occurred on 11th June, prior to 2010 (Rotschky et al., 2011), has now occurred earlier by 11 days (31st May) for the 2000–17 period for the entire Svalbard. The offset in SMO from north to south is 23 days, between 24th May in southern Spitsbergen (R1) and 16th June in Nordaustlandet (R7). Likewise, the north–south difference was just 9 days for MD. The northern and eastern regions (R2, R3, R6, and R7) experienced a generally lower number of MD (60–70 days), but exhibited a higher interannual variability in MD, as reflected by the higher SD (Fig. 13.4). The SD for SMO is larger in the eastern coastal areas, except for region R3. The average SD values for the entire glacier-covered area of Svalbard over the 18-year period are 15 days for MD and 11 days for SMO. These values are almost consistent with the findings of Rotschky et al. (2011) for 2000–08 (MD: 14.3 days and SMO: 12 days) and with the study of Sharp and Wang (2009) which reported 13.2 days for SMO during 2000–04 (MD is not reported). Overall, the mean values for SMO and MD for the entire Svalbard was 6th June and 64 days, respectively.

The spatial patterns of SMO and MD over 2000–17 are depicted in Figs. 13.5 and 13.6, respectively, and Table 13.3 provides the numerical values. The SMO and MD exhibited a general spatial pattern that reflects the climatic gradient across the archipelago. The warm West Spitsbergen Current influences the west coast of Svalbard, rendering a relatively mild climate and sea-ice-free ocean during most of the year. By June, more than half of the winter sea ice area in the Barents Sea melts due to solar heating, which results in a positive east–west T_a gradient of ~4°C. The earliest SMO was encountered at southern Spitsbergen (R1) on Julian day 114 (April 24, 2006) and the longest MD of 99 days in southeastern region (R4) in 2013. The delayed SMO occurred on July 4, 2008 at Nordaustlandet (R6), and the shortest MD was 29 days in 2003 at R4. We note that the SMO was advanced by 17% since 2015, across Svalbard, while the MD was shortened by 6% since 2014, across R1 to R5.

13.4.3 Interannual variability of SMO/MD

Though MD and SMO dates exhibit considerable interannual variability and regional differences, we found there were an increasing number of melt days and advancement in the SMO date over the 18-year period (Fig. 13.7). Linear trends were calculated using a least-squares regression analysis and relationships were quantified using linear Pearson's correlation analysis. There is uncertainty against using linear trends as a measure for climate change, because such alterations do not necessarily occur linearly

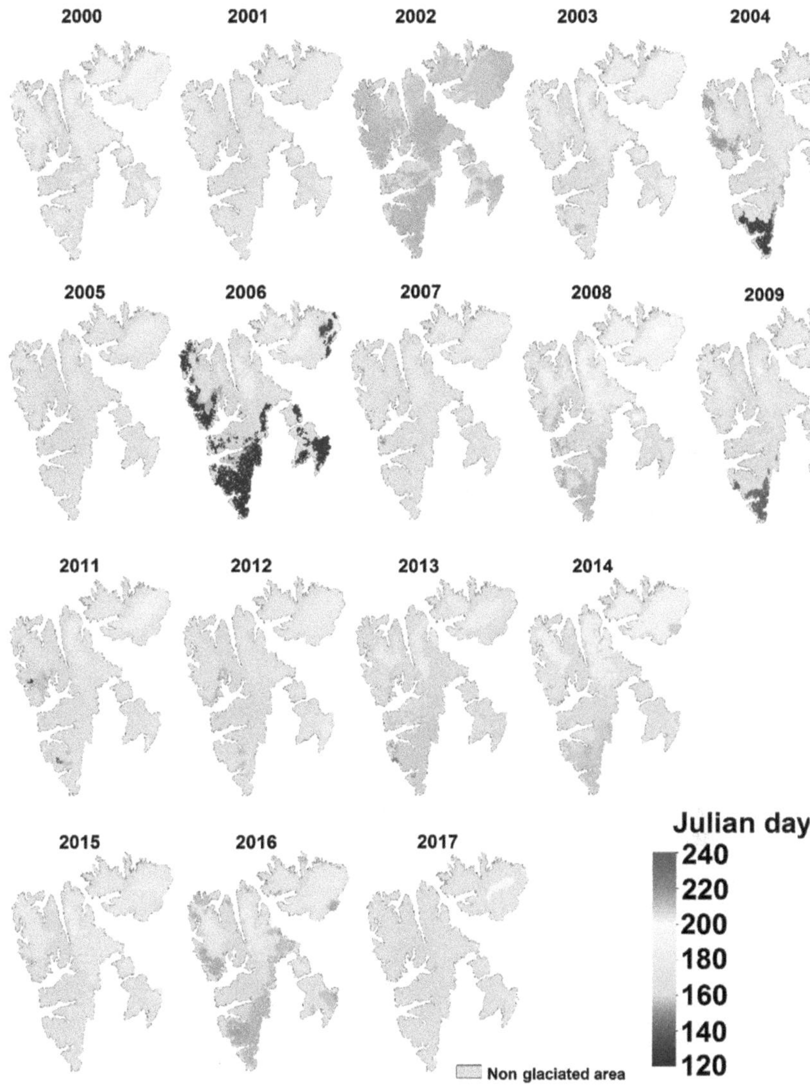

Figure 13.5 Spatial distribution of summer melt onset (SMO) date.

(Benestad, 2003). The statistical significance of trends and correlations were calculated using a two-tailed Student's *t*-test.

The regression of MD yielded the highest and lowest trends for R2 (Fig. 13.7A). It is pertinent to note that the MD mean for R6 and R7 yielded a highest correlation ($r = 0.96$, $P < .01$) for R1 and R2 ($r = 0.79$, $P < .01$), and least for R3 and R4 ($r = 0.53$, $P < .05$). The least squares fit

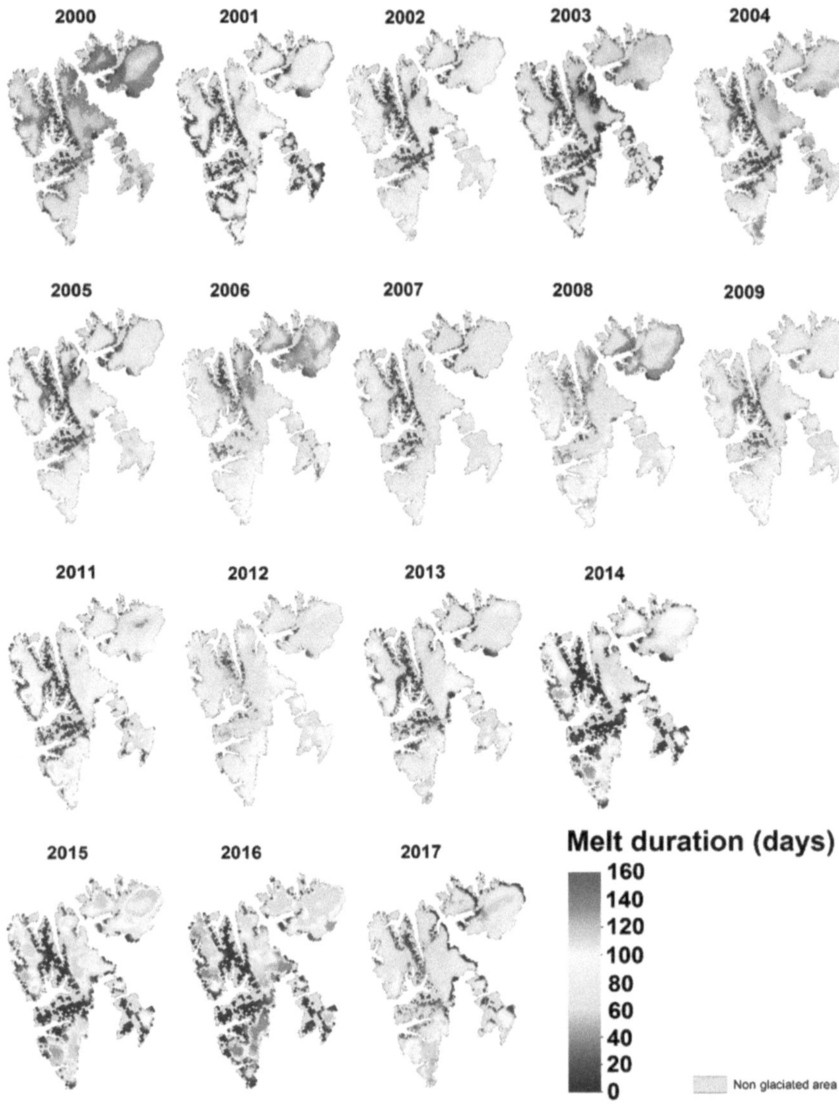

Figure 13.6 Spatial distribution of total number of melt days (MD) in summer.

to the SMO time series yielded a negative trend for R1 and R2. We also observed that SMO are highly correlated to R6 and R7 ($r = 0.98$, $P < .01$), to R1 and R2 ($r = 0.81$, $P < .01$), and to R3 and R4 ($r = 0.74$, $P < .01$). It is noted that the western (R1 and R2) and eastern regions (R6 and R7) experience a relatively stable climate in terms of SMO and MD due to the warm West Spitsbergen current off the west coast and the

Table 13.3 Correlation between ATMD and climate indices.

Region / Index	RI	R2	R3	R4	R5	R6	R7
IPO	−0.53 $P < .05$	−0.25	−0.10	−0.34	−0.52 $P < .05$	0.40 $P < .1$	0.42 $P < .01$
NINO3.4	−0.21	0.01	0.16	−0.05	−0.12	0.58 $P < .02$	0.53 $P < .05$
SOI	0.37	0.22	−0.03	0.18	0.19	−0.40	−0.36
NAO	−0.37	−0.12	−0.11	−0.14	−0.49 $P < .05$	0.15	0.21
PDO	−0.52 $P < .05$	−0.20	−0.03	−0.31	−0.51 $P < .05$	0.46 $P < .10$	0.48 $P < .10$
Mean sea-ice area (May–July) in Barents Sea	−0.38	−0.31	−0.62 $P < .01$	−0.56 $P < .02$	−0.46 $P < .1$	−0.26	−0.34

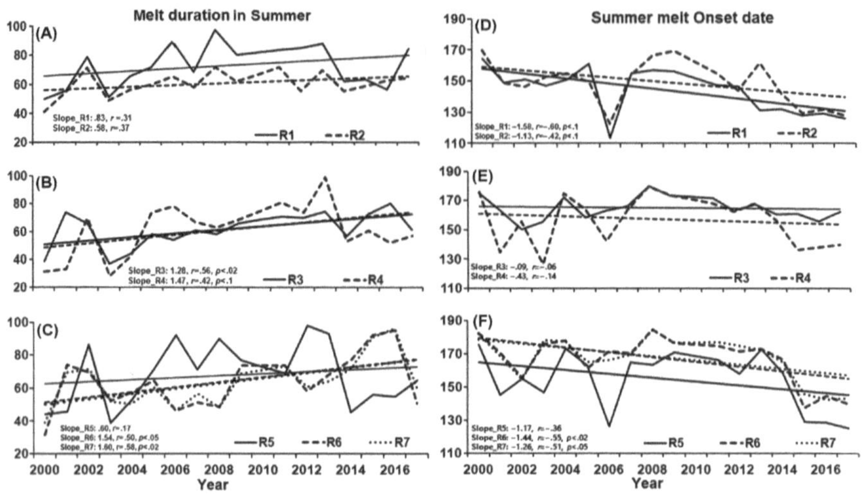

Figure 13.7 Regional averages of (A) total melt days in summer (MD) and (B) summer melt onset (SMO) date from 2000 to 2017.

influence of sea ice conditions in the Barents Sea in the winter, respectively.

13.4.4 Case studies for 2002, 2006, and 2013

We extend the analysis further using the time series for T_a, σ_0, and precipitation for 2002, 2006, and 2013 (Fig. 13.8), to assess the changes observed in Figs. 13.5 and 13.6. T_a induces changes in the surface energy budget and alters the atmospheric circulation, it is the most commonly cited climate change variable. Precipitation is a result of moisture convergence due to air circulation, which is influenced by orographic features.

The choice of particular years was based on the highest number of MD in all regions during 2002 and 2013, and an extreme melt event in 2006, wherein early melt that began at the end of April 2006 was preceded by a regional-scale temperature anomaly caused by long periods of strong southerlies (Isaksen et al., 2007). The other reason for selecting these years is the availability of meteorology data at the nearest stations. In 2002 the SMO, on average, for all the regions were found to be day of the year 154 (3rd June) and the number of melt days was 71 days. For the 2002 scenario, at the sample region (R2) the SMO occurred on 26th May, when the T_a was about 2°C and precipitation (mostly rain) was 3.5 mm. It was a continuous and longest melt event that extended to 71 days for this region (Table 13.3) with an average precipitation of 4.7 mm.

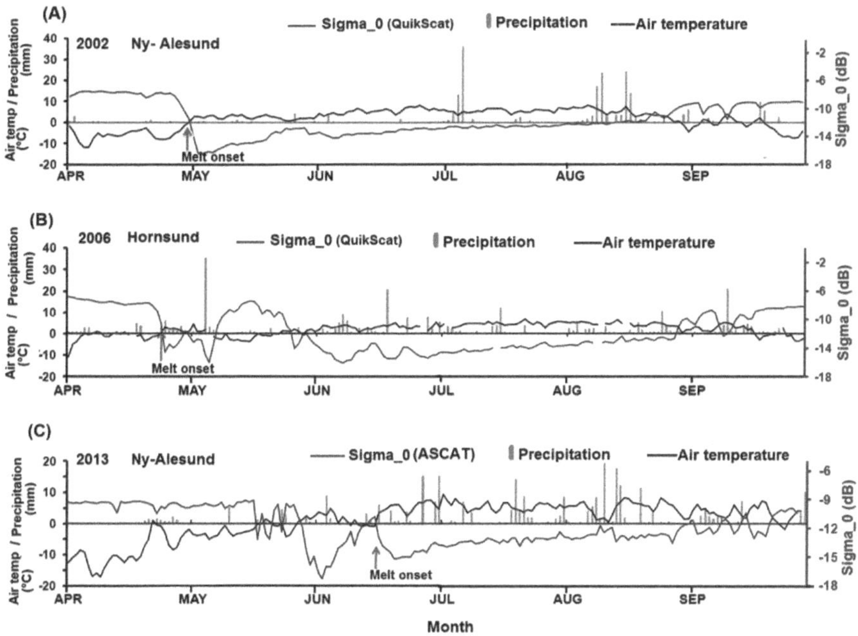

Figure 13.8 Time series of the region-average radar backscatter for (A) Ny-Ålesund in 2000, (B) Hornsund in 2006, and (C) Ny-Ålesund 2013 overlaid on air temperature and precipitation data from a local weather station. *Meteorological data source: Norwegian Meteorological Institute.*

With high precipitation (>10 mm) events to the end of summer from (July to September), no significant change in σ^0 is observed.

The melt onset date was day of the year 114 (24th April) for sample region R1, which coincided with positive degree T_a and a moderate precipitation event. A few days later, a high snowmelt was triggered by a high precipitation event (35 mm on 5th May). For the first melt event, the duration was about 2 weeks. The second melt event commenced on 25th May and lasted for 3 months, which is the longest in the QuikSCAT observation period. The conspicuous feature of the surface melt is that it is well correlated with precipitation events (likely rain). According to the studies, all through the year, low-pressure systems facilitated humid maritime air from the southwest leading to above-zero temperature events (Winther et al., 2003). Boike et al. (2003) suggested that melting processes due to warm weather events are dominated by sensible heat flux and latent heat of condensation due to rain, which concurs with our inferences. In 2013 the melt onset began on 11th June (delayed by

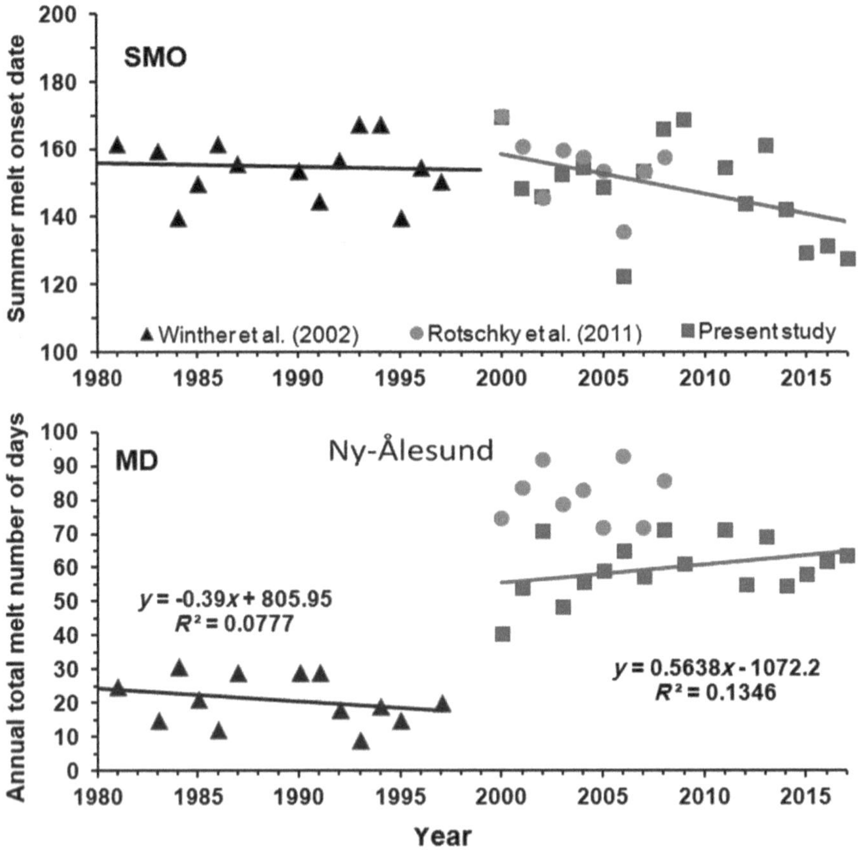

Figure 13.9 Summer melt onset (SMO in Julian days) and melt duration in summer (MD, days) at Ny-Ålesund derived from albedo study using pyranometers (▲) done by Winther et al. (2002), QuikSCAT-based backscatter study (●) (Rotschky et al., 2011), and those derived from the present study (■).

about 2 weeks compared to 2002) and the duration of melt lasted for 70 days for region R2. The average precipitation for MD was 3 mm. It is prudent to note that positive degree T_a and precipitation (mostly rain) dominated SMO and MD events.

13.4.5 Comparison of SMO/MD for Ny-Ålesund with previous studies

We now compare the values for SMO and MD backward in time using data derived from previous studies. Using an optical sensor (MODIS), Winther et al. (2002) determined melt onset dates at Ny-Ålesund (R2) for the period 1982–97 using in situ measurements of albedo, while Rotschky et al. (2011) used QuikSCAT-based σ^0 for 2000–08. Fig. 13.9 shows a scatterplot of MD and SMO from these two studies and supplemented by those values obtained in this study. Compared to Winther et al. (2002), our SMO values depict an enhanced negative trend (−1.1 day $year^{-1}$, $r = 0.4$, $P < .1$), possibly due to the availability of few cloud-free scenes (with only two passes per day) used in the former study. On the other hand, we find a positive trend for MD, as opposed to the negative trend observed by Winther et al. (2002). No inference could be drawn from a short-period study of Rotschky et al. (2011).

Within the record of 18 years, the exceptional early SMO on April 24, 2006 is worth highlighting. In general, SMO for R2 occurred on 2nd June with a SD of 9 days. The early date for SMO was 16th May, while the later date was 21st June. The higher SD for MD was due to low values reported by Winther et al. (2002). Two major differences could be attributed to the changes observed in Fig. 13.9. Two reasons could be attributed to the variation in the scatter distribution for SMO. First, Rotschky et al. (2011) estimates of SMO using 6–7 days threshold and allows 10% variation of c value. Second, grid cells with high or low values for MD/SMO have been removed from their analysis, which is expected to change the regional average. Moreover, our analysis for MD is restricted to the summer period only, so the count of melt days in autumn is excluded from our analysis.

13.4.6 Climatic control on ATMD and SMO

The signatures of atmospheric teleconnections (El Niño-Southern Oscillation—ENSO, North Atlantic Oscillation—NAO, Arctic Oscillation—AO, Interdecadal Pacific Oscillation—IPO) involved in the Arctic have been revealed in many studies (e.g., Hurrell, 1995; Thompson and Wallace, 2000). On the interannual time scale over the Arctic North Atlantic region, the occurrence frequency for cyclones and moisture advection toward high latitudes is associated with the AO/NAO and NAO can be considered to be the regional (North Atlantic) manifestation of the AO (Zhang et al., 1997). The IPO is the leading mode of 40–60-year variability in the Pacific sea surface temperatures (SSTs), which through the interactions with the atmosphere affects temperature, precipitation, and other climate fields in many land regions. Since the IPO-related SST anomalies exhibit spatial patterns similar to those of ENSO, the

IPO is naturally referred to as the ENSO-like interdecadal variability (Zhang et al., 1997).

The correlation coefficient and the level of significance (*P*-value) between these indices and SMO and ATMD are presented in Table 13.3. For R1, IPO and PDO showed a negative correlation ($P < .05$). Likewise, IPO, NAO, and PDO were found to be negatively correlated ($P < .05$) with ATMD for R5. However, ATMD for the eastern regions (R6 and R7) was found to be positively correlated with IPO, NINO3.4, and PDO at different levels of significance. SMO for R3, R4, and R5 was negatively correlated to the May–July mean sea-ice area in the Barents Sea, which suggests the dependence of ATMD on the sea ice conditions in the adjacent Barents Sea. As for SMO, there is no much difference in the geographic locations as far as its correlation with IPO and PDO is concerned. Except for the southern region (R1), the SMO date is positively correlated to NINO3.4, but negatively related to SOI. Summer mean sea ice in the Barents Sea seems to moderately influence the SMO date for R1 and R2. It is prudent to note that SMO was found to exhibit a high and negative correlation with IPO and PDO for R7.

13.5 Discussion

The Arctic region has attained a prominent role in scientific debate regarding global climate change. Arctic amplification (Fyfe et al., 2013) plays a dominant role in diminishing sea ice cover (http://nsidc.org/arcticseaicenews), promotes warmer winters (Isaksen et al., 2016), and has reduced springtime Eurasian snow cover over the past several decades (e.g., Derksen and Brown, 2012). These factors not only directly alter the heat budget of the atmosphere in the area but also cause modification of large-scale atmospheric circulation leading to positive feedback on the ocean and land environment. In this background scenario, we mapped MD and SMO over a period of 18 years using microwave scatterometer data covering the Svalbard Archipelago to overcome the deficiency of in situ point measurements. Our results generally agree with those of Sharp and Wang (2009) and Rotschky et al. (2011) for the overlapping period.

The literature indicates that the air temperatures in Svalbard are highly sensitive to the varying extent of sea ice surrounding the archipelago that impacts on both local albedo and heat transfer from the ocean back to the atmosphere (Benestad et al., 2002). Consequently, the large interannual variability observed in SMO and STMD can be attributed to the interannual variations in the sea-ice extent. Our results for the eastern regions showed anomalously late (early) SMO and a reduced (extended) MD associated with above (below) average concentrations of sea ice along the east Svalbard coast (Spreen et al., 2008), concurring with the findings of Sharp and Wang (2009). The late continuation of melt may reflect the fact that sea ice only disappears from the adjacent waters late in the melt season, so the latent heat released from the newly open ocean may have the effect of prolonging the melt season. For example, the anomalous early SMO anomaly in 2006 coincided with an unusually large extent of marine open water around Svalbard during winter, spring, and summer 2005–06, which was anomalous compared to the previous-observed lower ice extent (https://nsidc.org/arcticseaicenews/category/daily-image/).

One of the challenges in monitoring melt/freeze processes on a land surface is the strong dependency of the σ^0 on the incidence angle and frequencies of the sensor, which was significantly different for ASCAT. In this work we have ignored the role of different incidence angles on σ^0. The changes in σ^0 significantly vary over the image swath, and from acquisition to acquisition, making the interpretation of the time series difficult. With a longer wavelength of $\sim$5 cm the penetration depth increases, and so the volume scattering from the subsurface layers become dominant, which adds up to σ^0, as compared to lower σ^0 for the *Ku*-band due to relatively shallower penetration. Many studies have addressed this dependency by fitting a linear function to the σ^0 measurements spanning a long period such as a month (Mougin et al., 1995; Schmullius, 1997). This allowed studies of the relationship between the σ^0 observations and seasonal freeze/thaw processes in high-latitude areas. Nevertheless, this monthly fitting procedure suppresses short-term signal fluctuations caused, for example, by changes in melt and freeze events, or changes in snow morphology, rendering a more physically based interpretation of the σ^0 data impossible.

A general warming has been documented in the Arctic associated with a significant decline in sea-ice extent taking place since satellite observations began in 1978 (Johannessen et al., 1999). In Svalbard, warming has been observed since 1960 (Førland and Hanssen-Bauer, 2003), particularly for the spring months (Christensen et al., 2007). Other studies have reported a positive trend in June–August mean temperature at the rate of 0.178°C/a for the period 1997–2006. The increasing MD trend during the above period and beyond that we find in this study for some regions is consistent with this observed warming trend.

With the air temperature rising regionally in the Arctic, it is expected that surface melt rates will increase. According to the Clausius–Clapeyron relationship, saturation vapor pressure increases with air temperature and, therefore, the water-holding capacity of the atmosphere increases. Therefore the winter storm track which brings warm and moist air into the Arctic from low latitudes (Barnett et al., 2006) can intensify precipitation under warmer conditions. The snow season duration for the Svalbard land areas has decreased over the period 1958–2017 by approximately 20 days (Adakudlu et al., 2019). The future snow conditions up to 2071–2100 strongly depend on the emission scenario. The median simulation with RCP4.5 still shows areas with the same, or slightly more, snow storage, while the median simulation with RCP8.5 shows a reduction in maximum snow storage of 50% or more over most of Svalbard. The fine-scale simulation gives increased maximum snow storage (up to 100% for certain areas) for large areas in the northeast and reduced amounts in western and southern areas. The number of days with snow cover will be reduced all over Svalbard for both RCP4.5 and RCP8.5. In some areas, the snow cover duration is reduced marginally, while in other areas it is drastically reduced. With increasing winter temperatures, the snow line will increase and the timing of maximum snow storage will shift from June (1971–2000) to May (2071–2100) in the fine-scale RCP8.5 simulation. All of these changes in snow cover and resultant alterations of thermal contrast due to receding Arctic sea ice are expected to have strong implications for the atmospheric circulation.

13.6 Conclusion

We have used the publicly available 18-year record of scatterometer-based backscatter to infer snowmelt onset and MD from Svalbard. Using the backscatter, the melt onset and duration were validated by using in situ air temperature records derived from local weather stations available but restricted to southern and western Spitsbergen. However, the few and geographically sparse distribution of weather stations in the inaccessible terrain and measuring problems hamper analyses of climatic conditions in Svalbard. Our results capture the timing and geographical extent of the short-lived melt events caused by the warm air temperature and precipitation events, as a consequence of the influx of moist, mild air from the Norwegian and Barents seas. Because of the short temporal span, past studies could not infer an increasing (decreasing) trend in SMO (MD), which suggests the delayed melt onset and the reduced number of melt days.

The early SMO date and the longest MD occurred in its southernmost and western parts (R1 and R2), in response to the influence of warm and cold oceanic currents flowing along the west and east coastline of Svalbard, respectively. The large interannual variability in the melt onset and duration is a result of a general warming trend, possibly connected to the annual changes in sea ice extent in the Barents Sea and an overall sea ice decline in the Arctic, respectively. We infer that the early SMO and the shorter MD are often dictated by the arrival of warm air flow, accompanied by the regional-scale rainfall. To reinforce the inferences that emerged in this study, it is recommended to monitor Svalbard using the follow-up missions for extending the time series.

Acknowledgments

QuikSCAT and OSCAT data were downloaded from http://www.scp.byu.edu. OSCAT-2 was downloaded from ftp://ftp.mosdac.gov.in/. Most of the indices were also downloaded from the NOAA CPC website http://www.cpc.ncep.noaa.gov/products/precip/CWlink/daily_ao_index. The multivariate ENSO index (MEI V2) was obtained from https://www.esrl.noaa.gov/psd/enso/mei/. The Arctic sea ice area was distributed by NSIDC via their website: https://nsidc.org/data/G02135/. We thank the Director of NCPOR for providing the infrastructural facilities to carry out this work. This is NCPOR contribution B-8/ 2020-21.

References

Abdalati, W., 2007. Greenland Ice Sheet Melt Characteristics Derived from Passive Microwave Data. National Snow and Ice Data Center, Boulder, CO.

Abdalati, W., Steffen, K., 2001. Greenland ice sheet melt extent: 1979–1999. J. Geophys. Res. 106, 33983–33988.

Adakudlu, M., et al., 2019. In: Hanssen-Bauer, I., Førland, E.J., Hisdal, H., Mayer, S., Sandø, A.B., Sorteberg, A. (Eds.), Climate in Svalbard 2100—a knowledge base for climate adaptation. Norwegian Environment Agency (Miljødirektoratet), Report No 2387–3027.

AMAP, 2017. Snow, Water, Ice and Permafrost in the Arctic (SWIPA) 2017. Arctic Monitoring and Assessment Programme (AMAP)), Oslo, NO, xiv + pp. 269.

Anderson, M.R., 1987. Snow melt on sea ice surfaces as determined from passive microwave satellite data. Large Scale Effects of Seasonal Snow Cover. International Association of Hydrological Sciences (IAHS Publication, Wallingford, England, pp. 329–342, No. 166.

Ashcraft, I.S., Long, D.G., 2006. Comparison of methods for melt detection over Greenland using active and passive microwave measurements. Int. J. Remote Sens. 27, 2469–2488.

Barnett, D.N., Brown, S.J., Murphy, J.M., Sexton, D.M.H., Webb, M.J., 2006. Quantifying uncertainty in changes in extreme event frequency in response to doubled CO_2 using a large ensemble of GCM simulations. Clim. Res. 26, 511–889.

Benestad, R.E., 2003. What can present climate models tell us about climate change? Clim. Change 59 (3), 311–331.

Benestad, R.E., Førland, E.J., Hanssen-Bauer, I., 2002. Empirically downscaled temperature scenarios for Svalbard. Atmos. Sci. Lett. 3, 71–93.

Bintanja, R., van der Linden, E.C., 2013. The changing seasonal climate in the Arctic. Sci. Rep. 3, 1556. Available from: https://doi.org/10.1038/srep01556.

Boike, J., Roth, K., Ippisch, O., 2003. Seasonal snow cover on frozen ground: energy balance calculations of a permafrost site near Ny-Ålseund, Spitsbergen. J. Geophys. Res. (Atmospheres) 108, 8163. Available from: https://doi.org/10.1029/2001JD000939.

Box, J.E., Colgan, W.T., Christensen, T.R., Schmidt, N.M., Lund, M., Parmentier, F.-J. W., et al., 2019. Key indicators of Arctic climate change: 1971–2017. Environ. Res. Lett. 14, 04501.

Box, J.E., Colgan, W.T., Wouters, B., Burgess, D.O., O'Neel, S., Thomson, L.I., et al., 2018. Environ. Res. Lett. 13, 125012.

Brown, R., Derksen, C., Wang, L., 2007. Assessment of spring snow cover duration variability over northern Canada from satellite datasets. Remote Sens. Environ. 111, 367–381.

Christensen, J., Hewitson, B., Busuioc, A., Chen, A., Gao, X., Held, I., et al., 2007. Regional climate projections. In: Solomon, S., et al., (Eds.), Climate Change 2007. The Physical Science Basis. Contribution of Working Group I to the Fourth Assessment Report of the Intergovernmental Panel on Climate Change. Cambridge University Press, New York, pp. 849–940.

Cohen, J., Screen, J.A., Furtado, J.C., Barlow, M., Whittleston, D., Coumou, D., et al., 2014. Recent Arctic amplification and extreme mid-latitude weather. Nat. Geosci. 7 (9), 627–637.

Day, J.J., Hargreaves, J.C., Annan, J.D., Abe-Ouchi, A., 2012. Sources of multi-decadal variability in Arctic sea ice extent. Environ. Res. Lett. 7, 034011.

Derksen, C., Brown, R., 2012. Spring snow cover extent reductions in the 2008–2012 period exceeding climate model projections. Geophys. Res. Lett. 39, L19504.

Divine, D.V., Dick, C., 2006. Historical variability of sea ice edge position in the Nordic Seas. J. Geophys. Res. 111, C01001. Available from: https://doi.org/10.1029/2004JC002851.

Early, D.S., Long, D.G., 2001. Image reconstruction and enhanced resolution imaging from irregular samples. IEEE Trans. Geosci. Remote Sens. 39, 291–302.

Fetterer, F., Knowles, K., Meier, W.N., Savoie, M., Windnagel, A.K., 2017. Updated daily. Sea Ice Index, Version 3. [Indicate subset used]. Boulder, Colorado USA. NSIDC: National Snow and Ice Data Center. doi: https://doi.org/10.7265/N5K072F8.

Førland, E.J., Hanssen-Bauer, I., 2003. Past and future climate variations in the Norwegian Arctic: overview and novel analyses. Polar Res. 22, 113–124.

Førland, E.J., Benestad, R., Hanssen-Bauer, I., Haugen, J.E., Skaugen, T.E., 2011a. Temperature and precipitation development at Svalbard 1900–2100. Adv. Meteorol. 2011, 1–14. Available from: https://doi.org/10.1155/2011/893790.

Førland, E.J., Benestad, R.E., Hanssen-Bauer, I., Haugen, J.E., 2011b. "Future climate development in the NorACIA-Region. Combined results from simulations with different climate models." Brief Report Series, No. 023, Norwegian Polar Institute (In Norwegian).

Førland, E.J., Benestad, R., Hanssen-Bauer, I., Haugen, J.E., Skaugen, T.E., 2011c. Temperature and precipitation development at Svalbard 1900–2100. Adv. Meteorol. 2011, 893790. Available from: https://doi.org/10.1155/2011/893790.

Forster, R.R., Long, D.G., Jezek, K.C., Drobot, S.D., Anderson, M.R., 2001. The onset of Arctic sea-ice snowmelt as detected with passive- and active-microwave remote sensing. Ann. Glaciol. 33, 8593.

Fung, A., 1994. Microwave Scattering and Emission Models and Their Applications. Artech House, p. 573.

Fyfe, J.C., von Salzen, K., Gillett, N.P., Arora, V.K., Flato, G.M., McConnell, J.R., 2013. One hundred years of Arctic surface temperature variation due to anthropogenic influence. Sci. Rep. 3, 2645. Available from: https://doi.org/10.1038/srep02645.

Garrity, K., 1992. Characterization of snow on floating ice and case studies of brightness temperature change during the onset of melt. In: Carsey, F.D., et al., (Eds.), Microwave Remote Sensing of Sea Ice. AGU, Washington, DC, pp. 313–328. vol. 68.

Hagen, J.O., Melvold, K., Pinglot, F., Dowdeswell, J.A., 2003. On the net mass balance of the glaciers and ice caps in Svalbard, Norwegian Arctic, Arctic, Antarctic. Alp. Res. 35 (2), 264.

Hanssen-Bauer, I., Førland, E.J., 1998. Long-term trends in precipitation and temperature-in the Norwegian Arctic: can they be explained by changes in atmospheric circulation patterns? Clim. Res. 10, 143–153.

Hicks, B.R., Long, D.G., 2011. Inferring Greenland melt and refreeze severity from Sea Winds scatterometer data. Int. J. Remote Sens. 32, 8053–8080.

Hisdal, V., 1985. Geography of Svalbard, 2nd edition Norsk Polar Institute, p. 75.

Hurrell, J.W., 1995. Decadal trends in the North Atlantic Oscillation: regional temperatures and precipitation. Science 269, 676–679.

Isaksen, K., Benestad, R.E., Harris, C., Sollid, J.L., 2007. Recent extreme near-surface permafrost temperatures on Svalbard in relation to future climate scenarios. Geophys. Res. Lett. 34, L17502. Available from: https://doi.org/10.3402/polar.v30i0.5963.

Isaksen, K., Nordli, Ø., Førland, E.J., èupikasza, E., Eastwood, S., Niedźwiedź, T., 2016. Recent warming on Spitsbergen—influence ofatmospheric circulation and sea ice cover. J. Geophys. Res. Atmos. 121, 11913–11931. Available from: https://doi.org/10.1002/2016JD025606.

Johannessen, O.M., Shalina, E.V., Miles, M.W., 1999. Satellite evidence for an Arctic sea ice cover in transformation. Science 286, 1937–1939.

Joshi, M., Merry, C.J., Jezek, K.C., Bolzan, J.F., 2001. An edge detection technique to estimate melt duration, season and melt extent on the Greenland Ice Sheet using passive microwave data. Geophys. Res. Lett. 28 (5), 3497–3500. Available from: https://doi.org/10.1029/2000GL012503. issn: 0094-8276.

Kaser, G., Cogley, J.G., Dyurgerov, M.B., Meier, M.F., Ohmura, A., 2006. Mass balance of glaciers and ice caps: consensus estimates for 1961–2004. Geophys. Res. Lett. 33, L19501. Available from: https://doi.org/10.1029/2006GL027511.

Kunz, L.B., Long, D.G., 2006. Melt detection in antarctic ice shelves using scatterometers and microwave radiometers. IEEE Trans. Geosci. Remote Sens. 44 (9), 2461–2469. 10.1109/TGRS.2006.874138.

Kunzi, K.F., Patil, S., Rott, H., 1982. Snow-covered parameters retrieved from NIMBUS-7 SMMR data. IEEE Trans. Geosci. Remote Sens. GE-20, 452–467.

Legrésy, B., Rémy, F., Legresy, B., 1998. Using the temporal variability of satellite radar altimetric observation to map surface properties of the Antarctic ice sheet. J. Glaciol. 44, 207–213.

Li, X., Zhang, Y., Liang, L., 2017. Snowmelt detection on the Greenland ice sheet using microwave scatterometer measurements. Int. J. Remote Sens. 38, 796–807.

Liu, H.W., Wang, L., Jezek, K., 2006. Spatio-temporal vari-ations of snow melt zones in Antarctic ice sheet derived from satellite SMMR and SMM/I data (1978–2004). J. Geophys. Res. 111, 1111–1120.

Long, D.G., 2010. Standard BYU QuikSCAT and Seawinds land/ice image products, report, BYU Center for Remote Sensing, Microwave Earth Remote Sensing Laboratory, Provo, Utah.

Long, D.G., Drinkwater, M.R., 1994. Greenland ice sheet surface-properties observed by the SEASAT-A scatterometer at enhanced resolution. J.Glaciol 40, 213–230.

Long, D.G., Hicks, B.R., 2005. Standard BYU QuikSCAT/SeaWinds Land/Ice Image Products, Report, Brigham Young University, Provo, Utah.

Long, D.G., Drinkwater, M.R., Holt, B., Saatchi, S., Bertoia, C., 2001. Global ice and land climate studies using scatterometer image data. EOS Trans. AGU 82, 503and electronic supplement available at. Available from: http://www.agu.org/eos_elec/010126e.html.

Mätzler, C., Hüppi, R., 1989. Review of signature studies for microwave remote sensing of snowpacks. Adv. Space Res. 9, 253–265.

Mortin, J., Howell, S.E.L., Wang, L., Derksen, C., Svensson, G., Graversen, R.G., et al., 2014. Extending the QuikSCAT record of seasonal melt-freeze transitions over arctic sea ice using ASCAT. Remote Sens. Environ. 141, 214–230.

Mougin, E., Lopes, A., Frison, P.L., Proisy, C., 1995. Preliminary analysis of ERS-1 wind scatterometer data over land surfaces. Int. J. Remote Sens. 16 (2), 391–398.

Nghiem, S.V., Steffen, K., Kwok, R., Tsai, W.-Y., 2001. Detection of snow melt regions on the Greenland Ice Sheet using diurnal backscatter change. J. Glaciol. 47, 539–547.

Nghiem, S.V., Steffen, K., Neumann, G., Huff, R., 2007. Snow accumulation and snowmelt monitoring in Greenland and Antarctica. Int. Assoc. Geodesy Symp. 130, 31–38.

Nordli, Ø., Rajmund, P., Ogilvie, A.E.J., Isaksen, K., 2014. Long-term temperature trends and variability on Spitsbergen: the extended Svalbard Airport temperature series." 1898–2012. Polar Res. 33 (1). Available from: https://doi.org/10.3402/polar.v33.21349.

Nuth, C., Kohler, J., König, M., Deschwanden, A., 2013. Decadal changes from a multitemporal glacier inventory of Svalbard. Cryosphere 7 (5), 1603–1621.

Østby, T.I., Schuler, T., Hagen, J.O.M., Hock, R., Kohler, J., 2017. Diagnosing the decline in climatic mass balance of glaciers in Svalbard over 1957–2014. Cryosphere 11 (1), 191–215.

Overland, J.E., Wang, M., 2010. Large-scale atmospheric circulation changes are associated with the recent loss of Arctic sea ice. Tellus A 62 (1), 1–9.

Panday, P.K., Frey, K.E., Ghimire, B., 2011. Detection of the timing and duration of snowmelt in the Hindu Kush-Himalaya using QuikSCAT, 2000–2008. Environ. Res. Lett. 6, 1–13. Available from: https://doi.org/10.1088/1748-9326/6/2/024007.

Rotschky, G., Schuler, T.V., Haarpaintner, J., Kohler, J., Isaksson, E., 2011. Spatiotemporal variability of snowmelt across Svalbard during the period 2000–08 derived from QuikSCAT/Sea Winds scatterometry. Polar Res. 30 (1), 5963. Available from: https://doi.org/10.3402/polar.v30i0.5963.

Rott, H., Sturm, K., Miller, H., 1993. Active and passive microwave signatures of Antarctic firn by means of field measurements and satellite data. Ann. Glaciol. 17, 337–343.

Sand, K., Winther, J.-G., Maréchal, D., Bruland, O., Melvold, K., 2003. Regional variations of snow accumulation on Spitsbergen, Svalbard, 1997. Nordic Hydrol. 32, 17–32.

Schmullius, C.C., 1997. Monitoring Siberian forests and agriculture with the ERS-1 wind scatterometer. IEEE Trans. Geosci. Remote Sens. 35 (5), 1363–1366.

Serreze, M.C., Crawford, A.D., Barrett, A.P., 2015. Extreme daily precipitation events at Spitsbergen, an Arctic Island. Int. J. Climatol. 35, 4574–4588.

Sharp, M., Wang, L., 2009. A five-year record of summer melt on Eurasian Arctic ice caps. J. Clim. 22, 133–145.

Smith, L.C., Sheng, Y., Forster, R.R., Steffen, K., Frey, K.E., Alsdorf, D.E., 2003. Melting of small Arctic ice caps observed from ERS scatterometer time series. Geophys. Res. Lett. 30, 2034. Available from: https://doi.org/10.1029/2003GL017641.

Spreen, G., Kaleschke, L., Heygster, G., 2008. Sea ice remote sensing using AMSR-E 89 GHz channels. J. Geophys. Res. (Ocean) 113, . Available from: https://doi.org/10.1029/2005JC003384C02S03.

Steffen, K., Nghiem, S.V., Huff, R., Neumann, G., 2004. The melt anomaly of 2002 on the Greenland Ice Sheet from active and passive microwave satellite observations. Geophys. Res. Lett. 31 (1–5), L20402. Available from: https://doi.org/10.1029/2004GL020444.

Stiles, W.H., Ulaby, F.T., 1980. "The active and passive microwave response to snow parameters. 1. Wetness. J. Geophys. Res. - Ocean 85, 1037–1044.
Svendsen, H., Beszczynska-Møller, A., Hagen, J.O., Lefauconnier, B., Tverberg, V., Gerland, S., et al., 2002. The physical environment of Kongsfjorden–Krossfjorden, an Arctic fjord system in Svalbard. Polar Res. 21 (1), 133–166. 10.3402/polar.v21i1.6479L.
Tedesco, M., 2007. Snowmelt detection over the Greenland ice sheet from SSM/I brightness temperature daily variations. Geophys. Res. Lett. 34, L02504. Available from: https://doi.org/10.1029/2006GL028466.
Thompson, D.W., Wallace, J.M., 2000. Annular modes in extratropical circulation, Part II: Trends. J. Clim. 13, 1018–1036.
Turetsky, M.R., Abbott, B.W., Jones, M.C., Anthony, K.W., Olefeldt, D., Schuur, E.A. G., et al., 2019. Permafrost collapse is accelerating carbon release. Nature 569, 32–34.
Ulaby, F.T., Moore, R.K., Fung, A.K., 1982. Microwave Remote Sensing: Active and Passive. Volume 2 - Radar Remote Sensing and Surface Scattering and Emission. Addison-Wesley Publ. Co., Reading, MA, p. 624, Research Support by NASA, NSF, U.S. Department of Defense.
van Pelt, W., Pohjola, V., Pettersson, R., Marchenko, S., Kohler, J., Luks, B., et al., 2019. A long-term dataset of climatic mass balance, snow conditions, and runoff in Svalbard (1957–2018). Cryosphere 13, 2259–2280. Available from: https://doi.org/10.5194/tc-13-2259-2019.
Vihma, T., Screen, J., Tjernstrom, M., Newton, B., Zhang, X.D., Popova, V., et al., 2016. The atmospheric role in the Arctic water cycle: a review on processes, past and future changes, and their impacts. J. Geophys. Res. 121, 586–620.
Wang, L., Sharp, M.J., Rivard, B., Marshall, S., Burgess, D., 2005. Melt season duration on Canadian Arctic ice caps." 2000-2004. Geophys. Res. Lett. 32, 1–4.
Wang, L., Sharp, M., Rivard, B., Steffen, K., 2007. Melt season duration and ice layer formation on the Greenland ice sheet, 2000-2004. J. Geophys. Res. Earth Surf. 112, 2000–2004.
Wang, L., Derksen, C., Brown, R., 2008. Detection of pan-Arctic terrestrial snowmelt from QuikSCAT, 2000-2005. Remote Sens. Environ. 112, 3794–3805.
Winther, J.-G., 1993. Short- and long-term variability of snow albedo. Nordic Hydrol. 24, 199–212.
Winther, J.-G., Godtliebsen, F., Gerland, S., Isachsen, P.E., 2002. Surface albedo in Ny-Ålesund, Svalbard: variability and trends during 1981–1997. Glob. Planet Change 32, 127–139.
Winther, J.G., Bruland, O., Sand, K., Gerland, S., Marechal, D., Ivanov, B., et al., 2003. "Snow research in Svalbard—an overview. Polar Res. 22 (2), 125–144.
Wiscombe, W.J., Warren, S.G., 1980. A model for the spectral albedo of snow. I. Pure snow. J. Atmos. Sci. 37, 2712–2733.
Wismann, V.R., 2000. Monitoring of seasonal snowmelt on Greenland with ERS scatterometer data. IEEE Trans. Geosci. Remote Sens. 38, 1821–1826.
Zhang, Y., Wallace, J.M., Battisti, D.S., 1997. ENSO-like interdecadal variability: 1900–93. J. Clim. 10, 1004–1020.

CHAPTER 14

Arctic sea ice variability and trends in the last four decades: role of ocean-atmospheric forcing

Avinash Kumar, Juhi Yadav, Rohit Srivastava and Rahul Mohan
National Centre for Polar and Ocean Research, Ministry of Earth Sciences, Government of India, Vasco-da-Gama, India

14.1 Introduction

The Arctic is an integral part of the global climate system, which controls the global surface energy balance and ocean-atmospheric circulation by exchanging heat between the ocean, atmosphere, and sea ice (Petty et al., 2018). Arctic sea ice has always been a significant indicator of climate change; however, sea ice has declined at an unprecedented rate in recent decades due to the impact of Arctic amplification (Comiso et al., 2017). The reduction in the Arctic sea ice has increased the exchange of heat between the ocean and atmosphere, thus influencing the ocean circulation processes (Onarheim et al., 2018). Climate models suggest that 50%–60% of the recent changes in the Arctic sea ice is due to anthropogenic global warming, whereas rest are attributable to internal climate variability (Stroeve and Notz, 2018; Castruccio et al., 2019). Earlier studies have investigated Arctic sea ice variability and identified various physical processes. The possible factors of sea-ice variability include influence of atmospheric thermodynamic on sea ice melting (Polyakov et al., 2012), atmospheric forcing on sea ice drifting (Vihma, 2014), northward ocean heat transport in the Atlantic basin (Mahajan et al., 2011), and the Arctic amplification in the warming climate (Screen and Simmonds, 2010) associated with the ice-albedo, water vapor, and cloud radiative feedbacks (Colman, 2015). Since late 1978, satellite-based measurements have helped the cryosphere community in understanding and monitoring the state of Arctic sea ice (Cavalieri and Parkinson, 2012; Bliss et al., 2019; Comiso et al., 2017). The satellite-based studies reported a precipitous

Understanding Present and Past Arctic Environments
DOI: https://doi.org/10.1016/B978-0-12-822869-2.00010-4

decline in the Arctic sea ice at a rate of approximately -14% decade^{-1} (Serreze and Stroeve, 2015; Bushuk et al., 2017). In addition, the high-resolution satellite and in situ measurements showed that Arctic sea ice is becoming thinner (Kwok et al., 2009). Sea ice loss in the Arctic is more pronounced in the coastal regions (Peng and Meier, 2018; Comiso et al., 2017; Onarheim et al., 2018); however, the amount of sea ice reduction in Arctic regional seas varies seasonally from region to region. The Arctic Ocean, Baffin Bay, and the Barents Sea experience significant summer and autumn sea ice losses, while the Greenland Sea, the Barents Sea, and the Seas of Okhotsk and Japan experience significant winter sea ice losses. (Cavalieri and Parkinson, 2012).

The Arctic sea ice trend is generally associated with recent shifts in the climate, including pronounced warming over sub-Arctic land and the Arctic Ocean, and a decrease in winter mean sea level pressure (Serreze et al., 2003). In the Arctic, the trend of sea level pressure is positive across the Arctic Ocean and negative over the Atlantic Ocean and mid-latitudes (Ogi et al., 2015). However, the positive sea level pressure anomalies are associated with the anticyclonic circulation over the Siberian coast and Europe. Recent studies emphasized that anticyclones in summer play a major role in generating seasonal variation in atmospheric circulation anomalies, which leads to a longer melting season (Perovich et al., 2008), resulting in transition from a multiyear sea ice to the first-year sea ice (Rigor and Wallace, 2004; Stroeve et al., 2011). Earlier studies have revealed comparatively stronger correlation between the occurrence of summer anticyclones in the Arctic and the loss of summer sea ice (Castruccio et al., 2019). Therefore, Arctic sea ice is now more susceptible to an extensive summer retreat, delayed autumn freeze-up, and less extensive winter growth (Bliss et al., 2019). In the recent years, sea ice decline is more pronounced in winters when the temperature difference between sea ice and open water is the largest. Sea ice declining trends in the Arctic are robustly linked to the rise in surface air temperature and sea surface temperature in the Arctic (Kay et al., 2015; Kwok, 2018).

Previous studies indicate that the interannual Arctic sea ice variation is consistent with decadal changes in dominant atmospheric circulation patterns (Li and Bates, 2007; Proshutinsky and Johnson, 1997). Atlantic Multidecadal Oscillation (AMO) is a long-term series of oscillations correlated with the thermohaline cycle and plays a significant role in melting the sea ice (Mokhov et al., 2008; Semenov et al., 2015). The heat flux anomalies correlated with AMO extremes are demonstrated in the 20th century by the anomalous heat fluxes in the North Atlantic (Gulev et al., 2013) as well as by anomalous heat flows in the Arctic from the Atlantic, with the most important anomalies being found in the winter. Such variations are related to changes in the sea ice layer, which are induced by fluctuations in the ocean heat inflow (Koenigk et al., 2019).

In this study, we have examined the large-scale annual and seasonal variability and trends in sea ice extent, especially its association with the recent changes in sea surface temperature, air temperature, mean sea level pressure, and wind speed. The study aims to identify the impacts of annual and seasonal changes in air–ice–ocean interaction on the long-term variability in the sea ice extent compared to the 39-year period (1979–2017). The time series study was carried out for the Arctic as a whole and for each of nine regional seas: (1) Seas of Okhotsk and Japan, (2) the Bering Sea, (3) the Hudson Bay, (4) the Baffin Bay, (5) the Greenland Sea, (6) the Kara and Barents Sea, (7) the Arctic Ocean, (8) the Canadian Archipelago, and (9) the Gulf of St. Lawrence (Fig. 14.1A).

14.2 Materials and methods

The sea ice extent and sea ice concentration data were analyzed for the period of 1979–2017. The data were obtained from the series of satellite passive microwave brightness temperature from the NASA's Nimbus-7 Scanning Multichannel Microwave Radiometer, Special Sensor Microwave Imager, and Special Sensor Microwave Imager/Sounder instruments on Defense Meteorological Satellite Program satellites.

The calculation of sea ice extent (area having at least 15% sea ice concentration in a grid cell) is based on the sea ice concentration datasets available from National Snow and Ice Data Center (NSIDC). The threshold of 15% of ice concentration for estimating sea ice extent is widely used (Parkinson et al., 1999). Sea ice concentration datasets are taken from the Met Office Hadley Centre Sea Ice and Sea Surface Temperature dataset (HadISST2; https://www.metoffice.gov.uk/hadobs/hadisst2/) (Titchner and Rayner, 2014). Sea ice concentration is estimated from brightness temperature data based on the different temperature signatures of sea ice and water. For example, the polarized signature of water is higher in the vertical channel than in the horizontal, while the sea ice temperature is lower in brightness.

In this study, the variations in the sea ice extent and sea ice concentration are ducumented and addressed in the Arctic region as a whole and in

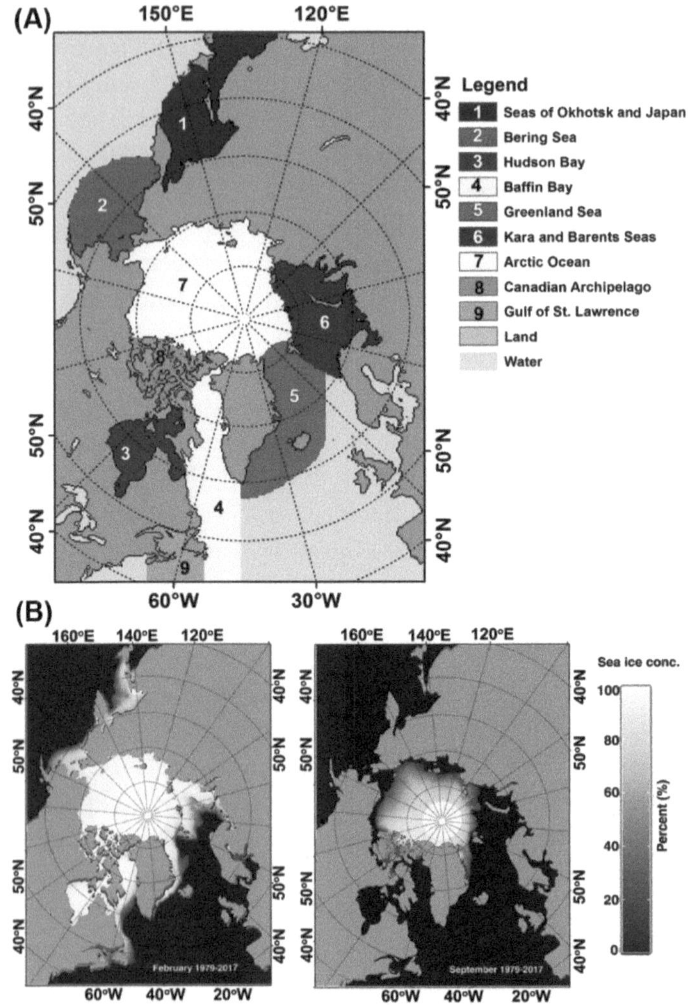

Figure 14.1 (A) Map showing different sectors studied for the northern hemisphere (NH); (B) average sea ice concentrations (1979–2017) of the NH for the boreal winter maximum in February and the boreal summer maximum in September derived from SMMR, SSMI, and SSMIS satellite observations. *SMMR*, Scanning Multichannel Microwave Radiometer; *SSMI*, Special Sensor Microwave Imager; *SSMIS*, Special Sensor Microwave Imager/Sounder.

each of its nine regional seas such as Seas of Okhotsk and Japan, the Bering Sea, the Hudson Bay, the Baffin Bay, the Greenland Sea, the Kara and Barents Sea, the Arctic Ocean, the Canadian Archipelago, and the Gulf of St. Lawrence (Fig. 14.1). The regional seas are consistent with the geographical definitions provided by the NSIDC (ftp://sidads.colorado.edu/DATASETS/NOAA/G02186/ancillary/).

The sea ice extent variabilities and trends for the Arctic as a whole and then for each of the nine Arctic regions were examined. The annual, monthly, and seasonal sea ice was determined using the mean extent value from the climatology year (1981–2010). A simple linear regression approach is applied for the trend calculation is provided in both $km^2\ yr^{-1}$ and in % $decade^{-1}$ (Table 14.1). Standard deviations of the trends were calculated and statistically significant trends were determined, computing the trend ratio R to its standard deviation (Taylor, 1997) identifying a trend as significant at a 95% and 99% confidence level if R values exceed 2.02 2.7, respectively. These statistically significance levels are obtained using the Student's t-test with the zero trends as the null hypothesis and assuming degrees of freedom as 37 ($n = 39 - 2$).

To understand the atmospheric forcing, the Global atmospheric reanalysis data of monthly mean 2 m air temperature, sea surface temperature, mean sea level pressure, and wind speed at the altitude of 10 m from January 1979 to December 2017 were utilized to compute anomalies and trends. These reanalysis data were obtained from the ERA-Interim produced by the European Centre for Medium-Range Weather Forecasts (https://www.ecmwf.int/en/forecasts/datasets/reanalysis-datasets/era-interim). The reanalysis data provides a multivariate record of the global atmospheric circulation which are produced by the data assimilation system (Dee et al., 2011). Annual and seasonal analyses of these atmospheric forcing parameters have been performed for the 39-year period to understand the influence of atmospheric forcing on the Arctic sea ice. The anomaly and trends have been calculated for the total Arctic region (55°N-90°N, 180°W-180°E).

The impacts of ocean warming on the Arctic sea ice were analyzed using the AMO mode of climatic oscillation. The AMO datasets are obtained from the National Oceanic and Atmospheric Administration Earth System Research Laboratory (https://psl.noaa.gov/data/timeseries/AMO/). The AMO index is defined as the long-term spatial patterns of low-pass filtered SST anomaly over the North Atlantic Ocean from which the effects of the anthropogenic forcing on temperature have been removed (Trenberth and Shea, 2006). It is a multidecadal periodic oscillation index calculated from the updated monthly Kaplan SST dataset. The AMO positive and AMO negative phases correspond to the cases in which the AMO index is above and below zero,

Table 14.1 The trends in annual and seasonal sea ice extent (with standard errors) were computed for the Arctic Barents Sea and Kara Sea sectors over the period 1979–2017.

Regions	Yearly			Summer			Autumn			Winter			Spring		
	10^3 km^2 yr^{-1}	*R*	% Decade	10^3 km^2 yr^{-1}	*R*	% Decade	10^3 km^2 yr^{-1}	*R*	% Decade	10^3 km^2 yr^{-1}	*R*	% Decade	10^3 km^2 yr^{-1}	*R*	% Decade
Total Arctic	−54.5 ± 3.1	***− 17.4***	−4.7 ± 0.3	−76.5 ± 6.1	***− 12.6***	− 10.3 ± 0.8	−57.7 ± 5.0	***− 11.5***	−5.5 ± 0.5	−45.3 ± 3.3	***− 13.9***	−3.0 ± 0.2	−39.8 ± 2.9	***− 14***	−3.0 ± 0.2
Seas of Okhotsk	−3.5 ± 0.8	***− 4.4***	−9.5 ± 2.2	0.1 ± 0.0	**2.5**	-	−0.9 ± 0.6	***−1.6***	−6.5 ± 4.1	−9.0 ± 2.1	***− 4.3***	−9.2 ± 2.1	−4.6 ± 1.2	***− 4***	−12.4 ± 3.4
Bering Sea	−0.5 ± 0.7	***−0.7***	−1.7 ± 2.6	0 ± 0	***− 3.6***	−21.9 ± 6.0	− 1.740.7	**−2.4**	−11.9 ± 5.0	0.2 ± 1.6	0.1	0.3 ± 2.4	−0.6 ± 1.4	−0.4	−1.7 ± 3.9
Hudson Bay	−3.2 ± 0.6	***− 5.9***	−4.2 ± 0.7	−4.7 ± 0.8	***− 5.8***	−25.5 ± 4.4	−5.5 ± 1.4	***− 4.0***	−10.9 ± 2.7	−0.1 ± 0.1	−1.2	−0.1 ± 0.1	−2.5 ± 0.5	***− 5***	−2.2 ± 0.4
Baffin Bay	−5.6 ± 1.2	***− 4.7***	−7.4 ± 1.6	−4.6 ± 0.8	***− 5.9***	−24.2 ± 4.1	−5.4 ± 1.2	***− 4.6***	−10.1 ± 2.2	−7.1 ± 2.3	***− 3.0***	−5.5 ± 1.8	−5.6 ± 1.5	***− 4***	−5.4 ± 1.4
Greenland Sea	− 6.0 ± 0.8	***− 7.8***	− 9.6 ± 1.2	−4.4 ± 1.1	***− 3.9***	−11.7 ± 3.0	−5.4 ± 0.9	***− 5.7***	−9.5 ± 1.6	−9.2 ± 1.3	***− 7.3***	−11.4 ± 1.6	−5.1 ± 0.9	***− 6***	− 6.8 ± 1.2
Kara and Barents Seas	−15.7 ± 1.8	***− 8.9***	−13.4 ± 1.5	−14.9 ± 2.0	***− 7.5***	−32.0 ± 4.3	−17.3 ± 2.4	***− 7.3***	−16.6 ± 2.3	−14.1 ± 1.9	***− 7.5***	−8.4 ± 1.1	−16.8 ± 2.3	**− 7**	−11.0 ± 1.5
Arctic Ocean	−16.0 ± 1.6	***− 9.9***	−2.4 ± 0.2	−44.9 ± 4.7	***− 9.6***	−7.8 ± 0.8	−16.0 ± 2 2	***−7.2***	−2.3 ± 0.3	−0.5 ± 0.1	**− 4.2**	−0.1 ± 0.0	−2.5 ± 0.4	***− 5.4***	−0.4 ± 0.1
Canadian Archipelago	−1.0 ± 0.3	***− 3.6***	−1.5 ± 0.4	− 2.9 ± 0.9	***− 3.4***	− 5.4 ± 1.6	−1.1 ± 0.3	***− 3.9***	−1.6 ± 0.4	0 ± 0	−0.1	0±0	−0.2 ± 0.1	−1.9	−0.3 ± 0.1
Gulf of St. Lawrence	−0.7 ± 0.2	***− 3.2***	−11.3 ± 3.6	0 ± 0	-	-	0 ± 0	***0.9***	2.6 ± 3.0	−2.4 ± 0.7	**− 3.7**	−14.0 ± 3.8	−0.4 ± 0.2	−1.8	−7.6 ± 4.3

Assuming a null hypothesis of zero trend and 37 degrees of freedom, *R* values in bold are significant at 95% and above, and those in italics and bold are significant at 99% and above.

respectively. Our analysis focuses on the annual and seasonal variations over the Arctic.

14.3 Results and discussion

14.3.1 Spatiotemporal changes in sea ice variabilities and trends

14.3.1.1 Changes in total Arctic region

The overall Arctic sea ice extent decline is twice as large in September, as in the month of February (Fig. 14.1). Monthly sea ice extent anomaly (from January 1979 to December 2017) and seasonal (winter and summer) anomalies in the total Arctic are shown in Fig. 14.2. The trend lines of linear regression fit for monthly and seasonal mean sea ice extent values for the total Arctic are shown in Fig. 14.2. The slopes of linear regression fits for each of the nine regions: (1) Seas of Okhotsk and Japan, (2) the Bering Sea, (3) the Hudson Bay, (4) the Baffin Bay/Labrador Sea, (5) the Gulf of St. Lawrence, (6) the Greenland Sea, (7) the Kara and Barents Seas, (8) the Arctic Ocean, and (9) the Canadian Archipelago are given in Table 14.1. The downward trend of $-55{,}100 \pm 1600\ km^2\ yr^{-1}$ has been observed in the last 39 years; however, the present rate is very high compared to the earlier reported values for the anomalies of Arctic sea ice (Parkinson et al., 1999; Parkinson and Cavalieri, 2008; Cavalieri and Parkinson, 2012), and the annual mean of total Arctic sea ice extent has decreased by $-1.93 \times 10^6\ km^2$ since 1979 (Fig. 14.2A). The total Arctic

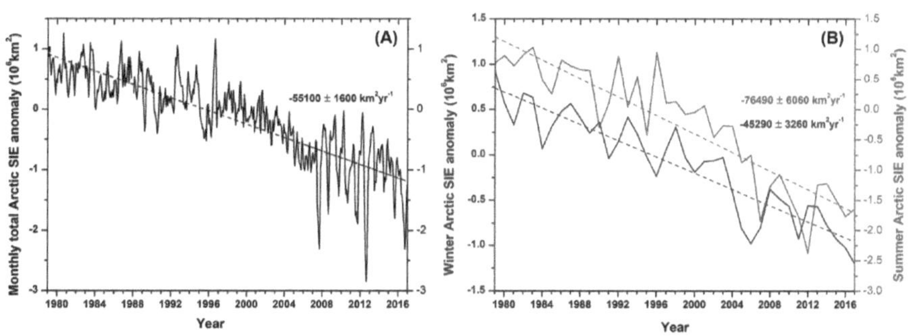

Figure 14.2 Timeseries of sea ice extent (SIE) derived from SMMR, SSMI, and SSMIS satellite data during 1979–2017. (A) Monthly anomaly of SIE with the corresponding lines of least-square fit of the total Arctic region. (B) Seasonal averages of SIEs, with the corresponding lines of least-square fit during boreal winter (January–March) and boreal summer (July–September). *SMMR*, Scanning Multichannel Microwave Radiometer; *SSMI*, Special Sensor Microwave Imager; *SSMIS*, Special Sensor Microwave Imager/Sounder.

trend for the annual mean, $-54,500 \pm 3100\ km^2\ yr^{-1}$, is similar to the monthly anomaly rate, but it has a higher standard deviation, representing the lower number of data points (37 years vs 444 + 2 months). The decline in sea ice extent is significant throughout the year (Fig. 14.2; Table 14.1), with the largest change observed in summer, which is in agreement with previous studies (Stroeve et al., 2007; Cavalieri and Parkinson, 2012).

In the last 39 years, the sea ice extent is characterized by large seasonal variations and decreased at the rate of $-4.7 \pm 0.3\%$ per decade, whereas the sea ice extent decline occurred during winter and spring are significantly lower ($-3.0 \pm 0.2\%$ per decade) than the summer ($-10.3 \pm 0.8\%$ per decade) and autumn ($-5.5 \pm 0.5\%$ per decade) seasons (Table 14.1). However, the magnitude of sea ice extent trend is recorded high during the summer ($-76,500 \pm 6100\ km^2\ yr^{-1}$), with the next strongest trend being for autumn ($-57,700 \pm 5000\ km^2\ yr^{-1}$). Further, the weakest seasonal trends during spring ($-39,800 \pm 2900\ km^2\ yr^{-1}$) and winter ($-45,300 \pm 3300\ km^2\ yr^{-1}$) are statistically significant at a 99% confidence level (Table 14.1; Figs. 14.4J and 14.5J).

14.3.1.2 Regional changes in sea ice extent and trends

The study of regional sea ice extent variabilities is important to understand the cumulative distribution of sea ice concentration and the influence of ocean-atmospheric forcing at the different timescales. In this study, over the 39-year period, monthly and seasonal (boreal winter and summer) sea ice anomalies and their trends for each of the nine Arctic regions and for the total Arctic as a whole are shown in Fig. 14.3 and Figs. 14.4 and 14.5, respectively. The monthly sea ice extent anomaly and trends show each of nine Arctic regions recorded negative trends, whereas the four regions, such as the Arctic Ocean ($-16.6 \pm 2.9 \times 10^3\ km^2\ yr^{-1}$), the Kara and Barents Seas ($-16.5 \pm 2.2 \times 10^3\ km^2\ yr^{-1}$), the Greenland Sea ($-6.2 \pm 0.8 \times 10^3\ km^2\ yr^{-1}$), and the Baffin Bay region ($-6.5 \pm 1.9 \times 10^3\ km^2\ yr^{-1}$), are majorly responsible to contribute the negative sea ice extent trend for the total Arctic as a whole (Fig. 14.3). However, the yearly average data trends computed for 39 years also recorded the same magnitude of sea ice changes (Table 14.1). The sea ice extent variability on monthly and seasonal scales of each nine regions, the trends are discussed in this section.

The Seas of Okhotsk and Japan recorded the sea ice extent decline over the last four decades at the rate of $-9.5 \pm 2.2\%\ decade^{-1}$, while in 1997–2001, it marked the increase in sea ice extent (Fig. 14.3A). The

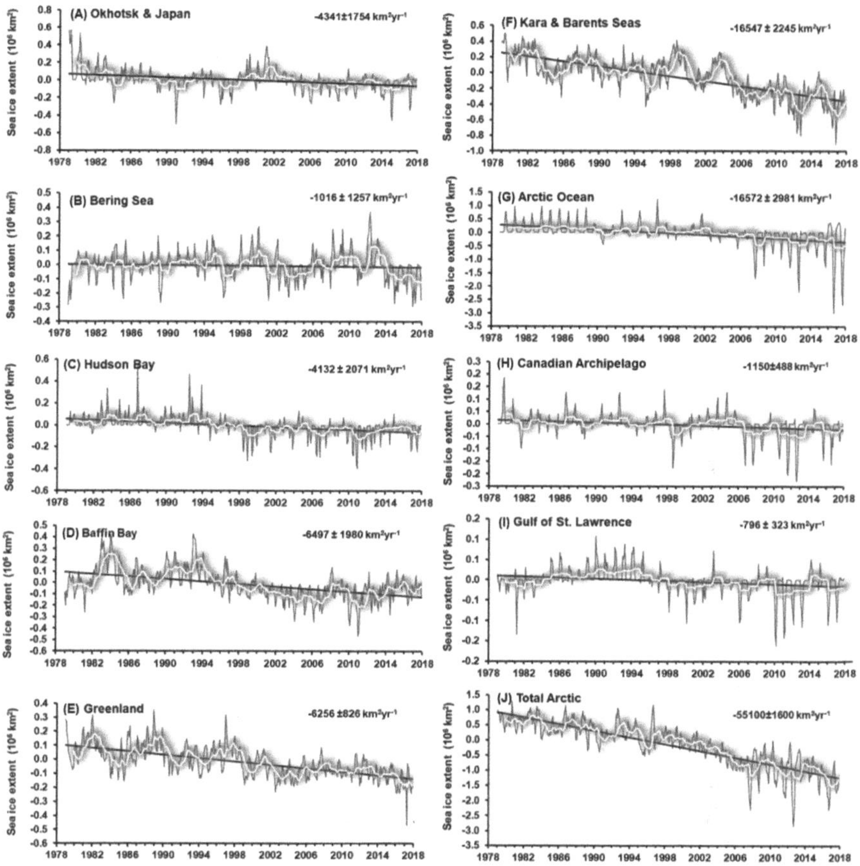

Figure 14.3 Monthly anomalies and trends of nine sectors: (A) Seas of Okhotsk and Japan, (B) Bering Sea, (C) Hudson Bay, (D) Baffin Bay/Labrador Sea, (E) Gulf of St. Lawrence, (F) Greenland Sea, (G) Kara and Barents Seas, (H) Arctic Ocean, (I) Canadian Archipelago, and (J) total Arctic. Gray shaded lines show the 12 months running average.

remarkable changes are observed mainly in the winter ($-9.0 \pm 2.1 \times 10^3$ km^2 yr^{-1}) and spring ($-4.6 \pm 1.2 \times 10^3$ km^2 yr^{-1}) seasons, whereas the summer trend recorded neutral value due to low sea ice cover in summer season(Figs. 14.4A and 14.5A; Table 14.1). The Bering Sea exhibits statistically insignificant trends and the interannual variability is dominated by long-term sea ice extent trends (Fig. 14.4B; Table 14.1). In summer and winter, sea ice recorded minor changes in negative and positive trends, respectively (Table 14.1). The Seas of Okhotsk and Japan recorded the strongest negative trend in April, whereas the Bering Sea observed the

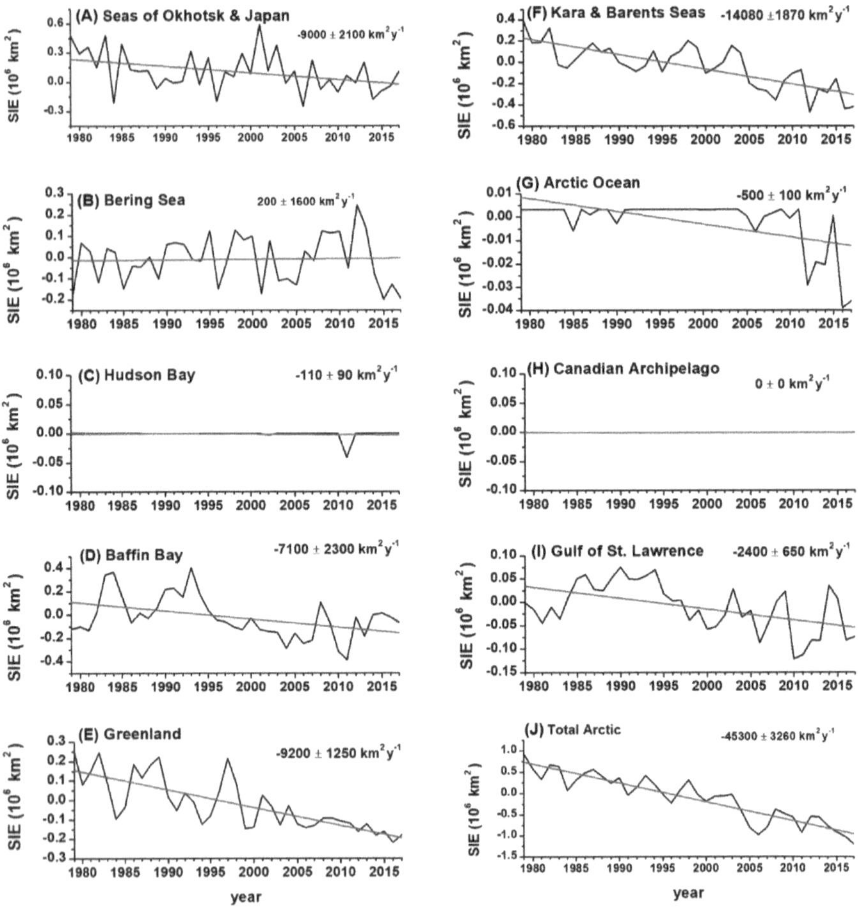

Figure 14.4 Winter (January–March) anomalies and trends of nine sectors: (A) Seas of Okhotsk and Japan, (B) Bering Sea, (C) Hudson Bay, (D) Baffin Bay/Labrador Sea, (E) Gulf of St. Lawrence, (F) Greenland Sea, (G) Kara and Barents Seas, (H) Arctic Ocean, (I) Canadian Archipelago, and (J) total Arctic.

strongest positive (negative) trend in April (November) over the last four decades (Fig. 14.6A and B).

The results from Hudson Bay show statistically significant sea ice decline in the intraannual anomalies, autumn, summer, spring, and the annual mean. The lowest in autumn is followed by summer, annual mean, and spring values of the 39 years (1979–2017) (Figs. 14.3C, 14.4C, and 14.5C; Table 14.1). The winter months trend shows neutral values (Fig. 14.4C) because the Hudson Bay has recorded to be majorly covered with at least 15% sea ice concentration. The Baffin Bay region observed

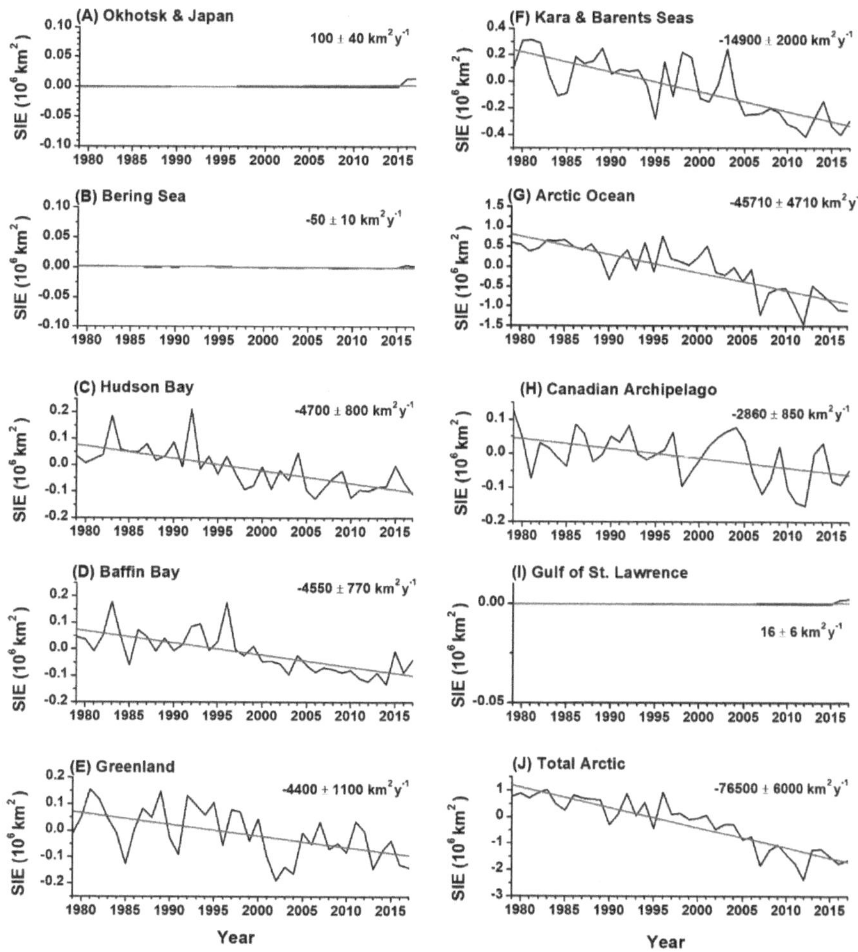

Figure 14.5 Boreal summer (July–September) anomalies and trends of nine sectors: (A) Seas of Okhotsk and Japan, (B) Bering Sea, (C) Hudson Bay, (D) Baffin Bay/ Labrador Sea, (E) Gulf of St. Lawrence, (F) Greenland Sea, (G) Kara and Barents Seas, (H) Arctic Ocean, (I) Canadian Archipelago, and (J) total Arctic.

oscillating patterns of sea ice variability, but overall the region experienced negative trends of sea ice extent (Figs. 14.3D, 14.4D, and 14.5D). The oscillating patterns result in an increase in sea ice extent in some years, but due to larger decreasing magnitude, it could not result as a positive trend. The Baffin Bay region, exhibits statistically significant negative trend in the monthly anomaly, seasonal means and annual averages (Table 14.1). The Hudson Bay exhibited the strongest negative trend in November,

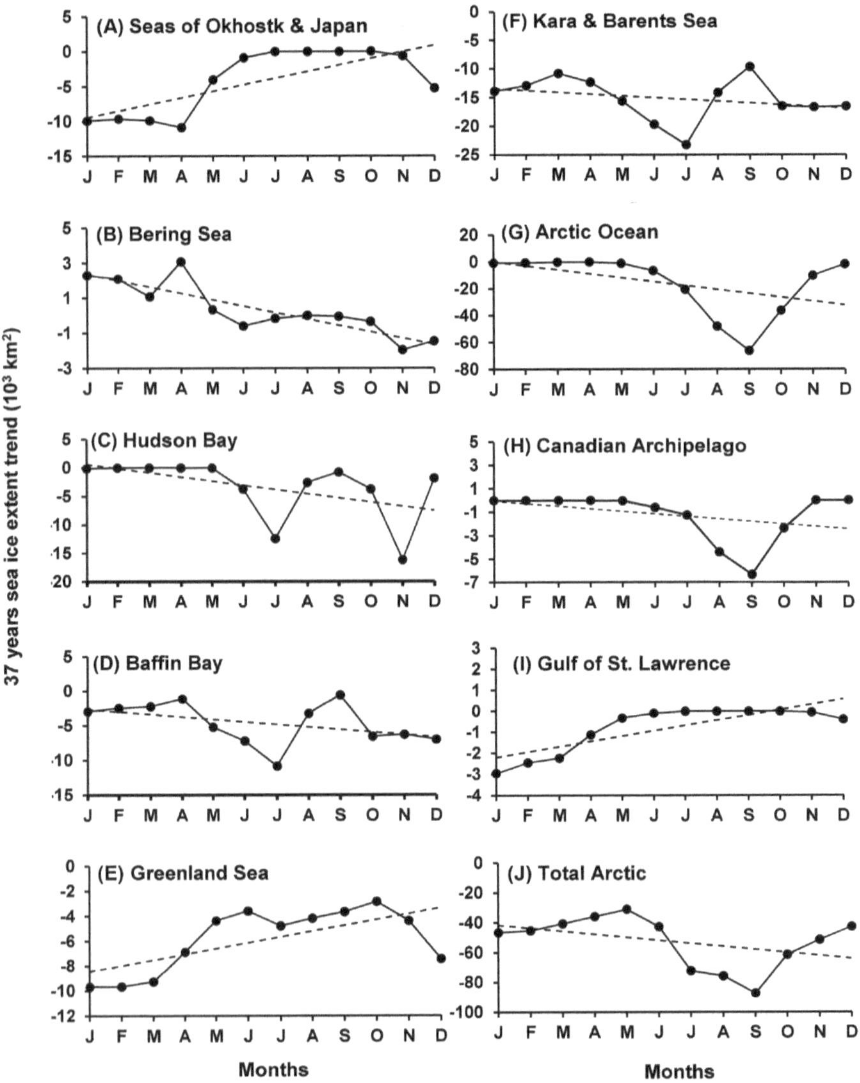

Figure 14.6 Monthly sea ice extent trends for 39 years (1979–2017) of following regions: (A) Seas of Okhotsk and Japan, (B) Bering Sea, (C) Hudson Bay, (D) Baffin Bay/Labrador Sea, (E) Gulf of St. Lawrence, (F) Greenland Sea, (G) Kara and Barents Seas, (H) Arctic Ocean, (I) Canadian Archipelago, and (J) total Arctic.

whereas the Baffin Bay region showed the strongest negative trend in July over the last four decades (Fig. 14.6C and D).

The Greenland Sea has maximum and minimum values in interannual variability, but significant negative trends throughout the study period

(Fig. 14.3E). The seasonal data shows maximum sea ice decline during winter ($-9.2 \pm 1.2 \times 10^3$ km^2 yr^{-1}) followed by autumn, spring, and summer (Table 14.1). The Greenland Sea shows negative trends in every month and season throughout the year. Similarly, the Kara and Barents seas illustrate a strong interannual variability and show the second-largest sea ice decline ($-15.7 \pm 1.8 \times 10^3$ km^2 yr^{-1}) among the nine regions of the total Arctic (Fig. 14.3 and Table 14.1). The Kara and Barents seas region recorded maximum sea ice loss in autumn and spring, followed by summer and winter (Table 14.1). In this region, the rate of sea ice extent shows strong negative trends in the summer (-32.5 ± 4.3% decade^{-1}, i.e, almost four times higher), compared to the winter trends (-8.4 ± 1.1% decade^{-1}). The strongest negative trend is recorded over the Greenland Sea in the month of February, whereas in the Kara and Barents seas region observed the strongest negative trend in the month of July over the last four decades (Fig. 14.6E and F).

The Arctic Ocean exhibits significant declining trends ($-16 \pm 1.6 \times 10^3$ km^2 yr^{-1}) in the last 39 years record and the rate of sea ice loss has increased that resulted in strong negative trends, for the last two decades (Fig. 14.3G; Table 14.1). However, the winter sea ice extent trends recorded constant close to 0, since the maximum area is covered with sea ice during winter (at least 15% sea ice concentration; Fig. 14.7). The significant strong negative trend values were observed during summer ($-44.9 \pm 4.7 \times 10^3$ km^2 yr^{-1}) followed by autumn ($-16 \pm 2.2 \times 10^3$ km^2 yr^{-1}) (Figs. 14.4G and 14.5G). The sea ice extent trends in the Canadian Archipelago are not statistically significant throughout the monthly and seasonal timescales (Figs. 14.3H, 14.4H, 14.5H; Table 14.1). In this region, September is the month when the strongest negative trend was recorded over the last four decades (Fig. 14.6G and H).

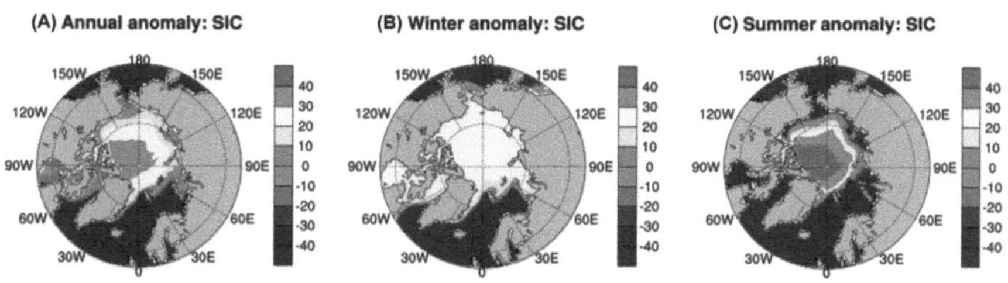

Figure 14.7 Sea ice concentration anomaly map (1979–2017) showing (A) annual, (B) winter (January–March), and (C) summer (July–September).

The Gulf of St. Lawrence recorded the least sea ice coverage among the nine regions and absence of sea ice in the summer, however, during winter maximum increase was reported from 1985 to 1995 and overall the winter showed a negative trend in all the four seasons. The total result suggests overall statistically insignificant negative trends in sea ice extent monthly anomalies, seasonal values (Figs. 14.3I, 14.4I, 14.5I; Table 14.1). In the Gulf of St. Lawrence, January is the month that recorded the highest negative trend over the last four decades (Fig. 14.6I).

14.3.2 Role of ocean-atmospheric forcing on Arctic sea ice changes

14.3.2.1 Spatial variabilities

The Okhotsk and Japan Sea is the southernmost ocean in the northern hemisphere where sea ice has decreased since 1979, which is attributed to the rising temperatures and weak wind speeds with low mean sea level pressure (Fig. 14.8). The sea surface temperature, air temperature, and wind speed anomalies are high during summer and in the winter, whereas low mean sea level pressure in this area is correlated with sea ice melting and drifting. The region is influenced by global atmospheric circulation

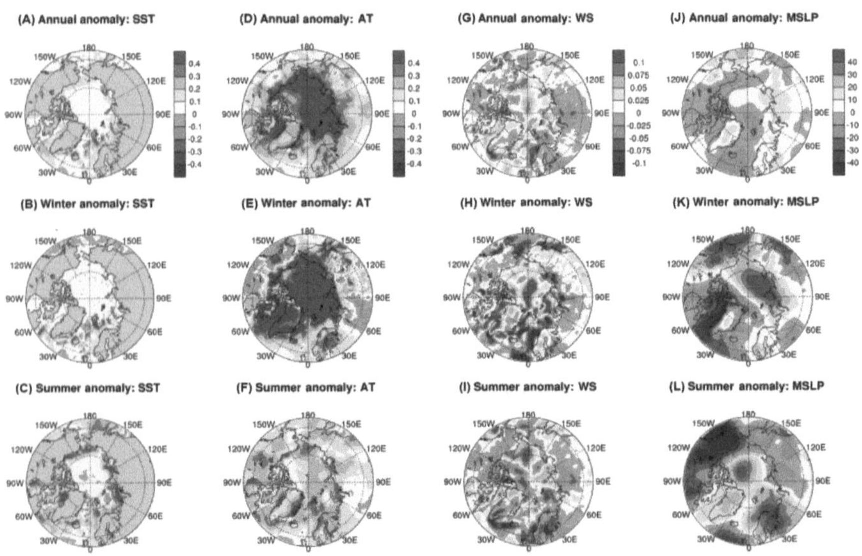

Figure 14.8 Map showing the spatial anomalies (1979–2017) of sea surface temperature, 2 m air temperature, wind speed, and mean sea level pressure, computed by removing the climatological mean for the period 1981–2010: (A) annual, (B) winter, and (C) summer.

causing anomalous fluxes of heat resulting in the sea ice variability (Honda et al., 1999).

The Bering Sea is the only gateway to the Pacific Ocean with the Arctic Ocean acting as a transition zone between the cold dry air mass of the Arctic. The study shows positive sea surface temperature anomalies over the past 39 years in summer, although air temperature has recorded consistent positive anomalies throughout the annual and seasonal datasets (Fig. 14.8). Wind speed in the Bering Sea shows a negative trend, while mean sea level pressure recorded high positive anomalies during the annual and winter period except for low mean sea level pressure in the summer. The reduction of sea ice in this region is linked with the warm Pacific water entering the Arctic from the narrow passage of the Bering Strait and leading to a higher melting of sea ice (Peng and Meier, 2018). Sea ice variability in this region is largely related to the atmospheric circulation, particularly the Siberian high and the Aleutian Low, which generates a strong pressure gradient across Bering Strait and leads to ice transport (Stroeve et al., 2014).

Hudson Bay is the largest inland sea in the world, surrounded by land from all sides with relatively narrow passages that connect it to the ocean of its surrounding world. This sector is isolated from the open ocean circulation, hence, variations in sea ice extent of Hudson Bay are closely tied to the changes in seasonal sea surface temperature and winds. In the last 39 years, sea surface temperature records indicate a gradual rise in temperature in annual and summer while winter also recorded a positive trend. The results show air temperature and wind speed higher throughout the study period, whereas the mean sea level pressure is recorded lowest during the summer followed by winter and annual (Fig. 14.8). Sea ice concentration shows strong seasonal cycle during winter (fully covered with sea ice) and summer (near-zero ice coverage) (Fig. 14.7). The Hudson Bay becomes ice-free from early summer to early winters and the changes in the ice cover pattern are expected to be mainly the result of atmospheric forcing (Wang et al., 1994; Gough et al., 2004).

The Baffin Bay is a pathway of fresh and cold Arctic water to the Labrador Sea resulting into a deep convention process and thermohaline circulation of the North Atlantic. The Baffin Bay has a negative sea ice extent (Fig. 14.3D) and positive anomalies of sea surface temperature, air temperature, and wind speed were observed on annual and seasonal scale, whereas the air temperature recorded a strong positive trend in winter (Fig. 14.8). However, the mean sea level pressure exhibits a very strong positive anomaly in summer, whereas it shows very low pressure in the winter (Fig. 14.8). Ice formation starts in the open water in immediate post-summer, which steadily increases from north to south and in late winter, ice cover reaches its maximum (Fig. 14.7). Warm Atlantic water enters through the Davis Strait and cold water flows toward the south in the western side of this region. The ice season eastern half of the bay has relatively low sea ice cover than the western part due to inflow of warm West Greenland Current (Tang et al., 2004).

The Greenland Sea is a region with a very wide passage to the Central Arctic region that allows Atlantic Waters to interact with waters in the Greenland Sea and its vicinity. This region experiences low summertime ice followed by low wintertime ice (Fig. 14.7) associated with a warm water influx from the Atlantic (Wu et al., 2003). The eastern part of the Greenland Sea is free of ice all the year round because of the Norwegian Atlantic current, which prevents the local sea ice formation (Onarheim et al., 2018). Annual and seasonal data show positive sea surface temperature and air temperature anomalies, whereas consistently very high air temperature was recorded in the northern part of the Greenland sea during winter (Fig. 14.8). However, wind speed and mean sea level pressure show inverse relations, in summer mean sea level pressure exhibits very high pressure and weak wind forcing, whereas the opposite relation was observed from winter and annual datasets (Fig. 14.8).

The Kara and Barents seas are the northernmost Arctic shelf sea that lies on the eastern side of the main gateway between the Arctic and North Atlantic Ocean and it is influenced by the heat transport occurring in the ocean and the atmosphere (Yang et al., 2016). The strong negative trends of sea ice concentration were observed in this sector throughout the year (Fig. 14.7). The temperature data suggest that sea surface temperature was maximum during summer, followed by winter and annual especially in the southern part of Kara and Barents Seas while air temperature shows maximum in the northern part in winter (Fig. 14.8). However, the mean sea level pressure and wind speed show positive anomalies during winter and summer, especially in the northern part of the Kara and Barents Seas (Fig. 14.8). The northern part of the Kara and Barents Seas became colder, possibly due to block effects of Novaya Zemlya and maintained frozen conditions for a much longer duration, hence sea ice variability is found lowest during winters (Parkinson et al., 1999). In the Arctic region, warm Atlantic water enters from Farm Strait and creates a temperature difference in the ocean heat transport, therefore the southern part of this sector remains ice-free year-round (Wu et al., 2003; Petoukhov and Semenov, 2010).

The Arctic Ocean is the largest sector in the northern hemisphere that recorded the highest negative trends since 1979 in summer and annual datasets (Fig. 14.7). Positive anomalies of sea surface temperature were mainly noticed along the periphery of the Arctic Ocean, especially during the summer when melting processes were dominant, whereas in winter this sector is fully covered with sea ice (Figs. 14.7 and 14.8). Air temperature recorded an increasing trend throughout the year, whereas very strong positive anomalies were observed in the annual mean and in winter (Fig. 14.8). The mean sea level pressure in this sector was recorded low in the last 39 years annual mean, whereas strong positive anomalies were observed in the winter and summer (Fig. 14.8). Over the season, changes and variability in the sea ice cover are due to atmospheric forcing in the Arctic Ocean, because the winds have a direct impact on both the movement of the ice and warm air into the region (Comiso and Nishio, 2008).

The annual variation of sea ice in the Arctic as a whole is usually associated with the annual amount of ice melt, though some ice is lost annually by advection from the Arctic through the Fram Strait and the Canadian Archipelago channels (Peng and Meier, 2018). The reduction in sea ice concentration throughout the year since 1979 over Canadian Archipelago and the Gulf of St. Lawrence are in accordance with the Arctic Ocean (Fig. 14.7). The Canadian Archipelago Sea surface temperature is not significantly varying throughout the year except a slight increase was observed during the summer, whereas air temperature shows positive anomalies throughout the year (Fig. 14.8). However, the wind and mean sea level pressure variation in this sector show high speed and low pressure, respectively. In annual and winter datasets except in summer, pressure anomaly shows strong positive anomalies (Fig. 14.8). Large distance advection in the ocean is associated with anomalously high pressure over the eastern Arctic and increased Beaufort gyre, leading to decreased sea ice thickness in the Canadian Archipelago (Kauker, 2003). The reduction of sea ice in September has led to a substantial possible effect on navigation in recent years. This is possibly due to the reduction in Archipelago sea ice cover and hence, the Northwest Passage has become accessible (Parkinson and Cavalieri, 2008). The Gulf of Lawrence sea ice is highly influenced by the warm Atlantic water; therefore, it has strong positive anomalies of sea surface temperature, air temperature, and wind speed. In this sector, the negative anomalies of mean sea level pressure are observed throughout the year.

14.3.2.2 Temporal variabilities

The role of ocean-atmospheric forcing on the Arctic sea ice variability can be understood through the interactions and feedback among the atmosphere, ocean, and sea ice. The physical linkages of ocean-atmospheric parameters with the sea ice from 1979 to 2017 were studied (Fig. 14.9). The highest and lowest sea ice extent were recorded respectively in the year 1982 (12.45 million km^2) and 2016 (10.16 million km^2). The negative temperature anomalies in the last two decades have impacted the formation of sea ice in the Arctic (Fig. 14.9). The average sea surface temperature and air temperature observed in the last 39 years were 0.23°C and −8.8°C, respectively, whereas 2016 was the warmest recorded year, with a sea surface temperature of 0.57°C and an air temperature of −6.4°C. Due to the formation of two cyclones, one fed by warm moist air over the Barents Sea and another over the Siberian region, sea ice loss was maximum in the year 2016. These cyclones produced strong winds that accelerated the breakdown of the ice cover and drifted it north of the Atlantic (Petty et al., 2018). Since the year 1979, the highest sea ice concentration of 69% was accounted for the year 1982 when

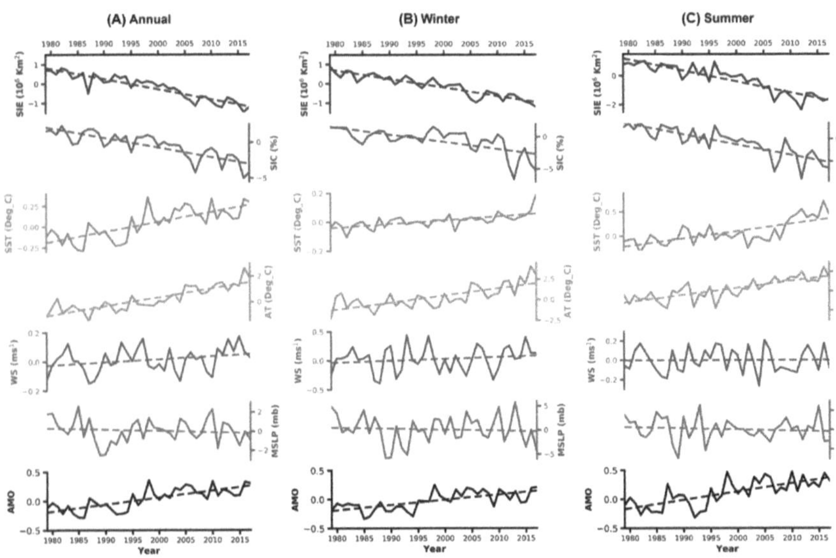

Figure 14.9 Anomalies and trends of sea surface temperature, 2 m air temperature, wind speed and mean sea level pressure, and Atlantic Multidecadal Oscillations over the period of 1979–2017 of the total Arctic (averaged 55°N to 90°N and 0°E to 360°E): (A) annual period; (B) winter period; and (C) summer period. The dotted lines are showing the linear trends at a significance level greater than 95%.

the lowest sea surface temperature (0.2°C) was recorded. However, in the year 1987, the lowest air temperature was recorded (−10.5°C) with second-highest sea ice concentration is 68.5% (Fig. 14.9). The highest mean sea level pressure was observed in the year 1985 (1015.6 mb) whereas since the year 2002, positive trend in temperature is observed and recorded negative pattern of mean sea level pressure.

The interannual sea ice extent variability is associated with large-scale atmospheric as well as local-scale changes. The analysis of seasonal variability suggests that the average winter sea ice extent (14.95 million km^2) is almost double of the average summer sea ice extent (7.44 million km^2) observed in the last 39 years. The loss of sea ice in the summer becomes more pronounced due to changing atmospheric circulation patterns including changes in sea level pressure and wind speed (Wu et al., 2006; Ogi et al., 2010). Minimum sea ice extent (13.87 million km^2) and sea ice concentration (72%) were recorded in 2017 during the winter period (1979–2017). In 2016 and 2017, however, the maximum air temperature of −17°C and sea surface temperature of −0.44°C were observed, respectively (Fig. 14.9). Since 1979, the decline in sea ice extent decrease is larger in summer (Table 14.1) with an average sea surface temperature (1.63°C) about three times higher than the sea surface temperature of the winter (−0.62°C). The lowest sea ice extent (5.32 million km^2) and sea ice concentration (43.5%) were recorded in the summer of 2012. In addition, the maximum temperature in summer was recorded in the year 2016, sea surface temperature was 2.29°C and the air temperature was 4.79°C followed by the sea surface temperature of 2.12°C and air temperature of 4.55°C in 2012 (Fig. 14.9). Over the last four decades, wind and mean sea level pressure have recorded significant negative relation, whereas positive correlations were observed between wind speed and temperature. However, a strong positive trend of the air temperature persists during the winter, whereas the sea surface temperature persists during the summer because the low concentration of sea ice enables more heat exchange from the ocean to the atmosphere and increased ice-albedo feedback (Mahajan et al., 2011). During the past two decades, AMO signals have been closely associated with the North Atlantic sea surface temperature, indicating a transition from the negative (cold) to the positive (warm) phase corresponding to the lower sea ice concentration (Fig. 14.9). The positive phase of the AMO is consistent with the widespread warming over the Arctic, which reduces the formation of sea ice in winter (Fyfe et al., 2013; Li et al., 2018).

14.4 Summary

The northern hemisphere is gaining the attention of the scientists during the last few decades, as the cyclic process of freezing and melting of Arctic sea ice is one of the key indicators of the on-going climate change across the world. There are a number of studies focusing on monitoring the sea ice extent in the northern hemisphere, due to the fact that the Arctic sea ice has experienced consistent and sharp decline owing to influence of the global heating. This study examined several key findings for the Arctic region as a whole and its nine regional seas from the year 1979–2017. In this study, it has been observed that sea ice extent loss in the last 39 years is taking place at the rate of −4.7% per decade, whereas sea ice loss during summer is (−10.38% per decade) which is three times more than the loss in winter (−3% per decade). In the last few years, extreme temperature anomalies in the Arctic have affected the sea ice concentration in summer as well as winter. Since the year 1979, the lowest sea ice concentration was recorded in 2016, due to increased intrusions of cyclones as well as the highest sea surface temperature and air temperature. The sea ice loss is accelerated during summer in the Arctic possibly due to the thinner ice which is more prone to the fractures and deformations and open seas, which allow absorption of more solar radiation in the Arctic climate system.

The study about the cumulative distribution of sea ice concentration over nine regional seas in the Arctic depends on their geographical location as well as different seasonalities experienced by the region. It has been observed that the Arctic region is experiencing a negative trend in sea ice extent. However, there are four major regions, the Arctic Ocean, Kara and Barents Seas, the Greenland Sea, and the Baffin Bay, largely influencing the trend over the total Arctic. The freezing and melting of the sea ice vary from one season to another in all the regions. For example, the regions that are ice-free during summer, including the Seas of Okhotsk and Japan, the Bering Sea, and the Gulf of St. Lawrence, do not exhibit any significant trend. Hence, the regions, experiencing faster melting due to increasing temperature, are vulnerable to further sea ice loss.

The study about the influence of ocean-atmospheric forcing over the Arctic sea ice shows a significant (significance level > 99%) positive trend of sea surface temperature and air temperature during the last two decades. As a result, if global temperatures continue to rise, the regions most responsible for the negative sea ice trend in the Arctic may potentially become the area with no sea ice in the summer. The regions situated in the periphery of the Arctic Ocean are warming and may be free from ice, similar to the Kara Barents Sea region. The heat exchange due to large-scale AMO leads to the widespread warming over the Arctic which results in increase in the loss of sea ice during winter, especially in the eastern Arctic region. The study of fast-declining Arctic sea ice is very important to understand the role of the ocean–atmosphere coupled mechanism on the global climate change.

Acknowledgments

We gratefully acknowledge Dr. M. Ravichandran, Director, National Centre for Polar and Ocean Research (NCPOR) and Ministry of Earth Sciences, for the continuous support and encouragement. Juhi Yadav thanks the University Grants Commission, New Delhi, India, for the award of Junior Research Fellowship [UGC Ref No.: 3358/(NET-DEC 2018). We greatly acknowledge various organizations such as National Snow and Ice Data Center, National Oceanic and Atmospheric Administration, UK Hadley Centre, National Centre for Atmospheric Research, and European Centre for Medium Range Weather Forecast, Technische Universitat Dresden, and the Polar Science Center, Applied Physics Laboratory for making various datasets available in their portals. Authors thank the Editor/s of the book, for their insightful comments and suggestions. This is NCPOR contribution no B-7/2020-21.

References

Bliss, A.C., et al., 2019. Regional variability of Arctic sea ice seasonal change climate indicators from a passive microwave climate data record. Environ. Res. Lett. Available from: https://doi.org/10.1088/1748-9326/aafb84. IOP Publishing.

Bushuk, M., et al., 2017. The seasonality and interannual variability of Arctic sea ice reemergence. J. Clim. 30 (12), 4657–4676. Available from: https://doi.org/10.1175/JCLI-D-16-0549.1.

Castruccio, F.S., et al., 2019. Modulation of Arctic Sea ice loss by atmospheric teleconnections from Atlantic multidecadal variability. J. Clim. Am. Meteorol. Soc. 32 (5), 1419–1441. Available from: https://doi.org/10.1175/JCLI-D-18-0307.1.

Cavalieri, D.J., Parkinson, C.L., 2012. Arctic sea ice variability and trends, 1979-2010. Cryosphere 6 (4), 881–889. Available from: https://doi.org/10.5194/tc-6-881-2012.

Colman, R.A., 2015. Climate radiative feedbacks and adjustments at the Earth's surface'. J. Geophys. Res. Atmos. 120 (8), 3173–3182. Available from: https://doi.org/10.1002/2014JD022896.

Comiso, J.C., Nishio, F., 2008. Trends in the sea ice cover using enhanced and compatible AMSR-E, SSM/I, and SMMR data. J. Geophys. Res. 113 (C2), C02S07. Available from: https://doi.org/10.1029/2007JC004257.

Comiso, J.C., Meier, W.N., Gersten, R., 2017. Variability and trends in the Arctic Sea ice cover: results from different techniques. J. Geophys. Res. Oceans. 122 (8), 6883–6900. Available from: https://doi.org/10.1002/2017JC012768. John Wiley & Sons, Ltd.

Dee, D.P., et al., 2011. The ERA-Interim reanalysis: configuration and performance of the data assimilation system. Q. J. R. Meteorol. Soc. 137 (656), 553–597. Available from: https://doi.org/10.1002/qj.828.

Fyfe, J.C., et al., 2013. One hundred years of Arctic surface temperature variation due to anthropogenic influence. Sci. Rep. 3, 1–7. Available from: https://doi.org/10.1038/srep02645.

Gough, W.A., Cornwell, A.R., Tsuji, L.J.S., 2004. Trends in Seasonal Sea Ice Duration in Southwestern Hudson Bay, ARCTIC.

Gulev, S.K., et al., 2013. North Atlantic Ocean control on surface heat flux on multidecadal timescales. Nature 499 (7459), 464–467. Available from: https://doi.org/10.1038/nature12268.

Honda, M., et al., 1999. Dynamic and thermodynamic characteristics of atmospheric response to anomalous sea-ice extent in the Sea of Okhotsk. J. Clim. 12 (12), 3347–3358. Available from: https://doi.org/10.1175/1520-0442(1999)012<3347:DATCOA>2.0.CO;2.

Kauker, F., 2003. Variability of Arctic and North Atlantic sea ice: a combined analysis of model results and observations from 1978 to 2001. J. Geophys. Res. 3182. Available from: https://doi.org/10.1029/2002JC001573.

Kay, J.E., et al., 2015. The Community Earth System Model (CESM) large ensemble project: a community resource for studying climate change in the presence of internal climate variability. Bull. Am. Meteorol. Soc. 96 (8), 1333–1349. Available from: https://doi.org/10.1175/BAMS-D-13-00255.1.

Koenigk, T., et al., 2019. Impact of Arctic sea ice variations on winter temperature anomalies in northern hemispheric land areas. Clim. Dyn. 52 (5–6), 3111–3137. Available from: https://doi.org/10.1007/s00382-018-4305-1.

Kwok, R., 2018. Arctic sea ice thickness, volume, and multiyear ice coverage: losses and coupled variability (1958–2018). Environ. Res. Lett. 13 (10). Available from: https://doi.org/10.1088/1748-9326/aae3ec.

Kwok, R., et al., 2009. Thinning and volume loss of the Arctic Ocean sea ice cover: 2003–2008. J. Geophys. Res. 114 (C7), C07005. Available from: https://doi.org/10.1029/2009JC005312.

Li, S., Bates, G.T., 2007. Influence of the Atlantic Multidecadal Oscillation on the winter climate of East China. Adv. Atmos. Sci. 24 (1), 126–135. Available from: https://doi.org/10.1007/s00376-007-0126-6.

Li, F., et al., 2018. Atlantic Multidecadal Oscillation modulates the impacts of Arctic sea ice decline. Geophys. Res. Lett. 45 (5), 2497–2506. Available from: https://doi.org/10.1002/2017GL076210.

Mahajan, S., Zhang, R., Delworth, T.L., 2011. Impact of the atlantic meridional overturning circulation (AMOC) on arctic surface air temperature and sea ice variability. J. Clim. 24 (24), 6573–6581. Available from: https://doi.org/10.1175/2011JCLI4002.1.

Mokhov, I.I., et al., 2008. Connection between Eurasian and North Atlantic climate anomalies and natural variations in the Atlantic thermohaline circulation based on long-term model calculations. Doklady Earth Sci. 419 (2), 502–505. Available from: https://doi.org/10.1134/S1028334X08030331.

Ogi, M., Yamazaki, K., Wallace, J.M., 2010. Influence of winter and summer surface wind anomalies on summer Arctic sea ice extent. Geophys. Res. Lett. 37 (7), n/a–n/a. Available from: https://doi.org/10.1029/2009GL042356.

Ogi, M., et al., 2015. Summer-to-winter sea-ice linkage between the Arctic Ocean and the Okhotsk Sea through atmospheric circulation. J. Clim. 28 (12), 4971–4979. Available from: https://doi.org/10.1175/JCLI-D-14-00297.1.

Onarheim, I.H., Eldevik, T., Smedsrud, L.H., Stroeve, J.C., et al., 2018. Seasonal and regional manifestation of Arctic sea ice loss. J. Clim. 31 (12), 4917–4932. Available from: https://doi.org/10.1175/JCLI-D-17-0427.1.

Parkinson, C.L., Cavalieri, D.J., 2008. Arctic sea ice variability and trends, 1979-2006. J. Geophys. Research: Ocean. 113 (7), 1979–2006. Available from: https://doi.org/10.1029/2007JC004558.

Parkinson, C.L. et al., 1999. Arctic sea ice extents, areas, and trends, 1978-1996, 104, pp. 1978–1996.

Peng, G., Meier, W.N., 2018. Temporal and regional variability of Arctic sea-ice coverage from satellite data. Ann. Glaciol. 59 (76pt2), 191–200. Available from: https://doi.org/10.1017/aog.2017.32.

Perovich, D.K., et al., 2008. Sunlight, water, and ice: extreme Arctic sea ice melt during the summer of 2007. Geophys. Res. Lett. 35 (11)), L11501. Available from: https://doi.org/10.1029/2008GL034007.

Petoukhov, V., Semenov, V.A., 2010. A link between reduced Barents-Kara sea ice and cold winter extremes over northern continents. J. Geophys. Res. Atmos.s 115 (21), 1–11. Available from: https://doi.org/10.1029/2009JD013568.

Petty, A.A., et al., 2018. The Arctic sea ice cover of 2016: a year of record-low highs and higher-Than-expected lows. Cryosphere. 12 (2), 433–452. Available from: https://doi.org/10.5194/tc-12-433-2018.

Polyakov, I.V., et al., 2012. Warming of the intermediate Atlantic water of the Arctic Ocean in the 2000s. J. Clim. 25 (23), 8362–8370. Available from: https://doi.org/10.1175/JCLI-D-12-00266.1.

Proshutinsky, A.Y., Johnson, M.A., 1997. Two circulation regimes of the wind-driven Arctic Ocean. J. Geophys. Res. Ocean. 102 (C6), 12493–12514. Available from: https://doi.org/10.1029/97JC00738.

Rigor, I.G., Wallace, J.M., 2004. Variations in the age of Arctic sea-ice and summer sea-ice extent. Geophys. Res. Lett. 31 (9). Available from: https://doi.org/10.1029/2004GL019492.

Screen, J.A., Simmonds, I., 2010. The central role of diminishing sea ice in recent Arctic temperature amplification. Nature. 464 (7293), 1334–1337. Available from: https://doi.org/10.1038/nature09051.

Semenov, V.A., Mokhov, I.I., Koltermann, K.P., 2015. Influence of the Atlantic Multidecadal Oscillation on settling anomalous climate regimes in Northern Eurasia based on model simulation Tracer Oceanography View project Social and economic risks of hazardous hydrological phenomena View project. Artic. Doklady Earth Sci. Available from: https://doi.org/10.1134/S1028334X14120307.

Serreze, M.C., Stroeve, J., 2015. Arctic sea ice trends, variability and implications for seasonal ice forecasting. Philos. Trans. Ser. A Math. Phys. Eng. Sci. 373 (2045). Available from: https://doi.org/10.1098/rsta.2014.0159.

Serreze, M.C., et al., 2003. A record minimum arctic sea ice extent and area in 2002. Geophys. Res. Lett. 30 (3), 1110. Available from: https://doi.org/10.1029/2002GL016406.

Stroeve, J., Notz, D., 2018. Changing state of Arctic sea ice across all seasons. Environ. Res. Lett. 13 (10), 103001. Available from: https://doi.org/10.1088/1748-9326/aade56.

Stroeve, J., et al., 2007. Arctic sea ice decline: faster than forecast. Geophys. Res. Lett. 34 (9), 1–5. Available from: https://doi.org/10.1029/2007GL029703.

Stroeve, J.C., et al., 2011. Sea ice response to an extreme negative phase of the Arctic Oscillation during winter 2009/2010. Geophys. Res. Lett. 38 (2), n/a–n/a. Available from: https://doi.org/10.1029/2010GL045662.

Stroeve, J.C., et al., 2014. Changes in Arctic melt season and implications for sea ice loss. Geophys. Res. Lett. Available from: https://doi.org/10.1002/2013GL058951.

Tang, C.C.L., et al., 2004. The circulation, water masses and sea-ice of Baffin Bay. Prog. Oceanogr. 63 (4), 183–228. Available from: https://doi.org/10.1016/j.pocean.2004.09.005.

Taylor, J.R., 1997. An Introduction to Error Analysis: The Study of Uncertainties in Physical Measurements. University Science Books, Sausalito, California, second.

Titchner, H.A., Rayner, N.A., 2014. The Met Office Hadley Centre sea ice and sea surface temperature data set, version 2: 1. Sea ice concentrations. J. Geophys. Res. Atmos. 119 (6), 2864–2889. Available from: https://doi.org/10.1002/2013JD020316.

Trenberth, K.E., Shea, D.J., 2006. Atlantic hurricanes and natural variability in 2005. Geophys. Res. Lett. 33 (12), L12704. Available from: https://doi.org/10.1029/2006GL026894.

Vihma, T., 2014. Effects of Arctic sea ice decline on weather and climate: a review. Surv. Geophys. 35 (5), 1175–1214. Available from: https://doi.org/10.1007/s10712-014-9284-0.

Wang, J., Mysak, L.A., Ingram, R.G., 1994. Interannual variability of sea-ice cover in Hudson bay, Baffin bay and the Labrador sea. Atmos. Ocean. 32 (2), 421–447. Available from: https://doi.org/10.1080/07055900.1994.9649505.

Wu, B., Wang, J., Walsh, J., 2003. Possible feedback of winter sea ice in the Greenland and Barents Seas on the local atmosphere.

Wu, B., Wang, J., Walsh, J.E., 2006. Dipole anomaly in the winter arctic atmosphere and its association with sea ice motion.

Yang, X.-Y., et al., 2016. Dynamical link between the Barents–Kara Sea ice and the Arctic oscillation. J. Clim. 29 (14), 5103–5122. Available from: https://doi.org/10.1175/JCLI-D-15.

CHAPTER 15

Space Weather Phenomenon of Polar Ionosphere Over the Arctic Region

A.K. Gwal[1], Suryanshu Choudhary[1] and Himanshu Chaurasia[2]
[1]Department of Physics, Rabindranath Tagore University, Bhopal, India
[2]Department of Physics, Govt. College Rehti, Sehore, India

15.1 Introduction

Sun is the primary source of all natural phenomena and space weather disturbances can cause severe fluctuations in the Earth's electrical and magnetic fields. Several types of storms and their effects are observing various geostationary satellites, and these storms are significant source of disturbances in the space plasma. The space weather causes mainly two types of effects—(1) induced effect of charged particle and (2) electrodynamical effects that are involved to change magnetic and electric fields of magnetosphere. During disturb days, space weather condition can change composition of thermosphere (95 to 500 km altitude) and also the composition of the upper atmosphere. It alters the recombination rate of the ion and electron of ionosphere (60 to beyond 1000 km altitude) and can change the electron density drastically in the polar ionosphere. Estimation rate of electron density would be allowing Global Positioning System (GPS) navigation system to correct the errors in signal propagation. During disturb days, space weather condition can also change the ionospheric F-region and perturb Critical Frequency (foF2) of F2 layer (Mendillo and Schatten, 1983). Many ionospheric parameters are governed and controlled by solar geophysical conditions and it is partially affected by geomagnetic storms in the high-latitude region. A diurnal and seasonal variation study of Total Electron Content (TEC) for space weather effect observation near the Earth magnetosphere plays a vital role to understand the dynamics of polar ionosphere (Titheridge, 1966; Walker and Ting, 1972). Sometimes, space weather forecasting comprises long- and short-term predicting capability under various conditions like solar wind interaction with the magnetosphere, ring current interaction

Understanding Present and Past Arctic Environments
DOI: https://doi.org/10.1016/B978-0-12-822869-2.00016-5

with the ionosphere, and energetic charged particles. Observations of L1 and L2 signals of GPS satellites at time intervals of 30 seconds by using dual-frequency receivers have been widely used for the study of polar ionospheric irregularities at high-latitudes region (Aarons et al., 1997; Aarons and Lin, 1999; Mendillo et al., 2000). Many GPS-based studies have been carried out to analyze the effects of geomagnetic storms and solar flares by measurement of Slant Total Electron Content (STEC) value at high-latitude region (Basu et al., 2001; Mendillo et al., 2000). Effects of magnetic storm at equatorial latitudes by using GPS signal reported that the measurements at Fortaleza (Dip Latitude = 4.5 degrees) show a precipitous fall in polar ionospheric TEC value (Basu et al., 2001; Doherty et al., 2001). Basic characteristics of night-time enhancement phenomena of TEC value have been already studied and concluded that solar and magnetic disturbances generate high latitudinal variation (Jain et al., 1995). During the winter season, TEC variations have been observed at low solar activity in the conjugate hemisphere stations (Bailey et al., 1992).

To understand the space weather effect at high-latitude region especially in Arctic having a great importance, its dynamic nature can change the whole ecosystem of poles. Many studies have been conducted on high-latitude polar ionosphere during disturbed space weather conditions, especially during extreme to moderate storms (Kelly et al., 1979; Abdu et al., 1990; Fejer, 1991; Sobral et al., 1997).

The solar wind blows across the magnetic equator and transports space plasma along the near-horizontal field of Earth magnetic lines and impinged the magnetosphere (Fejer et al., 1999; Sobral et al., 1997). Due to this phenomenon, ionospheric Critical Frequency (foF2), increases and produce fountain effect, and this physical mechanism can be responsible for the occurrence of adiabatic heating at the Arctic region. The amount of energy carried by the solar wind influences Earth's magnetosphere, and Earth's strong magnetic field can trap solar wind particles in the limited region of magnetosphere. It is always clear that solar wind particle or terrestrial particles are accelerated by solar energy, and the contribution of high latitudes for particle fluxes is very important because solar wind is one of the significant sources of energy at high-latitude ionosphere. The geomagnetic field creates an obstacle to the solar wind flow and produced intense interaction between the energetic charged particles and the magnetospheric plasma. A unit mass of particle and its momentum can transfer entire energy from heliosphere to the Earth's magnetosphere. There are two important mechanisms behind entire phenomena, one is large-scale convection of the magnetic field lines due to solar wind interaction and the other is precipitating particles that come from the collision retardation of magnetosphere energetic particles. The space weather impacts are highly pronounced at high-latitude ionospheric parameter, which may affect the high-frequency communication as well as trance-ionospheric communications. During ionospheric storm, changes start in F-region either by using foF2 or TEC (Sharma et al., 2005) or IEC data (Mendillo et al., 1970). It has been also reported that the day-to-day variability of TEC at high latitude can follow the variations in equatorial electrojet (EEJ) strength (Dabas et al., 1984; Rama Rao et al., 2006). The extended work from equator to 25°N dip latitude observed the same variability pattern in the equatorial peak region.

15.2 Loss of lock

Loss of lock means that the signal accurately and degrading positioning accuracy are no longer tracked. Doppler change and GPS signal recording in the Arctic region may occur as a result of GPS signal phase variations. It exceeds the bandwidth of phase lock loop (PLL), which is mainly responsible for loss of lock in GPS signal. In addition to this, fade amplitude can cause signal-to-noise ratio (SNR), which drops the threshold level of receiver. Tracking loops employ codeless and semicodeless technologies. In codeless and semicodeless tracking loops, losses are 27–30 and 14–17 dB, respectively. The effects of amplitude fading are more susceptible with respect to code correlation, L2 phase lock loop (PLL) employs narrow bandwidth (≈ 1 Hz, compared to ≈ 15 Hz for L1) to eliminate excess noise. A PLL receiver recovers short duration of loss of lock within a short time-period of 1–4 minutes. Due to GPS satellite angle, severe ionospheric scintillation activities and scintillation may be moderate to high at high latitude equator due to high electron density and large-scale irregularities. During the scintillation processes radio signals at L-band frequencies may vary at crest region and ionospheric scintillation anomaly may further increase electron density. Loss of lock produces signal gradients, but small scale irregularities contribute to strong scintillation that may severely affect the L-band GPS signals. Due to the presence of scintillation, GPS signal is degraded and it can also degrade the performance of GPS receiver. Measurement of polar ionospheric parameter through GPS signal requires dual-frequency GPS signal data through arrangements of wide lane ambiguities. Once the loss of lock occurs, the phase count breaks and connects losses of GPS signal to equatorial plasma irregularities, which occurs when the electron density in the ionosphere's F-region undergoes large and rapid changes. These offsets are quite common with the orbiting satellite data such as GNSS and GPS satellite system.

15.3 Instrumentation and data analysis at the Arctic region

An operation of GISTM system model of GSV4004A GPS Silicon Valley polar ionospheric instrument has been installed at Indian Station at "HIMADRI" [Ny-Ålesund (78.9°N Lat. and 11.9°E Long.)], Norway for navigation and measurement of TEC value at Arctic region. The GPS satellite TEC data of high latitude are collected by the GPS receiver during the period of June 2008 to July 2008. A Space Physics Interactive Data Resource (SPIDR), Solar and Heliospheric Observatory (SOHO), Solar X-ray Observatory (SXOS), Geostationary Operational Environmental Satellite (GOES), Advance Composition Explorer (ACE),http://swdcwww.Kugi.Kyoto-u.ac.jphttp://pss.swpc.noaa.gov.in identified for the measurement of real-time space weather condition by using geomagnetic, solar, and magnetic parameters. A modification of the NovAtel Euro-4 GPS receiver by a high-performance software system for monitoring all GPS satellites at the L1 (1575.42 MHz) and L2 (1227.6 MHz) frequencies. Instrument also measures the phase and amplitude (at 50 Hz rate) and code/carrier divergence (at 1 Hz rate) of GPS satellite. ISMRB is a data logger that collects TEC data from visible satellites and executes it. Finally the data are received in the interval of 60 seconds and converted into STEC to vertical TEC (VTEC) and S4 index for ionospheric scintillation purpose. The average of STEC for all the visible satellites is done at every 60 seconds and cut-off elevation of 30 degrees. Average STEC is measured at each 60-second interval after conversion into VTEC. It is assumed that ionosphere may be horizontally stratified and spatially uniform for simplifying the thin layer of ionosphere at average altitude of 380 km above the Earth's surface, and it is known as thin shell model. The thin shell concept is to reduce each 3D space measurement to data point (the IPP) on a two-dimensional map. The single-layer concept of atmosphere assumes that the free electrons present in the ionosphere are highly concentrated in an infinitely thin imaginary spherical shell surrounding the Earth at a height of 300–400 km, which is the expected maximum electron density (Klobuchar, 1987). Mostly, GPS signal data have been taken at every 60 seconds and the cut-off elevation for ionospheric scintillation is kept to 15 degrees.

15.4 Extra-terrestrial effect at the Arctic region

A solar variability and its effect on high latitude Arctic region are correlatively expressed as magnetic field variation during the month of June and July 2008. Fig. 15.1 shows the overall geomagnetic condition at Ny-Ålesund at Arctic; it is well-suitable solar and geomagnetic parameters which reflected the magnetospheric condition. The TEC of polar ionosphere is coupled with the geomagnetic index (Kp), the amplitude scintillation index (S4), the phase scintillation, and the Number of PRNs locked. Now Fig. 15.2 represents the AE index variation that is taken from the online facility of World Data Centre (WDC) during the operational time. The geomagnetic field rises at 0300 and 1200 UT to the value 4 during both the month, and change in the AE index could also be seen with some time lag of about 1–2 hours, although not much of the effect of this increase is seen in the Kp index value on the polar ionosphere. Mild effect is seen on the amplitude scintillation, but considerable amount of variation is seen on the phase scintillation. The numbers of PRNs locked have been also shown in the bottom of Fig. 15.1, and different PRNs are also showing the amplitude scintillation, and it has remarkable effect on phase scintillation. Solar radiation of 2800 MHz is observed during the observation process, and it is one of the possible reasons of ionization, especially over the polar ionospheric region. It is also a well-known phenomenon of solar EUV radiation. Geomagnetic parameters are also used for their contribution to create disturbances on Polar region. Solar EUV is mainly responsible for ionization over the latitudes of the ionosphere, which is quite similar to diurnal variations of TEC due to solar activity. With the local sunrise, the ionization starts and built up around 06:30 am local. The rate of change of ionization is quite large in the morning hours or even sunset hours. A reverse process is operative after the termination of the solar radiation and change in geomagnetic activity.

A strong correlation has been observed between ionosphere parameters and solar variations especially during the autumn to winter season, and fast decay in TEC value is observed in the evening hours, which is followed by local sunset (Oinats et al., 2008). Though the diurnal variation

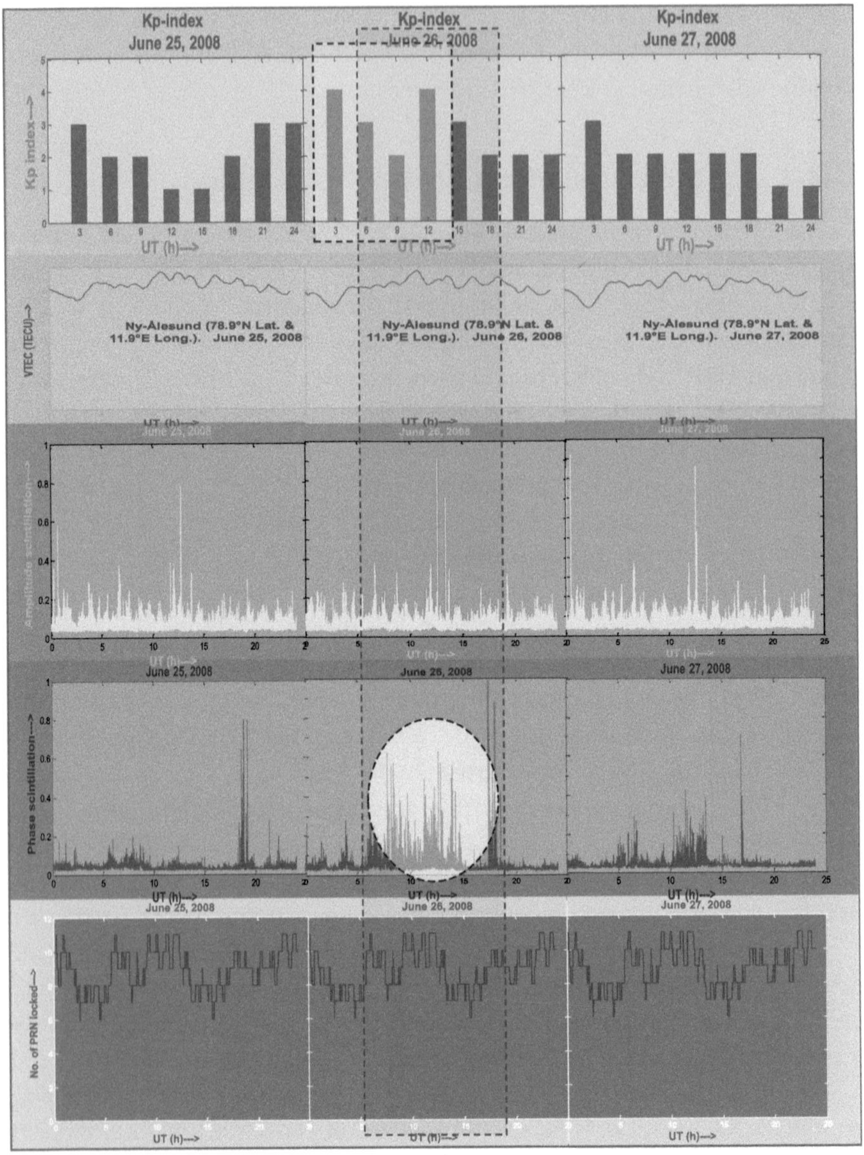

Figure 15.1 Geomagnetic condition at Ny-Ålesund, Arctic region.

of the TEC closely follows the sun, the monthly mean TEC values have shown a mixed effect of its variation with solar flux. Since most of the months have shown a positive dependence, the presence of contradictory results indicates the fact that there may be some other processes playing

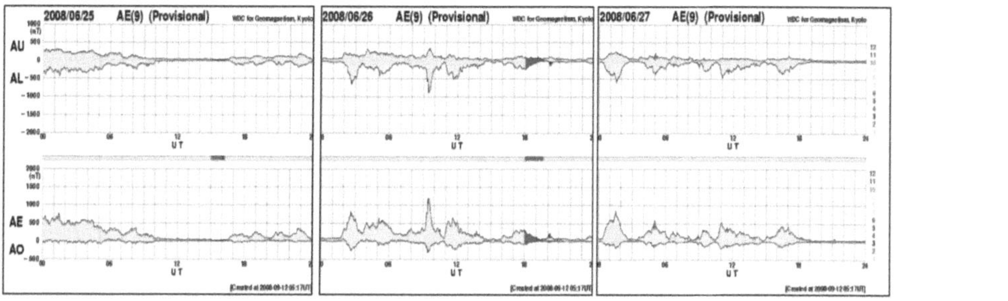

Figure 15.2 Variation of AE index at Ny-Ålesund, Arctic region.

the role for such effect. Hence it could be said that the daily variation of TEC has reasonable dependence on the solar flux. The randomness in the day-to-day variability in TEC is attributed to the changes in the activity of the sun and the associated changes in the intensity of the incoming radiations, and the zenith angle (χ) at which they impinge on the Earth's atmosphere, in addition to the changes that take place in its magnetic field and the EEJ strength, added to the effects due to the dynamics of the neutral winds. Added to all these, the plasma flow associated with the EIA may also play a significant role in the day-to-day variability of the observed diurnal variations in the TEC. The ions produced from the striking process of the ultraviolet light have an active role in reflecting and observing the Earth's radiowaves. This layer also is an unstable medium because it is vulnerable to several distortions, which affects its physical condition.

15.5 Total electron content (TEC) variation at high-latitude region (Arctic region)

The diurnal variations of TEC over the Polar ionospheric region during the month of June and July 2008 at Ny-Ålesund, Arctic (78.9ºN and 11.9ºE) are shown in three consecutive Figs. 15.3, 15.4, and 15.5. The mean diurnal TEC has been also studied for calculation and justifications of diurnal daily variation of TEC data. It is clearly shown that the TEC variation during the daytime goes down and observes small variations. A noticeable feature in the results of high-latitude TEC data is to form valley pattern in the plotted graph; the TECU value starts from 19 in between the time period of 01:30 UT and 02:10 UT. The value of TEC at high latitudes remains constant in between 24 and 28 TECU. The TEC value decreases with the steep gradient and remains almost constant

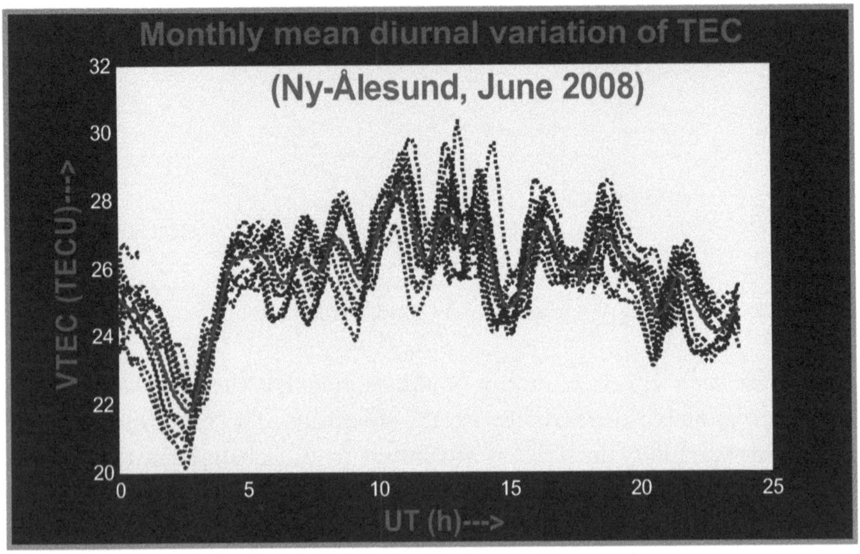

Figure 15.3 TEC diurnal variation in June at high latitude, Ny-Ålesund, Arctic region.

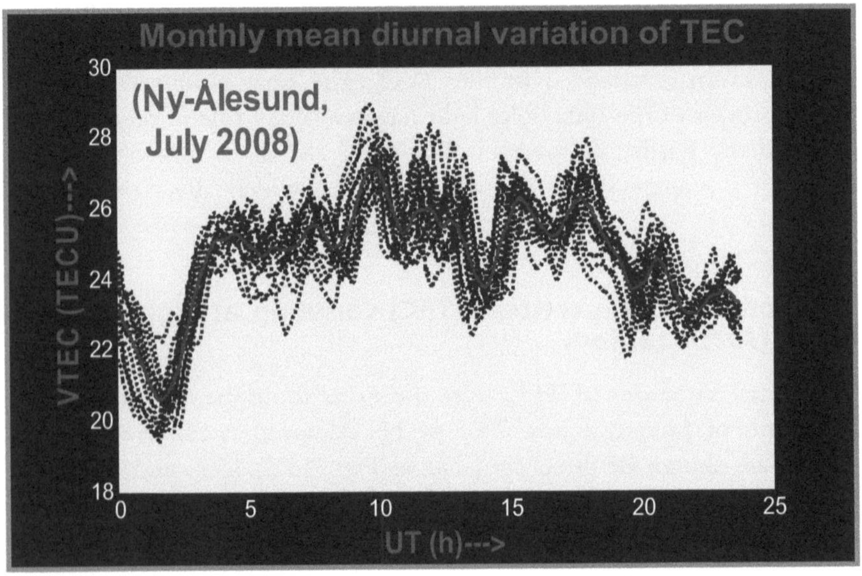

Figure 15.4 TEC diurnal variation in July at high latitude, Ny-Ålesund, Arctic region.

during the nighttime with the value of about 25 TECU being reported. Comparison between daily variations and mean TEC value graphs gives an idea that the polar ionosphere remains quite stable during night and

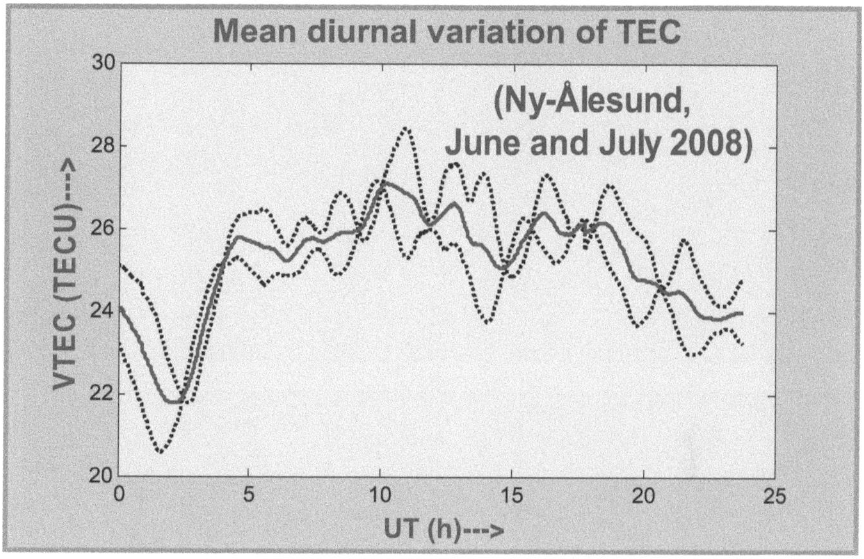

Figure 15.5 Mean TEC diurnal variation in June–July at high latitude, Ny-Ålesund, Arctic.

highly variable during the daytime. The phenomenon is also visible at the time of event, which holds the minimum TEC value shifting and the variability of the TEC value. Fig. 15.6 shows the diurnal variations of TEC value during the quiet days at Ny-Ålesund, Arctic region in the month of June and July 2008. A VTEC (given in TECU) goes down, and the value is 19 TECU during nighttime on both the months but gradually goes up to the maximum value of 29 TECU during daytime especially in the quiet days at Arctic region. Fig. 15.7 shows the diurnal of TEC value during the disturb days at Arctic region in the same month of different dates. A VTEC (given in TECU) goes down to 25 TECU in nighttime, but during the daytime, its value remains the same like quiet days. It is very interesting that during disturb days, polar ionosphere is more perturb and shows nighttime TEC variation. The differences in the TEC value clearly show that the anomalous behavior of polar ionosphere at Arctic region is governed by extraterrestrial effect. Fig. 15.4 shows the diurnal variations of TEC value during the quiet days at Ny-Ålesund, Arctic region in the month of June and July 2008. A VTEC (given in TECU) goes down value is 19 TECU during nighttime on both the months but gradually goes up to the maximum value of 29 TECU during daytime especially in the quiet days at Arctic region. Fig. 15.5 shows the diurnal variation of

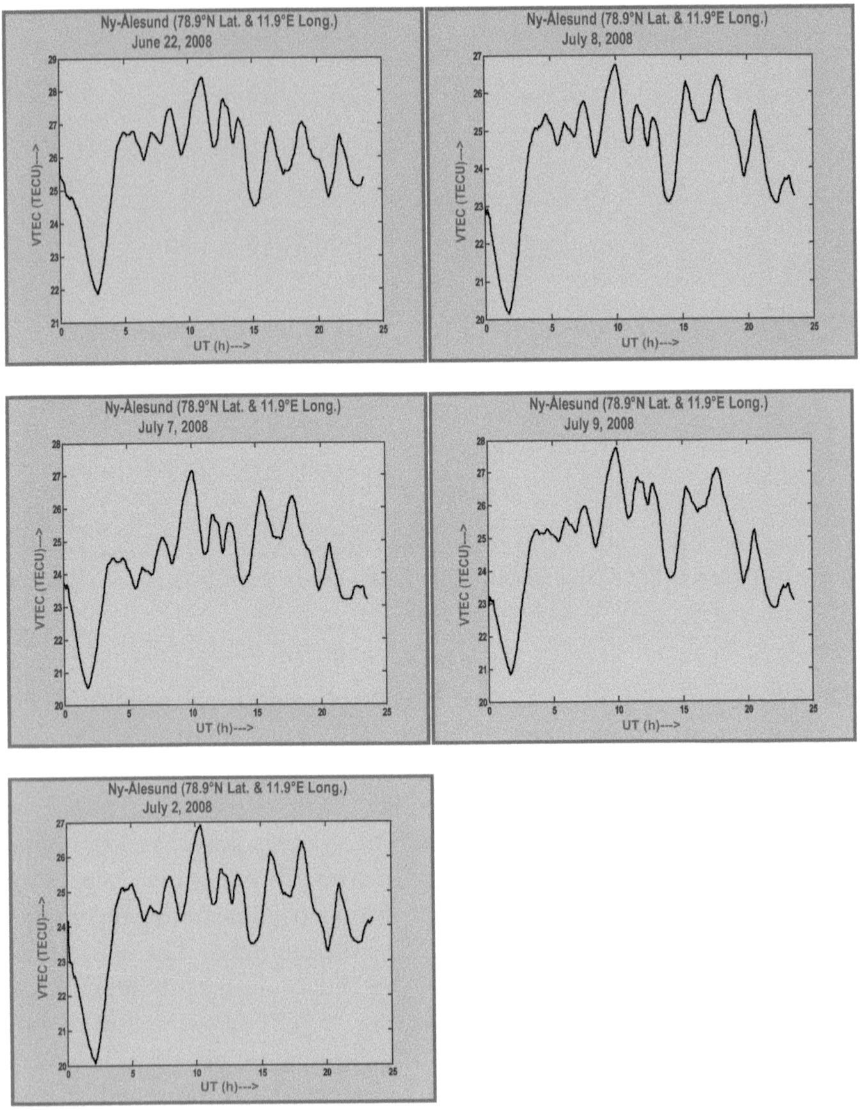

Figure 15.6 Diurnal variation of the TEC at Ny-Ålseund on quiet days during the month of June 2008 and July 2008.

TEC value during the disturb days at Arctic region in the same month of different dates. A VTEC (given in TECU) goes down up to 25 TECU in nighttime, but during the daytime its value remains same like quiet days. It is very interesting that during disturb days, polar ionosphere is more

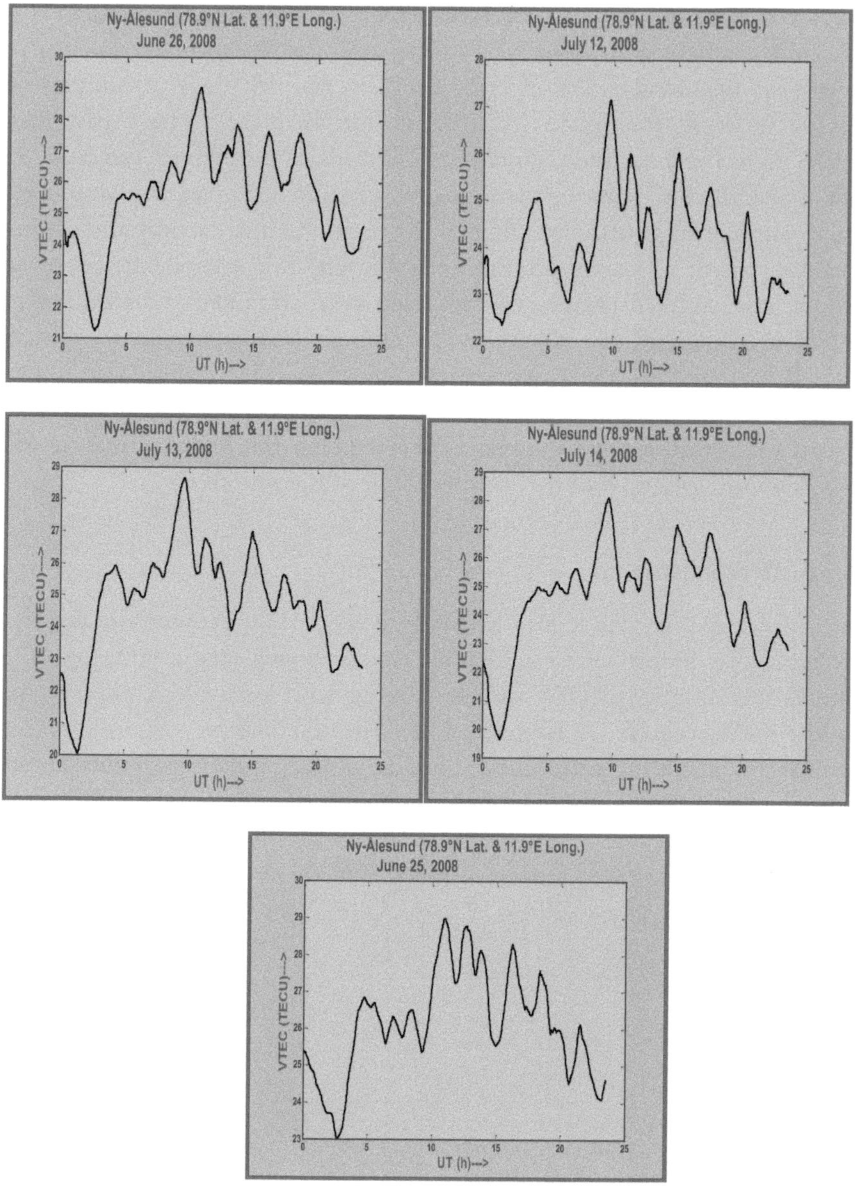

Figure 15.7 Diurnal variation of the TEC at Ny-Ålseund on disturb days during the month of June 2008 and July 2008.

perturb and shows the nighttime TEC variation. The differences in the TEC value clearly show that the anomalous behavior of polar ionosphere at Arctic region is governed by extraterrestrial effect.

To investigate the loss of lock effect at high-latitude station Ny-Ålesund, Arctic region, data are collected for analysis and have been presented in the consecutive Figs. 15.8, 15.9, and 15.10. The number of PRNs showing the phase scintillation but its reverse effect has been observed in the amplitude scintillation. Thus it could be well said that the phase scintillation is more predominant phenomena than the amplitude scintillation at high latitudes while at low latitudes it is the amplitude scintillation which is more pronounced. During this period the satellites which have suffered the loss of lock have been compiled and scintillation is leading to loss of lock, together while the satellite path traces have been also shown in Fig. 15.11 with zoom image of the satellite path trace. Thus from the satellite path trace could be inferred that the presence of irregularities in the polar ionosphere overhead at the high-latitude station Ny-Ålesund goes toward the southern side of the station.

15.6 Discussion

At Ny-Ålesund, Arctic region is mainly governed by extraterrestrial activity and polar ionospheric scintillation effects, which can identify by GPS signals, less number of GPS satellites having with loss of lock phenomena in the GPS receiver. It happens when the scintillation value exceeds a limiting value, and it is found that the loss of lock is more at higher elevation angles. Moderate geomagnetic activity is found during this time

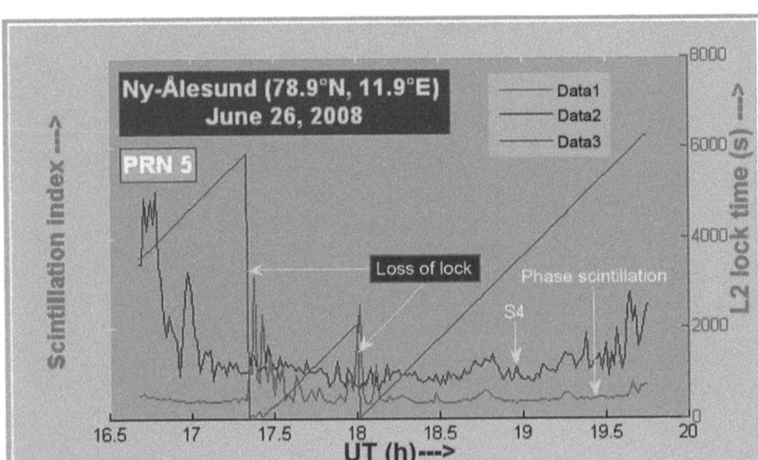

Figure 15.8 PRN 5 loss of lock effect and scintillation index observed at Ny-Ålesund, Arctic region.

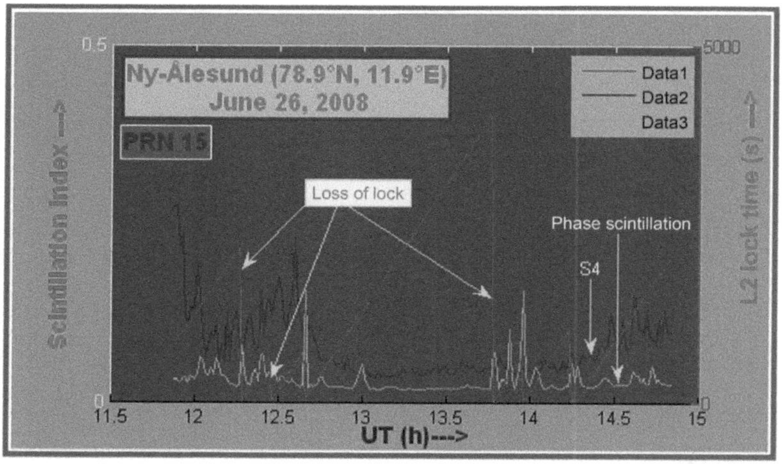

Figure 15.9 PRN 15 loss of lock effect and scintillation index observed at Ny-Ålesund, Arctic region.

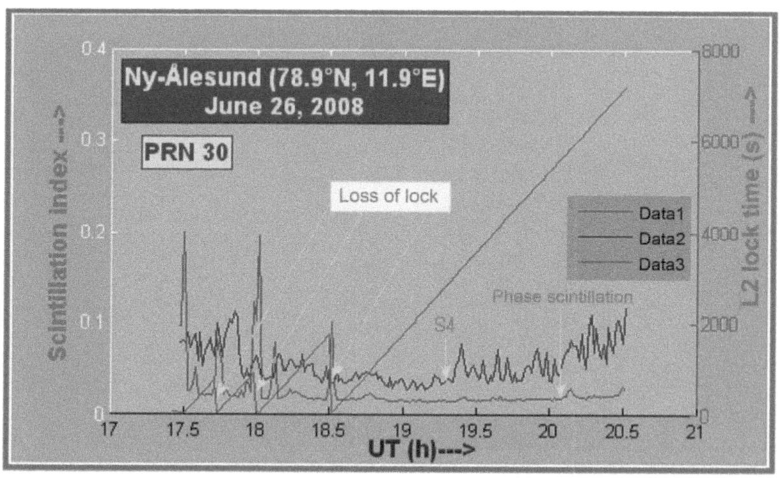

Figure 15.10 PRN 30 loss of lock effect and scintillation index observed at Ny-Ålesund, Arctic region.

period, and solar activity also shows the cycles of 11 years that are approaching for minimum face, which means that disruption in the GPS signals is very low (Klobuchar, 1996; Aarons and Basu, 1994). These high-latitude effects obtained due to loss of lock in the GPS receiver are broadcasted by geostationary satellites (Loh et al., 1995). The code-delay and error translation are rectified through carrier-phase ionospheric

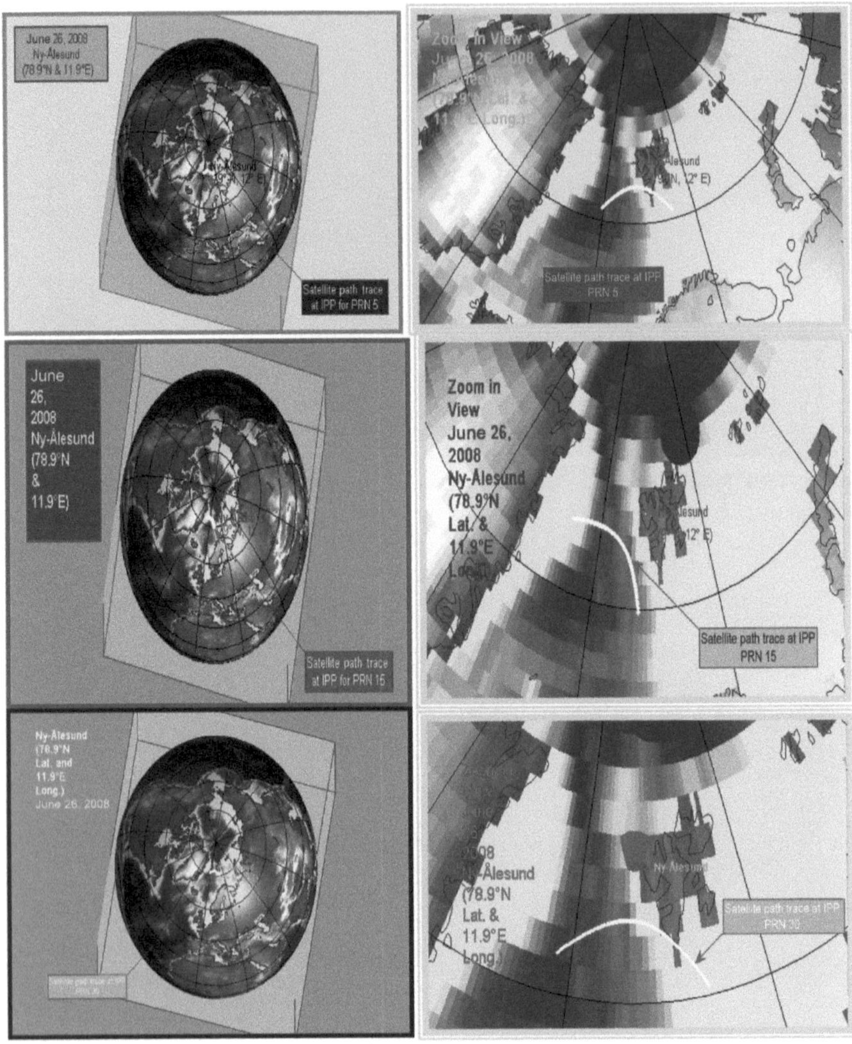

Figure 15.11 Satellite path with PRNs during the month of June and July 2008 at Ny-Ålesund at Arctic region.

observation, which is also known as "Levelling Process" that reduces carrier-phase ambiguities in the GPS data (Ya'acob et al., 2008). The TEC morphology over polar ionosphere at high-latitude Arctic station is very complex and sometimes advanced technology also fails to monitor the effect of ionospheric scintillation. In present result, daily TEC variations and diurnal variations proved that scintillation activity at high latitude has negative correlation with Kp index (geomagnetic activity). The value of TEC remains variable as compared to the nighttime values at Ny-Ålesund, Arctic region, and this variability shows the diurnal variation with sharp peak during afternoon hours and minimum around pre-midnight hours. The variability of TEC is high during the daytime and comparatively very low during nighttime, but the behavior of TEC value was different at quiet and disturb days. The phase and amplitude scintillation are dominant at high latitudes, and the loss of lock occurs if the scintillation exceeds a limit.

The responses of TEC value during moderate geomagnetic condition were studied here, and it was found that the response that came from ionosphere was negative (decrease of electron density) in the high-latitude region. The prime cause of ion formation was radiation that comes from Sun, and day-to-day variability in TEC is naturally governed by extraterrestrial effect (Batista et al., 1994; Doherty et al., 2000). The effect of geomagnetic storms on the polar ionosphere had been studied by many polar scientists, and they have suggested that during storm period, the composition of ionospheric plasma is very dynamic and may give rise to loss coefficients, thus decreasing electron density. The uplifting of plasma due to $E \times B$ action in vertical direction and eventually diffusion along the magnetic field lines to the subtropical latitudes are inhibited, thereby the anomaly weakens. This result proved the decrease in ionization at subtropical region (± 20 degree from the geomagnetic equator), and prestorm studies showed enhancement in electron density at mid-latitudes (Buresova and Lastovicka, 2008). The comparison of TEC value with AE index indicates a good relationship in the study, and it is clearly observed in the graph that during the main phase of storms, AE index increases. It concludes that an AE index increases due to injection of plasma particles into the outer radiation belts, which prevents the ring current. However, it should be noted that plasma sheet current intensifications or earthward motions produced variation in the solar fluxes and shift in the aurora oval having magnetospheric disturbances. When magnetic disturbances started, the direction of solar fluxes may change aurora oval toward equator. The neutral winds also cause the weakening of electro-jet (the electro-jet strength is basically the product of the east–west conductivity of the ionosphere and the electrostatic field) during geomagnetic storm, give rise to decrease in ionization at low latitudes have worked on this problem and have shown that during magnetic disturbances, the wind (which they term storm time wind) opposes the transportation of ionization from equatorial anomaly. This influences the latitudinal variation of electron concentration at lower and middle latitudes. Further they have shown that this storm time wind produces negative effects during summer days in the daytime, while in winter, it depends on the magnitude of the storm. For weak storm, these winds produce positive effects, and for strong and major disturbances, a negative effect is seen. Therefore it could be said that during the storm, the electric field structure in the ionosphere is affected, which may be due to some sort of coupling with Polar region or magnetosphere, which in turn complicates the effect at low latitude ionization anomaly.

Acknowledgments

We would like to give our sincere thanks to Indian Space Research Organisation (ISRO) Department of Space, Govt. of India, Bangalore for providing the financial support under the project entitled, "Space weather studies of the ionosphere and its impact on ground and space-based systems." We also acknowledged the NCAOR, Goa for giving logistic support and opportunity to our group to work under the Ministry of Earth Science, Govt. of India for enduring us with the golden opportunity.

References

Aarons, J., Basu, S., 1994. Ionospheric amplitude and phase fluctuations at the GPS frequencies, Vol. Proceedings of ION GPS-94, Institute of Navigation, Arlington, Va., pp. 1569–1578.

Aarons, J., Lin, B., 1999. Development of high latitude phase fluctuations during the January 10, April 10-11, and May 15, 1997 magnetic storms. J. Atmos. Solar-Terr. Phys. 61 (3–4), 309–327.

Aarons, J., Mendillo, M., Yantosca, R., 1997. GPS phase fluctuations in the equatorial region during sunspot minimum. Radio Sci. 32, 1535–1550.

Abdu, M.A., Walker, G.O., Reddy, B.M., Sorbral, J.H.A., Fejer, B.B.G., Kikuchi, T., et al., 1990. Major phenomena of the equatorial ionosphere-thermosphere system under disturbed conditions. Annal. Geophys. 8, 419–430.

Bailey, G.J., Sellek, R., Balan, N., 1992. Nighttime enhancements in TEC during winter at solar minimum: the effect of the conjugate hemisphere. Adv. Space Res. 12, 309–312.

Basu, S., Basu, S., Valladares, C.E., Yeh, H.C., Su, S.Y., Mackenzie, E., et al., 2001. Ionospheric effects of major magnetic storms during the international space weather period of September and October 1999: GPS observations, VHF/UHF scintillations, and in situ density structures at middle and equatorial latitudes. J. Geophys. Res. 106 (A12), 30389–30413.

Batista, I.S., De-Souza, J.R., Abdu, M.A., De-Paula, E.R., 1994. Total electron content at low latitudes and its comparison with the IRI90. Adv. Space Res. 14 (12), 87–90.

Buresova, D., Lastovicka, J., 2008. Pre-storm electron density enhancements at mid latitudes. J. Atmos. Sol.-Terr. Phys. 70 (15), 1848–1855.

Dabas, R.S., Bhuyan, P.K., Tyagi, T.R., Bhardwaj, R.K., Lal, J.B., 1984. Day-to-day changes in ionospheric electron content at low latitudes. Radio Sci. 19, 749–756.

Doherty, P.H., Klobuchar, J.A., Kunches, J.M., 2000. The correlation between solar 10.7 cm radio flux and ionospheric range delay. J. GPS Solut. 3, 75–79.
Doherty, P.H., Boston, M.A., Essex, E.A., 2001. A dynamic diffusive equilibrium model of the ion densities along plasmaspheric magnetic flux tubes. J. Atmos. Solar-Terr. Phys. 63 (11), 1249–1260.
Fejer, B.G., 1991. Low latitude electrodynamic plasma drifts: a review. J. Atmos. Sol.-Terr. Phys. 58, 677–693.
Fejer, B.G., Scherliess, L., De Paula, E.R., 1999. Effects of vertical plasma drift velocity on the generation and evolution of equatorial spread F. J. Geophys. Res. 104, 19859–19869.
Jain, S., Vijay, S.K., Gwal, A.K., Huang, Y.N., 1995. Nighttime enhancements in ionospheric electron content: seasonal and solar cycle variation. Annal. Geophys. 13, 256–261.
Kelly, M.C., Fejer, B.G., Gonzales, C.A., 1979. An explanation for anomalous equatorial ionospheric electric fields associated with a northward turning of the interplanetary magnetic field. Geophys. Res. Lett. 6, 301–304.
Klobuchar, J.A., 1987. Ionospheric time-delay algorithm for single frequency GPS users. IEEE Trans. Aerosp. Electron. Syst. 23 (3), 325331.
Klobuchar, J.A., 1996. Ionospheric effects on GPS. In: Parkinson, B.W., Spilker, J.J. (Eds.), Global Positioning Systems: Theory and application, vol. 1. American Institute of Aeronautics and Astronautics INC, Washington.
Loh, R., Wullschleger, V., Elrod, B., Lage, M., Haas, F., 1995. The US Wide Area Augmentation System (WAAS). Navigation 52 (3), 435–465.
Mendillo, M., Schatten, K., 1983. Influence of solar sector boundaries on ionospheric variability. J. Geophys. Res. 88, 9145–9153.
Mendillo, M., Papagiannis, M.D., Klobuchar, J.A., 1970. Ionospheric storms at midlatitudes. Radio Sci. 5, 895.
Mendillo, M., Lin, B., Aarons, J., 2000. The application of GPS observations to equatorial aeronomy. Radio Sci. 35, 885–904.
Oinats, A.V., Ratovsky, K.G., Kotovich, G.V., 2008. Influence of 27-day solar flux variations on the ionosphere parameters measured at Irkutsk in 20032005. Adv. Space Res. 42 (4), 639–644.
Rama Rao, P.V.S., Gopikrishna, S., Niranjan, K., Prasad, D.S.V.V.D., 2006. Temporal and Spatial variations in TEC using simultaneous measurements from the Indian GPS network of receivers during low solar activity period of 2004-2005. J. Annal. Geophys. 24, 3279–3292.
Sharma, N., Dabas, R.S., Pillaii, M.G.K., Gwal A.K., 2005. Day-to-day variability of equatorial and low latitude F-region ionosphere in the Indian zone. In: Proceedings of XXVIII General Assembly of International Union Radio Science (URSI), New Delhi, 23-29 October 2005 [GPIG03.4(0659)].
Sobral, J.H.A., Abdu, M.A., Gonzalez, W.D., Tsurutani, B.T., Batista, I.S., Gonzalez Clua de, A.L., 1997. Effects of intense storms and substorms on the equatorial ionosphere/thermosphere system in the American sector from ground based and satellite data. J. Geophys. Res. 102 (14), 305313.
Titheridge, J.E., 1966. Continuous records of total electron content of ionosphere. J. Atmos. Sol.-Terr. Phys. 28, 1135–1150.
Walker, G.O., Ting, S.D., 1972. Electron content and other related measurements for a low-latitude station obtained at sunspot maximum using geostationary satellite. J. Atmos.-Terr. Phys. 34, 283.
Ya'acob, N., Abdullah, M., Ismail, M., 2008. Determination of GPS total electron content using single layer model (SLM) ionospheric mapping function. IJCSNS Int. J. Comp. Sci. Netw. Security 8 (9), 154.

CHAPTER 16

Optical properties of brown carbon in aerosols and surface snow at Ny-Ålesund during the polar summer

Archita Rana[1], Supriya Dey[1] and Sayantan Sarkar[1,2]
[1]Department of Earth Sciences, Indian Institute of Science Education and Research Kolkata, Mohanpur, India
[2]School of Engineering, Indian Institute of Technology Mandi, Kamand, India

16.1 Introduction

As compared to mid-latitudes, the Arctic climate is more sensitive to emissions of light-absorbing aerosols such as black carbon (BC) with a warming estimate of 1.24K, which is comparable to a 10-fold increase in BC concentration at mid-latitudes (Stjern et al., 2017). Long-range transport events to the Arctic directed from Europe and Russia are more frequent and significant compared to those from Asia and North America and have contributed significantly to the Arctic temperature rise over the past century (AMAP Assessment, 2011). After the polar sunrise, the effect of photooxidation of aerosol and sea-to-air emissions of marine organics is common (Fu et al., 2015). Organic-rich particles significantly contribute to the total aerosol mass in the atmosphere over the Arctic after the polar sunrise (Willis et al., 2016). Overall, carbonaceous aerosols are known to be one of the major contributing factors to Arctic climate change due to their light absorption and scattering properties (Seinfeld and Pandis, 2012). Along with BC, the light-absorbing fraction of organic carbon, that is, brown carbon (BrC), has recently emerged as a species of interest. Light absorption by non-BC constituents can be as high as 25% of the total light absorption in Svalbard and mainland Norway (Doherty et al., 2010).

Optical properties of BrC depend heavily on its origin, that is, the relative importance of primary emissions versus secondary formation, and therefore, on the nature of atmospheric processing. Biomass burning (BB)

Understanding Present and Past Arctic Environments
DOI: https://doi.org/10.1016/B978-0-12-822869-2.00022-0

and fossil fuel combustion are the major primary sources of BrC, whereas secondary pathways could be low-temperature oxidation of biogenic materials and their polymerization products, heterogeneous reactions of dienes, and atmospheric processing of anthropogenic/biogenic volatile organic compounds (VOCs) in the presence of NO_x (Andreae and Gelencser, 2006; Laskin et al., 2015). BrC shows strong wavelength dependence of light absorption ($\lambda^{-2}-\lambda^{-6}$), affects the surface UV budget, and has implications for tropospheric chemistry (Laskin et al., 2015). According to Feng et al. (2013), BrC has radiative forcing of $+0.1-0.25$ W m^{-2} over the globe, which is 25% of that contributed by BC and is also responsible for 20%–50% of total aerosol light absorption in the 300–400 nm wavelength range (Liu et al., 2013). BrC has two components—a water-soluble fraction with most absorption near UV, and a methanol-soluble fraction with higher absorption due to the presence of large molecules with conjugated aromatic rings attached to polar functional groups (Sun et al., 2007). Water-soluble BrC can account for 20%–70% of total organic aerosol mass and can act as cloud condensation nuclei due to its ability to increase the hygroscopicity of particles (Kirillova et al., 2014). They contain aromatic rings or aliphatic structures with carboxyl, hydroxyl, carbonyl, or methoxy functional groups (Fu et al., 2015). A significant fraction of BrC is constituted by high-molecular-weight polycyclic as well as monocyclic compounds called humic-like substances (HULIS), which also plays an important role in climate forcing by direct absorption of UV–Vis radiation (Lin et al., 2012; Laskin et al., 2015).

BB emissions, especially wildfires, can contribute significantly to BrC aerosol in Ny-Ålesund, Svalbard via long-range transport as shown in studies such as Moroni et al. (2017, 2020). BrC from BB produces greater atmospheric heating compared to BC (Zhang et al., 2019), and its optical properties vary depending on the biomass type, combustion conditions, and the degree of atmospheric processing (Forrister et al., 2015). Primary emissions can affect aerosol characteristics on the day of the BB event, whereas secondary aerosol formation can continue on subsequent days in the Arctic atmosphere during the polar sunrise and summers (Moroni et al., 2017).

Aerosol-induced snow melting is also a common phenomenon in the Arctic. Dry and wet deposition of aerosol on surface snow and glaciers leads to the reduction of snow albedo thereby accelerating snow melting. In the visible range, snow albedo depends on the presence of impurities

Table 16.1 Absorption coefficient and Angstrom exponent (AE) for water and methanol extracts of aerosol samples as functions of trajectory clusters.

Clusters	Extracts	250 m agl b_{abs300} (M m^{-1})	b_{abs365} (M m^{-1})	AE	5000 m agl b_{abs300} (M m^{-1})	b_{abs365} (M m^{-1})	AE
Cluster 1	Water	0.30 ± 0.39	0.38	4.78 ± 0.11	0.30 ± 0.39	0.38	4.78 ± 0.11
	Methanol	4.05 ± 1.78	0.70 ± 0.13	6.69 ± 0.22	4.05 ± 1.78	0.70 ± 0.13	6.69 ± 0.22
Cluster 2	Water	0.44 ± 0.47		6.06 ± 0.38	0.53	0.16	6.29 ± 0.18
	Methanol	2.47 ± 0.12	0.58 ± 0.11	6.70 ± 0.24	3.91	0.66	6.45 ± 0.18
Cluster 3	Water	0.53	0.16	6.29 ± 0.18	0.44 ± 0.47		6.06 ± 0.38
	Methanol	3.91	0.66	6.45 ± 0.18	2.47 ± 0.12	0.58 ± 0.11	6.70 ± 0.24

Figure 16.1 Study location at Ny-Ålesund, Svalbard.

such as dust, volcanic ash, and carbonaceous aerosols (BC, BrC), whereas in the infrared spectrum, it is affected by the size of snow grains

(Pedersen et al., 2015). Carbonaceous aerosols are also known to affect photochemical reaction rates in snowpacks (Grannas et al., 2004) and non-BC constituents can contribute about 25%–50% of particulate light absorption in Arctic snow (Bond et al., 2013).

In view of this, this study investigates the optical properties of BrC in fine-mode aerosol and surface snow samples collected in Ny-Ålesund, Svalbard during the Arctic summer. The role of long-range transport and potential source regions of BrC to the Norwegian Arctic are also elucidated.

16.2 Materials and methods

16.2.1 Study area

The field campaign for monitoring BrC light absorption properties for air and snow samples was conducted at Ny-Ålesund, Svalbard (78°55′N, 11° 56′E) (Fig. 16.1) as part of Indian Arctic Expedition 2019. The $PM_{2.5}$ sampler was installed on the rooftop of Gruvebadet Laboratory, around 1 km outside the village. Aerosol samples were collected from mid-August to early October 2019 and fresh snow samples were collected mostly toward the end of September 2019.

16.2.2 Sampling and analysis

16.2.2.1 Aerosol sampling and extraction

A low-volume fine particulate sampler (NPM-FDS2.5, Netel India Ltd.) with a flow rate of 16.7 L min^{-1} was used to collect 72 hours $PM_{2.5}$ filter samples on pretreated quartz fiber filters. A total of nine samples were collected during August 21–October 2, 2019. Filters were stored at −80°C and were transported to the laboratory in India under frozen conditions. One-half of each quartz filter was extracted with 20 mL of ultrapure water and another one-fourth fraction was used for extraction using methanol by ultrasonication for 30 minutes to get water- and methanol-extractable fractions of BrC aerosol, respectively (Varga et al., 2001). The extracts were filtered through 0.22 μm PVDF syringe filters and were then analyzed for BrC optical characteristics on a UV–Vis spectrophotometer from 250 to 700 nm wavelength (Evolution 201, Thermo Fisher) and a fluorescence spectrometer (FluoroMax-4, Horiba) for an excitation wavelength (λ_{ex}) of 280 nm (Mukherjee et al., 2020). In this study, light absorption coefficients at a given wavelength ($b_{abs\lambda}$) were calculated as follows:

$$b_{abs\lambda} = \frac{(A_\lambda - A_{700})(V_l)}{V_a.l}\ln 10$$

where A_λ is the absorbance at any given wavelength, A_{700} is the absorbance at 700 nm to account for baseline drift, V_l is the volume of the liquid filter extract (20 mL for water and 10 mL for methanol), and V_a is the total volume of air passed through the filter (for a sampling period of 72 hours) and l is the absorbing path length (1 cm) (Liu et al., 2013). These $b_{abs\lambda}$ were used to estimate Angstrom exponent (AE) values of water- and methanol-extractable fractions for 300–700 nm (Hecobian et al., 2010):

$$b_{abs\lambda} \propto \lambda^{-AE}$$

16.2.2.2 Snow sampling

A total of 14 samples of surface snow (~5 cm) were collected (fresh snow) in late September and early October of 2019 in four different locations over the tundra in Ny-Ålesund, Svalbard. Aluminum spatula was used to shovel snow into zip lock bags and clean disposable latex gloves were used during the sampling. The sample sizes were limited to approximately 500–1000 g and samples were kept frozen until processing. The samples were preserved at −80°C before being cryo-shipped to India. In the laboratory, snow was transferred to clean glass beakers and melted quickly in a microwave oven to avoid contamination. The meltwater was passed through a prebaked 47-mm quartz filter using a vacuum pump and the volumes of filtrates were measured. The filter was dried and extracted

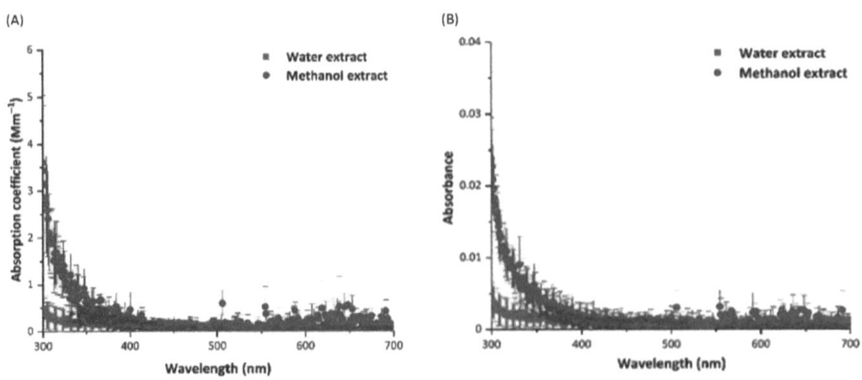

Figure 16.2 Water and methanol extract absorption coefficient for the filter (A) and absorbance for snow (B) samples.

using the same procedure as mentioned in Section 16.2.2.1 to generate aqueous and methanol extracts.

16.2.2.3 Air mass back trajectories

To identify aerosol transport pathways and potential source regions over the Arctic, daily air mass backward trajectories were calculated using meteorological data from the Global Data Assimilation System of the National Ocean and atmospheric administration, and the *Hybrid Single Particle Lagrangian Integrated Trajectory* (HYSPLIT) transport model. HYSPLIT was run for 168 hours backward to calculate trajectories for three starting heights, that is, 250, 1000, and 5000 m above ground level (Moroni et al., 2017). Thereafter, trajectory clusters were calculated as per standard HYSPLIT clustering protocol. The numbers and directions of clusters were similar for all three starting heights, and therefore, the 250 m and 5000 m clusters are discussed hereafter to comment on the transport of aerosol to the Arctic atmosphere.

16.3 Results

16.3.1 Light absorption by water- and methanol-soluble brown carbon

Filter samples for late August to early October 2019 ($n = 9$) were extracted in water and methanol, and the absorbance of extracted samples was measured using UV–VIS and finally converted to absorption coefficient following Liu et al. (2013). A plot of absorption coefficient versus wavelength is shown in Fig. 16.2A. Although both extracts absorb in the entire spectrum (300–700 nm), the absorption increases toward the lower wavelengths, that is, near-UV, with the methanol-soluble fraction absorbing more compared to the water-soluble one. The absorption coefficient at 365 nm (b_{abs365}) was 0.18 $\pm$ 0.19 M m^{-1} for the water extract and 0.65 $\pm$ 0.12 M m^{-1} for the methanol extract. This b_{abs365_water} is higher compared to the b_{abs365} in the Canadian High Arctic (mean: 0.07 $\pm$ 0.04 M m^{-1}; range: 0.02–0.14 M m^{-1}) (Yue et al., 2019). Unlike the methanol extract, only a few samples showed absorption at 365 nm for the water extract. Below 315 nm, there was visible absorption by all samples for both fractions, and therefore, we compared absorption at 300 nm, which was approximately nine times for the methanol extract (3.61 $\pm$ 1.43 M m^{-1}) compared to the water extract (0.39 $\pm$ 0.34 M m^{-1}). The AE for the water extract was 5.8 $\pm$ 0.1, whereas for the methanol extract

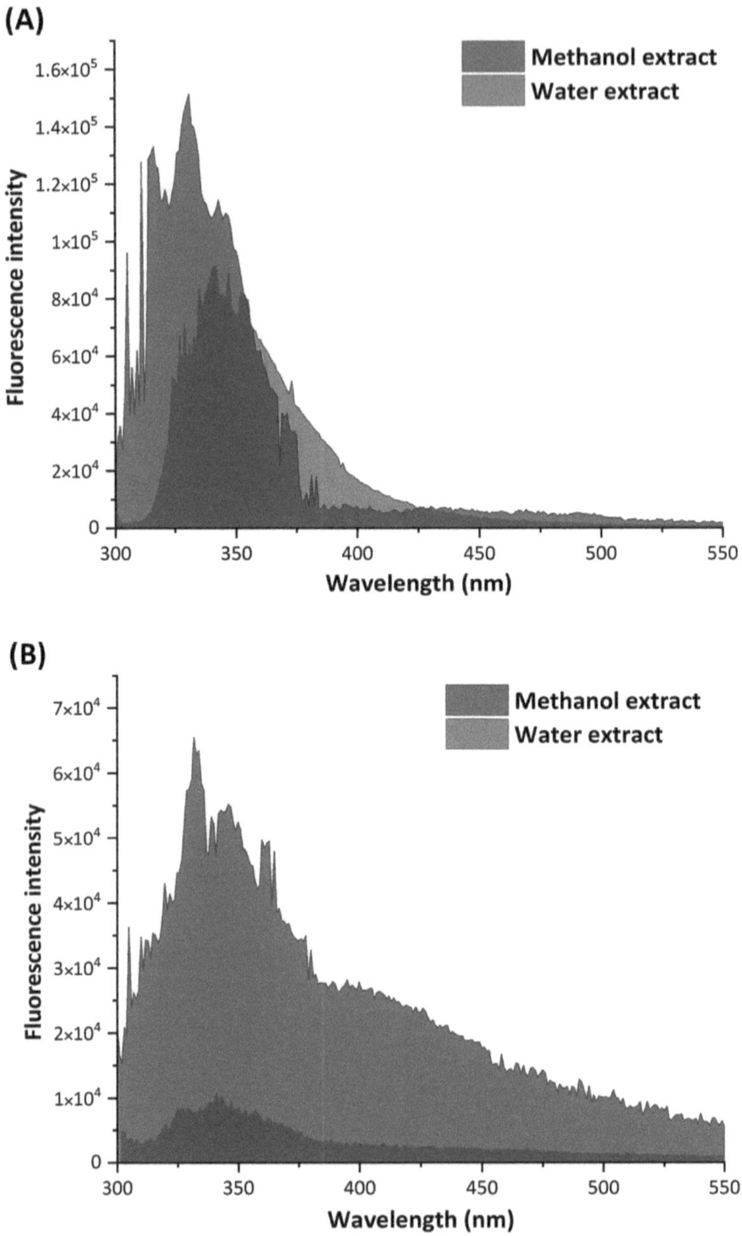

Figure 16.3 Fluorescence spectra for water and methanol extract for the filter (A) and snow (B) samples.

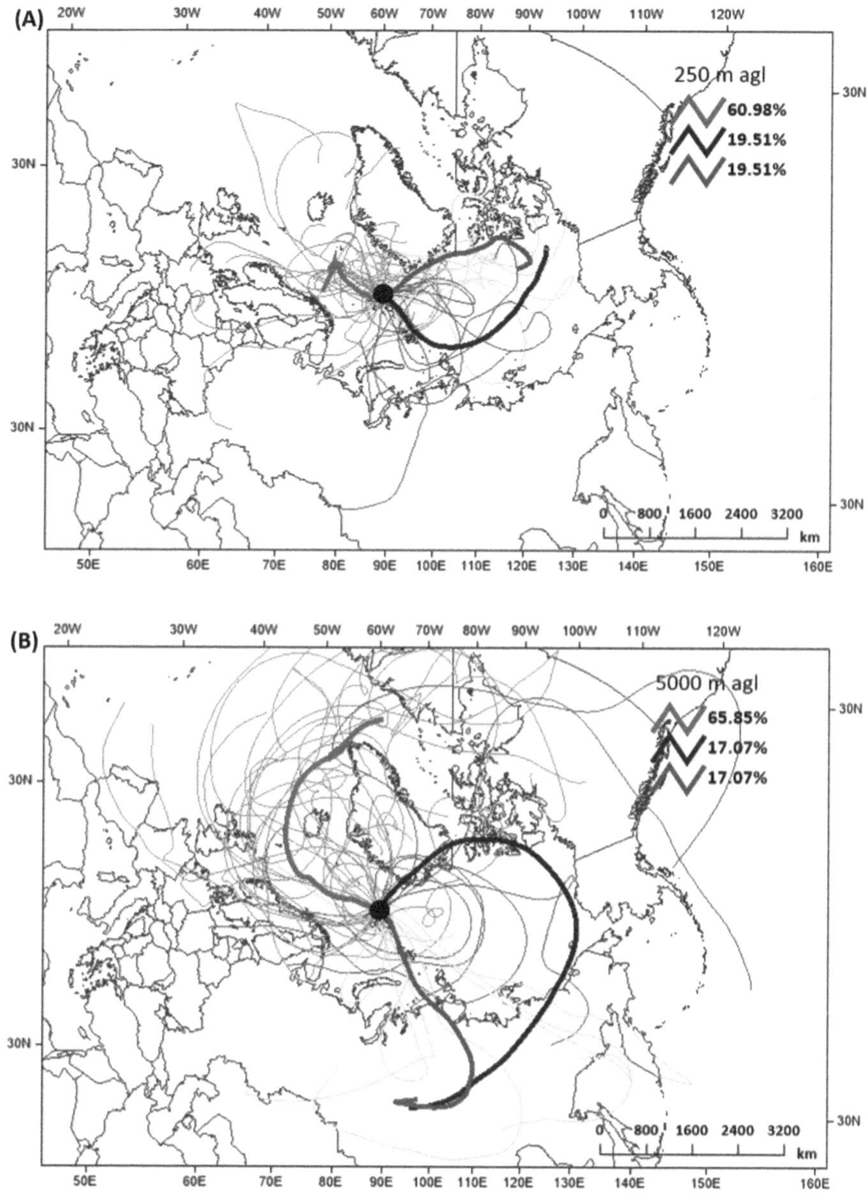

Figure 16.4 Clusters of air mass back trajectories over Svalbard for a study period at 250 m agl (A) and 5000 m agl (B).

it was 6.6 ± 0.2 and was lower compared to that reported by Yue et al. (2019) which is about 8.5 ± 4.2 for the Canadian Arctic.

Surface snow samples ($n = 14$) were extracted and absorbances were plotted against wavelength (Fig. 16.2B) similar to Voisin et al. (2012). Similar to air samples, the absorbance for snow samples increased toward lower wavelengths with the methanol-soluble fraction absorbing more compared to the water-soluble fraction. For water extracts, the absorbance at 365 nm was 0.0017 ± 0.0015 and was almost half of the methanol extract (0.0033 ± 0.0019). For 300 nm the absorbance of the methanol extract (0.024 ± 0.0057) was almost six times than that of the water extract. The AE for the water extracts of snow samples was 7.3 ± 0.2, whereas for the methanol extract it was 6.8 ± 0.2.

16.3.2 Fluorescence spectra for air and snow samples

Fluorescence spectra for air and snow samples for $\lambda_{ex} = 280$ nm are shown in Fig. 16.3A and B. Methanol extract shows higher intensity compared to water extract for both air and snow samples. Water-soluble fraction shows peak at similar wavelength at about 342 nm and 341 nm with a fluorescence intensity (FI) of 9.2×10^4 and 1.1×10^4 for air and snow samples, respectively. The peak for methanol extract for both air and snow samples is at similar wavelength of ~ 332 nm. For an excitation of 280 nm, the emission at 340 nm shows that there could be the presence of protein-like fluorophores in the aerosol in the Arctic as it is similar to excitation–emission matrix (EEM) peak at around 320 nm for excitation of 280 nm from Fu et al. (2015) during polar sunrise. There is also FI between 400 and 450 nm, which could be because of the presence of HULIS fraction of BrC aerosol, but it is not the primary contributor to the light absorption during August–September 2019.

16.3.3 Air mass transport and source sectors during the Arctic summer

Trajectory clusters are shown in Fig. 16.4A and B for heights of 250 m agl and 5000 m agl for the sampling period. Three clusters were identified and were used to comment on the potential source regions contributing to carbonaceous aerosol in the Arctic summer. The major clusters contributing 61%–66% arrived from over the Arctic Ocean near Greenland and Iceland. The other two clusters contributing 17%–19% arrived from Alaska and Russia over the ocean. With the increasing height of air trajectories, the effect of terrestrial and marine emissions becomes apparent, which otherwise are not very clear from trajectories near ground height

(250 m agl). Although the contribution by cluster 1 was about 60% to the total number of trajectories, the average absorption coefficient for the water extracts by trajectories in cluster 1 was the lowest among the three clusters (Table 16.1).

16.4 Discussion

The methanol-soluble fractions show greater light absorption ability for both air and snow samples compared to the water-soluble fraction, which could be because methanol can extract a much larger fraction of organics from aerosol ($\sim$92%) (Chen and Bond, 2010). Significant absorption by the water-soluble fraction at lower wavelengths could be because of secondary organic aerosol (SOA) formation via photochemical gas-to-particle conversion during the polar summer (Yue et al., 2019). The b_{abs} for methanol for air samples and absorbance for snow samples decreased relatively slowly with increasing wavelength compared to water, which might be due to the presence of unsaturated polyconjugated compounds (Sun et al., 2007) and/or due to the presence of water-insoluble compounds, for example, nitroaromatic compounds (NACs) and nitro-polycyclic aromatic hydrocarbons (PAHs) with 4–6 aromatic rings fused together (Lin et al., 2018). The sources of atmospheric NACs include gas-phase oxidation and reactions of phenolic compounds (Zangrando et al., 2013) in the presence of NO_x and oxidants. The AE of about 6 for water extracts shows that there might be BrC aerosols from BB plumes and fresh SOA (Hecobian et al., 2010). Also, high absorption coefficients of water extracts for clusters 2 and 3 indicate contribution of terrestrial emission and forest fires from Alaska and regions in Russia (Fig. 16.1).

FIs for both snow and air samples show peak near 340 nm for excitation at 280 nm, which could be because of the presence of protein-like fluorophores (Fu et al., 2015). It is also similar to the EEM fluorescence peak at around $\sim$354 to 364 nm for protein-like substances reported previously (Xie et al., 2020; Wu et al., 2020). This contribution by protein-like fluorophores could be because of enhanced sea-to-air emissions of marine organic matter and the production of SOA from the photooxidation of anthropogenic and biogenic VOCs during polar summer (Fu et al., 2015; Yue et al., 2019). Major trajectory clusters coming over the Arctic Ocean from Iceland and Greenland suggest marine organics being an important contributor to protein-like fluorophores (Yue et al., 2019) along with a contribution from terrestrial aerosol. BB aerosol possibly from wildfires in Alaska and Russia during summer 2019 could also be one of the contributors to the BrC aerosols (Moroni et al., 2017, 2020).

16.5 Conclusion

During the polar summer, protein-like fluorophores contribute significantly to BrC light absorption in the Norwegian Arctic. These fluorophores are mostly likely from sea-to-air emission of marine organics or secondary formation of aerosol by photooxidation as suggested by trajectory clusters. BB aerosol from forest fires in Alaska and regions in Russia could also be a possible contributor to the BrC aerosol in this region.

Acknowledgements

The authors would like to acknowledge the Ministry of Earth Sciences, Government of India for funding, the National Centre for Polar and Ocean Research (NCPOR) for allowing and coordinating participation in the Indian Arctic Expedition 2019 and for providing instrumentation facility at Ny-Ålesund, Svalbard. The authors are thankful to Dr. Rohit Srivastava of NCPOR and Dr. Sunil Sonbawne of Indian Institute of Tropical Meteorology, Pune for help with aerosol sampling. Indian Institute of Science Education and Research, Kolkata is acknowledged for providing laboratory space and instrumentation facilities for analysis.

References

AMAP Assessment, 2011. Mercury in the Arctic. Arctic Monitoring and Assessment Programme (AMAP), Oslo, Norway.

Andreae, M.O., Gelencser, A., 2006. Black carbon or brown carbon? The nature of light absorbing carbonaceous aerosols. Atmos. Chem. Phys. 6, 3131–3148.

Bond, T.C., Doherty, S.J., Fahey, D.W., Forster, P.M., Berntsen, T., DeAngelo, B.J., et al., 2013. Bounding the role of black carbon in the climate system: a scientific assessment. J. Geophys. Res. Atmos. 118 (11), 5380–5552.

Chen, Y., Bond, T.C., 2010. Light absorption by organic carbon from wood combustion. Atmos. Chem. Phys. 10, 1773–1787.

Doherty, S.J., Warren, S.G., Grenfell, T.C., Clarke, A.D., Brandt, R.E., 2010. Light-absorbing impurities in Arctic snow. Atmos. Chem. Phys. 10 (23), 11647–11680.

Feng, Y., Ramanathan, V., Kotamarthi, V.R., 2013. Brown carbon: a significant atmospheric absorber of solar radiation? Atmos. Chem. Phys. 13, 8607–8621.

Forrister, H., Liu, J., Scheuer, E., Dibb, J., Ziemba, J., Thornhill, K.L., et al., 2015. Evolution of brown carbon in wildfire plumes. Geophys. Res. Lett. 42, 4623–4630.

Fu, P., Kawamura, K., Chen, J., Qin, M., Ren, L., Sun, Y., et al., 2015. Fluorescent water-soluble organic aerosols in the High Arctic atmosphere. Sci. Rep. 5, 1–8.

Grannas, A.M., Shepson, P.B., Filley, T.R., 2004. Photochemistry and nature of organic matter in arctic and antarctic snow. Glob. Biogeochem. Cycles 18 (1).

Hecobian, A., Zhang, X., Zheng, M., Frank, N., Edgerton, E.S., Weber, R.J., 2010. Water-soluble organic aerosol material and the light-absorption characteristics of aqueous extracts measured over the South eastern United States. Atmos. Chem. Phys. 10, 5965–5977.

Kirillova, E.N., Andersson, A., Tiwari, S., Srivastava, A.K., Bisht, D.S., Gustafsson, Ö., 2014. Water-soluble organic carbon aerosols during a full New Delhi winter: isotope-based source apportionment and optical properties. J. Geophys. Res. Atmos. 119, 3476–3485.

Laskin, A., Laskin, J., Nizkorodov, S.A., 2015. Chemistry of atmospheric brown carbon. Chem. Rev. 115, 4335–4382.

Lin, P., Rincon, A.G., Kalberer, M., Yu, J.Z., 2012. Elemental composition of HULIS in the Pearl River Delta Region, China: results inferred from positive and negative electrospray high resolution mass spectrometric data. Environ. Sci. Technol. 46, 7454–7462.

Lin, P., Fleming, L.T., Nizkorodov, S.A., Laskin, J., Laskin, A., 2018. Comprehensive molecular characterization of atmospheric brown carbon by high resolution mass spectrometry with electrospray and atmospheric pressure photoionization. Anal. Chem. 90, 12493–12502.

Liu, J., Bergin, M., Guo, H., King, L., Kotra, N., Edgerton, E., et al., 2013. Size-resolved measurements of brown carbon in water and methanol extracts and estimates of their contribution to ambient fine-particle light absorption. Atmos. Chem. Phys. 13, 12389–12404.

Moroni, B., Cappelletti, D., Crocchianti, S., Becagli, S., Caiazzo, L., Traversi, R., et al., 2017. Morphochemical characteristics and mixing state of long range transported wildfire particles at Ny-Ålesund (Svalbard Islands). Atmos. Environ. 156, 135–145.

Moroni, B., Ritter, C., Crocchianti, S., Markowicz, K., Mazzola, M., Becagli, S., et al., 2020. Individual particle characteristics, optical properties and evolution of an extreme long-range transported biomass burning event in the European Arctic (Ny-Ålesund, Svalbard Islands). J. Geophys. Res. Atmos. 125 (5), 1–17.

Mukherjee, A., Dey, S., Rana, A., Jia, S., Banerjee, S., Sarkar, S., 2020. Sources and atmospheric processing of brown carbon and HULIS in the Indo-Gangetic Plain: insights from compositional analysis. Environ. Pollut. 267, 115440.

Pedersen, C.A., Gallet, J.C., Ström, J., Gerland, S., Hudson, S.R., Forsström, S., et al., 2015. In situ observations of black carbon in snow and the corresponding spectral surface albedo reduction. J. Geophys. Res. Atmos. 120, 1476–1489.

Seinfeld, J.H., Pandis, S.N., 2012. Atmospheric Chemistry and Physics: From Air Pollution to Climate Change. John Wiley & Sons.

Stjern, C.W., Samset, B.H., Myhre, G., Forster, P.M., Hodnebrog, Ø., Andrews, T., et al., 2017. Rapid adjustments cause weak surface temperature response to increased black carbon concentrations. J. Geophys. Res. Atmos. 122 (21), 11,462–11,481.

Sun, H., Biedermann, L., Bond, T.C., 2007. Colour of brown carbon: a model for ultraviolet and visible light absorption by organic carbon aerosol. Geophys. Res. Lett. 34, Article L17813.

Varga, B., Kiss, G., Ganszky, I., Gelencser, A., Krivacsy, Z., 2001. Isolation of water-soluble organic matter from atmospheric aerosol. Talanta 55, 561–572.

Voisin, D., Jaffrezo, J.L., Houdier, S., Barret, M., Cozic, J., King, M.D., et al., 2012. Carbonaceous species and humic like substances (HULIS) in Arctic snowpack during OASIS field campaign in Barrow. J. Geophys. Res. Atmos. 117 (9), 1–17.

Willis, M.D., Burkart, J., Thomas, J.L., Köllner, F., Schneider, J., Bozem, H., et al., 2016. Growth of nucleation mode particles in the summertime Arctic: a case study. Atmos. Chem. Phys. 16, 7663–7679.

Wu, G., Wan, X., Ram, K., Li, P., Liu, B., Yin, Y., et al., 2020. Light absorption, fluorescence properties and sources of brown carbon aerosols in the Southeast Tibetan Plateau. Environ. Pollut. 257, 113616.

Xie, X., Chen, Y., Nie, D., Liu, Y., Liu, Y., Lei, R., et al., 2020. Light-absorbing and fluorescent properties of atmospheric brown carbon: a case study in Nanjing, China. Chemosphere 251, 126350.

Yue, S., Bikkina, S., Gao, M., Barrie, L.A., Kawamura, K., Fu, P., 2019. Sources and radiative absorption of water-soluble brown carbon in the high Arctic Atmosphere. Geophys. Res. Lett. 46 (24), 14881–14891.

Zangrando, R., Barbaro, E., Zennaro, P., Rossi, S., Kehrwald, N.M., Gabrieli, J., et al., 2013. Molecular markers of biomass burning in Arctic aerosols. Environ. Sci. Technol. 47 (15), 8565–8574.

Zhang, A., Wang, Y., Zhang, Y., Weber, R.J., Song, Y., Ke, Z., et al., 2019. Modeling global radiative effect of brown carbon: a larger heating source in the tropical free troposphere than black carbon. Atmos. Chem. Phys. 20, 1901–1920.

CHAPTER 17

Measurements of carbon monoxide at Polar regions

B.C. Arya[1], Sachin D. Ghude[2], Y.N. Ahammed[3], Arun Kumar[1], Devesh Kumar Shukla[4] and S.L. Jain[1]
[1]Radio and Atmospheric Sciences Division, National Physical Laboratory, New Delhi, India
[2]Indian Institute of Tropical Meteorology, Pashan, Pune, India
[3]Department of Physics, Yogi Vemana University, Kadapa, India
[4]Environmental Science and Bio-medical Metrology Division, National Physical Laboratory, New Delhi, India

17.1 Introduction

Carbon monoxide (CO) is a chemically important gas in the atmosphere. It is mainly removed from the atmosphere by reaction with hydroxyl radical (OH) (Thompson, 1992). Information about concentrations of OH radical and other trace gases (e.g., ozone, CO, nitrogen oxides, and formaldehyde) help to provide a better understanding of the chemistry of the atmosphere. Carbon monoxide, as a major sink of OH radical, controls the lifetime of many gases which are relevant for climate change studies (Logan et al., 1981; Crutzen, 1995). As a result, CO distribution and trend should influence OH concentration and act in a significant manner on the oxidation capacity of the atmosphere. Therefore an anthropogenic or natural disturbance in the CO concentration could perturb OH concentration and have a significant impact on many aspects of atmospheric chemistry (Thompson and Cicerone, 1986; Khalil and Rasmussen, 1990).

The global budget of atmospheric CO is moderately estimated (Seiler and Conrad, 1987; Crutzen, 1995; Houghton et al., 1995; Holloway et al., 2000); however, it is uncertain. It is generally believed that atmospheric CO comes from fossil fuel combustion, biomass burning, and gas-phase oxidation of natural hydrocarbon (McConnell et al., 1971; Erickson, 1989; Khalil and Rasmussen, 1990). However, there is lack of information about the CO production and distribution processes. Measurements have shown that natural dissolved organic matter (DOM) in lake, wetland, surface marine and fresh water, and rain water leads to

Understanding Present and Past Arctic Environments
DOI: https://doi.org/10.1016/B978-0-12-822869-2.00014-1

the production of CO (Zuo and Jones, 1995, 1996, 1997; Vaughan and Biough, 1998).

Dissolved organic carbon in seawater represents one of the largest reservoirs of carbon on the Earth. Sea ice is often enriched with DOM. Photochemical degradation of DOM from sea-ice, snow-covered regions and seawater by sunlight [ultraviolet visible (UV) radiation] produces carbonyl compounds and carbon gases (primarily CO). Greenland Ice core also showed high CO values suggesting CO production process within the ice.

Measurements showed that the CO production process is found to be radiation dependent and active in the surface snow layer. Strong absorption of UV and visible radiations within the Arctic Sea has been observed in spite of the usually high albedos of sea ice and overlying snow. Studies over Franklin Bay, Arctic region during early March to mid-July show that the total amount of CO emitted from sea ice to the Arctic atmosphere would be 1.4×10^{10} moles a^{-1}, taking the main sea-ice area to be 12×10^{6} km^2 in the Northern Hemisphere from March to mid-July. This value represents 3% of the global oceanic CO flux to the atmosphere.

Measurements made from the snow samples (Rasmussen et al., 1982; Haan et al., 1996) showed a large amount of CO and the abundant of CO observed was attributed to the adsorption of atmospheric CO molecules on snow crystal. Recent investigations (Haan et al., 2001) have shown that CO production processes are likely to be involved in freshly fallen snow and significant release of CO flux from the snowpack. This CO production process found to be radiation dependent and active in the surface snow layer, suggesting HCHO photolysis to be partly responsible for CO production in snow (Haan et al., 2001). Therefore, to define the role of snow- and ice-covered region in the atmospheric chemistry and CO budget, information about CO production process from snow are important.

To measure the CO concentration and variability in the surface air over the snow-covered region of Antarctica, measurements were carried out at Maitri (Indian Research Base in Antarctica) during summer campaign of XXII Indian Antarctic Scientific Expedition (2003–04). Measurements were also carried out in the first Arctic winter phase expedition (March 3–31, 2008). The measurements of CO and surface ozone were also carried out at Indian Arctic station "Himadri," Ny-Ålesund, Svalbard (78°54′N, 11°53′E) from March 20 to April 19, 2010, which coincide with the episode of tropospheric ozone depletion that occurred from March 29, 2010, 2200 hours to April 01, 2010, 1700 hours. Here, we report the results obtained from these measurements, which show the diurnal cycle of carbon monoxide and daytime increase along with solar actinic radiation.

17.2 Experimental site and setup

Measurement of surface-level CO concentration has been carried out at a high latitude site (70.46S, 11.44E), Maitri, located in the Schirmacher region of east Antarctica. Schirmacher Oasis is a rocky hill range of about 35 sq km. Maitri, is situated on the ice-free rocky area of Schirmacher Oasis. The entire area is surrounded by the adulatory shelf ice and polar ice sheet (Ravendra et al., 2001), and snow deposition is widespread on the leeward side of the hillocks. The northeastern and north-western corners of the area are on shelf ice, while the south-western extremity is on polar ice sheet. The southernmost ice sheet is bare "blue ice." The northernmost part of the area comprises shelf ice. The central part of the area is occupied by Lake B7 (500 m $\times$ 300 m) (Ravendra et al., 2001) where the Maitri (140 masl) station is located. It is a lake of glacial origin. On both the northern and southern sides of Maitri, ice region exists at about 500 m from the station. Snowfall is quite frequent during the winter months, but gale force winds scrub the rocky surfaces clean; however, snow deposition is widespread on the leeward side of the hillocks. The single-storey building of Maitri station and other structures such as generator huts, incinerator, were about 150–200 m away in the northeast from the site of the measurements so that any pollutant released from them had little or no chance of reaching the site of measurements with the prevalent winds, except during few exceptional episodes.

CO measurements were carried out using ultrasensitive gas filter correlation (GFC) nondispersive infrared absorption CO analyzer of Advanced Pollution Instrument Inc. (API, Model 300), which is described in detail by Dickerson and Delancy (1988) and Cogan and Lobert (1998). This instrument is equipped with an internal zero and span option (IZA), which allows for internal creation of zero gas that effectively removes ambient level of CO from the sample gas. Interfering effects of H_2O are minimized by the GFC wheel yielding a rejection ratio of H_2O of the order of 5×10^{-6}:1.

Therefore for an ambient water vapor of about 2800 ppm (observed at observation site during February) the water vapor interference would be less than 1 ppb. To minimize zero drift of the instrument the laboratory temperature was maintained at 20°C. The main air inlet was installed on top of the sidewall of the observatory 2.5 m above the ground. The Teflon tube between the inlet and the instrument was around 2 m long with an inner diameter of 4 mm. Great care has been taken to avoid any error throughout the measurement period. The zero and span calibration were performed regularly for data accuracy and to enhance its stability for low-level detection. Regular automatic zero checking, every 2 hours, was carried out and daily span checks of the instrument were performed with 1 ppm (1%) commercial gas standard (Scott specialty gases, San Bernardino, CA). This calibration frequency was found to be adequate as no significant difference was observed in the measurement made before and after the calibration. The technical specification of the CO analyzer is given in Table 17.1. This instrument measures the CO concentration continuously and averages it for time intervals programmed by the user. The analyzer stored data separately in the data record format of 5-minute average and 1-hour average (5-minute average is based on 1-minute average calculated with sampling rate of 10 seconds and 1 hour average is calculated by 5-minute average). We have used 5-minute data record to calculate hourly average and SD.

17.3 Observations and results

Continuous measurement of surface CO concentration at Maitri, Antarctica was carried out over a 6-week period during February and March 2003 (austral summer), as a part of the trace gases measurement project. Simultaneously, columnar ozone and water vapor measurements using MICROTOPS sun photometer were also carried out. Direct solar UVB irradiance at 300 and 305 nm was derived using the irradiance

Table 17.1 Technical specifications of carbon monoxide analyzer.

Zero noise	<0.01 ppm RMS
Span noise	<0.5% of reading above 5 ppm (RMS)
Lower detectable limit (LDL)	0.02 ppm
Zero drift	<10 ppb/24 h, 20 ppb/7 days
Span drift	<0.5% of reading/24 h, 1% of reading/7 days
Linearity	1% of full scale
Precision	0.5% of reading
Sample flow rate	1800 scc min^{-1} ± 20%

calibration constant provided by the MICROTOPS instrument, and direct UV irradiance flux between the wavelength 280 and 420 nm has been calculated using the total UV visible (TUV) radiation model (Madronich, 1993), which we used to compare diurnal variation of CO with solar UVB radiation.

Temporal variation of daily mean atmospheric CO concentration along with sunlight hours of day for the period February 1, 2003 to March 14, 2003 is depicted in Fig. 17.1. The daily mean atmospheric CO concentration at Maitri varied significantly from one day to another, ranging from about 22 to 80 ppb. This day-to-day variability reflects the influence of local photochemistry and prevailing meteorological conditions at site (Gros et al., 2001). However, as depicted by the linear fit curve on the set of data shown in Fig. 17.1, CO concentration showed an increasing tendency, in spite of large day-to-day variability. Fig. 17.1 also displays the temporal variation of period of sunlight hour at observational site Maitri. On February 1 the period of sunlight hours was about 2130 hours (time between sunrise and sunset) which gradually decreased to about less than 1330 hours on March 14. Measurement of water vapor (one of the major sources of OH radicals) for the same location also showed a tendency of decreasing water vapor concentration for the same period (Jain et al., 2005).

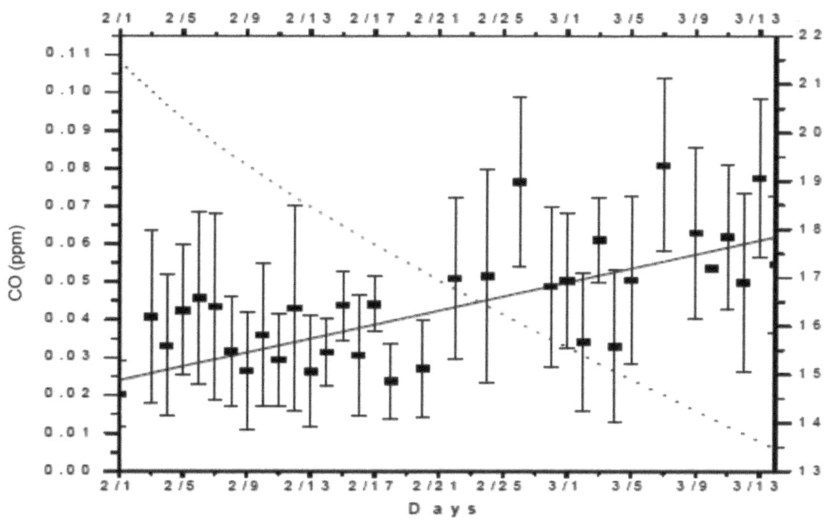

Figure 17.1 Comparison of daily mean CO concentration (dots), observed for February 1, to March 14, 2003 period, together with the daylight hours (*dashed line*). The straight line shows the linear fit cure on daily mean CO concentration.

Fig. 17.1 reveals that the increasing trend of CO concentration coincides with the decreasing trend of the amount of daily solar actinic flux (Correlation $R = -0.62$), as the photochemical oxidization of CO by OH (Spivakovsky et al., 1990; Thompson, 1992) is the main sink of CO. This leads to the slow accumulation of CO in the boundary layer as the amount of solar radiation decreases. Inverse CO correlation with the amount of daily solar actinic flux similar to those observed in our measurement and inverse CO correlation with estimated OH concentration have already been reported by earlier studies (Seiler, 1974; Novelli et al., 1994). Novelli et al. (1998a, b) and Matsueda et al. (1998) have also pointed out that the timing of maximum and minimum CO concentration in the southern hemisphere is delayed by about 2 months from the estimated period of the minimum amount in OH.

In the presence of sunlight, photochemical oxidation of CO by OH considerably decreases its residence time in the atmosphere. The solar actinic flux (Beaglehole and Carter, 1992; Ghude et al., 2005) and OH concentration (Mount, 1995; Lee et al., 2006) over Antarctica showed a marked diurnal variability during the summer season. As shown in Fig. 17.2, the data in the thick line, it has been observed that as the solar elevation increases, the CO concentration increases. To identify this process, measured CO values were compared with direct solar UVB radiation at 300 and 305 nm (measured by MICROTOPS-II sun photometer) and direct UV irradiance flux (280–420 nm) calculated from the TUV model (Fig. 17.2). A rapid increase in CO concentration with the increase of solar UV radiation has been observed. CO was found to increase rapidly between 6 and 13 hours and reached a maximum of about 62 ppb in the early afternoon when the solar irradiation was at its maximum. A direct UV irradiation calculated from TUV model (280–420 nm) was taken to investigate its covariance with observed CO concentration between 4 and 15 hours, as shown in Fig. 17.3. The results from linear regression of CO with UV radiation showed a significant positive correlation.

Fig. 17.4A, B and C, D show the hourly mean CO values for clear sunny days and cloudy days, respectively. During cloud-free sky days, on February 5 and 10, 2003, the weather conditions were mostly characterized by wind speed below 10 m s^{-1} and clear sky during, before, and after the days of observation. While, on cloudy days (18 February and 5 March), the wind speed was above 20 m s^{-1} and sky conditions were cloudy. However, bright intervals occurred sparsely and only for a very short time on the days of observation. As seen from Fig. 17.4A and B,

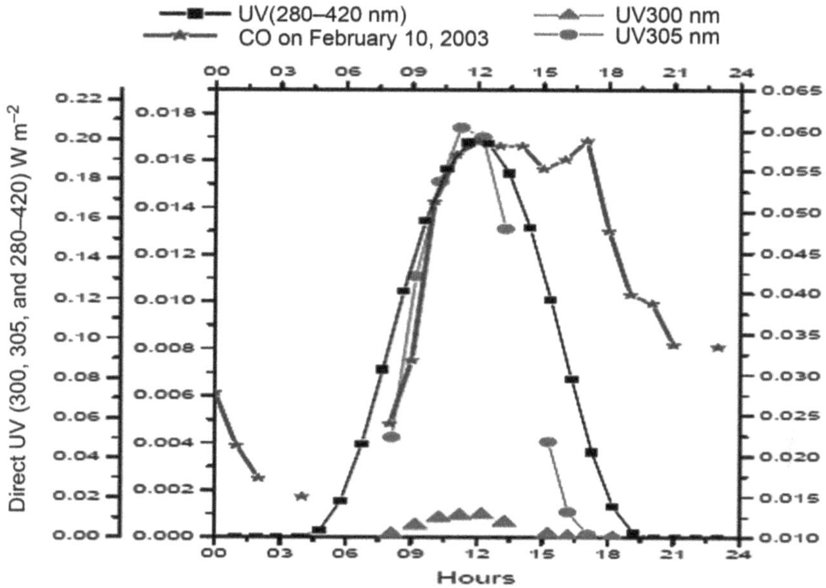

Figure 17.2 Comparison of diurnal variation of CO, on February 10, 2003, together with the direct UV (300 and 305 nm) obtained from MICROTOPS sun photometer and direct UV for the wavelength interval 280–420 nm derived from the TUV model [Columnar ozone = 289 DU; AOD = 0.012 (Gadhavi and Jayaraman, 2004); altitude: 0.140 m]. *TUV*, Total ultraviolet visible.

diurnal cycle was apparent during clear sky conditions with a higher CO mixing ratio during daytime. On the clear sky days, CO concentration has been observed to increase with a rate of approximately 5.1 ppb h^{-1} between the interval 06 and 13 hours. A maximum of about 62 ppb has been observed in the afternoon (from 12 to 18 hours) and then gradually decline to a minimum of about 20 ppb in the early morning. On the cloudy days however, CO concentration decreased during daytime, and cloudy condition and no diurnal variation were observed.

To study the mean of diurnal variability during the observational period of 6 weeks (from February 01–March 11, 2003), hourly mean values were plotted in Fig. 17.5. The vertical error bars show the one sigma standard deviation. Standard deviation has been observed relatively higher during the daylight hours (about $\pm$ 0.023 ppb) and reduces (about $\pm$ 0.013 ppb) during the dark hours. This signifies that more changes of CO occur during the day than nighttime hours. The variability in CO concentration has been observed with hourly mixing ratio ranging from 30 ($\pm$ 0.014) to 65 (+0.023) ppb. This shows about twofold diurnal

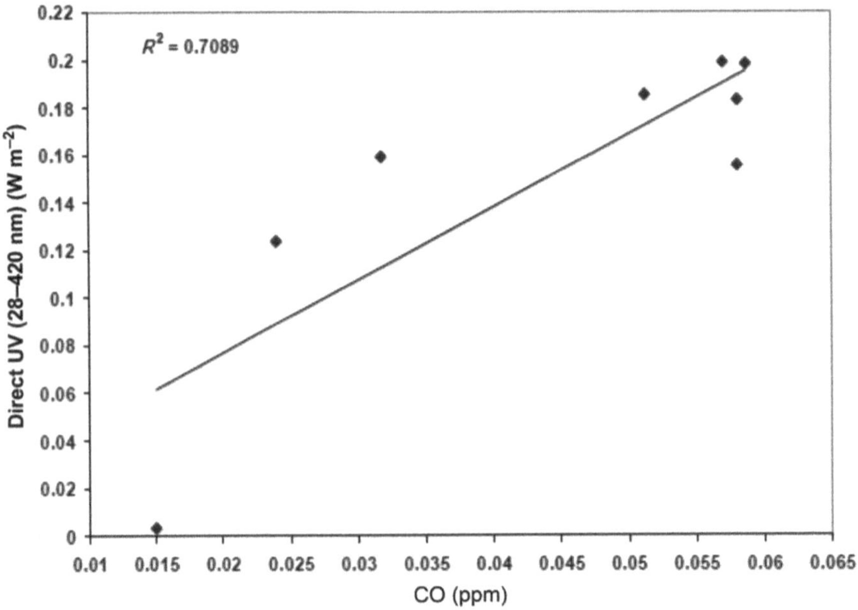

Figure 17.3 A scattered plot between direct UV irradiation calculated from TUV model (280–420 nm) and observed CO concentration between 4 and 15 h. *TUV*, Total ultraviolet visible.

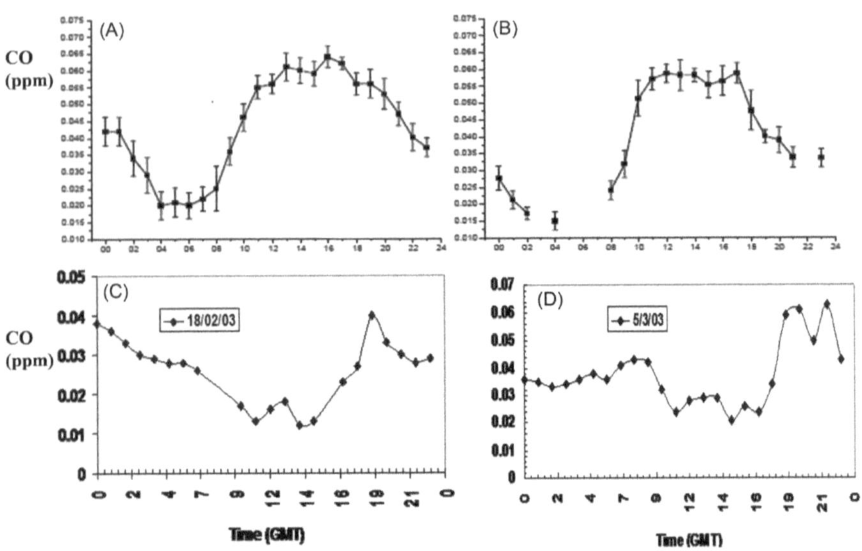

Figure 17.4 Diurnal variation of CO on clear sky days: (A) February 5, 2003 and (B) February 10, 2003; and on cloudy days: (C) February 18, 2003 and (D) March 5, 2003. The vertical bars show the one sigma standard deviation.

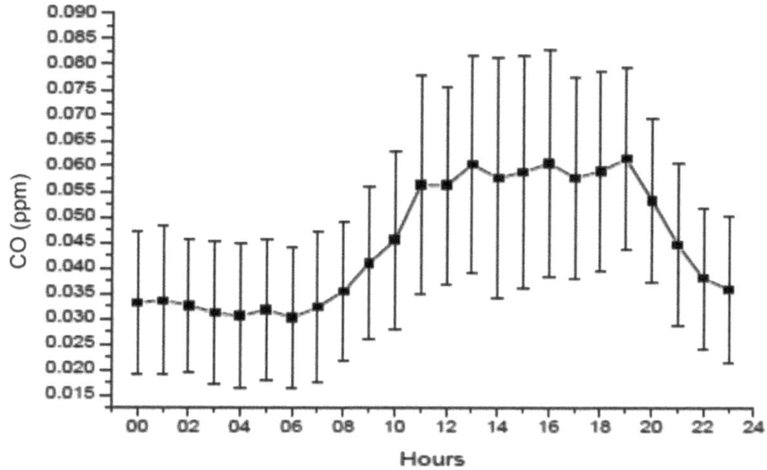

Figure 17.5 Diurnal cycle of CO concentration, observed from February 1–March 14, 2003. The vertical bar shows the one sigma standard deviation.

change that corresponds to about 30 ppb increases in the afternoon. The mean rate of increase in CO concentration has been observed 4.2 ppb h^{-1} (6–13 hours). The maximum CO concentration was detected in the afternoon when the solar elevation angle was high. During nighttime, CO concentration declined to minimum.

In view of the above, CO measurements were also carried out at Indian Arctic station "Himadri," Ny-Ålesund, International Research and Monitoring Facility, Svalbard (78°54′N, 11°53′E), in the Norwegian Arctic, during the first Indian Arctic winter phase expedition (March 3–31, 2008) to study the snow-pack production of carbon monoxide and its diurnal variability. Ny-Ålesund is a village located on the southern shore of Kongsfjorden, on the west coast of Spitsbergen (78°54′N, 11° 53′E) which is the largest of the Svalbard Islands. The facility is surrounded by steep mountains, among which the Zeppelin mountain is ~2 km from Ny-Ålesund, peaks at 474 masl. The sampling site was located at the foot of the Zeppelin mountain.

Continuous measurements of carbon monoxide (CO) have been carried out at Ny-Ålesund, Arctic using Horiba (APMA 370) Ambient CO monitor during March 9–27, 2008 with a 5-minute interval. CO analyzer was precalibrated at NPL using zero air and calibration standard of 10 ppmv CO concentration. CO concentration at measurement site varies between 147 and 201 ppbv with an average value of 164 ppbv, whereas at Maitri, Antarctica CO concentration varies between 20 and 90 ppbv with an average value of 60 ppbv (Fig. 17.6).

During the measurement period, that is, from March 10 to 24, 2008, there were two events of snowfall at an observation site. We attempted to correlate UV radiation with ambient CO concentration of measured data of CO and UV radiation from March 10 to 14, 2008 and March 24 to 27, 2008 (Fig. 17.7). The measurements of both these periods were

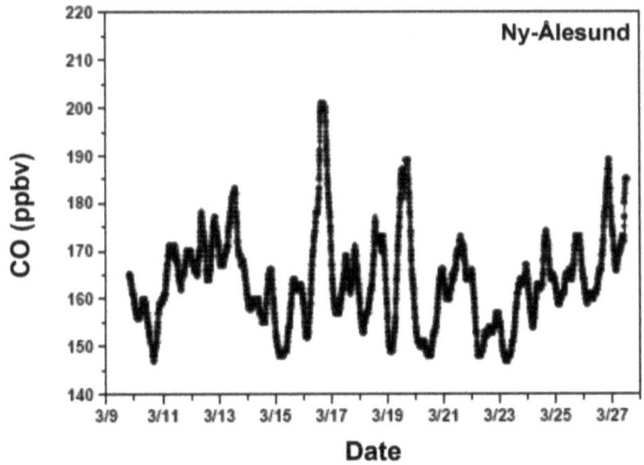

Figure 17.6 Temporal variation of carbon monoxide (CO) in ppbv at Ny-Ålesund, Arctic during March 9–27, 2008.

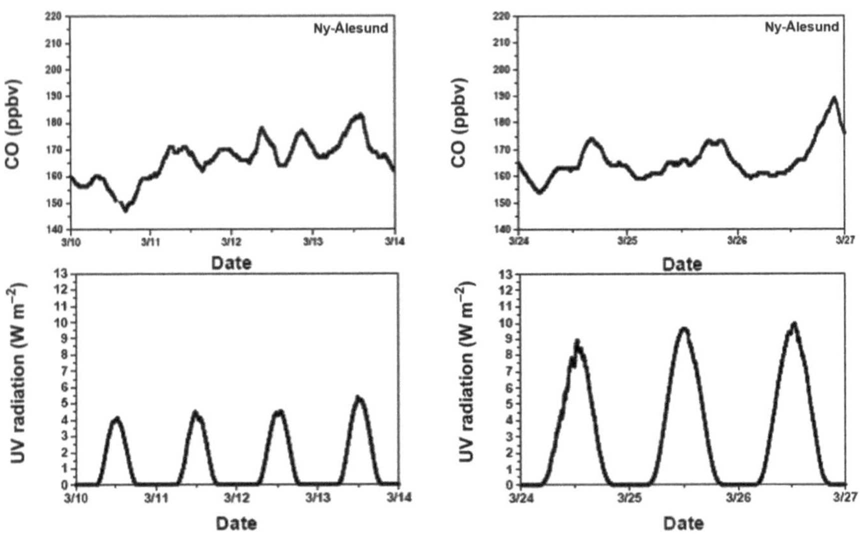

Figure 17.7 Diurnal variation of CO concentration and UV radiation at Ny-Ålesund during and after snowfall events.

characterized by atmospheric condition with calm wind condition and clear sunny days which leads to a very stable boundary layer. A positive correlation was observed between CO concentration and UV radiation. As UV radiation increases, the concentration increase in CO is also apparent. The integrated solar irradiation (measured at Koldewey station in Ny-Ålesund) of the UV radiation flux ($300\ \text{nm} < \lambda < 370\ \text{nm}$) was used to correlate UV radiation with ambient CO concentration. CO concentration attains comparatively a high value (190 ppbv) when fresh fallen snow is exposed to UV radiation on March 27, 2008.

The dramatic loss of O_3 and Hg over a period of days to week in Arctic boundary layer during polar sunrise has been investigated and analyzed by several investigators. This depletion process is related to the presence of bromine. The main source of reactive bromine species (Br and BrO) is bromide which is released via photochemical and heterogeneous chemical reactions from sea salt. Most of the ozone depletion events (ODEs) have been observed from coastal sites when is frozen and snow covered. The duration of ODEs at coastal sites was found typically between 1 and 3 days depending on meteorology.

$$\mathrm{Br} + \mathrm{O_3} \rightarrow \mathrm{BrO} + \mathrm{O_2} \tag{17.1}$$

$$\mathrm{BrO} + \mathrm{BrO} \rightarrow 2\mathrm{Br} + \mathrm{O_2} \tag{17.2}$$

$$\mathrm{BrO} + h\nu \rightarrow \mathrm{Br} + \mathrm{O}$$

Another ozone destruction reaction involves reactions of halogen oxides with HO_2:

$$\mathrm{BrO} + \mathrm{HO_2} \rightarrow \mathrm{HOBr} + \mathrm{O_2}$$
$$\mathrm{HOBr} + h\nu \rightarrow \mathrm{Br} + \mathrm{OH}$$

During this sunlit period, series of field and laboratory experiments have also shown that many chemical species such as carbonyl compounds (HCHO, CH3COH, acetone, etc.), NO_x, C_2H_2, and CO are produced photochemically from snowpack, as a result of which the ozone destruction chain is reduced by reaction of Br atoms with aldehydes.

$$\mathrm{Br} + \mathrm{HCHO} \rightarrow \mathrm{HBr} + \mathrm{HCO} \tag{17.3}$$

$$\mathrm{Br} + \mathrm{CH3CHO} \rightarrow \mathrm{HBr} + \mathrm{CH_3C(O)} \tag{17.4}$$

The rate constant for the reaction of formaldehyde with Br atoms is relatively fast ($1.7 \times 10-11\ \exp\,(-800/T)$).

The photolysis and oxidation of HCHO or other aldehydes yield either hydrogen atoms and HCO radicals or hydrogen molecules and carbon monoxide (CO) depending on the wavelength. Aldehydes especially HCHO are an important source of CO and HO_2 radicals in the clean atmosphere.

Any reactions that produce H or HCO in the troposphere act as HO_2 and CO sources:

$$HCHO + h\nu(\lambda < 325\ nm) \rightarrow H + HCO \tag{17.5}$$

$$HCHO + h\nu(325\ nm < \lambda < 360\ nm) \rightarrow H_2 + CO \tag{17.6}$$

$$H + O_2 \rightarrow HO_2 \tag{17.7}$$

$$HCO + O_2 \rightarrow HO_2 + CO \tag{17.8}$$

In addition to Br, chemically induced Cl reaction also takes place during ODEs. Due to relatively high rate constants of Cl with hydrocarbons as compared to oxidation by OH, Cl readily reacts with hydrocarbon, which leads to the production of CO.

Fig. 17.8 showed a time series of O_3 and CO measurements carried out at Ny-Ålesund at Indian Arctic station Himadri from March 28 to April 19, 2010, which coincide with the episode of tropospheric ozone depletion that occurred from March 29, 2010, 22:00 hours to April 01, 2010, 17:00 hours. Ozone measurements were carried out using Horiba (APOA-370) Ambient Ozone Monitor. Zeppelin station data are also shown in the figure for comparison. A back trajectory analysis (not shown) showed that episodes of ozone depletion are found associated with air masses advected from frozen Arctic Ocean site. A signature of slightly gradual increase (approx. 215 ppb to 240 ppb) of CO levels was observed during depletion event when the ozone depleted from 55 to 25 ppb, from March 29, 2010, 22:00 hours to April 01, 2010, 17:00 hours. The elevated level of CO may be attributed to the production of CO as a result of reaction of Cl and Br with nonmethane hydrocarbons.

17.4 Discussion

The results obtained from the field measurement of CO from a high latitude site (Maitri, Antarctica) for a 6-week period and CO measurements

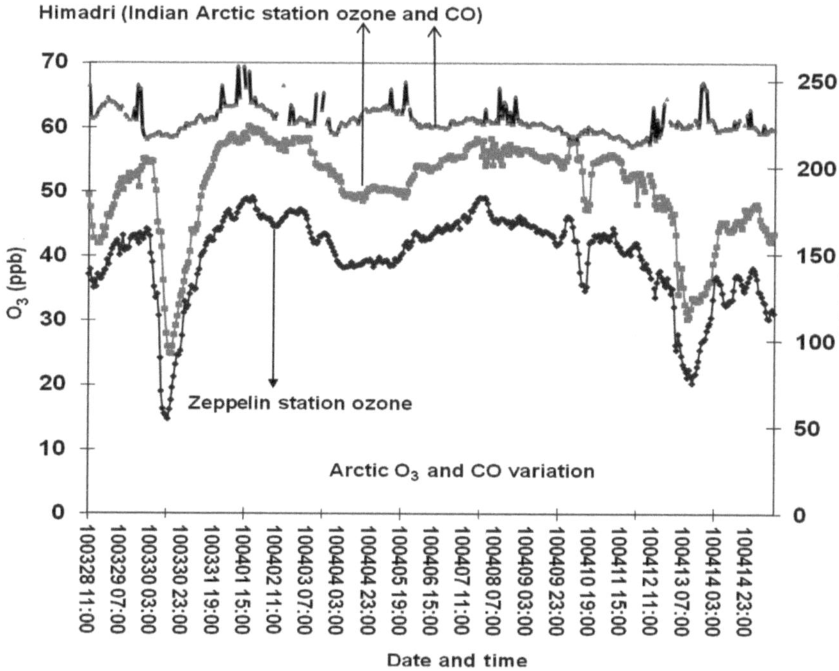

Figure 17.8 Time series of O_3 and CO measurements carried out at Ny-Ålesund at Indian Arctic station "Himadri" from March 28 April 19, 2010.

carried out at Indian Arctic station "Himadri," Ny-Ålesund, Svalbard (78° 54′N, 11°53′E) during the first Indian Arctic winter phase expedition (March 3–31, 2008) and from March 20 to April 19, 2010 are presented here. Diurnal changes in CO concentration were observed with higher CO mixing ratios in the daytime during clear sky at Maitri, Antarctica. On cloudy days, no diurnal variation was observed. This observation suggests that the increase in CO concentration (with increase in solar elevation) probably takes place through photochemical mechanisms that are activated by sunlight or some other unknown process that is correlated with the sunlight. Haan et al. (2001) conducted chamber experiments on freshly fallen snow from Mount Sonnblick in Austria and found evidence for light dependent. He noticed that CO generation was a function of snow total organic compound concentration and suggested that it is related to the oxidation of organic compounds. Similarly Grannas et al. (2002) found that ice pellets made from melted Arctic and Antarctic snow produced CH_2O and CH_3CHO when irradiated in the laboratory. Production rates were enhanced by the addition of nitrate to the melted snow, perhaps due to enhanced OH production. Sumner and Shepson (1999) suggested that carbonyl compounds may be generated by reactions of DOM with condensed phase OH radicals. It has recently been shown that direct photochemical transformation (295–340 nm) of DOM present in seawater results in the production of carbonyl compounds and alkenes, which are then released to the atmosphere. If this chemistry also occurs in the snowpack, photooxidation of DOM could serve as a source of carbonyl compounds in the Arctic snowpack. The presence of several organic compounds in snow has already been reported by Legrand and Mayewski (1997) and by other investigators (Li and Wsnclsester, 1993; Shepson et al., 1996; Hutterli et al., 1999; Couch et al., 2000). The result of closed chamber experiment conducted by Haan et al. (2001) at Mt. Sonnblick, Austria observed an increase of CO concentration of 3.8–9.5 ppb min^{-1} in the box and a decrease of 6 ppb min^{-1} when the box is darkened, which they attributed to loss of CO through porous layer. From this experiment, they derived that the diurnal variation of CO in snow is controlled by photochemical processes leading to large amount of CO production in the freshly fallen snow and release into the atmosphere when snow was exposed to sunlight during daytime.

A number of field and modeling studies indicate that a source of gas-phase organic compounds exists in polar snowpacks and that the evolution and emission of these species may significantly impact the overlying boundary layer. The snow-pack chemistry of H_2O_2 and volatile carbonyls (as well as other organic compounds) is likely to have a close interaction with hydroxyl radical. Because of its very high reactivity, OH is probably the major oxidant for organic compounds and other contaminants on sun-lit snow grains (Anastasio et al., 2007; France et al., 2007). The most significant role of hydrogen peroxide in snow grain chemistry is a photo-chemical source of OH radical. Calculations based on laboratory measurements showed that H_2O_2 photolysis is the dominant source of OH on Antarctic and Arctic snow grains (Chu and Anastasio, 2005). Away from coastal reigns with elevated sea salt levels, the greatest sinks for snow grain OH is possibly organic compounds (Anastasio et al., 2007). In this case, most photo-formed OH will react with organics and these reactions probably account for a significant portion of the carbon-containing products that are emitted from snow. In addition to CH_2O and other carbonyls such as those described earlier, the products formed from OH reactions with organics may also include CO and carboxylic acids. So it is important, that other mechanisms such as direct photoreactions of organic compounds and indirect photoreactions involving other oxidants are probably also significant in the release of CO and volatile organic compounds (VOCs).

Low molecular weight aldehyde (RH(C = O)) and ketones (RR′ (C = O)) are ubiquitous VOCs in the Arctic atmosphere (Barrie et al., 1988; Bottenheim et al., 1990; De Serves, 1994; Shepson et al., 1996). Recent observations that formaldehyde (HCHO), acetaldehyde (CH_3CH), and acetone ($(CH_3)_2CO$) were elevated in snow-pack interstitial air (Couch et al., 2000), and the NO_x was photochemically produced in the snowpack indicate that the snowpack is chemically active and that it can release carbonyl compounds to the troposphere. Such a source may explain why their Arctic troposphere concentrations cannot be explained by considering only (1) gas-phase oxidation of hydrocarbons as the main source and (2) photolysis and radical oxidation as the main sinks (Shepson et al., 1996; Sumner and Shepson, 1999). Hydrocarbon precursors were measured both at Narwhal and Alert in April 1994. The dominant reactive hydrocarbons were CH_4, C_2H_6, C_3H_8, n-C_4H_{10}, i-C_4H_{10}, n-C_5H_{12}, and i-C_5H_{12}.

The photolysis and oxidation of HCHO or other aldehydes yield either hydrogen atoms and HCO radicals or hydrogen molecules and carbon monoxide (CO) depending on the wavelength. Aldehydes especially HCHO are an important source of CO and HO_2 radicals in the clean atmosphere. Any reactions that produce H or HCO in the troposphere act as HO_2 and CO sources. The photolysis of higher aldehydes, RCHO, also forms HCO and hence HO_2.

The ozone depletion and halogen chemistry have a significant impact on nonmethane hydrocarbons photochemistry. Jobson et al. (1994) and Yokouchi et al. (1994) first observed that light hydrocarbons are rapidly consumed (mostly by CI atoms) during ODEs. They showed that reactive alkanes such as butane and *n*-isopentane are almost completely removed during ODEs. During ODEs relatively high rate constants of Cl with hydrocarbons as compared to oxidation by OH, Cl readily reacts with hydrocarbons and thus extensive chlorine atoms processing of VOCs take place. For example, during ODEs, there is a substantial consumption of propane, and a concomitant increase in acetone was observed. As discussed, HO_2 is produced in significant part by CI atom reaction with nonmethane hydrocarbons to produce HO_2 directly, or through photolysis of the product HCHO.

Haan et al. (2001) suggested that photolysis of the formaldehyde could be partially responsible for observed CO increase. Very recently, some studies have reported high HCHO concentration in the snowpack (Hutterli et al., 1999; Katja et al., 1999; Sumner and Shepson, 1999; Couch et al., 2000; Riedel et al., 2005) and favored the degassing process. Formaldehyde is rapidly destroyed by sunlight producing hydroperoxyle radical and CO through the following photolysis reaction:

1. $HCHO + h\nu \rightarrow CO + H_2, J_1 = 3.73 \times 10^{-05}\ s^{-1}$
2. $HCHO + h\nu \rightarrow H + CHO, J_2 = 2.17 \times 10^{-05}\ s^{-1}$
3. $HCO + O_2 \rightarrow HO_2 + CO$

The product of the second reaction reacts rapidly with molecular oxygen for HO_2 and CO.

For example, if we take a measurement of HCHO at Antarctica's Neumayer station concentration (400 ppt), a simple calculation of HCHO photolysis leading to CO production rate of about 85 ppt/h^{-1}. By taking two photodissociation rate coefficient (J) obtained using TUV model for clear sky condition at noontime for 11.44E, 70.46S, K represented the volumetric CO production rate ($K = (J_1 + J_2) \times [HCHO]$). This CO production rate is much lower (ignoring other CO production mechanism) than that of the observed rate of CO increase (approximately 5.1 ppb h^{-1}) during daytime. This suggests and as discussed earlier, there are various physical and chemical processes that determine the concentration of carbon monoxide in polar atmosphere. During nighttime, CO concentration declined to minimum. The decrease in CO concentration during nighttime and early morning conditions may be determined by the deposition of CO through porous snow layer. The upper part of the snowpack exhibits a high porosity, supports the diffusion of CO to deeper layers, and increases the surface area that interacts with the atmosphere (Haan et al., 2001).

17.5 Conclusion

On the basis of measurements made at Maitri, Antarctica, and Indian Arctic station "Himadri" a detailed analysis of the surface CO concentration was presented. A distinct diel variation was observed coinciding with the diurnal cycle of solar actinic radiation at Maitri, Antarctica. This observation implies that the diurnal variation in CO concentration has probably activated the process that is correlated with the sunlight, photochemical processing of organic matter trapped in snow. However, there are various process mechanisms in Polar regions, which account for observed diurnal variability of CO.

Acknowledgments

We express our sincere thanks to the Director of NPL for the encouragement and taking a keen interest in the progress of the work. The authors are grateful to the National Center for Antarctic and Ocean Research, Goa for the logistic support during Antarctic and Arctic expeditions. Thanks are also due to the leaders of 21st and 22nd Indian Scientific Expedition to Antarctica for providing the infrastructure for the experiment. We are also grateful to Sasha Madronich for the TUV model and valuable suggestions.

References

Anastasio, C., Galbavy, E.S., Hutterli, M.A., Burkhart, J.F., Friel, D.K., 2007. Photoformation of hydroxyl radical on snowgrains at Summit, Greenland. Atmos. Environ. 41 (24), 5110–5121. Available from: https://doi.org/10.1016/j.atmosenv.2006.12.011.

Barrie, L.A., Bottenheim, J.W., Schnell, R.C., Crutzen, P.J., Rasmussen, R.A., 1988. Ozone destruction and photochemical reactions at polar sunrise in the lower Arctic atmosphere. Nature 334, 138–141.

Beaglehole, D., Carter, G.G., 1992. Antarctic skies 1. Diurnal variations of the sky irradiance and UV effects of the ozone hole, spring 1990. J. Geophys. Res. 97 (D2), 2589–2596.

Bottenheim, J.W., Barrie, L.A., Atlas, E., Heidt, I., Niki, H., Rasmussen, R.A., et al., 1990. Depletion of lower tropospheric ozone during Arctic spring: The Polar Sunrise Experiment. J. Geophys. Res. 95, 18555–18568.

Chu, L., Anastasio, C., 2005. Formation of hydroxyl radical from the photolysis of frozen hydrogen peroxide. J. Phys. Chem. A 109 (28), 6264–6271.

Cogan, M., Lobert, J.M., 1998. Instrumentation for trace – level measurement of carbon monoxide in pristine environments. In: Proceedings of the 43rd Annual ISA Analysis Division Symposium, vol. 31, pp. 229–236.

Couch, T.L., Sumner, A.L., Dassau, T.M., Shepson, P.B., 2000. An investigation of the interaction of carbonyl compounds with the snowpack. Geophys. Res. Lett. 27, 2241–2244.

Crutzen, P.J., 1995. Overview of tropospheric chemistry: developments during the past quarter century and a look ahead. Faraday Discuss. 100, 1–21.

De Serves, C., 1994. Gas phase formaldehyde and peroxide measurements in the Arctic atmosphere. J. Geophys. Res. 99, 25391–25398.

Dickerson, R.R., Delancy, A.C., 1988. Modifications of a commercial gas filter correlation CO detector for enhanced sensitivity. J. Atmos. Oceanic Technol. 5, 424–431.

Erickson III, D.J., 1989. Ocean to atmosphere carbon monoxide flux: Global inventory and climate implications. Global Biogeochem. Cycles 3, 305–314.

France, J.L., King, M.D., Lee-Taylor, J., 2007. Hydroxyl (OH) radical production rates in snowpacks from photolysis of hydrogen peroxide (H_2O_2) and nitrate (NO_3^-). Atmos. Environ. 41, 5502–5509.

Gadhavi, H., Jayaraman, A., 2004. Aerosol characteristics and aerosol radiative forcing over Maitri, Antarctica. Curr. Sci. 86, 296–304.

Ghude, S.D., Jain, S.L., Arya, B.C., Kulakarni, P.S., Kumar, A., 2005. Direct solar ultraviolet irradiance (305 nm, 312 nm, 320 nm) at Schirmacher Region of East Antarctica and Delhi: observations, comparison and model calculations. In: *URSI XXVIII General Assembly*, F10P.20 (0096).

Gros, V., Bräunlich, M., Rbckmann, T., Jbckel, P., Bergamaschi, P., Brenninkmeijer, C. A.M., et al., 2001. Detailed analysis of the isotopic composition of CO and

characterization of the air masses arriving at Mt. Sonnblick (Austrian Alps). J. Geophys. Res. 106, 3179–3193.
Haan, D., Martinerie, P., Raynaud, D., 1996. Ice core data of atmospheric carbon monoxide over Antarctica and Greenland during the last 200 years. Geophys. Res. Lett. 2235–2238.
Haan, D., Zuo, Y., Gros, V., Brenninkmeijer, C.A.M., 2001. Photochemical production of carbon monoxide in snow. J. Atmos. Chem. 40 (3), 217–230.
Holloway, T., Levy II, H., Kasibhatla, P., 2000. Global distribution of carbon monoxide. J. Geophys. Res. 105D, 12,123–12,147.
Radiative forcing of climate change and an evaluation of the IPCC IS92 emisssions scenarios. In: Houghton, J.T., et al., (Eds.), Intergovernmental Panel on Climate Change (IPCC) Climate Change (1995). Cambridge University Press, New York, p. 339.
Hutterli, M.A., Rothlisberger, R., Bales, R.C., 1999. Atmosphere-to-snow-to-firn transfer studies of HCHO at Summit, Greenland. Geophys. Res. Lett. 26, 1691–1694.
Jain, S.L., Ghude, S.D., Arya, B.C., Kumar, A., Bajaj, M.M., 2005. Increase in total water vapour column and surface temperature at Maitri, Antarctica. Curr. Sci. 89 (11), 1917–1922.
Jobson, B.T., Niki, H., Yokouchi, Y., Bottenheim, J., Hopper, F., Leaitch, R., 1994. Measurements of C2-C6 hydrocarbons during the polar sunrise 1992 experiment: evidence for Cl atom and Br atom chemistry. J. Geophys. Res. 99, 25355–25368.
Katja, R., Weller, R., Schrems, O., 1999. Variability of formaldehyde in the Antarctic troposphere. Phys. Chem. Chem. Phys. 1, 5523–5527.
Khalil, M.A.K., Rasmussen, R.A., 1990. The global cycle of carbon monoxide: trends and mass balance. Chemosphere 20, 227–242.
Lee, J., Bloss, W., Heard, D., Bauguitte, S., Salmon, R., Roscoe, H., et al., 2006. Measurement of OH and HO_2 in costal Antarctica using laser induced fluorescence. Geophys. Res. Abstr. 8, 01483.
Legrand, M., Mayewski, P., 1997. Glaciochemistry of polar ice cores: a review. Rev. Geophys. 35 (3), 219–243.
Li, S.M., Wsnclsester, J.W., 1993. Water soluble organic constituents in Arctic aerosols and snow pack. Geophys. Res. Lett. 20, 45–48.
Logan, J.A., Prather, M.J., Wofsy, S.C., McElroy, M.B., 1981. Tropospheric chemistry: a global perspective. J. Geophys. Res. 86, 7210–7254.
Madronich, S., 1993. UV radiation in the natural and perturbed atmosphere. In: Tevini, M. (Ed.), Environmental Effects of UV (Ultraviolet) Radiation. Lewis Publisher, Boca Raton, FL, pp. 17–69.
Matsueda, H., Inoue, H.Y., Sawa, Y., Tsutsumi, Y., Ishii, M., 1998. Carbon monoxide in the upper troposphere over the western Pacific between 1993 and 1996. J. Geophys. Res. 103 (D15), 19093–19110.
McConnell, J.C., McElroy, M.B., Wofsy, S.C., 1971. Natural sources of atmospheric CO. Nature 233, 187–188.
Mount, G.H., 1995. New insight and question resulting from recent ion-assisted OH measurements. J. Atmosphere. Sci. 52, 3337–3341.
Novelli, P.C., Masarie, K., Tans, P., Lang, P., 1994. Recent changes in atmospheric carbon monoxide. Science 263, 1587–1590.
Novelli, P.C., Masarie, K.A., Lang, P.M., 1998a. Distributions and recent changes of carbon monoxide in the lower troposphere. J. Geophys. Res. 103, 19,015–19,033.
Novelli, P.C., Connors, V.S., Reichle, J.H.G., Anderson, B.E., Brenninkmeijer, C.A.M., Brunke, E.G., et al., 1998b. An internally consistent set of globally distributed atmospheric carbon monoxide ratios developed using results from an intercomparison of measurements. J.Geophys. Res. 103D, 19,285–19,293.

Rasmussen, R.A., Khalil, M.A.K., Hoyt, S.D., 1982. Methane and carbon monoxide in snow. APCA J. 32, 176–178.
Ravendra, R., Chaturvedi, A., Beg, M.J., 2001. Melt water lakes of Schirmacher Oasis—their genetic aspects and classification. In: Sahu, D.B., Pandey, P.C. (Eds.), Advances in Marine and Antarctic Science, Dariyaganj, New Delhi, pp. 301–313.
Riedel, K., Allan, W., Weller, R., Schrems, O., 2005. Discrepancies between formaldehyde measurements and methane oxidation model predictions in the Antarctic troposphere: An assessment of other possible formaldehyde sources. J. Geophys. Res. 110 (D15308). Available from: https://doi.org/10.1029/2005JD0058592995.
Seiler, W., 1974. The cycle of atmospheric CO. Tellus 26, 118–135.
Seiler, W., Conrad, R., 1987. Contribution of tropical ecosystems to the global budgets of trace gases, especially CH4, H2, CO, and N20. In: Dicnson, R.E. (Ed.), *Geophysiology of* Amazonia. John Wiley & Sons, New York, pp. 219–244.
Shepson, P.B., Sirju, A.P., Hopper, J.F., Barrie, L.A., Young, V., Niki, H., et al., 1996. Sources and sinks of carbonyl compounds in the arctic ocean boundary layer: polar ice floe experiment. J. Geophys. Res. 101, 2177–2193.
Spivakovsky, C.M., Yevich, R., Logan, J.A., Wofsy, S.C., McElroy, M.B., Prather, M.J., 1990. Tropospheric OH in a three-dimensional chemical tracer model: an assessment based on observations of CH3CCl3. J. Geophys. Res. 95, 18441–18471.
Sumner, A.L., Shepson, P.B., 1999. Snowpack production of formaldehyde and its effect on the Artic troposphere. Nature 398, 230–233.
Thompson, A.M., 1992. The oxidizing capacity of the Earth's atmosphere: probable past and future changes. Science 256, 1157–1165.
Thompson, A.M., Cicerone, R.J., 1986. Possible perturbations to atmospheric CO, CH_4, and OH. J. Geophys. Res. 91, 10853–10864.
Vaughan, P.P., Biough, N.V., 1998. Photochemical formation of hydroxyl radical by constituents of natural water. Environ. Sci. Technol. 32, 2947–2953.
Yokouchi, Y., Akimoto, H., Barrie, L.A., Bottenheim, J.W., Anlauf, K., Jobson, B.T., 1994. Serial gas-chromatographic mass-spectrometric measurements of some volatile organic-compounds in the arctic atmosphere during the 1992 Polar Sunrise Experiment. J. Geophys. Res. 99, 25379–25389.
Zuo, Y., Jones, R.D., 1995. Formation of carbon monoxide by photolysis of dissolved marine organic material and its significance in the carbon cycling of the oceans. Naturwissenschaften 82, 472–474.
Zuo, Y., Jones, R.D., 1996. Photochemical production of carbon monoxide in authentic rainwater. Geophys. Res. Lett. 23, 2769–2772.
Zuo, Y., Jones, R.D., 1997. Photochemistry of natural dissolved organic matter in lakes and wetland waters production of carbon monoxide. Water Res. 31, 850–858.

Further reading

Grannas, A.M., Shepson, P.B., Guimbaud, C., Domine, F., Albert, M., Simpson, W., et al., 2002. A Studyof photochemical and physical processes affecting carbonyl compounds in the Arctic marine boundary layer. Atmos. Environ. 36, 2733–2742.
Haan, D., Raynaud, D., 1998. Ice core record of CO variations during the last two millennia: atmospheric implications and chemical interactions within the Greenland ice. Tellus Ser. B 253–262.
König-Langlo, G., King, J.C., Pettré, P., 1998. Climatology of the three coastal Antarctic stations Dumont d'urville, Neumayer, and Halley. J. Geophys. Res. 103, 10,935–10,946.

Rogers, J.D., 1990. Ultraviolet absorption cross sections and atmospheric photo dissociation rate constants of formaldehyde. J. Phys. Chem. 94, 4011–4015.
Warneck, P., 1998. Chemistry of the Natural Atmosphere. Academic Press, San Diego, CA, p. 237.

CHAPTER 18

Spatial and temporal variability of physical parameter around Ny-Alesund region of Arctic

S.M. Pednekar
National Centre for Polar and Ocean Research, Ministry of Earth Sciences, Headland Sada, Vasco-da-Gama, India

18.1 Introduction

The Arctic circle (Fig. 18.1) is a comparatively small part of the global surface where the center of the sun can remain continuously above or below the horizon for 24 hours, but plays a crucial role in determining global climate change due to the complex couplings between cryosphere, atmosphere, ocean, and land (Serreze et al., 2007). Environmental change in the Arctic Ocean and Norwegian Seas is firmly connected toward the North Atlantic subpolar gyre (Østerhus et al., 2005). Presently the Arctic is undergoing significant environmental changes and has been the focal point of the climate sciences society (Winton, 2008; Overland, 2016). The surface temperature inside the Arctic Ocean and Norwegian Seas has been increased since the mid-1980s and keeps on expanding (Comiso, 2003) and early 1990s (Quadfasel et al., 1991; Carmack et al., 1995) due to the positive North Atlantic Oscillation (NAO) in the 1980s. The seawater has various physical properties that depend upon temperature, salinity, and pressure which interned controlled by the changes in atmospheric condition. The sea surface temperature (SST) is one of the most significant physical factors influencing the Earth's climate system (Lindsay and Rothrock, 1994; Reynolds et al., 2002, 2007; Deser et al., 2010), and SSTs performed a boundary condition in oceanic models (Reynolds and Smith, 1994).

The small spatial changes in the density of the seawater results in variation in physical properties. The properties of ocean water fluctuate with the measure of heat content (temperature), the dissolved matter (salinity), and ambient pressure of the water column. The variation of temperature and salinity in space and time forms the water mass, which is used to map

Understanding Present and Past Arctic Environments
DOI: https://doi.org/10.1016/B978-0-12-822869-2.00021-9

Figure 18.1 Arctic is an ocean surrounded by landmasses, and the basins is rimmed by a narrow continental shelf and land on the Greenland, Canadian, and Alaskan side and rimmed by a broad, shallow continental shelf on the Eurasian (Norway and Russia) side. The deep opening is between Greenland and the Norwegian archipelago of Svalbard connecting it to the North Atlantic. This narrow passage, called the Fram Strait along the Mid-Atlantic Ridge. The seafloor had drawn using GeoMap App.07/07/2020.

the ocean circulation (Pawlowicz, 2013). The global conveyor belt is liable for the thermohaline circulation which redistributes heat take-up by the Pacific through sea circulatory pathways (Fig. 18.2). Through thermohaline circulation, ocean heat transports over the large geographic distances pathways referred to as the ocean conveyer belt. It takes approximately 1000 years to complete one around trip.

Seawater in the Arctic region has been reported warming by many authors (Comiso, 2003; Chepurin and Carton, 2012). In the early 20th century, Bengtsson et al. (2004) gave a detailed discussion of the anomalously high

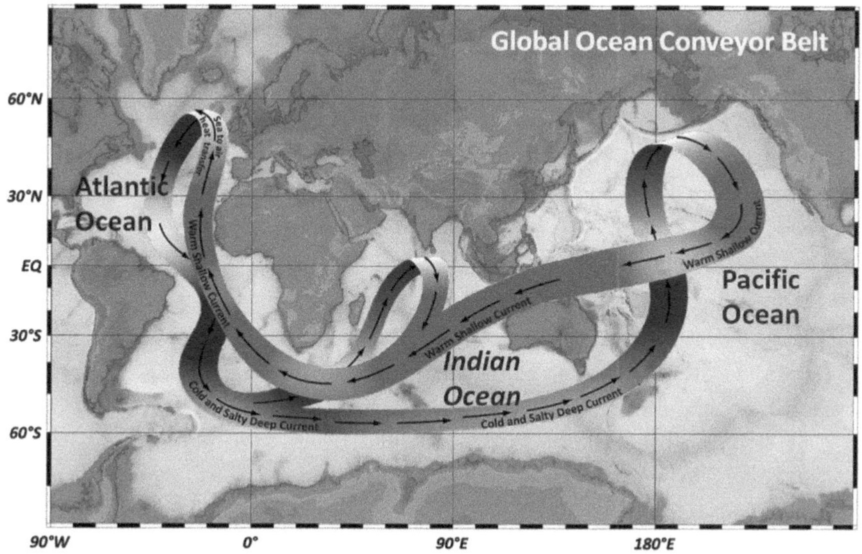

Figure 18.2 The global ocean conveyor belt (global thermohaline circulation) helps to redistribute ocean temperature and compositional properties throughout the global ocean. Gray/Blue arrows represent cold and salty deep water currents, while dark gray/Red arrows represent surface warm currents.

Arctic warming and suggested that the warming was a result of natural variations in wind-driven oceanic flow patterns, which causes a reduction in sea-ice cover. The annual cycle of sea-ice freeze/thaw and growth/melt, and the energy exchange between oceans and atmosphere, depending on SSTs (Key et al., 1997).

Chepurin and Carton (2012) studied the association between SST variation in subpolar gyres (Norwegian Seas) and those occurrences further north in the Arctic area by using the Pathfinder SST data from the AVHRR and OI-SST products from NOAA and the UK Meteorological Office. Arctic SSTs are also affected by the NAO and the Arctic Oscillation (AO) (Visbeck et al., 2001; Comiso, 2003; Cohen and Barlow, 2005; Chepurin and Carton, 2012; Qu et al., 2012; and others), and similarly, southern ocean SSTs are affected by the El Nino Southern Oscillation (Comiso, 2000). SSTs significantly correlated with NAO fluctuation in the Arctic, clarifying 33% of the Northern Hemisphere interannual SSTs variability in winter (Hurrell and Dickson, 2004), and Qu, et al. (2012) found a positive correlation between SSTs and the NAO in their study region (in Greenland). Based on the ice-sea mixing layer model, Deser et al. (2002) evaluated the role of atmospheric forcing in constraining anomalous ice and SSTs and found that atmospheric forcing contributed for winter to winter ingenuity and spatial development of sea-ice and anomaly of SST.

In this chapter, spatial and temporal variability of physical parameter namely SSTs around Ny-Alesund (Spitsbergen) presented and described using long-term OI-SST data. The following sections are described.

1. Background
2. Data and methodology
3. Temporal variability in SSTs on the global scale
4. Spatial variability in SSTs on the global scale
5. Seasonal variability in SSTs on the regional scale
6. Annual variability in SSTs on the regional scale
7. Decadal variability in SSTs on regional scale: 1982–91, 1992–2001, 2002–11, and 2012–19 (the 2012–19 data are for 8 years only)

18.2 Background

In recent decade, the major challenges among researchers are to understand changes in the climate system, and SSTs is one of the major physical parameters contributing to the World Ocean. Long-term temperature and sea level changes are also contributing to climate change. Fig. 18.3 shows an increasing trend with the time of two time series, ocean heat content for 0–700 m layer of the World Ocean, updated from Ishii et al. (2006) and Levitus et al. (2005) for 1955–2005, and a time series for 0–750 m for 1993–2005 updated from Willis et al. (2004). Many papers on altimetry show a current rate of sea level rise of 3.1 ± 0.7 mm yr^{-1} over 1993–2003 (Cazenave and Nerem, 2004; Leuliette et al., 2004; Fig. 18.4). A significant rate of change of 3 mm yr^{-1} has been shown to arise in the Southern Ocean (Cabanes et al., 2001). The global sea level changes derived from the tide gauges are presented in Fig. 18.5. Based on the above results the ongoing higher trend in recent years shown by altimetry, the authors assess the rate for 1961–2003 as 1.8 ± 0.5 mm yr^{-1} and for the 20th century as 1.7 ± 0.5 mm yr^{-1}.

The above studies demonstrated that the sea level and heat content in the ocean shown rising trends, which in turn, contributes to climate change. SST is one of the physical parameter controlled directly by the incoming solar radiation. Hence in this chapter, spatial and temporal variability of SSTs presented which is directly or indirectly connected with the worldwide atmosphere changes. Statistical analysis was conducted on long-term SSTs to

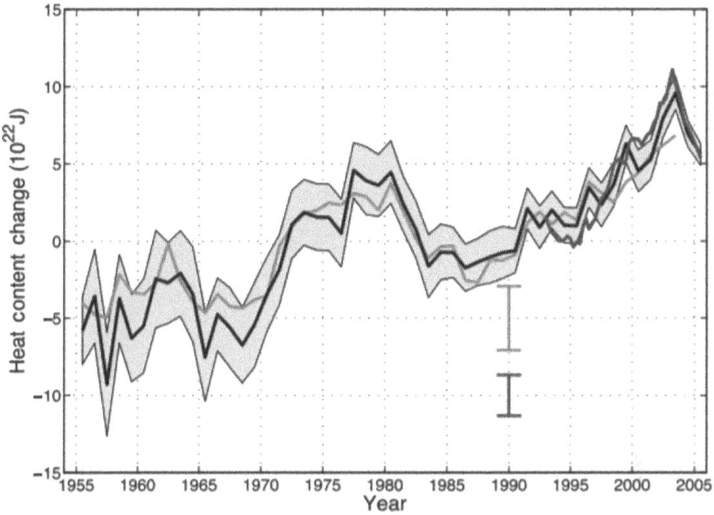

Figure 18.3 Annual time series of global ocean heat content for 0–700 m layer (1022 J). Black curve is updated from Levitus et al. (2005), with the shading representing the 90% confidence interval. The gray/red and dark gray/green curves are updated from the analyses by Ishii et al. (2006) and Willis et al. (2004), over 0–750 m, respectively. The error bars denoting the 90% confidence interval. The black and gray/red curves denote the deviation from the 1961 to 1990 average, and the shorter dark gray/green curve denotes the deviation from the average of the black curve for the period 1993–2003. Figure 5.1 from Bindoff et al. (2007).

find correlation coefficient, linear regression, and standard deviation (SD) from a reference period. The data and methodology used are described in the next section.

18.3 Data and methodology

It is well established that the SST is directly or indirectly linked to global climate change. To understand the spatial and temporal variability of the region, long-term series is required. The optimum interpolated (OI) SST are derived from the NOAA, Version 2 dataset (Reynolds et al., 2002) fulfilled the requirement. The monthly averages of SST for the period January 1982 to December 2019 were downloaded from NOAA, Oceanic and Atmospheric Research, Earth Science Research Laboratory, Physical Sciences Division (NOAA/OAR/ESRL/PSD), and Boulder, Colorado, USA (http://www.esrl.noaa.gov/psd/). In this chapter the analyses product having a resolution of $1° \times 1°$ grid obtained by merging

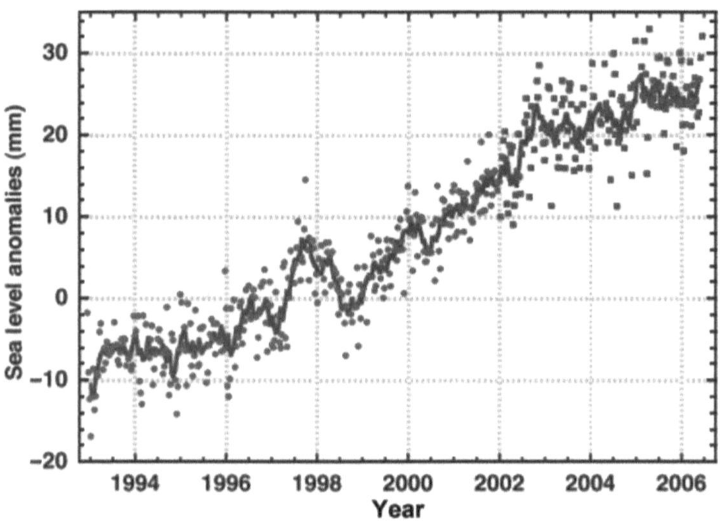

Figure 18.4 Variations in global mean sea level (difference to the mean 1993 to mid-2001) computed from satellite altimetry from January 1993 to October 2005, averaged over 65°S to 65°N. Dots are 10-day estimates (from the TOPEX/Poseidon satellite in gray/red and from the Jason satellite in dark gray/green). The black/blue solid curve corresponds to 60-day smoothing. Updated from Cazenave and Nerem (2004) and Leuliette et al. (2004). Figure 5.14 from Bindoff et al. (2007).

satellite and in situ SST observations (Reynolds and Smith, 1994). The OI-SST dataset contains monthly SST fields, obtained by averaging daily field obtained by linear interpolation of weekly OI fields. The examination utilizes in situ and satellite SSTs and simulated SSTs utilizing sea-ice information over the Marginal Ice Zonem (MIZ). In MIZ in situ and satellite data are inadequate because of cloud cover and navigation risks, respectively. The temperature at 5 m is also referred to as the model SST and is strongly nudged to the daily OI-SST (Reynolds et al., 2007). Wang et al. (2011) showed that the mean CFSR SST is about 0.05°C–0.1°C warmer than the daily OI-SST in the tropical Indian Ocean, the western tropical Pacific, and tropical North Atlantic, and they proposed that the warm SST biases are largely due to too strong shortwave radiation in those regions. The OI-SST dataset represents a modest improvement over the Version 1 dataset (Reynolds and Smith, 1994), based on bias corrections performed using in situ data. The OI-SST product utilizes a straight relationship with sea-ice concentration to induce SST, with SST limits to −1.8°C (the freezing point of seawater with the salinity of 33 PSU at the ocean surface) where ice spread is 100% (Reynolds et al., 2007).

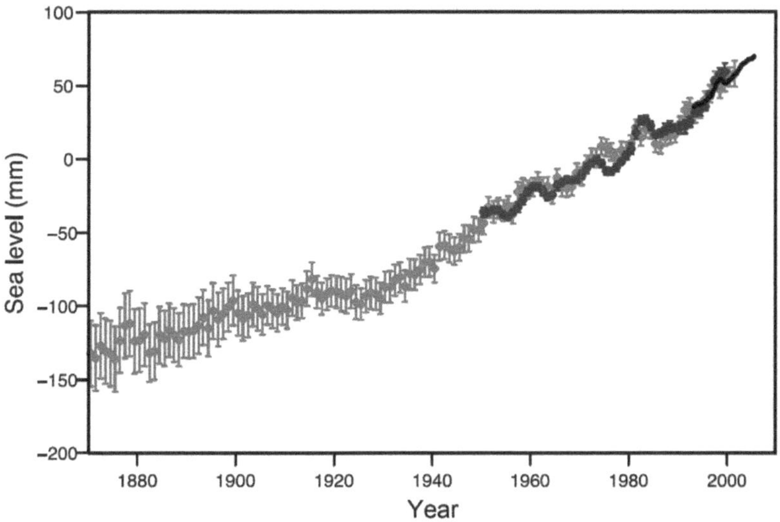

Figure 18.5 Annual averages of the Global mean sea level (mm). The gray/red curve shows reconstructed sea level fields since 1870 (updated from Church and White, 2006); the dark gray/blue curve shows coastal tide gauge measurements since 1950 (from Holgate and Woodworth, 2004), and the black curve is based on satellite altimetry (Leuliette et al., 2004). The gray/red and dark gray/blue curves are deviations from their averages for 1961–90, and the black curve is the deviation from the average of the gray/red curve for the period 1993–2001. Error bars show 90% confidence intervals. Figure 5.13 from Bindoff et al. (2007).

To study the temporal and spatial variability using global time series of SST analysis of monthly, interannual, and decadal variability, conducted using data from the period 1982 to 2019 compared changes between maps, respectively. We then performed a regression analysis to compute the trends, correlation, and standard deviation (SD) on a pixel-by-pixel basis over monthly averages, annual averages, and decadal averages.

The simple linear regression model is calculated from its observed values v_i at the time t_i, $i = 1, 2, 3, \ldots, n$, by fitting (t_i, v_i) to a linear relation through least squares methods.

$$v_i = b_0 + b_1 t_i + \varepsilon_i \quad i = 1, 2, \ldots, n \tag{18.1}$$

where the observed values of the v_i's constitute the responses or values of the dependent variable, the t_i's are the settings of the independent variable, b_0 and b_1 are the intercept and slope (trend) parameters, respectively, and the ε_i's are independently distributed normal errors each with mean zero and variance as s^2 based on a method described by Emery and Thomson, 2001, see also Appendix A (Fang et al., 2006; Pednekar, 2015).

Monthly average values of indices of AO and NAO were downloaded from the National Weather Service, Climate Prediction Center, NOAA (http://www.ncep.noaa.gov) for computing the time series running mean. The AO is an atmospheric phenomenon that can govern weather and climate patterns across mid and high latitude (North America, Europe and Asia) during winter. Basically air pressure near to the surface is measured, which is coupled with the polar vortex measured higher up in the atmosphere.

The AO is a large-scale climate pattern that influences weather throughout the Northern Hemisphere. It alternates between a positive phase (Fig. 18.6A) and a negative phase (Fig. 18.6B). In its positive phase, the AO tends to bring warmer weather to the middle latitudes, whereas in its negative phase, winter storms are more common in the Eastern United States and Europe (Katherine, 2012). Strong wind flows counter clockwise around the pole when the air pressure is low over the pole and higher over the mid latitudes as shown in Fig. 18.6A. Since 1997 the AO has regularly substituted among positive and negative stages. In the negative stage, the pressure is high in the far North, keeping the Arctic cold and dry, and pushing freezing air southwards over North America and Eurasia (AMAP, 2012).

The NAO is a fluctuation in pressure pattern over the North Atlantic Ocean. It affects weather over the globe. The NAO has two pressure systems, one over the Azores (higher pressure) and one close to Iceland (low pressure). Barnston and Livezey (1987) have shown that the NAO having

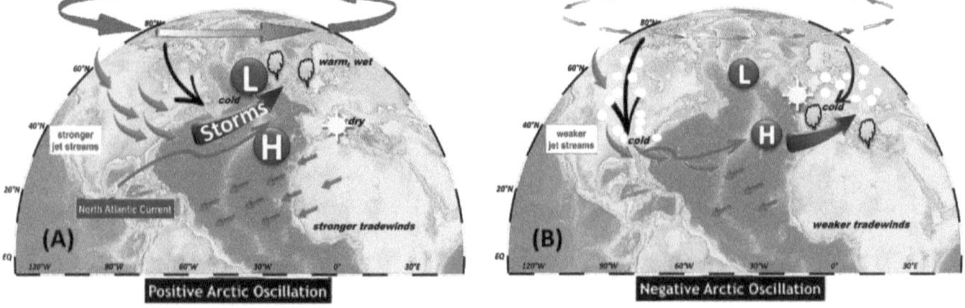

Figure 18.6 Schematic diagram of the Arctic Oscillation (AO) and its effects (adapted from AMAP, 2012). The positive AO denoted by (A) and negative AO as (B). Accordingly, the centers of low (gray/red-encircled L) and high (dark gray/blue-encircled H) pressure systems over the North Atlantic indicate the corresponding North Atlantic Oscillation phases (A: positive, B: negative).

prominent teleconnection throughout the year in the northern hemisphere. It is having the strongest signature in the winter months (December to March) when it's positive (negative) phase shows an enhanced (diminished) Iceland Low and Azores High (Hurrell et al., 2003). Positive index of NAO in winter is associated with a northeastward shift in the Atlantic storm activity with improved movement from Newfoundland into northern Europe and an unassuming diminishing toward the south (Hurrell and van Loon, 1997; Alexandersson et al., 1998). In winter, positive NAO is characterized by the stronger and frequent storms in the surrounding area of the Norwegian Sea and Iceland (Serreze et al., 1997; Deser et al., 2000).

Earlier studies have shown that AO and NAO play a major role in changing climate pattern over the Arctic circle. The data sets were further analyzed on a global scale and regional scale north of 65°N. The next sections described temporal and spatial analyses on a global scale.

18.4 Temporal variability in sea surface temperatures on the global scale

The temporal variability was visualized using long-term data. The time series plot of monthly averages of SSTs in the Arctic region north of 65°N up to 85°N and AO, NAO was filtered through 3-months running mean and 12-months running mean from 1982 to 2019. Since the calculation method is based on the 12-month running means, the seasonal variation is not present in the time series profiles, whereas in 3-month running mean, seasonal variation is present in the time series. The linear regression fit is calculated on 3-month and 12-month running mean time series of SSTs, and a linear relation through least squares method is shown in Fig. 18.7 as a straight solid line. The *x*-axis symbolizes time in months, and the *y*-axis gives the magnitude of SST. The solid line in Fig. 18.7A (Fig. 18.7B) is showing positively increasing trend in OI-SST of 0.00095 (0.001)°C/month. The SSTs varies from −1.210°C to 1.746°C (−0.537°C−0.258°C) for 3-month (12-month) running mean, and the anomalies varies between −0.293°C and 0.470°C (Fig. 18.7B). The regression sum of squares was estimated as 6.598°C (8.399°C) and residual mean square as 0.664°C (0.008°C) for 3-months (12-months) running mean, respectively. The coefficient of determination *R*-squared is 0.0220°C (0.691°C) for 3-months (12-months) running mean, respectively. Similarly, AO and NAO indices filtered through 3-months and 12-months running mean are

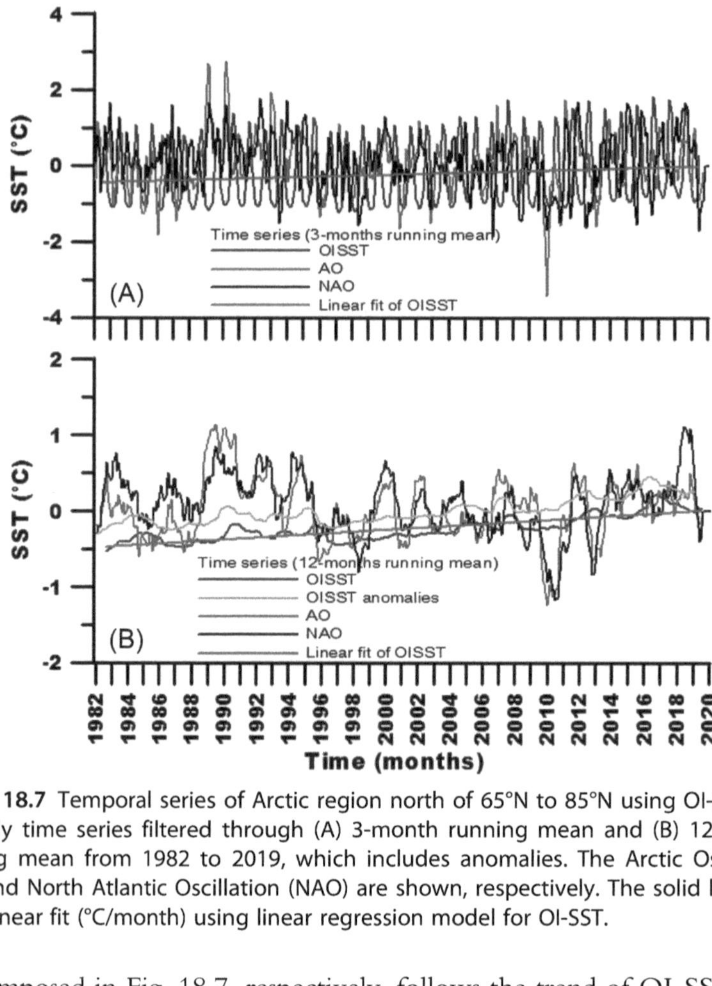

Figure 18.7 Temporal series of Arctic region north of 65°N to 85°N using OI-SST. The monthly time series filtered through (A) 3-month running mean and (B) 12-months running mean from 1982 to 2019, which includes anomalies. The Arctic Oscillation (AO) and North Atlantic Oscillation (NAO) are shown, respectively. The solid line indicates linear fit (°C/month) using linear regression model for OI-SST.

superimposed in Fig. 18.7, respectively, follows the trend of OI-SST. The AO fluctuates from −3.421°C to 2.754°C (−1.252°C–1.153°C) for 3-month (12-month) running mean, respectively. NAO varies from −1.829°C to 1.781°C (−1.170°C–1.121°C) for 3-month (12-month) running mean, respectively. The next section shown analyses conducted on the larger spatial scale to understand the variability in SST.

18.5 Spatial variability in sea surface temperatures on the global scale

The spatial analysis of variability based on 38 years period of SST from 1982 to 2019. The spatial correlation of SST with time based on the

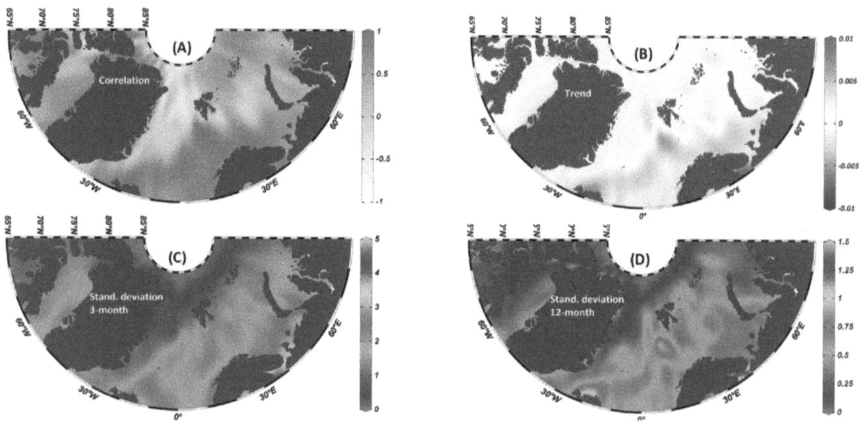

Figure 18.8 Spatial (A) correlation coefficient, (B) trend (°C/month), standard deviation using (C) 3-month running mean, and (D) using 12 months running mean from 1982 to 2019.

running mean is shown in Fig. 18.8A. The correlation ranges in between ± 1°C, higher the coefficient value is the good correlation. The correlation is shown positive (negative) relation south (west) of Spitsbergen. The entire east coast of Greenland is showing negative relation with time. There are some patches of positive and negative relation east of Spitsbergen. The Norwegian Sea and Iceland Sea is completely covered with good correction. The correlation between the NAO index and cyclone activity is highly negative in eastern Canada and positive in western Canada (Wang et al., 2006).

The spatial linear trend is shown in Fig. 18.8B, calculated using the linear regression method on a pixel by pixel bases using 12-month running mean. The scale of trend ranges between ± 0.01°C/month, ± 0.1°C/year, and ± 0.1°C/decade. The correlation and the trends are the mirror image of each other in term of spatial features. The positive trend associated with positive correlation and vice versa. The Norwegian Sea is completely covered with positive correction and trend, which indicates the warming trend over much of the Arctic Ocean and the rest of the region is the cooling trend in the northern Barents Sea (Blunden and Arndt, 2020). The annually averaged northern Barents Sea SSTs exhibit a warming trend, which has been attributed to changes in Atlantic Water influence in the region (see e.g., Barton et al., 2018). The positive correlation and the warming trends of SSTs spread the entire selected region except in the Greenland Sea and some pockets in the Barents Sea where the correlation is negative and SSTs trends are cooling. The spatial patterns of increasing/decreasing trends closely follow those observed annually; however, the rate of warming observed in the seasonal trends is apparently greater than that observed in the nonseasonal trends. The upward trend toward more positive NAO index winters from the mid-1960s to the mid-1990s has been associated with increased wave heights over the northeast Atlantic and decreased wave heights south of 40°N (Carter, 1999; Wang and Swail, 2001). The annual averaged of SST over the period 1982–2019 as an indication intrusion of warm water in the Norwegian Sea toward the north where the trend showed positive increase and vice versa.

The SD from the mean is calculated using 3-month (12-month) running mean is shown in Fig. 18.8C (Fig. 18.8D). The SD of SSTs in Fig. 18.3C and D ranges from 0°C to 5°C and 0°C to 1.5°C, respectively. The variability is observed higher in Fig. 18.3C as compared to Fig. 18.3D could be due to the inclusion of seasonal effects. The variability is stronger in the coastal region north of Iceland and Kolsky peninsula and reduces in open seas. The Norwegian Sea and the Barents Sea experiencing weaker variability; however, small patches of higher variability noticed south of Spitsbergen (Fig. 18.8D). The margin of the warming and cooling region had shown strong SD.

18.6 Seasonal variability of sea surface temperatures on the regional scale

The seasonal climatological monthly mean fields of the SSTs are described, respectively, as a background field. The climatological spatial fields are given in Fig. 18.9 for the SSTs from 38 years averages from January to December. The climatological spatial SSTs scale ranges from −2°C to 15°C is kept constant in all the maps for consistency. The seasonal pattern of the bounded region (65°N to 85°N and 40°W to 60°E) described as per the geographic location. Ny-Alesund, Spitsbergen, Svalbard experiencing polar climate, and southern region experiencing Norwegian climate change. The climatological month of the seasons also differed in these regions.

There are four seasons in Norwegian region, summer (June to August), autumn (September to November), winter (December to February), and spring (March to May). The upper North Atlantic Ocean varies primarily in response to changes in the surface winds, air-sea heat exchanges, and freshwater fluxes associated with NAO variations (Cayan, 1992a,b).

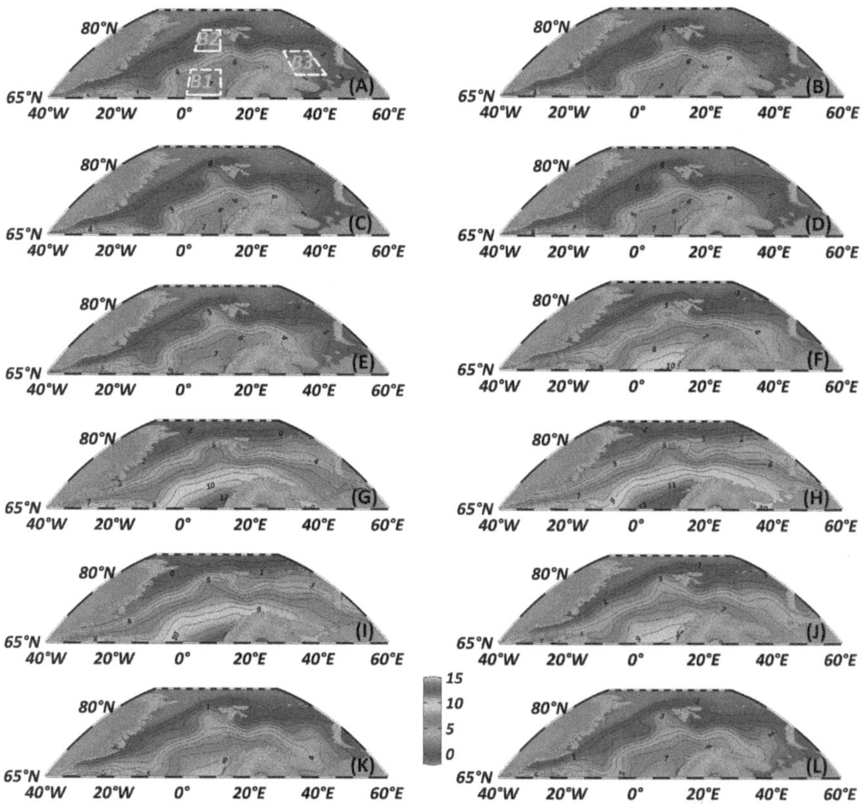

Figure 18.9 Climatological mean fields of sea surface temperature during 1982 and 2019 for (A) January, (B) February, (C) March, (D) April, (E) May, (F) June, (G) July, (H) August, (I) September, (J) October, (K) November, and (L) December. Square boxes selected for averaging SSTs in the Norwegian Sea as B1 (Box-1), west of Spitsbergen as B2 (Box-2), and the Barents Sea as B3 (Box-3).

In Norwegian seas, warmer (cooler) core of temperature is evaluated from June (December) with peak SST core of >13°C (<7°C) in summer (winter) during August (February). The amount of incoming solar radiation absorbed by the sea surface in the Arctic Ocean drives the SST in summer. The intrusion of warm North Atlantic current in summer also contributes to the warmer SST in the Norwegian Sea. The Arctic surface ocean is influence by the distribution of sea ice. The warming is greater in ice-free regions. As the space of sea-ice cover is reduced, more incoming solar radiation is absorbed by the ocean, and in turn, the warmer ocean melts more sea ice. In the Barents Sea, warm water from North Atlantic and North Pacific Oceans contributes additional heat by advection, respectively.

The Polar region, summer (May to September), winter (October to April), northern light winter (October to February), and sunny winter (March to May). Further in more detail the polar seasons in Spitsbergen, Svalbard is subdivided accordingly as follows. Polar summer from 17th May to 30th September, midnight sun from 17th May to 23rd August, and Golden autumn from 24th August to 30th September. Northern light winter 1st October to 28thFebruary, polar night from 11th November to 30th January and twilight from 1st October to 19th November and 1st to 28th February, summer winter from 1st March to 16th May and Pastel winter from 1st March to 19th April, and midnight sun from 20th April to 16th May.

In the Polar seas, the summer period occurred longer as compared to the Norwegian seas. The warm (cold) core, which is identified during summer (winter) in Norwegian seas, and its intrusion into the Polar seas noticed in Fig. 18.9 west of Spitsbergen during August–September (March–April) SSTs becomes less warmer (more cooler). The boundaries between warm and cold waters form the strong frontal region due to the mixing of two different water masses. In winter, cold water spreads and its extend entered the Norwegian Sea making SSTs cooler.

Further to understand better monthly variability, 5° × 10° grid boxes averages in selected boxes as shown in Fig. 18.9 bounded by Box-1 is (65.5°N to 70.5°N and 0.5°E to 10.5°E) in the Norwegian sea, Box-2 is (74.5°N to 79.5°N and 0.5°E to 10.5°E) near west of Spitsbergen, and Box-3 is (69.5°N to 74.5°N and 35.5°E to 45.5°E) in Barents sea, respectively. The monthly time series plot in three boxes is shown in Fig. 18.10. All the three profiles are showing peak (fall) during August (March) for summer (winter). The maximum average SST during August is 11.469°C, 5.420°C, and 7.658°C, respectively. The minimum average SST during March is 6.012°C, 1.498°C, and 1.529°C, respectively.

18.7 Annual variability of sea surface temperatures on the regional scale

The annual averages of SST maps prepared for the period 1982–2019 bounded by 65°N to 85°N and 40°W to 60°E to understand interannual variation in space and time. The scale of the SST range kept constant (−2°C–10°C). An average of each year analyzed from 1982 to 2019 is shown in Fig. 18.11. Warm core of SSTs >7°C noticed in Norwegian

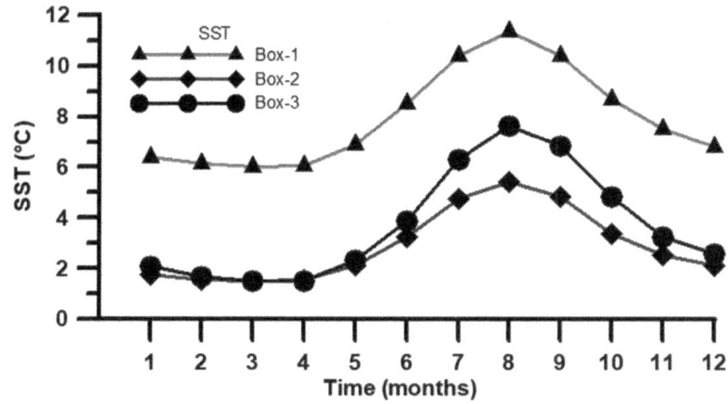

Figure 18.10 Climatological monthly averages of sea surface temperature in selected 5° × 10° Box-1, Box-2, and Box-3 are shown in Fig. 18.9.

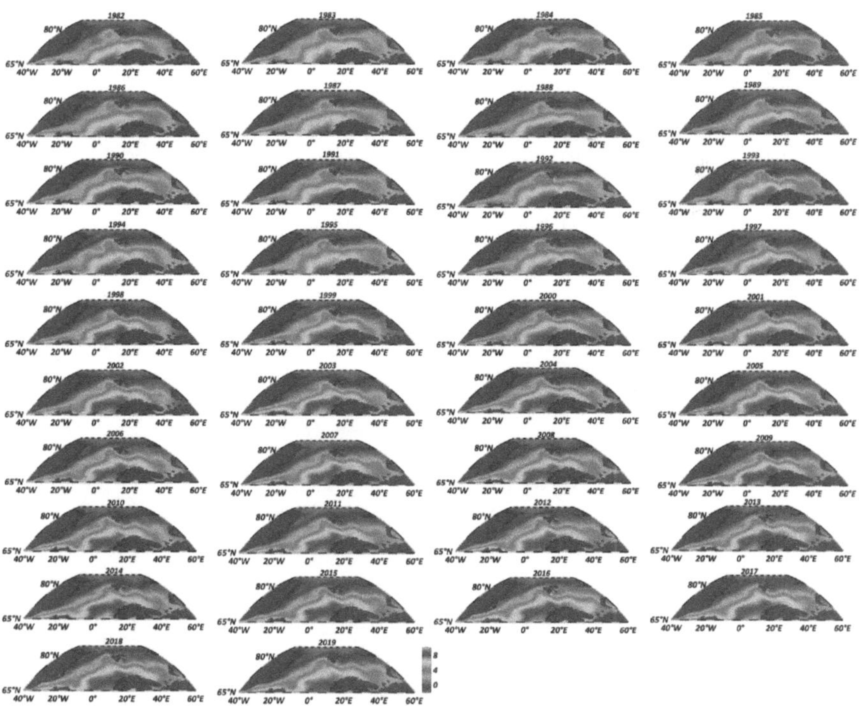

Figure 18.11 Arctic annual sea surface temperature maps for the period from 1982 to 2019. The year is mention in each map.

Sea in all the years between 0°E and 10.5°E wide and ~5° northward. The spatial spreading of warmer water of >8°C differed in each year, and it also extends toward Barents Sea. There are warming and cooling cycles according to the positive and negative phase of NAO/AO during 1982–2019.

Based on anomalies, Singh et al. (2013) identified the Barents and the Norwegian Sea, experiencing warming during the period 1983–86 (and low-amplitude warming in 1982), cooling during the period 1987–89, warming during the period 1990–94, small-amplitude cooling during the period 1995–97, and then warming from 1998 onwards until 2010. Thus as of 2019, the region is experiencing a warming trend.

Similar trends were observed by Comiso (2003) in his analysis of Arctic SSTs during the period 1981–2001. It should be noted that the periods of warming (1983–85, 1990–94, etc.) correspond to episodes of positive NAO indices, and the NAO may have contributed to the observed temperature changes in this region. The high NAO/AO index in 1990 corresponds to warmer SSTs in the region. Similar patterns of SST warming and high NAO/AO indices are observed in Fram Strait.

The average linear trend (Fig. 18.12) identified in Box-1 is 0.0365°C/year, Box-2 is 0.0034°C/year, and Box-3 is 0.0290°C/year, respectively, from 1982 to 2019. The trends have shown positive rising with the maximum (minimum) slope in the Norwegian Sea and Barents Sea (west of Spitsbergen). The warmer SSTs were associated with positive AO indices, except during the years 1998 and 2004/10, when negative AO indices

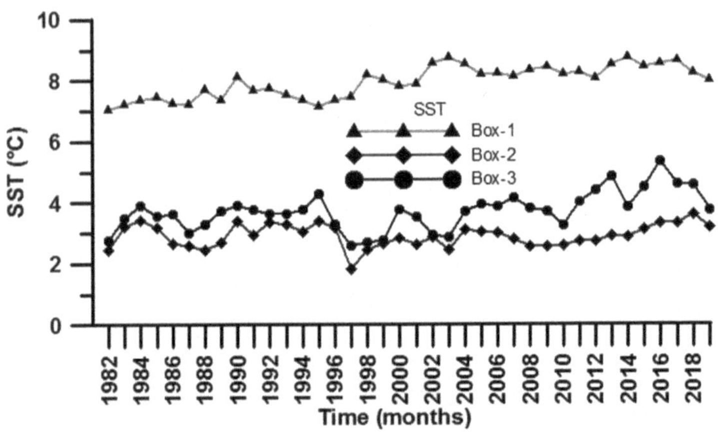

Figure 18.12 Time series of annual averages of sea surface temperature in selected 5° × 10° Box-1, Box-2, and Box-3, respectively, for the period from 1982 to 2019.

dominated. Other parameters may affect SSTs, such as air temperature, wind speed (Qu et al., 2012), etc., which can thus explain the observed warming during the years of negative AO indices. In December 2009 and January 2010, and again in December 2010 and January 2011, the AO was in a very negative mode, indicating higher pressure in the High Arctic. Central Europe was extremely cold, but the high Arctic stayed relatively warm.

18.8 Decadal variability of sea surface temperatures on the regional scale

To understand decadal variation in SSTs, the available data is divided into four subperiod, consisting of 1982–91 (top row), 1992–2001 (second row), 2002–11 (third row), and 2012–19 (bottom row) (Fig. 18.13). The extreme left panel represents correlation, SD, interdecadal trends, and the decadal average of SSTs (except for the last row, which shows the analysis for the 2012–19 period).

In correlation, the variables are correlated, and their strength of association is understood. The strength of correlation and the decadal trend which is the functional relationship between two variables established to make future projection events resemble each other. The decadal correlation of SSTs ranges between ± 1°C, and decadal trend of SSTs ranges between ± 0.2°C/year. The negative (positive) correlation is associated with a negative (positive) trend. The positive (Negative) correlation east (west) of 20°E where trends are warming (trends are cooling). The Norwegian Sea shows a negative correlation

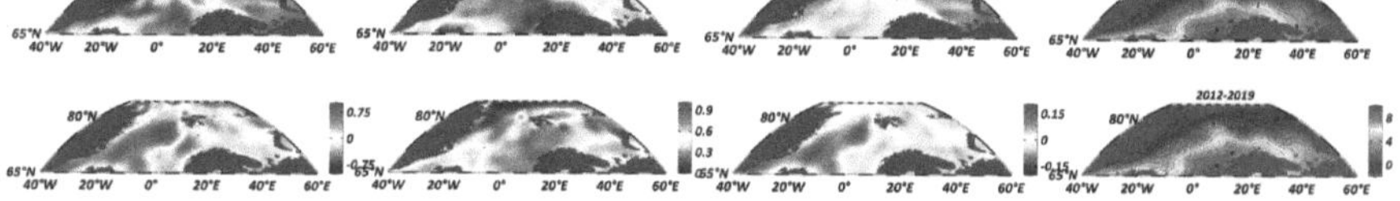

Figure 18.13 Decadal variations in Arctic sea surface temperatures (SSTs). The extreme left panels show in each correlation, standard deviation, linear trends in Arctic SSTs for the periods shown, whereas right panels show decadal averages for the period from 1982 to 1991, 1992 to 2001. 2002 to 2012, and 2013 to 2019 as mentioned in each panel.

and cooling trend. During 1982–91 and 1992–2001, the decadal average period indicated positive relationship and the warming trend in the whole region south of 74°N in 1982–91, up to 40°E in 1992–2001 average period and spotted negative connection and cooling trend of SSTs toward the north of 74°N. From the trend analysis of the recent period 2002–11 and 2012–19, it can be seen that the sea east of the Greenland coast experiencing cooling and warming in recent years.

The decadal SD of SSTs range from 0°C to 1°C, and the decadal average of SSTs ranges from −2°C to 10°C, respectively. The strong (weak) variability exists due to a strong (weak) gradient in SSTs. The decadal SSTs are warming in the Norwegian Sea, Iceland Sea, and it also influences extended toward the southern Barents Sea due to the intrusion of warm SSTs. The warming is stronger in 2002–11 and 2012–19 period as compared to 1982–91 and 1992–2001 period. The SSTs in the Greenland Sea and north of Spitsbergen are much cooler as compared to the south. The southern and the western region of Spitsbergen are also partly warming situations. The Scandinavia coastal SSTs has shown warming temperature.

Warm Atlantic waters have also been observed to enter the Arctic as pulses via Fram Strait and then along the slope to the Laptev Sea (Polyakov et al., 2005); the increased heat content and increased transport in the pulses both contribute a net warming of the arctic waters (Schauer et al., 2004). Multidecadal variability in the temperature of the Atlantic Water core affecting the top 400 m in the Arctic Ocean has been documented (Polyakov et al., 2004). The Spatio-temporal evolution on decadal scale of winter Sea ice in the Labrador Sea and associated SST variations in the North Atlantic studied by Deser et al. (2002). They suggested that the advection of cold freshwater by the Labrador Current could possibly account for the enhanced winter ice cover because they found that periods of enhanced winter ice cover in the northern Labrador Sea tended to precede colder than normal SSTs east of Newfoundland. Satellites thermal infrared data of 20 years of record (1981–2001) from the Advanced Very High Resolution Radiometer (AVHRR) used to study Arctic SSTs and found an increasing SSTs trend (Comiso, 2003).

18.9 Summary

The examined spatial and temporal variations in SSTs during the period from January 1982 to December 2019, using the operational SST product from NOAA, OI.V2. The present analysis has been demonstrated that climate change in the Arctic is a predictable reality today. The spatially and temporal varying SSTs were observed over the Arctic in the time sequence map during 1982–2019. An increase in SST was observed in 38 year period on a regional scale north of 65°N. The warming trend is observed in the Norwegian Sea, the Barents Sea and the west of Spitsbergen, the Iceland Sea, etc. over most of the Arctic region; however, toward pole marginal, cooling trend was observed. The trend is positively showing an increasing trend of 0.00095 (0.001)°C/month. The scale of the trend ranges between $\pm$ 0.01°C/month, $\pm$ 0.1°C/year, and $\pm$ 0.1°C/decade. The correlation and the trends are the mirror image of each other in terms of spatial features. The variability is observed in the 3-months running mean is higher as compared to the 12-months running mean could be due to the inclusion of seasonal effects.

In the Norwegian seas, warmer (cooler) core of temperature is evaluated from in June (December) with the peak in SST core of >13°C (<7°C) in summer (winter) during August (February). The warming temperature in the Norwegian Sea has been primarily due to the rapid increased in incoming solar radiation during summer. In the Polar seas, the summer period occurred longer as compared to the Norwegian seas. The warm (cold) core is identified during summer (winter) in Norwegian seas, and its intrusion into the Polar seas noticed west of Spitsbergen during August–September (March–April) SSTs becomes less warmer (more cooler). The spatial spreading of warmer water of >8°C differed in each year, and it also extends toward the Barents Sea. There are warming and cooling cycles according to the positive and negative phase of NAO/AO during 1982–2019. The average linear trend which is identified in Box-1 is 0.0365°C/year, Box-2 is 0.0034°C/year, and Box-3 is 0.0290°C/year, respectively, from 1982–2019. The trends have shown positive rising with the maximum (minimum) slope in the Norwegian Sea and Barents Sea (west of Spitsbergen). The decadal analysis revealed that the central Arctic region is undergoing a cooling trend during the decade 1992–2001 and a warming trend observed during other decades (1982–1991 and 2002–2019). The warming is stronger in 2002–11 and 2012–19 period as compared to 1982–91 and 1992–2001 period.

Acknowledgements

The author is thankful to Secretary, Ministry of Earth Sciences and Director, National Center for Polar and Ocean Research for their continuous support. Thanks to Schlitzer,

Reiner, Ocean Data View, 2020 for providing a package for the interactive exploration, analysis, and visualization of oceanographic data. NOAA analyzed the SST data (OI-SST Version 2), which were obtained from NOAA/OAR/ESRL/PSD, Boulder, Colorado, USA, and their use is gratefully acknowledged. Also thanks to the National Weather Service, Climate Prediction Center, NOAA for providing Arctic Oscillation (AO) and North Atlantic Oscillation (NAO) indices.

References

Alexandersson, H., Torben, S., Knut, I., Heikki, T., 1998. Long term variation of the storm climate over NW Europe. Glob. Atmosphere Ocean. Syst. 6 (2), 97–120.

AMAP (Arctic Monitoring and Assessment Programme) 2012. Arctic Climate Issues 2011. Changes in Arctic snow, water, ice and permafrost. SWIPA 2011 Overview Report, Oslo. xi + 97.

Barton, B.I., Lenn, Y.-D., Lique, C., 2018. Observed Atlantification of the Barents Sea Causes the Polar Front to Limit the Expansion of Winter Sea Ice. J. Phys. Oceanogr. 48 (8), 1849–1866.

Barnston, Livezey, 1987. Classification, seasonality and persistence of low-frequency atmospheric circulation patterns. Monthly Weather. Rev. 115 (6), 1083–1126.

Bengtsson, L., Semenov, V.A., Johannessen, O.M., 2004. The early twentieth-century warming in the Arcticda possible mechanism. J. Clim. 17, 4045–4057.

Blunden, J., Arndt, D.S., 2020. State of the climate in 2019 (Eds) Bull. Amer. Meteor. Soc. 101 (8), S1–S429.

Bindoff, N.L., Willebrand, J., Artale, V., Cazenave, A., Gregory, J., Gulev, S., et al., 2007. Observations: Oceanic Climate Change and Sea Level. In: Solomon, S., Qin, D., Manning, M., Chen, Z., Marquis, M., Averyt, K.B., Tignor, M., Miller, H.L. (Eds.), Climate Change (2007) The Physical Science Basis. Contribution of Working Group I to the Fourth Assessment Report of the Intergovernmental Panel on Climate Change. Cambridge University Press, Cambridge, United Kingdom and New York, NY, USA.

Carter, D.J.T., 1999. Variability and trends in the wave climate of the North Atlantic: a review, in Proc. 9th ISOPE Conf., Brest Vol. III, 12-18.

Carmack, E.C., et al., 1995. Evidence for warming of Atlantic Water in the southern Canadian Basin of the Arctic-Ocean - Results from the Larsen-93 Expedition. Geophys. Res. Lett. 22 (9), 1061–1064.

Cabanes, C., Cazenave, A., Le Provost, C., 2001. Sea level change from Topex-Poseidon altimetry for 1993-1999 and possible warming of the southern oceans. Geophys. Res. Lett. 28 (1), 9–12.

Cayan, D.R., 1992a. Variability of latent and sensible heat fluxes estimated using bulk formulae. Atmos. Ocean. 30, 1–42.

Cayan, D.R., 1992b. Latent and sensible heat flux anomalies over the northern oceans: the connection to monthly atmospheric forcing. J. Clint. 5, 354–369.

Church, J.A., White, N.J., 2006. A 20th century acceleration in global sea-level rise. Geophys. Res. Lett. 33, L01602.

Cazenave, A., Nerem, R.S., 2004. Present-day sea level change: observations and causes. Rev. Geophys. 42 (3), RG3001.

Chepurin, G.A., Carton, J.A., 2012. Subarctic and arctic sea surface temperature and its relation to ocean heat content. J. Geophys. Res. 117, C06019.

Cohen, J., Barlow, M., 2005. The NAO, the AO, and global warming: how closely related? J. Clim 18, 4498–4513.

Comiso, J.C., 2003. Warming trends in the Arctic from clear sky satellite observations. J. Clim. 16 (21), 3498–3510.

Comiso, J.C., 2000. Variability and trends in Antarctic surface temperatures from in situ and satellite infrared measurements. J. Clim. 13, 1674–1696.

Deser, C., Walsh, J., Timlin, M., 2000. Arctic sea ice variability in the context of recent atmospheric circulation trends. J. Clim. 13 (3).

Deser, C., Holland, M., Reverdin, Timlin, M., 2002. Decadal variations in Labrador sea ice cover and North Atlantic sea surface temperatures. J. Geophys. Res. 107 (C5), 3–1-3-12.

Deser, C., Phillips, A.S., Alexander, M.A., 2010. Twentieth-century tropical sea surface temperature trends revisited. Geophys. Res. Lett. 37.

Emery, W.J., Thomson, R.E., 2001. Data Analysis and Methods in Physical Oceanography. Elsevier, New York.

Fang, G., Chen, H.Y., Wei, Z.X., Wang, Y.G., Wang, X.Y., Li, C.Y., 2006. Trends and interannual variability of the south China Sea surface winds, surface height, and surface temperature in the recent decade. J. Geophys. Res. 111, 1–16.

Holgate, S.J., Woodworth, P.L., 2004. Evidence for enhanced coastal sea level rise during the 1990s. Geophys. Res. Lett. 31, L07305.

Hurrell, J.W., Kushnir, Y., Ottersen, G., Visbeck, M., 2003. The North Atlantic Oscillation: Climatic Significance and Environmental Impact. Geophys. Monogr. Ser. 134.

Hurrell, J.W., Dickson, R.R., 2004. Climate variability over the North Atlantic. In: Stenseth, N.C., et al., (Eds.), Marine Ecosystem and Climate Variation: the North Atlantic: a Comparative Perspective. Oxford University Press, pp. 15–32.

Hurrell, J.W., Van Loon, H., 1997. Decadal variations in climate associated with the north Atlantic oscillation climatic change 36 (3/4), 301–326.

Ishii, M., Kimoto, M., Sakamoto, K., Iwasaki, S.I., 2006. Steric sea level changes estimated from historical ocean subsurface temperature and salinity analyses. J. Oceanogr. 62 (2), 155–170.

Katherine, L., 2012. Once again, strange US winter weather partly due to Arctic Oscillation, www.adn.com.

Key, J.R., Collin, J.B., Fowler, C., Stone, R.S., 1997. High-latitude surface temperature estimates from thermal satellite data. Rem. Sens. Environ. 61, 302–309.

Leuliette, E.W., Nerem, R.S., Mitchum, G.T., 2004. Calibration of TOPEX/Poseidon and Jason altimeter data to construct a continuous record of mean sea level change. Mar. Geodesy 27 (1-2), 79–94.

Levitus, S., Antonov, J.I., Boyer, T.P., 2005. Warming of the World Ocean, 1955-2003. Geophys. Res. Lett. 32, L02604.

Lindsay, R.W., Rothrock, D.A., 1994. Arctic sea ice surface temperature from AVHRR. J. Clim. 7, 174–183.

Overland, J.E., 2016. A difficult Arctic science issue: midlatitude weather linkages. Pol. Sci. 10, 210–216.

Østerhus, S., Turrell, W.R., Jónsson, S., Hansen, B., 2005. Measured volume, heat, and salt fluxes from the Atlantic to the Arctic Mediterranean. Geophys. Res. Lett. 32, L07603.

Pawlowicz, R., 2013. Key physical variables in the ocean: temperature, salinity, and density. Nat. Educ. Knowl. 4 (4), 13.

Pednekar, S.M., 2015. Trends and interannual variability of satellite-based wind and sea surface temperature over the Southern ocean in the recent decade. Int. J. Geosci. 6, 145–158.

Polyakov, I.V., Beszczynska, A., Carmack, E.C., Dmitrenko, I.A., Fahrbach, E., Frolov, I. E., et al., 2005. One more step towards a warmer Arctic. Geophys. Res. Let. 32 (17), L17605.

Polyakov, I.V., Alekseev, G.V., Timokhov, L.A., Bhatt, U.S., Colony, R.L., Simmons, H.L., et al., 2004. Variability of the intermediate Atlantic water of the Arctic Ocean over the last 100 years. J. Clim. 17, 4485–4497.

Qu, B., Gabric, A.J., Zhu, J., Lin, D., Qian, F., Zhao, M., 2012. Correlation between sea surface temperature and wind speed in Greenland Sea and their relationships with NAO variability. Water Sci. Eng. 5 (3), 304–315.

Quadfasel, D., Sy, A., Wells, D., Tunik, A., 1991. Warming in the Arctic. Nature 350 (6317), 385.

Reynolds, R.W., Smith, T.M., 1994. Improved global sea surface temperature analyses using optimum interpolation. J. Clim. 7, 929–948.

Reynolds, R.W., Smith, T.M., Liu, C., Chelton, D.B., Casey, K.S., Schlax, M.G., 2007. Daily High-resolution blended analyses for sea surface temperature. J. Clim. 20, 5473–5496.

Reynolds, R.W., Rayner, N.A., Smith, T.M., Stokes, D.C., Wang, W., 2002. An improved in situ and satellite SST analysis for climate. J. Clim. 15, 1609–1625.

Schauer, U., Fahrbach, E., Østerhus, S., 2004. Arctic warming through the Fram Strait - Oceanic heat transport from three years of measurements. J. Geophys. Res. 109, C06026.

Serreze, M.C., Carse, F., Barry, R.G., Rogers, J.C., 1997. Icelandic Low Cyclone Activity: Climatological Features, Linkages with the NAO, and Relationships with Recent Changes in the Northern Hemisphere Circulation'. J. Clim. 10, 453–464.

Serreze, M.C., Barrett, A.P., Slater, A.G., Michael Steele, Zhang, J., Trenberth, K.E., 2007. The large-scale energy budget of the Arctic. J. Geophys. Res. 112, D11122.

Singh, R.K., Maheshwari, M., Oza, S.R., Kumar, R., 2013. Long-term variability in Arctic sea surface temperatures. Polar Sci. 7, 233–240.

Visbeck, M.H., Hurrell, J.W., Polvan, L., Cullen, H.M., 2001. The North Atlantic Oscillation: past, present, and future. Proc. Natl. Acad. Sci. (PNAS) 98 (23), 12876–12877.

Wang, W., Xie, P., Yoo, S.H., Xue, Y., Kumar, A., Wu, X., 2011. An assessment of the surface climate in the NCEP climate forecast system reanalysis. Clim. Dyn. 37, 2511–2539.

Wang, X.L., Swail, V.R., 2001. Changes of extreme wave heights in Northern Hemisphere oceans and related atmospheric circulation regimes. J. Clim. 14 (10), 2204–2221.

Wang, Y., Fang, G., Wei, Z., Qiao, F., Chen, H., 2006. Interannual variation of the south China Sea circulation and its relation to El Nino, as seen from a variable grid global ocean model. J. Geophys. Res. 111 (C11), 1–15. S14.

Willis, J.K., Roemmich, D., Cornuelle, B., 2004. Interannual variability in upper-ocean heat content, temperature and thermosteric expansion on global scales. J. Geophys. Res. 109, C12036.

Winton, M., 2008. Sea Ice – Albedo feedback and nonlinear Arctic climate change. In: Et, D.W., Bitz, C.M., Tremblay, L.-B. (Eds.), Arctic Sea ice Decline: Observations, Projections, Mechanisms, and Implications. American Geophysical Union, Washington, DC, pp. 111–132.

CHAPTER 19

Geopolitical and geoeconomic implications of climate change in the Arctic region: the future of contestation and cooperation

Dhanasree Jayaram[1], Anand V.[2] and Rashmi Ramesh[3]
[1]Department of Geopolitics and International Relations and Co-coordinator, Centre for Climate Studies, Manipal Academy of Higher Education, Manipal, India
[2]Department of Geopolitics and International Relations, Manipal Academy of Higher Education, Manipal, India
[3]School of Conflict and Security Studies, National Institute of Advanced Studies, Bengaluru, India

19.1 Introduction

In recent times the Arctic has been associated with conceptions such as the "Arctic boom" and the "Arctic gold rush" in the wake of accelerating thaw being experienced in the region due to climate and environmental change. These phrases reflect and signify the growing geopolitical and geoeconomic prominence of the Arctic in light of the region being a reservoir of immense natural resources as well as opening up for commercial shipping opportunities. The Arctic is a classic example of the confluence of geophysical, geopolitical, and geoeconomic dynamics, rendering it one of the most critical theaters of competition and cooperation not only among the littoral states but also the non-Arctic states such as China that are now making their presence felt in the region through various means—scientific, economic, political, and so on. Although traditionally, the Arctic has been "imagined" as a cold, frigid region with sparse human settlements and exotic fauna and flora, it has always featured in the geopolitical landscape and strategies of mainly the Arctic (littoral) states. However, in an age of Anthropocene[1], the Arctic has emerged as a "laboratory" of this new geological epoch, representing the "Arctic paradox" as well—in terms of both

[1] The Anthropocene "defines Earth's most recent geologic time period as being human-influenced, or anthropogenic, based on overwhelming global evidence that atmospheric, geologic, hydrologic, biospheric and other earth system processes are now altered by humans." More information is available at http://www.anthropocene.info/.

Understanding Present and Past Arctic Environments
DOI: https://doi.org/10.1016/B978-0-12-822869-2.00004-9

the effects of climate change and the (resultant) geoeconomic opportunities, tracing these developments back to beginning of the Anthropocene itself—the Industrial Revolution (Finger, 2016: p. 121). The changes in the Arctic—geophysical, geopolitical, and geoeconomic—have far-reaching implications for the rest of the world.

In this context the chapter attempts at analyzing the various geopolitical and geoeconomic implications of climate change in the Arctic. Accordingly, the major drivers of these implications and dynamics are identified as—territorial contestations among the Arctic littoral states—the role of the Arctic Council as a forum for promoting cooperation (involving the indigenous communities too), the geoeconomic potential of the Arctic, and the increasing militarization of the Arctic. It delineates the geopolitical and geoeconomic discourses on the Arctic from historical and contemporary points of view, also reflecting on the Arctic security environment during the second World War and Cold War. This provides a background to the ways in which climate and environmental changes have reshaped the geopolitics of the region by reshaping the geography itself, thereby questioning the assumption of permanence of geography in classical geopolitics. The chapter looks into the role of various actors in this respect, apart from the Arctic states. In particular the analysis takes into consideration the implications of the growing role of the Asian countries in the region, within and outside the Arctic Council. Finally, it also explores India's options in the Arctic, keeping in view the geopolitical and geoeconomic developments (based on the analysis). The chapter argues that despite the emerging rivalries and competition among the various stakeholders, the Arctic is likely to see more signs of cooperation through diplomatic efforts initiated by them, at least for now. However, steps need to be taken to avoid conflict or disputes from escalating in the future.

19.2 Geopolitical and geoeconomic discourses on the Arctic

The Arctic region comprises the Arctic Ocean and the adjacent seas along with parts of several countries, including the United States (northern parts of Alaska), Russia (northern parts of Siberia's mainland), Canada, Greenland (Denmark), Finland, Norway (Svalbard archipelago), Sweden, and Iceland. The Arctic is the northernmost region of the Earth and, being a polar region, is characterized by harsh climatic conditions and sparse population. It is centered on the North Pole and consists of more ocean than land unlike Antarctica. Moreover, unlike the Antarctic region, the Arctic has permanent and indigenous population residing within the Arctic Circle[2]. The region is part of the Earth's cryosphere, with both frozen water and ground—usually seen in the form of ice, snow, glaciers, and permafrost. Therefore the Arctic has been at the center for geophysical research since the 19th century, especially since the First International Polar Year (1882–83) (Hunkins, 1969). The region is also home to rich biodiversity with more than 21,000 species, including the polar bear, narwhal, lichens, Arctic fox, seals, and walrus, some of which are endemic to the Arctic (Barry et al., 2018). With climate change and rising temperatures, the Arctic cryosphere is either receding or crumbling, affecting the biodiversity and human population. The changing climate and geographical features are giving rise to new social, economic, security, and political dynamics in the region.

Recent reports on the Arctic furnish a rather grim picture of the effects of climate change in the region. According to the Intergovernmental Panel on Climate Change (IPCC) (2019: p. 205), "Arctic surface air temperature has likely increased by more than double the global average over the last two decades." More importantly, the report states that "Arctic sea ice extent continues to decline in all months of the year...the strongest reductions in September...are unprecedented in at least 1000 years." These assertions have been made by scientists with "very high confidence," "high confidence," or "medium confidence." Consequently, shipping activity in the region has increased during summer months over the same time period. The opening up of Arctic shipping lanes—mainly Northern Sea Route (NSR) and potentially Northwest Passage (NWP)—transforms the geoeconomic outlook of not just the Arctic region but also the entire world (explained in the subsequent sections of this chapter). Some of the nonlinear effects of climate change in the Arctic include the thawing of the permafrost and tundra, which could potentially release methane into the atmosphere—a greenhouse gas (GHG) that is more potent that carbon dioxide. However, there is much less agreement on whether the high rate of permafrost thawing has caused additional methane or carbon emissions. The rapidly

[2] The Arctic Circle is an "imaginary line that circles the globe at approximately 66° 34′ N (dashed blue circle in the map at right)." It "marks the latitude above which the sun does not set on the summer solstice, and does not rise on the winter solstice." More information is available at https://nsidc.org/cryosphere/arctic-meteorology/arctic.html.

retreating ice in Greenland and Arctic glaciers is said to have contributed to global sea level rise. These changes could adversely affect the Arctic communities (particularly the indigenous peoples) as well as trigger shifts in biomes and habitats, threatening the existing ecosystems in the region (Meredith et al., 2019: pp. 206–208). The retreating sea ice and snow cover also lead to decrease in albedo effect[3] and thereby increased surface temperatures and more retrenchment in ice and snow extent.

The Arctic has been an integral part of geopolitics due to its significance for both security/defense purposes as well as resource use. However, most of the geostrategists of the 19th and 20th centuries (or earlier), including Mahan, Mackinder, and Spykman, never took into account environmental change while propounding their theories on "Sea Power"[4], "Heartland"[5], and "Rimland"[6], respectively. The reason why they could not fathom these changes was because they regarded "geography" as a constant factor that shaped politics and foreign policy. Therefore even while recognizing the richness and value of the Arctic in terms of resources to some extent, they deemed it to be largely impregnable and thereby, aiding Russia's containment. On the contrary, Russia is now looking northward to the Arctic that is gradually opening up year after year due to warming, especially relevant to post-Cold War era geopolitics, in which it has lost the Soviet Union's constituents in Central Asia, which formed the "geographical pivot" according to Mackinder. The Arctic is becoming relevant not only from the perspective of

3 The albedo effect refers to the reflection of sun's energy by shiny ice and snow. If ice/snow melts, more energy will be absorbed by the ground and water instead of being reflected into the atmosphere, thus increasing the temperature.

4 In his seminal work, "The Influence of Sea Power upon History, 1660–1783," Alfred Mahan (an American naval officer), emphasized the role of sea power as central to world domination, reflected in one of his statements—"The one who owns the sea possesses all."

5 The "Heartland theory" was propounded by Halford Mackinder (a 20th century British geographer), in his works, "The Geographical Pivot of History" (1904) and "Democratic Ideals and Reality: A Study in the Politics of Reconstruction" (1919). According to him, Eurasia and Africa (which he called the "World Island"), on account of their resources, population, and size, form the core of world domination. He considered the landlocked region of Central Eurasia as the heartland. Whoever controls the heartland has the potential to "command the world."

6 Nicholas Spykman countered Mackinder's "Heartland theory" with the "Rimland theory," according to which, the coastal areas of Eurasian littorals are critical to control of the World Island, and not the heartland. His works include, "The Geography of the Peace" (1944) and "America's Strategy in World Politics: The United States and the Balance" (1942).

resources such as oil, gas, minerals, and fisheries but also commercial maritime trade with the East Asian region (Antrim, 2010).

However, this does not imply that the geopolitical significance of the Arctic was never realized by any power. For instance, the former US Secretary of State, William Seward (under Abraham Lincoln), bought Alaska from Russia in 1867, and he was equally keen on acquiring Greenland and Iceland too, but failed to do so (Dyer, 1940). Therefore when Donald Trump expressed his desire to buy Greenland, this was not the first time that the United States voiced its imperialist agenda. The Arctic region was a theater of the Second World War too, with the Americans, Canadians, and the Scandinavians building and maintaining air bases, including in northern Canada and Greenland. However, the "winter war" was mainly fought between the Soviet Union and Germany (that also invaded Norway and Russia). While the Arctic states had already set up several scientific stations, Nazi Germany built polar stations in Svalbard to monitor the Arctic's weather conditions too. Recently, Russian scientists from the Russian Arctic National Park also discovered a secret Nazi base (Arctic weather station) on a remote Arctic island (Menshikova, 2019). The Cold War period in fact marked increased militarization of the Arctic. After the Second World War, the strategic importance of the Arctic was recognized even more by countries such as the United States, which established several meteorological stations on the northern Canadian islands, floating research stations, air installations, and ballistic missile early warning radar systems (Roucek, 1983: pp. 465–466).

The 21st century geopolitical analysis of the Arctic is shifting toward a broader acknowledgment of the fact that environmental and climate changes are reshaping geography, which in turn has given way for an assessment of the region, not just in terms of power politics but also, from what is considered, "low politics" in international relations (issues ranging from economic to socio-cultural affairs). Moreover, at present, climate change seems to be the main driver of the transformation of the Arctic "from an afterthought to a central front in the new geopolitical view of the world" (Antrim, 2010: p. 16). Chaturvedi (2000: p. 441) argues that the Arctic "needs to be approached and understood by a new view of geopolitics, which is turn can no longer be seen in terms of the impact of fixed geographical conditions and configurations (heartlands/rimlands, lifelines, choke-points, critical strategic zones, etc.) upon the activities of Great Powers engaged in the pursuit of primacy." The Arctic has therefore become a focal point in the "critical" turn in geopolitical theories and discourses, raising the relevance of integrating the ecological and environmental aspects of geography into geopolitics—both in terms of "planetary boundaries"[7] and position of the Arctic in the international political economy on account of the undiscovered or newly discovered resources.

The Arctic has also been analyzed significantly through the lens of geoeconomics (as already mentioned), wherein states practise power politics through economic means to achieve strategic objectives. Based on Luttwak's (1990) seminal work, "From geopolitics to geo-economics: logic of conflict, grammar of commerce," the foundations of geoeconomics were laid out, primarily to establish the continuing rivalry between the states based on economics rather than the military. Geoeconomics has been defined in multiple ways by authors. It could be seen as economically driven territorial control or capture done through economic means. It could also be perceived as a relative shift from military (security) concerns to economic security ones. Geoeconomics could be used an analytical approach to study how and why states use economic means to project power globally or regionally. Control of sea lanes of communication and vital strategic resources is often considered a part and parcel of geoeconomics, which applies to the Arctic too. While some scholars distinguish between geopolitics and geoeconomics, others propose a combination of military and economic capabilities, thus implying that geoeconomics is a fusion of economic and geopolitical objectives and strategies (Scholvin and Wigel, 2018). As geographical conditions shape political outcomes, they also mold economic ones. The interplay between geoeconomics and geopolitics is central to analyzing the Arctic's security environment, especially in the wake of geophysical changes that have accelerated in recent decades.

The geopolitics and geoeconomics of the Arctic can be understood in multiple ways. From the geoeconomic relevance of the region due to melting of ice and consequent opening up of shipping routes and potential exploitation of the region's resources to militarization and entry of extra-regional players in the region, the Arctic's security scenario is multifaceted. It is evident that the existing disputes in the Arctic are not just

[7] The concept of "planetary boundaries" was created by a group of scientists from Stockholm Resilience Center and Australian National University in 2009. It indicates the safe operating space for humanity and the disastrous effects of crossing certain biophysical limits. More on this concept can be found in the article—Rockström et al. (2009).

territorial or material in nature. The meanings associated with maritime spaces, security, and other factors are equally relevant in the discourse, as is explored further in the chapter. Similarly, the question of human security in the region, particularly in reference to the indigenous peoples also needs a special reference here as contested meanings of security provide a diverse understanding on the Arctic security environment. As some would argue, it is shaped not only by changing geographical conditions but also by the relationship of the people with the land, thereby integrating indigenous peoples (and their traditional knowledge) into the paradigm. Besides, the Arctic is not a self-contained region, and its prospects (in terms of resource exploration and exploitation) depend on the global energy markets (availability, price, etc.) (Brutschin and Schubert, 2016). Scholars such as Käpylä and Mikkola (2016) have argued that in the Arctic, states largely cooperate as geographical conditions dictate so, as otherwise everyone's economic goals would be in jeopardy. Furthermore, the geoeconomic potential of the region may be exaggerated as factors, including "external geopolitical dynamics," "economic viability," "difficult weather and ice conditions," "global economic dynamics," infrastructural and technological bottlenecks among others come in the way of its utilization and development.

Has the Arctic emerged as a theater of conflict due to territorial disputes, resource competition, and sea routes? Is the Arctic a theater of cooperation, as exemplified by the Arctic Council? There are enough reasons to support both these arguments but what cannot be overlooked is the rising geopolitical and geoeconomic significance of the region and how these aspects are expected to strengthen further—toward conflict or cooperation—with more signs of climate and environmental change; so much so that the region has been "internationalized" with more active engagement of extra-regional powers such as China (geoeconomic). The Arctic Council that was established in 1996 to address issues concerning the Arctic states has inducted Germany, France, the United Kingdom, and Spain among others as observer countries. Notably, the Council also granted observer status to Asian countries—China, India, South Korea, Japan, and Singapore—in 2013. These states do not have voting rights, and they are obliged to "recognize Arctic States' sovereignty, sovereign rights, and jurisdiction in the Arctic" (Farré et al., 2014): p. 9). The Arctic Council has so far facilitated three legally binding agreements—Agreement on Enhancing International Arctic Scientific Cooperation (signed 2017), Agreement on Cooperation on Marine Oil Pollution

Preparedness and Response in the Arctic (signed 2013), and Agreement on Cooperation on Aeronautical and Maritime Search and Rescue in the Arctic (signed 2011).[8] To explore these questions and narratives, the next sections look into various aspects with respect to what defines the geopolitics and geoeconomics of the Arctic region.

19.3 Territorial contestations in the Arctic

The opening up of the Arctic and its resources has led to competing territorial claims in the region. Although parts of Finland, Sweden, and Iceland fall under the Arctic, only five countries—the United States, Canada, Denmark, Norway, and Russia—have an Arctic coast. Then, there are countries from outside the region, which have claimed Arctic identities in various ways, even though not purely being Arctic states. France has termed itself a "Polar Nation" in its national roadmap of 2016 on the basis of its extensive exploration of the polar regions (Ministry of Europe and Foreign Affairs, 2016). The United Kingdom has identified itself as "the Arctic's nearest neighbor" due to its close location and its susceptibility to the environmental changes in the Arctic (House of Commons Environmental Audit Committee, 2018). China has stressed that it is a "Near Arctic State" and that the changes happening in the Arctic has immense implications for the country (The State Council Information Office of the People's Republic of China, 2018). Nevertheless, all the Arctic coastal states have territorial disputes among each other.

The main territorial issues are with regard to overlapping claims over the continental shelf[9], surface level border disputes, and control over transit routes. These territorial issues have a significant bearing over the peace and security of the Arctic, since they can serve as the sources of possible conflicts in the Arctic. Claims over the continental shelf are primarily driven by the abundance of resources in the Arctic seabed. The

[8] More details of these agreements can be found on the Arctic Council's website at https://arctic-council.org/index.php/en/our-work/agreements.

[9] The continental shelf is defined in the Article 76 of the UNCLOS as comprising "the seabed and subsoil of the submarine areas that extend beyond its territorial sea throughout the natural prolongation of its land territory to the outer edge of the continental margin, or to a distance of 200 nautical miles from the baselines from which the breadth of the territorial sea is measured where the outer edge of the continental margin does not extend up to that distance.' The continental margin as defined in the Article, constitutes 'the submerged prolongation of the land mass of the coastal State, and consists of the seabed and subsoil of the shelf, the slope and the rise."

continental shelf part in the Arctic occupies a third of the region, the remaining part being shared equally between dry land and high seas (Kriz and Chrastansky, 2012: p. 116). The continental shelf occupies half of the Arctic Ocean floor, the remaining half being a deep basin. According to the estimates of the US Geological Survey, the Arctic could be home to nearly a quarter of the global energy resources (Holmes, 2008: p. 326). The claims over the continental shelf are determined by the Commission on the Limits of the Continental Shelf (CLCS) under the United Nations Convention on the Law of the Seas (UNCLOS)[10]. Except the United States, all the other Arctic coastal states have signed and ratified the UNCLOS and have subsequently submitted their claims regarding the outer limits of their extended continental shelf in the Arctic to the CLCS. The United States did not ratify the UNCLOS due to concerns of the UN bureaucracy imposing itself on the US sovereign activities in the maritime domain like seabed mining (Bentzen and Hall, 2017: p. 7). Norway was the first Arctic coastal state to ratify the UNCLOS (in 1996). This was followed by Russia in 1997, Canada in 2003, and Denmark in 2004 (Holmes, 2008: p. 331). In 2001 Russia became the first Arctic state to submit continental shelf claims, and it subsequently gave a revised submission in 2015. Norway followed in 2006, and Denmark submitted its claims during 2012–14. Canada submitted its claim only in 2019 (submissions to the CCLS). The important aspect to note here is that Canada, Denmark, and Russia have overlapping claims over the North Pole, by virtue of their claims over the submarine Lomonosov ridge (Quinn, 2019).

There are many territorial disputes over the surface borders in the Arctic. Even though the Cold War has ended, the United States and Russia continue to have a maritime border dispute between them in the Bering Sea. The State Duma, the lower house of Russia's legislature, is yet to ratify the 1991 agreement reached between the United States and Russia on the resolution of the dispute. Apart from this border dispute with Russia, the US boundary with Canada in the Beaufort Sea is yet to

[10] According to the UNCLOS, each country's sovereign territorial waters extend to a maximum of 12 nautical miles (22 km). Every coastal country may establish an Exclusive Economic Zone (EEZ) extending 200 nautical miles (370 km) from shore. Under article 76 of the UNCLOS, within 10 years of joining the framework, countries can claim their continental shelf beyond the 200 mile Exclusive Economic Zone limit to the Commission on the Limits of the Continental Shelf (CLCS), provided it meet certain specific geological conditions.

be formally delineated. Canada, on the other hand, is also mired in a territorial dispute over a barren rock called Hans Island, located in the Davis Strait between the former's Ellesmere Island and the latter's Greenland territory (Congressional Research Service, 2019: pp. 42–43). Norway had reached an agreement with Russia in 2010 for the resolution of its boundary dispute on the Barents Sea. However, it is yet to resolve the status of the maritime areas related to Svalbard (Bentzen and Hall, 2017: p. 5). According to the 1920 Svalbard Treaty, although the territory belonged to Norway, nationals of all 39 signatory countries have the freedom of entry and stay as well as the right to equal use of resources as per nondiscriminatory rules, which could be set out by Norway. However, Norway has restricted the Treaty's application to the land area and territorial waters of the islands. This has led to disputes regarding the status of Exclusive Economic Zones (EEZs) and continental shelf around Svalbard, most specifically with Russia. Moreover, even though the Treaty bans the use of the territory for military purposes, there is a lack of clarity about it, leading to further disputes (Østhagen, 2018: pp. 105–106).

As far as the transit routes in the Arctic are concerned, Canada maintains that the NWP will be treated as an inland waterway subject to its sovereign rule, whereas the United States and the EU want to treat the passage as an international strait, where innocent passage[11] is possible (Lalonde and Lasserre, 2013: p. 29). Even though there are disputes between the United States and Canada about the status of the NWP, the former always asks permission from the latter for transit, based on a 1988 bilateral agreement, and the latter never objects. On the other hand, the NSR passes mostly through Russia's EEZ rather than its territorial waters and therefore has far less of a conflict potential than NWP, even though vessels will have to respect Russian rules and regulations (Kriz and Chrastansky, 2012: p. 124). With such varied types of territorial disputes present in the Arctic, there is a possibility of military build-up that could subsequently trigger conflicts in the region. There have been arguments that the UNCLOS system may not be sufficient to resolve the territorial disputes of the Arctic, and that the key for resolution could be multilateral agreements on the lines of the Antarctic Treaty (Holmes, 2008: p. 325). Nevertheless, countries are likely to expand their territories and assert

[11] The UNCLOS allows innocent passage of ships through the territorial seas and EEZs of other countries, according to its Articles 17–19 and Article 38 of the Convention.

their sovereignty, especially in a region where a new geography is being "created" through environmental change.

19.4 Achieving cooperation through the Arctic Council

While the previous section looked into territorial contestations and related expansionist and/or conflict paradigms from a geopolitical point of view, this section deals with the paradigm of cooperation that has been equally influential in describing the Arctic and problematizes it. According to Oran Young (2009), the post-World War period in the Arctic can be divided into three stages: geopolitical/security, militarization (1945–91), circumpolar cooperation and institutions (post-Cold War), and climate change, socio-economic changes (present). The second stage opened the gates for a number of cooperative mechanisms and witnessed the establishment of institutions in the region. Former President of the Soviet Union, Mikhail Gorbachev, termed the Arctic as a "zone of peace." With this, the idea of an Arctic regional organization was strengthened. The initial efforts for regional cooperation however began when Canada proposed the establishment of an institution, at the fag end of the Cold War (Nuttall, 1997).

The idea of cooperation in the Arctic was a serious matter of concern prior to 1990s. The height of the Cold War induced the militarization component, as outlined earlier in the chapter. While the Soviet Union had a testing ground in the Arctic, the United States controlled the Thule Air Base in Greenland. Particularly on the lines of cooperation and governance, the Arctic is very distinct from Antarctica. The 1959 Antarctic Treaty System is a classic example of science diplomacy during a period of sustained conflict, especially between the two superpowers. The tag of "global commons" was not applicable in the case of the Arctic also probably because the Arctic Ocean is the backyard of two biggest military powers (the United States and Soviet Union/Russia) as well as proximate to some of the most important European countries. The Arctic Council only came into the picture when the Soviet Union weakened and finally collapsed. It was established in 1996, after multiple rounds of negotiations in Rovaniemi, Yellowknife, Kiruna, and Ottawa.

The "Joint Communique of Governments of the Arctic Countries on the Establishment of the Arctic Council" or the Ottawa Declaration is the founding document of the Council. It must be noted that the Arctic Council is not a formal organization on the lines of international law.

Rather, it is an informal forum for achieving consensus and cooperation among the Arctic states. There is no requirement for participating states' Parliamentary ratification in this case. The United States was strongly opposed to the idea of establishment of another international organization with binding rules and agreements. The Ottawa Declaration states that the Council focuses on environment, trade, standard of living, economic development, and sustainable development. It is not surprising that the Council emphasizes environmental issues, given that the region faces immense impact of climate change and drastic changes in the cryosphere. Additionally, the Arctic Council is an offshoot of the Arctic Environmental Protection Strategy and therefore adheres to its agenda.

The Council is composed of eight members—Russia, the United States, Canada, Denmark, Norway, Iceland, Finland, and Sweden. It is headed by the Chair, supported by the Arctic Council Secretariat. The Chairmanship rotates among the members every 2 years. Along with the members, the Council consists of six permanent participants, six working groups, and 13 observer countries. The permanent participants are indigenous community councils and forums that have an important place in the Arctic affairs.[12] The Saami Council, Gwich'in Council International, Inuit Circumpolar Council, Arctic Athabaskan Council, Aleut International Association, and Russian Association of Indigenous Peoples of the North are included as the permanent participants. The role of indigenous communities in the Arctic is quite unique. The Ottawa Declaration, 1996 declared that the Arctic Council would be a "high-level" forum for interactions among member states, along with the indigenous communities and Arctic inhabitants. These communities, in their capacity as permanent participants, have influenced the decisions of the Arctic Council, particularly the ones that are concerned with economic development and shipping (Gassiy, 2018). At the international level, the Arctic indigenous communities successfully create an Arctic narrative, generate awareness, and influence international agreements. The Saami Council and the Inuit Circumpolar Council played an important role in incorporating the clause for limiting the average temperature rise to 1.5°C (Inuit Circumpolar Council, 2019). The latter, in the 2019 Conference of Parties (COP 25), held in Spain reiterated the seriousness of climate change and its effects on the Arctic as well as the world at large (Inuit Circumpolar Council, 2019).

[12] More information on the "permanent members" can be found on the Arctic Council's website at https://arctic-council.org/index.php/en/about-us/permanent-participants.

The Arctic Council is one of a kind of forum to give a high pedestal to the civil society in the decision-making process and the governance of the region. This sets it apart from other intergovernmental organizations across the world. It also gives the Council, the legitimacy for the initiatives taken towards environmental protection and conservation, mainly because there is an effort to take the most affected group into consideration, and preserve the traditional knowledge while adapting to climate change. The "Working Groups" and "Task Forces" focus on issues such as conservation, sustainable development, marine biodiversity protection, and emergency preparedness. The Council is successful in creating an Arctic narrative, thereby generating awareness about the impacts of climate change in the region. However, while making the Arctic and climate change a global concern, it also ensures that the region does not qualify as a "global common." It continues to be a closed group of eight countries and permanent participants. The Ottawa Declaration emphasized that the Arctic Council remains a forum or more of a "soft-law" organization, in which consensus is the key to decision-making (Dodds, 2012: p. 195). Although an informal organization, it emerged as the prime institution that could achieve circumpolar cooperation during the post-Cold War period (Bloom, 1999).

The third stage in the post-World War period in the Arctic (climate change and socio-economic changes) has generated a lot of interest in not only environment-related issues but also their cascading impacts in various sectors. The opportunities that have risen have dented the consensus of the Council (Young, 2009: p. 424). The Ilulissat Declaration was adopted at the Arctic Ocean Conference, held at Ilulissat, Greenland, in 2008. Denmark invited the other four coastal states of the Arctic Ocean—Russia, the United States, Norway, and Canada—to discuss the political developments of the region ("2008 Ilulissat Declaration," 2008). The conference is seen as an important event, where the five coastal states emphasized their roles and responsibilities toward the region, exclusive of Finland, Sweden, and Iceland. The divide between the five coastal states and the three non-coastal states was visible. It is an inherent contradiction within the Arctic Council and a counter-current to circumpolar consensus and unanimity.

The Ilulissat Declaration had many more implications. The first phase of the Arctic that had come to an end by 1991 is seeing a revival since Russian marine explorers placed the national flag on the seabed beneath the North Pole (Parfitt, 2007). The Arctic Council, established in 1996, has precisely kept away from the political issues and focused on environmental conservation and socio-economic development. The Council's mandate remains the same, and political issues are being deliberated upon, independent of the Council. The 2008 Arctic Ocean Conference was one of these meetings that seemed to have undermined the importance of the Council's role in questions of governance. Iceland, Finland, and Sweden stress upon their Arctic identities and history in their Arctic policies. They also prioritize the Arctic Council and inclusive governance of the region. The conference (and the Declaration) thus gave rise to a sort of identity politics among the Arctic countries.

The Arctic Council has expanded, especially with the induction of Asian countries. China, Japan, and South Korea have extensive involvement in the region, particularly in the area of shipping. It must be noted that the East Asian economies are the greatest beneficiaries of opening up of the NSR, predominantly because of the reduced distance, time, and fuel consumption when they trade with Europe. The route, when fully operational, has the potential to offer an alternative to Malacca Strait and the Suez Canal. Singapore's interest in the Arctic is also related to shipping industry, as it is concerned about the loss of revenue in the Malacca Strait and its status as a global shipping hub (The Arctic Institute, 2019). The rapid rise in Chinese activity (not just shipping but also investments in the Arctic Circle) in the region has raised a few eyebrows. China's participation in the Council is guided by "respect, cooperation, win-win result, and sustainability," according to its White Paper. Technically, it is not an Arctic state, but it proclaims itself as a "near-Arctic state" with the ambition of being a "polar power" (Descamps, 2019). It has also made it clear that issues such as "climate change and international shipping" involve the interests of non-Arctic states, and therefore they must be engaged further in the Arctic matters (Hong, 2018).

In 2018 Canada, Denmark, Russia, the United States, Norway, the EU, Iceland, China, South Korea, and Japan signed the "International Agreement to Prevent Unregulated High Seas Fisheries in the Central Arctic Ocean." The agreement not only prevents commercial fishing in the Central Arctic Ocean (CAO) for 16 years but also paves way for conservation mechanisms and scientific cooperation in the CAO (Fisheries and Ocean and Government of Canada, 2018). Notably, Finland and Sweden are not part of the agreement. On the other hand, the agreement sets a tone for the recognition of the Asian powers like China, South Korea, and Japan in the Arctic affairs. Issues of climate change require political solutions that are legally binding upon the participants/stakeholders. The Council is not a formal organization that can give rise to agreements and treaties. This can probably be counted as one of the weaknesses of the Arctic Council, as the countries negotiated and arrived at the CAO Agreement outside the purview of the Council.

With some critical loopholes in the Arctic Council, there has been a debate about the reforms that can be induced into the forum. First, the Council does not have its own financial source and is dependent on the member countries' contributions for its budget. Also, the members fund the working groups and task forces on the basis of projects and activities that are undertaken. Hence, there is no continuity in financial sources. Working toward this is one of the major reforms suggested by some scholars (Balton and Ulmer, 2019; Śmieszek and Koivurova, 2017: pp. 20–21). Second, reforming the structure of the Council has also been suggested by many scholars such as Rottem (2020) and Young (2016). While the number of members is unchangeable, the number of observers is on the rise. Permanent participants, as a rule must be less than eight (number of members) and have only consultation rights (not voting rights). In order to enhance and give more value to the participation of the indigenous communities, permanent participants ought to be given voting rights on issues concerning sustainable development in the region and indigenous peoples' lifestyles. The role of the observers and that of the northern regional governments also needs to be clearly spelled out and expanded with the changing dynamics of the region, such as Asian investments. Besides, the Council has also been criticized for not instituting an exclusive working group, task force, or expert group on climate change (Exner-Pirot et al., 2019). Rottem (2020: p. 73) proposes to provide a clear strategy and vision for the Council by establishing an Arctic Summit. Śmieszek (2019) analyses three types of reforms—legal, organizational, and functional—and concludes that the Council is not suited to a legally binding regulatory framework. The Arctic Council is therefore a forum for calling attention to nonmilitary facets such as scientific research and science diplomacy, emergency preparedness, climate change, and economy. However, the complexities of the Arctic region are on the rise, and there is a demand for reforms to face these emerging challenges.

19.5 The Arctic geoeconomics: what is in store?

Climate change has the ability to alter the geography of a region or a country to the extent where territorial demarcations can no more be the sole determinant of interactions among nations (Dalby, 2014: pp. 3–4). Changing geography can alter the economic profile of a region too, which is often correlated with geoeconomics. In the Arctic the melting sea ice has paved the way for a relatively ice-free Arctic Ocean during certain months (in summers), and countries have been speculating on shipping opportunities offered by these developments. The contracting sea ice has also increased the search for resources that might push the countries to look beyond traditional energy producers such as countries in the Middle East. Russia, Norway, and Canada are already on the energy map of the Arctic in terms of exploration and extraction. While some scholars have forecasted an energy-rush/oil-rush in the near future, few other scholars have downplayed the narrative, stating a number of challenges in the process, as also stated earlier in the chapter.

Canada was the first Arctic country to begin exploring and extracting oil in the region. The focus was mostly on the Northwestern territories. In 1960s 40 years after Canada began extraction, Russia discovered hydrocarbon fields in the Yamalo-Nenets region. Following this, the United States discovered oil fields in Brooks Range, Alaska (Rossiya Segodnya, n. d.). The US Geological Survey (USGS) report of 2008 can be considered a breaking point, from where the interest in the Arctic resources saw a new high. The Circum-Arctic Resource Appraisal (CARA) by the USGS stated that the study "aimed to discover the undiscovered stocks of conventional hydrocarbons and exclude non-conventional resources such as gas hydrates, shale oil, tar sand and coal bed methane" (US Geological Survey, 2008). The study was conducted in 33 specific areas of the Arctic, based on the speculation of the presence of resources. The report noted that 25 out of 33 study areas (such as East Barents Basin, Arctic Alaska, East Greenland Rift Basins, West Greenland-East Canada region, etc.), had a probability of at least 10 percent of accumulation of either oil, natural gas, or both (US Geological Survey, 2008).

Along with the hydrocarbons, the Arctic is known to have abundant resources of copper, gold, tungsten, nickel, uranium, coal, iron, silver, and zinc. There is also wide presence of marine resources and fossils. The Arctic commercial fishing mostly depends on cod, pollock, and salmon. Increasing water temperature and the acidification of the ocean are causing depletion of some marine species, whereas a few other species have migrated to the northern reaches of the Arctic. This is an emerging challenge to the traditional fishing communities of the region (Goldfarb, 2017). The distribution of the resources in the region is skewed toward the Russian coast. Russia has the largest coastline with the Arctic Ocean and is naturally at an advantageous position in terms of resources distribution.

Although the distribution of the resources is known, there are inherent challenges in accessing these resources. First, the findings of the CARA are an estimation of the resources present in the region. In order to save financial resources, companies might not plunge into proper extraction in new hydrocarbon/mineral fields, without further probing and confirmation. Second, companies/countries are yet to completely develop technology that can successfully penetrate through the thick sea ice. Hydrocarbons are covered by layers of permafrost, which further complicates the process. Third, even if technological impediments are overcome, economically, extracting resources in adverse environmental conditions may not be viable, at least in the near future. Fourth, there are genuine environmental concerns regarding such activities. Drilling and extraction can release methane, a GHG having adverse effects on the Arctic Ocean and the region's environment as a whole. These challenges have busted the myth of an immediate resource-rush and severe competition between stakeholders in the region (Chaturvedi, 2014; Käpylä and Mikkola, 2016).

The potential for commercial shipping in the Arctic Ocean, particularly the NSR, was first discovered in 1860s by a Russian merchant who owned gold mines near Yenisey. Since then, Russia gradually deepened its involvement in shipping and development of the NSR. In the year 1932, Chief Administration of the NSR was established by the Soviet Union, with an aim to develop the sea route as well as the northern reaches of the country (Drent, 1993). Today's interest in the Arctic shipping arises out of the changing climatic conditions and increased commercialization. The NSR along the Russian and Norwegian coasts and the NWP along the Canadian and American coasts are the two sea routes that are currently in use. However, the NSR is set to play a more pivotal role, simply because the erstwhile Soviet Union and the Russian Federation have invested extensively in infrastructural development. As a sea route that has catered to the communities in Siberia as well as to regional trade, NSR will continue to bolster trade and commerce across the Arctic (Ho, 2011).

According to the Arctic Marine Shipping Assessment (AMSA) prepared by the Protection of the Arctic Marine Environment (PAME), a Working Group of the Arctic Council, shipping will see a surge in the Norwegian and Russian waters. Kara Sea, Barents Sea, and Pechora Sea will be the important points in regional transportation. The report also notes that the traffic along the NSR might increase manifold by 2025 (Ho, 2011). The opening of sea routes has attracted not only European but also Asian countries. The NSR can reduce the distance between

Japan and the United Kingdom by approximately 5000 nautical miles, because it would avoid circumventing the Asian continent and passing through the Suez Canal. It would also result in the reduction of time taken for navigation (Lee and Song, 2014). This is one of the main reasons for the Asian countries, particularly China, Japan, and South Korea to show keen interest in the development of the NSR. In the 2018 White Paper (The State Council Information Office of the People's Republic of China, 2018), China highlighted that the Polar Silk Road, the polar version of the ambitious Belt and Road Initiative, will be developed in partnership with the Arctic countries. Much of the Polar Silk Road is in the Russian Arctic and coincides and connects to the NSR. By ensuring rail, road, and freight connectivity in the High North, China aims to overcome its "Malacca dilemma"[13].

However, there are a number of challenges and obstacles to Arctic shipping too. First, the issues concerning sovereignty and ownership render different meanings to routes, security, and territorial spaces. For instance, the ownership of the NSR is claimed by Russia, mainly due to the proximity and historical reasons. The establishment of NSR administration by Russia to regulate traffic along the sea route and a legislation (2012 NSR Act) that define sea route's dimensions (borders) reinforce its intent to establish its sovereignty over the NSR (Lagutina, 2019: p. 62). Similarly, as already explained earlier, Canada claims the ownership of the NWP, given that the sea route passes through its territorial waters. The United States has always asserted that the navigational routes do not belong to any one country and that the "freedom of navigation" must be ensured. Additionally, the United States has not ratified the UNCLOS and is not obligated to follow its provisions. Second, the Arctic presents adverse conditions for shipping. The sea routes remain frozen for a long period of time in a year, requiring the use of ice-class ships. In recent years, during summers, the waters mostly are ice-free, but have moving ice and icebergs that pose danger to shipping. Due to these risks, the insurance companies might take a back seat, thereby affecting shipping (Tseng and Cullinane, 2018).

Third, infrastructural constraints have also subdued the "Arctic boom" mood. The Arctic Ocean is the shallowest of the world's oceans, and hence, heavy cargo vessels cannot navigate through it. Polar conditions

[13] The Malacca dilemma refers to the vulnerabilities associated with China's overdependence on seaborne energy imports via the Malacca Strait.

require ice-class vessels. This is yet to be developed to its full capacity by the Asian countries, if they aim to improve shipping prospects in the region. As identified in the findings of a survey of the Japanese shipping companies, Arctic shipping ambitions are impeded by many factors, including "higher construction and operational costs, unpredictable ice-situation for some of the year, and limitations on vessel size due to shallow straits" (Moe and Stokke, 2019: p. 33). Fourth, capable support services must be developed to meet any emergency situations. Search and rescue capabilities must be developed in a comprehensive manner to ensure the safety of the mariners. Adequate training facilities at the hands of experienced mariners are also necessary for the sailors (Dalaklis et al., 2018). Fifth, the indigenous communities are a very important part of the Arctic. Their interests often collide with the commercial and economic interests of other stakeholders. There is a fear of loss of traditional settlements, livelihood, and lifestyles when the economic benefits are prioritized. One must not ignore that these communities are most vulnerable to climate change. Balancing between these interests is a necessity in the 21st century Arctic security environment.

19.6 Militarization of the Arctic

The strategic value of the Arctic was recognized as early as the 1930s, when the US Army General Billy Mitchell told the Congress that "in the future, whoever holds Alaska will hold the world ... (the Arctic) is the most strategic place in the world." Further, during the Second World War, the German occupation of Norway, Spitsbergen, and Greenland and the establishment of weather monitoring stations there, as well as the Japanese occupation of Aleutian Islands were testament to the strategic importance of the region. The Arctic was perhaps strategically the most significant region during the Cold War period. The Arctic became a frontline in the Cold War as the distance between the northernmost points of Eurasia and North America is just 2000 km (Beixi, 2016: p. 164). Hence, the region was witness to regular patrol missions of the strategic submarine and bomber fleets of the superpowers. However, the Arctic changed from a theater of military and strategic competition to political, economic, and environmental cooperation with the speech by Mikhail Gorbachev in October 1987 at Murmansk (Purver, 1988: pp. 153–156). Hence, with the end of the Cold War, the military value of the Arctic dipped significantly, and countries started working on demilitarizing the region. One of the significant instances of collaboration in demilitarizing the region was the Arctic Military Environmental Cooperation (AMEC) program. It was formalized in 1996 between the United States, Norway, and Russia to mitigate the environmental fallouts of Cold War strategic deployments in the Arctic through the decommissioning of such facilities. However, this trend lasted hardly for a decade, after which geopolitical tensions started rising again. Countries started coming up with their national strategic interests and visions for the Arctic, with a focus on developing military deployments. Coincidentally, one of the casualties of this remilitarization was the AMEC, which was terminated, triggered by the expulsion of Norwegian officials from Russia for alleged espionage activities (Jokela, 2015: pp. 59–60).

In the 21st century the rising geopolitical assertion of Russia and China have led to concerns about their geostrategic approach[14] to the Arctic. The 2007 cyberattacks by Russia on Estonia, Russia's 2008 intervention in Georgia, and the 2014 annexation of Crimea by Russia indicated to the rest of the Arctic countries the increasingly muscular trajectory of Russia's power projection in the region. Russia had placed its flag on the Arctic seabed corresponding to the North Pole during a 2007 scientific expedition. This had raised widespread concerns across the region about Russia's geopolitical intent, even though it cannot be legally treated as a territorial claim. Subsequently, in 2008 the Arctic coastal states reiterated their commitment to the UNCLOS framework for formalizing their maritime territorial claims (van Efferink, 2011). Since 2008, Russia has been adopting a series of strategic documents calling for increased deployment of its military capabilities in the Arctic. Russia has been refurbishing a number of redundant Soviet military facilities and is modernizing its naval fleet with a substantial focus on Arctic deployment. Russia's Arctic-based Northern Fleet is its largest one and has the most number of icebreakers and strategic submarines as well as majority of its nuclear forces. Russia has more than 40 icebreakers, and many of them are nuclear-powered. The country is seeking to construct new heavy duty icebreakers, given the fact that a large part of

[14] The United States' move to come out of the Anti-Ballistic Missile Treaty in 2002 and pursue its missile defense program is substantially rooted in the Arctic, and is seen to have disturbed the global strategic stability in the post-Cold War era from the standpoint of Russia and China. This, along with the US-led series of interventionist and regime change operations by the turn of the 21st century in Eurasia has played a key role in bringing back a certain level of Cold War mentality, where the rising geopolitical clout of Russia and China as well as their regime security got threatened.

its existing inventory belongs to the Soviet era, especially the nuclear-powered ones (Congressional Research Service, 2019: pp. 42–83).

Apart from the naval modernization, Russia is also upgrading its strategic and air/missile defense systems. Between 2004 and 2006, Russia resumed regular submarine patrol and missile tests in the Arctic (Conley and Rohloff, 2015: p. 20). Russia has been setting up a number of radar systems in the Arctic region to reinforce its air/missile defense systems. Since 2007, strategic bombers have resumed air patrols over the Arctic region, and there have been increasing number of cases of Russian fighters intruding into Scandinavian, the United States, and Canadian airspaces in the Arctic (Lasserre and Tetu, 2016: p. 306). It has also established in 2014 a new Arctic Joint Strategic Command in Murmansk Oblast. Since 2015, the country has started deploying Arctic brigades in this region (Levis, 2015: pp. 17–19). Russia's exploitation of its energy potential has been instrumental in the country finding enough financial resources to spend on its military build-up in the Arctic. The opening up of the Arctic has unraveled the possibility for Russia to break out of its continental geography and re-emerge as a significant maritime power like the Soviet Union in the 1960s to 70 s.

Meanwhile, China has also increased its forays into the Arctic. In 2013 China accomplished the first ever voyage of a container ship through the NSR, demonstrating its intent of a strategic presence in the region. Out of the 27 vessels that transited the Arctic in 2017, 11 had either their origin or destination in China. The country's collaboration with Russia is becoming increasingly visible in the Arctic. For instance, China has struck an agreement with Russia to create, what is called an "Ice Silk Road" in 2017 (Ying, 2017). China has been very much proactive in courting Greenland and Iceland. It has been exploiting the former's disenchantment with its sovereign authority, Denmark, and the latter's economic crisis in the wake of the Global Financial Crisis to establish facilities with possible dual-use purposes (Auerswald, 2019). For instance, China's setting up of an observatory in Iceland, supposedly to study the northern lights, have raised concerns of being a possible surveillance project (Schreiber, 2018). Of late, the Arctic littoral states seem to have drawn their own red lines with regard to allowing China's influence. For instance, Iceland terminated the sale of a large plot of land to a Chinese investor in 2011. Similarly, Denmark rejected the sale of a redundant naval base in Greenland to a Chinese company, and Norway canceled the setting up of a Chinese resort in Svalbard, both in 2016 (Pezard, 2018: pp. 6–7).

Hence, when the former US President Donald Trump expressed his interest in buying Greenland from Denmark, an offer which the Danish government quickly rejected, the move needs to be understood in the context of growing strategic foothold of Russia and China in the Arctic (Lau and Elmer, 2019). It has been observed that the security dynamics of the Arctic since the Cold War has evolved "from militarization to de-militarization, and to re-militarization" (Beixi, 2016: p. 166).

The US military strategy in the Arctic has been seen to be largely determined by global strategic requirements rather than regional aspects. For instance, the United States installed the Ground-based Midcourse Interceptors for its National Missile Defence program at Fort Greely in Alaska in 2003. This event has been instrumental in bringing the strategic rivalry back to the Arctic, independent of the environmental change happening in the region. The United States has also upgraded its radar systems in Thule, Greenland, in this context (Jokela, 2015: p. 61). With the growing challenge from Russia and China, the Trump administration came up with certain steps to maintain its strategic interests in the region. The United States re-established its second fleet in 2018, after closing it down in 2011. The fleet, which was established in 1950, was instrumental for the US Navy to counter the moves of the Soviet Navy in the North Atlantic. The United States has also been contemplating about the conduct of Freedom of Navigation Operations in the Arctic to reinforce its legal navigational rights in the region (Schreiber, 2019). For airborne threats, the United States has activated antiaircraft systems in the Arctic targeting Russian long range bomber patrol in the region (Konyshev and Sergunin, 2012: p. 39). The US Department of Defence came up with its 2019 Arctic Strategy, following its 2013 and 2016 versions to bolster its military presence in the region for countering the moves by Russia and China in the region independently as well as in collaboration with its allies (Office of the Undersecretary of Defence for Policy, 2019). Hence, the Cold War era rivalry seems to be warming up with the melting of the Arctic.

Five of the eight Arctic states (the United States, Canada, Denmark, Iceland, and Norway) are members of the North Atlantic Treaty Organization (NATO). The Arctic held strategic significance for NATO during the Cold War as the northernmost front of its geopolitical rivalry with the Warsaw Pact. The Arctic is important for the NATO as it is the gateway to the North Atlantic. The end of the Cold War led to the receding role of the Arctic in NATO's strategic matrix. However, with the renewal of great power rivalry in the contemporary scenario and the rising profile of the Arctic, the role of the region has grown back into significance for the NATO (Paul, 2019). The geopolitical rivalry between Russia and NATO is therefore rising. The NATO has a strategic interest in the GIUK gap, the maritime area between Greenland, Iceland, and the United Kingdom, through which Russian naval vessels have to pass through to enter the Northern Atlantic Ocean. Consequently, the NATO has been attempting to re-establish its Atlantic Command, which was closed in 2002 (Regehr, 2017: p. 1). However, NATO's prospects in the region are hampered by disputes among its member states such as the United States, Canada, Denmark, and Norway (Konyshev and Sergunin, 2012: p. 42). Nevertheless, the growing assertiveness of Russia and China will most certainly only reinforce the relevance of NATO in the region. For instance, Canada, which has been resisting NATO's role in the Arctic, has taken an about turn since 2014 in the context of Russia's growing military capabilities in the region and beyond (Jokela, 2015: p. 64). Norway, in this regard, represents a frontline state for the NATO. In fact, the NATO held the largest military exercise in the post-Cold War era, called the Trident Juncture 2018 exercise held in the waters near Norway. The exercise saw the United States sending its aircraft carrier to the Arctic waters for the first time since the end of the Cold War (Rogoway, 2018).

Norway has adopted a policy of "High North, Low Tension" in its relationship with other Arctic states (Bentzen and Hall, 2017: p. 5). Nevertheless, it is not stopping the country from taking steps to ensure its defense. Norway and Russia, even though having resolved their Barents Sea dispute, are caught up in an ongoing issue regarding fishing in the region. Moreover, Norway has been continuously objecting to Russian fishing activities around Spitsbergen. This has led to Russia's Northern Fleet starting to conduct regular patrols around the area, leading to objections being raised by Norway (Closson, 2018). With growing threats from Russia's militarization, Norway in 2010 moved its military headquarters to Bodo, which is located north of the Arctic circle (Barents Observer, 2009). However, Norway has to do a delicate balancing act between maintaining strong ties with the NATO and at the same time keeping its ties with Russia from going awry. Norway has increased the number of the US soldiers that it will host on a rotational basis. This will help it boost its defense, but at the same time retain its understanding with Russia to not station foreign troops in is territory during peacetime (Pezard, 2018: p. 5). Sweden and Finland have to strike a fine balance between partnering with NATO and playing a role in the EU's Common

Security and Defense Policy and not antagonizing Russia. The security scenario in the European Arctic is complicated by the fact that there is no overarching organizational architecture for regional security. The consequence has been a vicious cycle of militarization, which is bound to grow with the strengthening of maritime commerce in the region in the coming decades.

19.7 India's options and the road ahead

India's role so far in the region has been restricted to the scientific arena. Geographically, the Arctic is distant from the country and even the opening up of shipping routes are clearly not as much of interest to India as to China, Japan, or South Korea, especially in light of India's southern location. The traditional shipping route from Europe to Asia via the Suez Canal, on which India's trade is based, is not going to fade away (Sinha, 2013: p. 29). India has established considerable experience in polar science and research. India is currently an observer in the Arctic Council and has asserted that unlike other countries, its interest in the Arctic region is only driven by scientific interests—to enhance its scientific expertise, "particularly its polar research capabilities" (Menon and Dikshit, 2013). Following decades of experience in scientific research in Antarctica, India set up its first research station in the Arctic in 2008—"Himadri" in Ny-Alesund, Spitsbergen, and Norway—to conduct studies on "glaciology, atmospheric sciences, and biological sciences." Since many studies have pointed toward the connection between the Indian monsoon and the Arctic atmospheric processes, the importance of monitoring effects of climate change in the Arctic has been acknowledged by many Indian scientists (Nayak, 2016: pp. 28–29) (Rajan and Krishnan, 2016: p. 43). Scientific cooperation and science diplomacy will continue to be a formidable pillar of India's engagement with the Arctic region in the future as well.

Several Indian analysts, including Saran (2013), have always called for the declaration of the Arctic as a "global commons." The Arctic is not a part of the global commons, but as more and more countries show their interest in its resources and commercial utility, it could be argued that the only way in which the region's environmental integrity and sustainability could be maintained would be to declare it as a global commons. Saran (2013) in fact has been critical of India's inception into the Arctic Council as an observer as according to him, with this move, India has "succumbed to the temptation of sharing in the emerging opportunities for resource extraction as the Arctic continues to melt because of global warming." He argues that India should push for a "global commons" approach in the region to ensure that harmful activities are not carried out in the Arctic as climate change and its related effects in the region are expected to have repercussions around the world. These effects range from sea level rise (due to loss of land ice) and impacts of release of methane gas hydrates to uncertainties associated with the Arctic circulation (impacts on global climate) and even the spread of new diseases (Woodgate, 2013; Van Wormer et al., 2019). However, this is likely to be struck down by the Arctic states who have been upholding their sovereignty fervently all this while, as explained in the section on territorial contestations in the region. Countries such as Canada and Russia have even gone ahead and claimed their exclusive (territorial) rights over "territorial waters" through which the NWP and NSR pass.

Just like the Antarctic, scientific cooperation seems to be guiding cooperation in the region but with shifts in the geopolitical landscape (concerning the Arctic) such as Russia-China friendship, India-Russia cooperation, and China's inroads into Nordic countries, the Arctic Council's task of maintaining stability and sustainability in the region is likely to be strained. Increasing commercialization and militarization in the region are leading to a Cold War-like situation, so much so that one of the arguments put forth by Indian scholars is that India should rather press for a military-free zone in the Arctic (Sinha, 2013: p. 29), using its observer status in the Arctic Council. In addition, India should also promote the creation of "resource management regimes" aimed at sustainable use of resources and ecological protection in the region as well as greater coordination between the Asian countries through an "Asia Arctic Forum of Asian," focussing on scientific research (Sinha and Gupta, 2014: p. 883). At the same time, there is a consensus that India should also develop a comprehensive assessment of the opportunities arising in the Arctic and needs to adopt a strategy that allows India access to the region, even if it is for material gains. In this context, it should be noted that India, despite geographically being far removed from the Arctic, is not without any experience in military operations under similar conditions. This is because India has its armed forces stationed in the highest battlefield in the world, the Siachen glacier, battling the polar-like conditions.

India's geoeconomic interests in the Arctic have been largely downplayed due to its lack of resources to take the plunge with the same vigor and intensity as some of the other Asian countries, primarily China. However, India's aspirations to be a global power to reckon with are to an extent contingent on its capacity to place itself strategically in many theaters, including the Arctic. India is therefore focussing on cooperation with the Arctic states such as Russia and Norway. One of the ways that is often recommended by analysts for boosting India's engagement with the region is to involve its immense human resources and private sector in infrastructural development in the region in various sectors, including energy (Nanda, 2019; Lackenbauer, 2017; Sakhuja, 2010). India's cooperation with Russia is also critical to the former's energy diplomacy, which is driven by the need for diversifying its sources and investing more in natural gas to meet its burgeoning energy demand. More recent developments show that India's interest is no longer just restricted to scientific research, but has gradually expanded into more strategic aspects of the Arctic too. During Prime Minister Modi's visit to Vladivostok, Russia in 2019, explicit indications were made toward greater interaction between India and Russia on energy (Liquefied Natural Gas or LNG) cooperation. It is directed at augmenting LNG supplies from Arctic LNG 2 project and the future project on Arctic LNG 3; as well as possible "geological exploration, including on the shelf, as well as Russian energy resource deliveries to India with possible use of the Northern Sea Route" (Staalesen, 2019).

As an observer nation, India's options are definitely limited. However, there are some steps that can be taken in partnership with the Arctic countries. It has to be noted that improving bilateral relations is the key to access the Arctic in a holistic manner. The first step is to formulate and implement an "Arctic policy," not only to compete with other Asian countries but also to clearly state India's intentions in the Arctic and the manner in which it would carry out those stated goals (Ramesh, 2018). This has been partly fulfilled by the release of India's Arctic Policy Roadmap For Sustainable Engagement draft, in which the focus areas are "scientific research, economics and human development; connectivity; global governance and international cooperation; and development of Indian human resource capabilities" (Rej, 2021).[15] Building awareness is one of the primary steps in this regard. Speculations of why the Arctic

[15] At the time of writing this chapter, India's draft Arctic policy had not been released. Hence, it does not delve deeper into the draft policy.

is important for India can be answered through such initiatives. Encouraging a multidisciplinary approach to study the polar regions and simultaneously building collaborative academic networks is the need of the hour. While scientific research can continue to be the priority for India, in the medium and long terms, India can think of gaining access to the resources of the region—both renewable and nonrenewable. India can go to the Arctic by connecting the International North South Transport Corridor with the European-Baltic railway networks. This is probably the shortest route for India to reach the High North (Menezes and Sharma, 2018).

In a nutshell, there are primarily three pathways in which the Arctic's future could be visualized, when seen through the prism of climate change—"global commons," a zone of reinvigorated diplomacy amidst (and in recognition of) increasing geopolitical and geoeconomic competition, or a Cold War-like situation, reflective of protracted rivalries, and a zero-sum game. Out of the three, the second pathway seems most plausible at this stage as Arctic states, non-Arctic states and indigenous communities find common grounds to cooperate to prevent the existing competition and rivalries from blowing out. The Central Arctic Ocean agreement and the Russia-Norway Svalbard Treaty among others signify the intent among the concerned stakeholders to de-escalate tensions and reach agreements. This does not mean that enhanced maritime activity (resources and shipping) would not lead to disputes and tensions at all. Trigger points are many but the fact of the matter is that as of now, countries have more reasons to cooperate than confront due to the conditions in the Arctic. With heightened commercial, economic, and trade activities, issues such as prevention of oil spills, fisheries management, and search and rescue assume greater significance. These issues require cooperation, without which countries would find it difficult to utilize the Arctic's resources or the new shipping routes. However, with countries engaging in rebuilding military facilities in the region and beefing up capabilities targeted at the Arctic, miscalculated moves within the region escalating to a bigger conflict or spill-over effects of conflicts/disagreements elsewhere being transmitted to the Arctic cannot be ruled out. In such a situation, countries also would need to engage in confidence-building measures to avoid or prevent conflict in the long term. India may not be as big a player in the Arctic as others, but it can strive to create cooperative frameworks and promote sustainable development, as its footprint in the region increases.

References

Auerswald, D., 2019. China's multifaceted Arctic strategy. War on the rocks. Available from: https://warontherocks.com/2019/05/chinas-multifaceted-arctic-strategy/. Accessed on 5 January 2020.

Balton, D.A., Ulmer, F., 2019. A Strategic plan for the Arctic council: recommendations for moving forward. Belfer Center for Science and International Affairs and Wilson International Center for Scholar Paper. Available from: https://www.belfercenter.org/publication/strategic-plan-arctic-council-recommendations-moving-forward. Accessed on 5 January 2020.

Barents Observer, 2009. Norwegian army moves north. Available from: https://barentsobserver.com/en/node/18103. Accessed on 5 January 2020.

Barry, T., et al., 2018. Arctic biodiversity assessment. Conservation of Arctic flora and fauna, Borgir.

Antrim, C.L., 2010. The next geographical pivot. Nav. War. Coll. Rev. 63 (3). Available from: https://digital-commons.usnwc.edu/nwc-review/vol63/iss3/3. Accessed on 5 January 2020.

Beixi, D., 2016. Arctic security: evolution of Arctic security dynamics and prospect for a security regime in the Arctic. Adv. Polar Sci. 27 (3), 163–169.

Bentzen, N., Hall, M., 2017. Arctic continental shelf claims: mapping interests in the circumpolar north. European Parliament Members' Research Service Briefing PE 595.870. Available from: http://www.europarl.europa.eu/thinktank/en/document.html?reference = EPRS_BRI(2017)595870. Accessed on 5 January 2020.

Bloom, E.T., 1999. Establishment of the Arctic Council. Am. J. Int. Law 93 (3), 712–722. Available from: https://doi.org/10.2307/2555272.

Brutschin, E., Schubert, S.R., 2016. Icy waters, hot tempers, and high stakes: geopolitics and geoeconomics of the Arctic. Energy Res. Soc. Sci. 16, 147–159. Available from: https://doi.org/10.1016/j.erss.2016.03.020.

Chaturvedi, S., 2000. Arctic geopolitics then and now. In: Nuttall, M., Callaghan, T.V. (Eds.), The Arctic: Environment, People, Policy. Routledge, New York, pp. 441–458.

Chaturvedi, S., 2014. India's Arctic engagement: challenges and opportunities. Asia Policy 18 (1), 73–79. Available from: https://doi.org/10.1353/asp.2014.0037.

Closson S., 2018. Good fences make good neighbors: Russia and Norway's Svalbard. Kennan Cable No. 37. Kennan Institute. Available from: https://www.wilsoncenter.org/publication/kennan-cable-no-37-good-fences-make-good-neighbors-russia-and-norways-svalbard. Accessed on 5 January 2020.

Congressional Research Service, 2019. Changes in the Arctic: background and issues for congress. Report for Congress R41153. Available from: https://fas.org/sgp/crs/misc/R41153.pdf. Accessed on 5 January 2020

Conley H.A., Rohloff C., 2015. Ice curtain: Russia's strategic reach to the Arctic. CSIS Report. Available from: https://csis-prod.s3.amazonaws.com/s3fs-public/legacy_files/files/publication/150826_Conley_NewIceCurtain_Web.pdf. Accessed on 5 January 2020.

Dalaklis, D., Drewniak, M.L., Schröder-Hinrichs, J., 2018. Shipping operations support in the "High North": examining availability of icebreakers along the Northern Sea Route. WMU J. Marit. Aff. 17, 129–147. Available from: https://doi.org/10.1007/s13437-018-0142-7.

Dalby, S., 2014. Environmental geopolitics in the twenty-first century. Alternatives: Global, Local, Political 39 (1), 3–16.

Descamps, M., 2019. The ice silk road: is china a "Near-Arctic-State"? The Institute for Security and Development Policy, Focus Asia: Perspective and Analysis. Available from: https://isdp.eu/content/uploads/2019/02/The-Ice-Silk-Road-is-China-a-Near-Arctic-State.pdf. Accessed on 5 January 2020.

Dodds, K.J., 2012. Anticipating the Arctic and the Arctic Council: pre-emption, precaution and preparedness. Polar Rec. 49 (2), 193–203. Available from: https://doi.org/10.1017/s0032247412000198.

Drent, J., 1993. Commercial fishing on the northern sea route. North. Mariner/Le Mar. Du. Nord. 3 (2), 1–17.

Dyer, B., 1940. Robert J. Walker on acquiring greenland and iceland. Miss. Val. Historical Rev. 27 (2), 263–266.

Exner-Pirot, H., Ackrén, M., Loukacheva, N., Nicol, H., Nilsson, A.E., 2019. Form and Function: The Future of the Arctic Council. The Arctic Institute. Available from: https://www.thearcticinstitute.org/form-function-future-arctic-council/, Accessed on 5 January 2020.

Farré, A.B., Stephenson, S.R., Chen, L., Czub, M., et al., 2014. Commercial Arctic shipping through the Northeast Passage: routes, resources, governance, technology, and infrastructure. Polar Geography 37 (4), 298–324. Available from: https://doi.org/10.1080/1088937X.2014.965769.

Finger, M., 2016. The Arctic, laboratory of the anthropocene. In: Heininen, L. (Ed.), Future Security of the Global Arctic: State Policy, Economic Security and Climate. Palgrave Pivot, London, pp. 121–137.

Fisheries and Ocean, Government of Canada, 2018. International Agreement to Prevent Unregulated High Seas Fisheries in the Central Arctic Ocean. Available from: https://www.dfo-mpo.gc.ca/international/arctic-arctique-eng.htm. Accessed on 5 January 2020.

Gassiy, V., 2018. Indigenous communities in the Arctic change in socio-economic and environmental perspective. In: Kanao, M. (Ed.), Arctic Studies—A Proxy for Climate Change. IntechOpen. Available from: https://doi.org/10.5772/intechopen.80765.

Goldfarb, B., 2017. Feeling the heat: how fish are migrating from warming waters. Yale Environment 360. Available from: https://e360.yale.edu/features/feeling-the-heat-warming-oceans-drive-fish-into-cooler-waters. Accessed on 5 January 2020.

Ho, J., 2011. The Arctic meltdown and its implication for ports and shipping in Asia. In: Kraska, J. (Ed.), Arctic Security in an Age of Climate Change. Cambridge University Press, New York, pp. 33–44.

Holmes, S., 2008. Breaking the ice: emerging legal issues in Arctic Sovereignty. Chic. J. Int. Law 9 (1), 323–351.

Hong, N., 2018. China's Interests in the Arctic: Opportunities and Challenges. Institute for China-America Studies. Available from: https://chinaus-icas.org/wp-content/uploads/2018/03/2018.03.06-China-Arctic-Report.pdf, Accessed on 5 January 2020.

House of Commons Environmental Audit Committee, 2018. The changing Arctic. Twelfth Report of Session 2017–19. Available from: https://publications.parliament.uk/pa/cm201719/cmselect/cmenvaud/842/842.pdf. Accessed on 5 January 2020.

Hunkins, K., 1969. Arctic geophysics. Arctic 22 (3), 169–364. Available from: https://doi.org/10.14430/arctic3216.

2008 Ilulissat Declaration, 2008. Available from: https://cil.nus.edu.sg/wp-content/uploads/2017/07/2008-Ilulissat-Declaration.pdf. Accessed on 5 January 2020.

Inuit Circumpolar Council, 2019. Inuit at COP25—Our Ice is Melting, Our Oceans are at Risk, Our Planet is on Fire. Available from: https://www.inuitcircumpolar.com/news/inuit-at-cop25-our-ice-is-melting-our-oceans-are-at-risk-our-planet-is-on-fire/. Accessed on 5 January 2020.

Jokela, J., 2015. Arctic security matters. EU Institute for Security Studies Report 24. Available from: https://www.iss.europa.eu/sites/default/files/EUISSFiles/Report%2024.pdf. Accessed on 5 January 2020.

Käpylä, J., Mikkola, H., 2016. The promise of the geoeconomic Arctic: a critical analysis. Asia Europe J. 14, 203–220. Available from: https://doi.org/10.1007/s10308-015-0447-5.

Konyshev, V., Sergunin, A., 2012. The Arctic at the crossroads of geopolitical interests. Russian Politics Law 50 (2), 34–54.
Kriz, Z., Chrastansky, F., 2012. Existing conflicts in the Arctic and the risk of escalation: rhetoric and reality. Perspectives 20 (1), 111–139.
Lackenbauer, W., 2017. India and the Arctic: revisionist aspirations, Arcticrealities. Jindal Glob. Law Rev. 8 (1), 23–54. Available from: https://doi.org/10.1007/s41020-017-0040-4.
Lagutina, M.L., 2019. Russia's Arctic Policy in the Twenty-First Century: National and International Dimensions. Lexington Books, London.
Lalonde, S., Lasserre, F., 2013. The position of the United States on the northwest passage: is the fear of creating a precedent warranted? Ocean. Dev. Int. Law 44 (1), 28–72.
Lasserre, F., Tetu, P., 2016. Russian air patrols in the Arctic: are long-range bomber patrols a challenge to Canadian security and Sovereignty? Arctic Yearbook 2016. Available from: https://arcticyearbook.com/images/yearbook/2016/Scholarly_Papers/11.Lasserre-and-Tetu.pdf. Accessed on 5 January 2020.
Lau, S., Elmer, K., 2019. Did China's growing presence in Arctic prompt Donald Trump's offer to buy Greenland? South China Morning Post. Available from: https://www.scmp.com/news/china/diplomacy/article/3025207/did-chinas-growing-presence-arctic-prompt-donald-trumps-offer. Accessed on 5 January 2020.
Lee, S.W., Song, J.M., 2014. Economic possibilities of shipping though Northern sea route. Asian J. Shipping Logist. 30 (3), 415–430. Available from: https://doi.org/10.1016/j.ajsl.2014.12.009.
Levis, A., 2015. Russia and the Arctic. George Mason University Systems Architectures Laboratory Report. Available from: https://nsiteam.com/social/wp-content/uploads/2015/12/Arctic-GMU.pdf. Accessed on 5 January 2020.
Luttwak, E.N., 1990. From geopolitics to geo-economics: logic of conflict, grammar of commerce. Natl Interest. 20, 17–23.
Menezes, D., Sharma, H., 2018. The International North-South Transport Corridor (INSTC): India's Grand Plan for Northern Connectivity. The Polar Connection. Available from: http://polarconnection.org/india-instc-nordic-arctic/. Accessed on 1 February 2020.
Menon, M., Dikshit, S., 2013. India gets observer status in Arctic Council. The Hindu. Available from: https://www.thehindu.com/news/international/world/india-gets-observer-status-in-arctic-council/article4717770.ece. Accessed on 5 January 2020.
Menshikova, M., 2019. The Arctic during World War II: three polar stations bring victory closer. The Arctic. Available from: https://arctic.ru/infographics/20190621/864748.html. Accessed on 5 January 2020.
Meredith, M., Sommerkorn, M., Cassotta, S., Derksen, C., Ekaykin, A., Hollowed, A., et al. (2019) Polar regions. In: Pörtner, H.-O., Roberts, D.C., Masson-Delmotte, V., Zhai, P., Tignor, M., Poloczanska, E., Mintenbeck, K., Alegría, A., Nicolai, M., Okem, A., Petzold, J., Rama, B., Weyer, N.M. (Eds), IPCC special report on the ocean and cryosphere in a changing climate. In press.
Ministry of Europe and Foreign Affairs, 2016. The Great Challenge of the Arctic. National Roadmap for the Arctic. Available from: https://www.diplomatie.gouv.fr/IMG/pdf/frna_-_eng_-interne_-_prepa_-_17-06-pm-bd-pdf_cle02695b.pdf. Accessed on 5 January 2020.
Moe, A., Stokke, O.S., 2019. Asian countries and Arctic shipping: policies, interests and footprints on governance. Arct. Rev. 10, 24–52. Available from: https://doi.org/10.23865/arctic.v10.1374.
Nanda, D., 2019. India's Arctic potential. Observer Research Foundation Occasional Paper 186. Available from: https://www.orfonline.org/research/indias-arctic-potential-48263/. Accessed on 5 January 2020.
Nayak, S., 2016. Balancing development and environmental concerns in the Arctic. In: Sakhuja, V., Narula, K. (Eds.), Asia and the Arctic: Narratives, Perspectives and Policies. Springer, Singapore, pp. 27–32.

Nuttall, M., 1997. The Arctic Council: can cooperation be sustained? Polar Rec. 33 (185), 99–100. Available from: https://doi.org/10.1017/s0032247400014418.
Office of the Undersecretary of Defence for Policy, 2019. Arctic strategy. Department of defence report to congress. Available from: https://media.defense.gov/2019/Jun/06/2002141657/-1/-1/1/2019-DOD-ARCTIC-STRATEGY.PDF. Accessed on 5 January 2020
Østhagen, A., 2018. Managing conflict at sea: the case of Norway and Russia in the Svalbard zone. Arct. Rev. Law Politics 9, 100–123. Available from: http://doi.org/10.23865/arctic.v9.1084.
Parfitt T., 2007. Russia plants flag on North Pole seabed. The Guardian. Available from: https://www.theguardian.com/world/2007/aug/02/russia.arctic. Accessed on 5 January 2020
Pezard S., 2018. The New Geopolitics of the Arctic: Russia's and China's Evolving Role in the Region. RAND Corporation Testimonies. Available from: https://www.rand.org/content/dam/rand/pubs/testimonies/CT500/CT500/RAND_CT500.pdf. Accessed on 5 January 2020.
Purver, R., 1988. Arctic security: the Murmansk initiative and its impact. Curr. Res. Peace Violence 11 (4), 147–158.
Quinn E., 2019. Canada files submission to establish continental shelf outer limits in Arctic Ocean. The Barents Observer. Available from: https://thebarentsobserver.com/en/arctic/2019/05/canada-files-submission-establish-continental-shelf-outer-limits-arctic-ocean. Accessed on 5 January 2020.
Rajan, S., Krishnan, K.P., 2016. India's scientific endeavours in the Arctic. In: Sakhuja, V., Narula, K. (Eds.), Asia and the Arctic: Narratives, Perspectives and Policies. Springer, Singapore, pp. 43–48.
Ramesh, R. (2018) India's Arctic engagement: shifting from scientific to strategic interests? South Asian Voices. Available from: https://southasianvoices.org/indias-arctic-engagement-shifting-from-scientific-to-strategic-interests/. Accessed on 1 February 2020.
Regehr E., 2017. Shielding the Arctic from NATO's return to territorial defence. The Simons Foundation Arctic Security Briefing Papers. Available from: http://www.thesimonsfoundation.ca/highlights/shielding-arctic-natos-return-territorial-defence. Accessed on 5 January 2020.
Rej, A., 2021. India Releases Draft Arctic Policy. The Diplom. Available from: https://thediplomat.com/2021/01/india-releases-draft-arctic-policy/. Accessed on 20 March 2021.
Rockström, J., et al., 2009. A safe operating space for humanity. Nature 461, 472–475.
Rogoway T., 2018. U.S. carrier enters Arctic circle For the first time in 28 years as new Cold war heats up. The Drive. Available from: https://www.thedrive.com/the-war-zone/24373/u-s-carrier-enters-arctic-circle-for-the-first-time-in-28-years-as-new-cold-war-heats-up. Accessed on 5 January 2020.
Rossiya Segodnya, n.d. Natural resources. Available from: https://arctic.ru/resources/. The Arctic. Accessed on 11 January 2020.
Rottem, S.V., 2020. The Arctic Council: Between Environmental Protection and Geopolitics. Palgrave Pivot, London.
Roucek, J.S., 1983. The geopolitics of the Arctic. Am. J. Econ. Sociol. 42 (4), 463–471. Available from: https://doi.org/10.1111/j.1536-7150.1983.tb01733.x.
Sakhuja, V., 2010. China: breaking into the Arctic Ice. Indian Council of World Affairs Issue Brief. Available from: https://icwa.in/showfile.php?lang = 1&level = 3&ls_id = 944&lid = 887. Accessed on 11 January 2020.
Saran, S., 2013. India's date with the Arctic. The Hindu. Available from: https://www.thehindu.com/opinion/op-ed/indias-date-with-the-arctic/article4915241.ece. Accessed on 5 January 2020.
Scholvin, S., Wigel, M., 2018. Geo-Economics as Concept and Practice in International Relations:Surveying the State of the Art. Centre for Security Studies, ETH Zürich.

Available from: https://css.ethz.ch/en/services/digital-library/articles/article.html/dd73604f-ffee-44e2-8960-4ecd8927fa32/pdf, Accessed on 11 January 2020.

Schreiber, M., 2018. A new China-Iceland Arctic science observatory is already expanding its focus. Arctic Today. Available from: https://www.arctictoday.com/new-china-iceland-arctic-science-observatory-already-expanding-focus/. Accessed on 5 January 2020.

Schreiber, M., 2019. The US Navy's revived 2nd Fleet, with a focus on the North Atlantic and nearby Arctic, is now operational. Arctic Today. Available from: https://www.arctictoday.com/the-us-navys-revived-2nd-fleet-with-a-focus-on-the-north-atlantic-and-nearby-arctic-is-now-operational/. Accessed on 5 January 2020.

Sinha, U.K., 2013. India and the 'Age of the Arctic'. Indian Foreign Aff. J. 8 (1), 23–30. Available from: http://www.associationdiplomats.org/Publications/ifaj/Vol8/8.1/DEBATE.pdf. Accessed on 5 January 2020.

Sinha, U.K., Gupta, A., 2014. The Arctic and India: strategic awareness and scientific engagement. Strategic Anal. 38 (6), 872–885. Available from: https://doi.org/10.1080/09700161.2014.952945.

Śmieszek, M., 2019. Do the cures match the problem? Reforming the Arctic Council. Polar Rec 55 (3), 121–131. Available from: https://doi.org/10.1017/s0032247419000263.

Śmieszek, M., Koivurova, T., 2017. The Arctic council: between continuity and change. In: Lackenbauer, P.W., Nicol, H., Greaves, W. (Eds.), One Arctic. Canadian Arctic Resources Committee and Centre on Foreign Policy and Federalism. Canadian Arctic Resources Committee and the Centre on Foreign Policy and Federalism, Ottawa and Waterloo, pp. 1–26.

Staalesen, A., 2019. India looks towards Russian Arctic. The Barents Observer. Available from: https://thebarentsobserver.com/en/arctic-industry-and-energy/2019/09/india-looks-towards-russian-arctic. Accessed on 5 January 2020.

The Arctic Institute, 2019. Singapore. Available from: https://www.thearcticinstitute.org/countries/singapore/. Accessed on 5 January 2020

The State Council Information Office of the People's Republic of China, 2018. China's Arctic Policy. White Paper. Available from: http://english.www.gov.cn/archive/white_-paper/2018/01/26/content_281476026660336.htm. Accessed on 5 January 2020.

Tseng, P.H., Cullinane, K., 2018. Key criteria influencing the choiceof Arctic shipping: a fuzzy analytic hierarchy process model. Marit. Policy Manag. 45 (4), 422–438. Available from: https://doi.org/10.1080/03088839.2018.1443225.

US Department of Defence, 2019. Arctic Strategy. Report to Congress. Available from: https://media.defense.gov/2019/Jun/06/2002141657/-1/-1/1/2019-DOD-ARCTIC-STRATEGY.PDF. Accessed on 5 January 2020.

US Geological Survey, 2008. Circum-Arctic resource appraisal: estimates of undiscovered oil and gas north of the Arctic circle. Available from: https://pubs.usgs.gov/fs/2008/3049/fs2008-3049.pdf. Accessed on 5 January 2020.

Van Wormer, E., Mazet, J.A.K., Hall, A., et al., 2019. Viral emergence in marine mammals in the North Pacific may be linked to Arctic sea ice reduction. Sci. Rep. 9 (15569). Available from: https://doi.org/10.1038/s41598-019-51699-4.

Woodgate, R., 2013. Arctic ocean circulation: going around at the top of the world. Nat. Educ. Knowl. 4 (8), 1–13.

Ying, 2017. China, Russia agree to jointly build "Ice Silk Road". Xinhua News Agency. Available from: http://www.xinhuanet.com//english/2017-07/04/c_136417241.htm. Accessed on 5 January 2020.

Young, O., 2009. The Arctic in play: governance in a time of rapid change. Int. J. Mar. Coast. Law 24 (2), 423–442. Available from: https://doi.org/10.1163/157180809x421833.

Young, O., 2016. The Arctic council at twenty: how to remain effective in a rapidly changing environment. U.C. Irvine Law Rev. 6, 99–119. Available from: https://scholarship.law.uci.edu/ucilr/vol6/iss1/5/.

Further reading

van Efferink, L., 2011. Arctic geopolitics: Russia's territorial claims, UNCLOS, The Lomonosov Ridge. Exploring Geopolitics. Available from: https://exploringgeopolitics.org/publication_efferink_van_leonhardt_arctic_geopolitics_russian_territorial_claims_unclos_lomonosov_ridge_exclusive_economic_zones_baselines_flag_planting_north_pole_navy/. Accessed on 5 January 2020.

Paul, M., 2019. Polar power USA: full steam ahead into the Arctic. SWP Comment 42. Available from: https://www.swp-berlin.org/fileadmin/contents/products/comments/2019C42_pau.pdf. Accessed on 5 January 2020.

CHAPTER 20

Glacial mass change–induced earthquakes in the Himalayan region of South Asia and its bearing to understand Arctic glaciers dynamics: proxy of climate change

O.P. Mishra[1], Neloy Khare[2], Sweta Baidya Das[3], Vikas Kumar[1], Jagvir Singh[2], Vandana[1], Priya Singh[1], M. Ghatak[4], Shashank Shekhar[1], Anurag Tiwari[1], Sasi Kiran Gera[1], Ravikant Mahto[1] and Poorti Gusain[1]
[1]National Centre for Seismology, Ministry of Earth Sciences, New Delhi, India
[2]Ministry of Earth Sciences, Govt. of India, New Delhi, India
[3]National Institute of Disaster Management, New Delhi, India
[4]DGCO, Geological Survey of India, New Delhi, India

20.1 Introduction

Most interestingly, glaciers simply represent the existence of ice on a landscape for years, and it is defined as a thicker ice mass, which originates through the accumulation of ice masses that undergo subsequent compaction and recrystallization of snow. Different stages of ice with respect to glacial stratigraphy were found to occur in the region as shown in Fig. 20.1.

At present, climate change is one of the most discoursed topics, especially several factors that influence processes involved with climate change. The most anticipated consequence of climate change is manifested by a change in temperature, precipitation, sea-level rise, and global warming, which causes uncontrollable glacial melts in the Polar regions and the region which is associated with appreciable numbers of glaciers outside the Polar region, which is none other than the Himalayas, the third pole of the Earth as the standing water tower for South Asia (Mishra, 2014). It is a known fact that the Northern Hemisphere is land-dominated, the snow cover or the mountain glaciers that acts as a crucial factor in surface energy

Understanding Present and Past Arctic Environments
DOI: https://doi.org/10.1016/B978-0-12-822869-2.00025-6

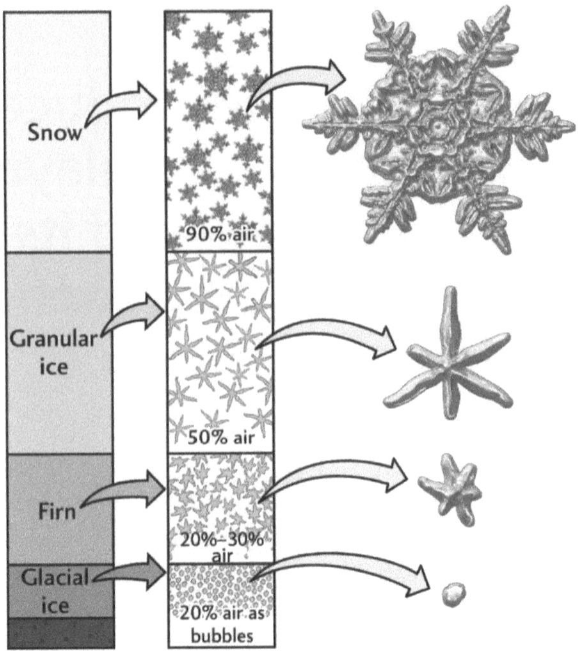

Figure 20.1 A schematic diagram shows the stratigraphy of the glacier. *Adapted from Rana, A.S., 2018. Summer School on Integrated Water Resources Management, PMD, Pakistan.*

budget and in the global hydrological system (Groisman et al., 1994). Snow is potentially sensitive to climate change (Barnett et al., 2005; Immerzeel et al., 2009) and regulates global climate variability (Frei et al., 2012). It is a proven fact that the livelihood of mountainous cities and villages is mostly dependent on the glaciers as it provides water for drinking and other household needs, irrigation, and also in some places, hydroelectricity is generated by damming the glacial meltwater (Bookhagen and Burbank, 2010; Lutz et al., 2014). Therefore any changes in the snow cover will directly influence the mountain river discharge (Beniston, 2003; Mukhopadhyay and Khan, 2015). Not only this, the snow cover of the Tibetan plateau regulates the atmospheric circulations and, therefore, the Asian monsoon system will be affected by that atmospheric circulation (Wu and Zhang, 1998; Bansod et al., 2003; Qian et al., 2011).

Coupled model intercomparison project (CMIP5) shows a future projection of snowfall reduction from 50% to 60% in the Ganges Basin, 50% to 70% in the Brahmaputra Basin, and 30% to 50% in the Indus Basin by 2071–2100 (Viste and Sorteberg, 2015). It is also projected that the snow depth of the Himalayas would reduce by 25%–50% and in the Hindu Kush–Karakoram by 17%–39% (Terzago et al., 2014). Future projections also show that the snow line elevations are also going to rise between 400 and 900 m at the rate varying from 4.4 to 10 m per year (Viste and Sorteberg, 2015), which is indeed an alarming situation for the survival of the flora and faunas. Further studies have shown that in Holocene temporal clustering of seismic events coincides with glacial advances (Ghazoui et al., 2019 thesis). Changes in ice cap volume alter the load acting on the lithosphere by significantly modifying its stress field; such stress field variations deform the earth's crust and especially the slip behavior of existing faults (Hetzel and Hampel, 2005; Hampel et al., 2009). They have shown that glacier retreat significantly increases the slip rate and vice versa in the case of normal and thrust faults located below the glacier/ice cap. High-frequency and low-amplitude climatic variations like variation in water storage and associated surface loading produce elastic deformation of the lithosphere (Argus et al., 2014; Amos et al., 2014; Borsa et al., 2014; Johnson et al., 2018), and it is found that seasonal water storage impacts and modulates the stress field and consequently the seismicity in California. Variation in climatic processes like changes in precipitation timing and rate will modify pore fluid pressure, resulting in an altered coefficient of friction of faults, which will affect the slip rate directly, ultimately resulting in an increased seismic activity (Vernant and Chery, 2006; Godard and Burbank, 2011).

Increased glacial melt vis-à-vis crustal unloading can potentially increase the frequency of low-to-medium magnitude earthquakes, and an integrated research is needed concerning this (Mishra, 2014). Studies on polar glaciers have already demonstrated the glacial melt inducing seismic activities or earthquakes, but there are very rare studies on the effects of Himalayan glacial melt on the Himalayan earthquakes. Initial study on Himalayan region has revealed the correlation between glacier melt and seismicity. At present, there is no detailed study on interrelationship between glacial melts in the Himalayan region and seismogenesis associated with modulation in the rate of change of seismicity in the glacial region of India. Therefore this paper aims to develop an understanding of the underlying processes of the complex climatological and geophysical interactions to assess the potential threat posed by climate change in the form of glacioisostasy-induced seismic activity in the third pole, The Himalaya. The outcome of this piece of research will enrich our understanding of intricate issues of seismogenesis in mountainous regions and the role of glacial melting in seismogenesis.

Different types of glaciers (Fig. 20.2) are found to be distributed all over the world, including the Asian region as shown in Fig. 20.3, which are having a different style and degree of melting under given climatologically and meteorological conditions that control the extent of crustal loading, causing the stress perturbation.

The glaciers, distributed in macroregions of the world, are well evident as shown in Fig. 20.3. These have different potentials of water resources, and the nature and extent of ice melting are also reported to be different because of varying conditions of natural and anthropological factors, including the changes in temperature and tectonic dynamism of the areas, which, in turn, demonstrates differential stress perturbation depending upon the extent of crustal loading due to mass changes associated with melting of ice sheets in different parts of the globe. As mentioned before, the cryospheric resources of South Asia and its adjacent region (Fig. 20.4) are corroborated with uneven distribution as shown in Table 20.1, indicating that the Himalaya is susceptible to glacial mass changed–induced subsurface earthquakes of varying frequency and intensity, depending upon the areal and

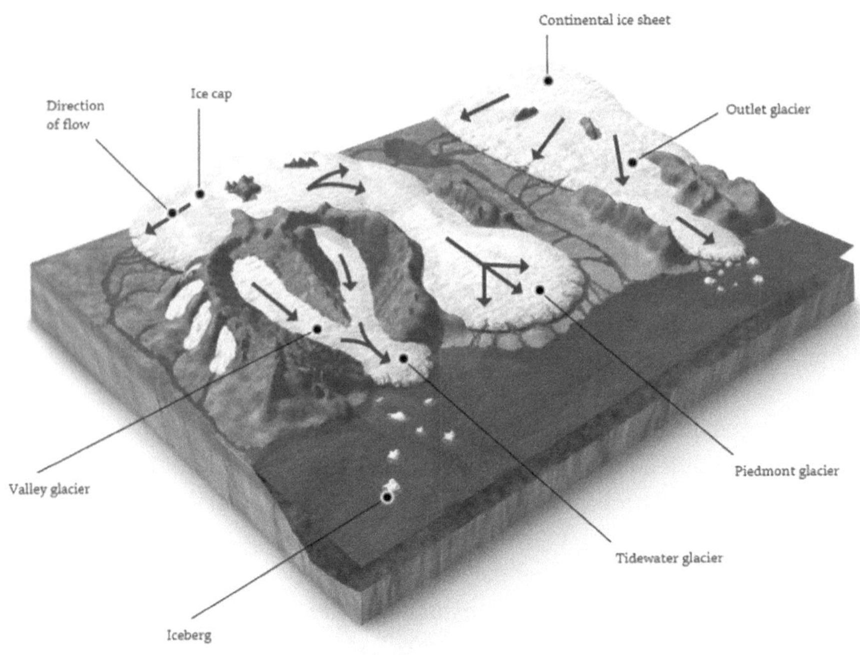

Figure 20.2 Different types of glaciers. *Adapted from http://www.dkfindout.com/us/earth/glaciers/types-glacier/.*

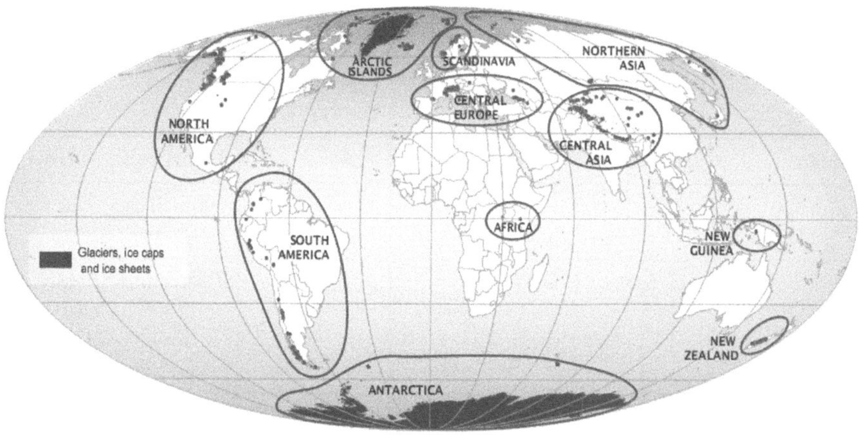

Figure 20.3 World's glaciated macroregions. *Adapted from Rana, A.S., 2018. Summer School on Integrated Water Resources Management, PMD, Pakistan.*

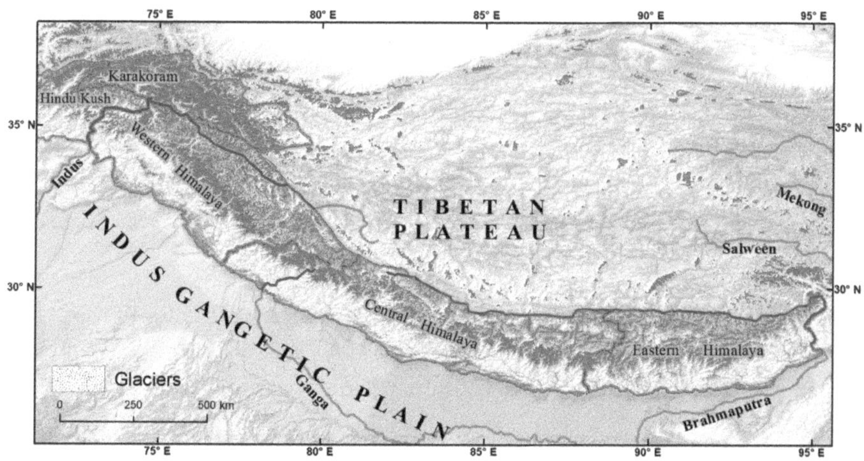

Figure 20.4 Distribution of glaciers in the Himalayan region of South Asia.

volumetric changes of ice masses either in terms of accumulation or of melting of ice sheets.

Some of the important glaciers located in South Asia and its adjacent region are shown in Fig. 20.5, demonstrating their impacts on locations having enough capability to change lithostatic stresses.

Due to the Indian and Eurasian Plate movement, the mountain building activity (orogenesis) of the Himalayas is continuously in progress. Therefore the area surrounding the Himalayas, like the northern states of

Table 20.1 Cryospheric resources of South Asia.

Country	Number of glaciers	Area of glaciers (km^2)	Volume of glaciers (km^3)	References
Bhutan	1583	~1487 ± 235	NA	Nagai et al. (2016)
	885	~642 ± 16.1	NA	Bajracharya et al. (2014)
India	9575	~37,466	~3371.94	Raina and Srivastava (2008), Geological Survey of India (1999)
Nepal	3252	~5323	~481	Bajracharya et al. (2006), Gurung (2005)
Tibet	1578	~2864	NA	Bajracharya et al. (2006)
China (including Tibetan Plateau)	46 298	~59,406	~5590	Liu et al. (2000)
Pakistan	5218	~15,040.7	~2738.5	Campbell and Pradesh (2005)

India and other neighboring countries like Nepal, Bhutan, Tibet, and China are continuously experiencing the seismic activity of the Himalayas. The purpose of this study is to understand the effect of deglaciation on the seismic activity of the Indian Himalayan region. Therefore the locations of the data used for the present study incorporate five North Indian states such as Jammu & Kashmir, Himachal Pradesh, Uttarakhand, Sikkim, and Arunachal Pradesh. Along with these two other countries, Nepal and Bhutan are also included to get an uninterrupted picture of the entire North Indian Himalayan Arc (see study area map). The geographical extent of each state and country is given in Table 20.2.

It is significant to explore the connections between cryosphere regions. Indubitably, no two regions have more to benefit from such a teleconnection than the Arctic and the Himalayas. Albeit, the two regions are geographically distinct and situated at far distance with their own unique characteristics. While the Arctic is an ocean surrounded by land and

Glaciers of South Asia

1-Baltaro Glacier (Hunza Valley, Pakistan); 2-Yala Glacier (Langtang Valley,Nepal); 3- Rongbuck Glacier (Tibet); 4- Gungrinchmezoe Glacier (Bhutan); 5- Siachen Glacier (karakoram, India); 6- Gangotri Glacier (India)

Figure 20.5 Snaps showing different glaciers located in different parts of the Himalayas and its adjoining region.

created by its high latitude, the Himalayas are entirely landlocked and created by its high altitude. Similarly, the Arctic is sparsely populated, with less than 4 million inhabitants compared to the Himalayas where about 750 million are estimated to live in the Himalayan watershed region. Both regions also face multiple national and cultural boundaries and challenges, with sources from one nation often having their primary impact on another. Having similar challenges despite the geographic differences, we need to understand these regions properly, especially in the present

Table 20.2 Location of the earthquake centers and the glaciers considered for the study.

State	Latitude	Longitude
Jammu & Kashmir	32.5–37N	73–80.5E
Himachal Pradesh	30–33.5N	75.5–79E
Uttarakhand	28.5–31.5N	77.5–81E
Nepal	26–30.5N	80–88E
Sikkim	27–29N	88–89E
Bhutan	27–29N	89–92E
Arunachal Pradesh	26.5–29.5N	91.5–97.5E

scenario of climate change and global warming, which is imposing severe adverse effects on the cryosphere regions of the Arctic and the Himalayas. The climate response of a large fraction of glaciers of the Arctic and Himalayas is significantly linked with atmospheric changes and ocean circulation. Snow and glaciers melt in the Arctic enhanced significantly due to large temperature changes, changes in sea ice cover, atmospheric flow patterns, and precipitation.

A detailed investigation on the Himalayas glaciers is scientifically important primarily because glaciers are widely known as one of the best indicators of climate change impacting the Earth over the past few centuries (Sharma et al., 2019). Globally, the glaciers are shrinking posing danger to the Arctic and Himalayan glaciers. The Svalbard glaciers and ice caps cover an area of 34,600 km^2 with a total ice volume of roughly 7000 km^3 while the Himalayas occupy nearly 38,000 km^2 (Sharma et al., 2019). Despite having contrast in their geographical settings, the glaciers in both the regions show a generally increased melting during the present decades.

The regions most affected by climate change, South Asia, are ought to be essentially studied to help unravel teleconnection with the climate-induced changes taking place over the Arctic region. India has been monitoring two representative glaciers Vestre Broggerbreen and Feiringbreen in the Svalbard region, to understand the dynamics and mass of Arctic glaciers in the context of a changing climate and also have focused research on the Himalayan glaciers. Since glaciers are an important source of freshwater and influence the climate of a region, the research is of immense significance to the well-being of the entire Indian subcontinent.

In this study, we attempted to address the issues of glacial mass changes to understand what lead to induce earthquakes or the modulation of seismicity in the Himalayan region of South Asia. The study also sheds light on the modulation of seismicity that can be treated as one of the proxies of the climate change, not only confined to the Himalayas but to other cryosphere regions in future. The study aimed at corroborating the glacial retreat as well as the gaining of ice with reference to assess its impacts on the Himalayan seismicity to understand how the rupture of the micro-to-moderate earthquakes in the Polar region get influenced by glacial mass change phenomena to generate glacial mass change–induced modulated seismicity (GMCIMS) in the region.

20.2 Data and results

In order to achieve the objectives, some of the selected earthquake data from the United States Geological Survey (USGS), International Seismological Centre (ISC), and Incorporated Research Institutions for Seismology (IRIS) have been considered for the present calculation. Earthquake data for only the Indian Himalayan areas are considered for the present study. The earthquake data used here range from 1974 to 2016 and belongs to the shallower depth, that is, surface to 100 km depth. The earthquake data presented in Table 20.3 are of a 10-km radius of the corresponding glaciers. The common incidence periods (from two data) are counted as one event. Data of average change in glacial ice thickness from 1974 to 2016 have been consolidated from NASA data repository for the Indian Himalayan region. The glacial ice thickness data were available as a DEM image in NSIDC, which has been converted into the numbers using R software. Earthquake and glacial ice thickness change data of same time frame (1974–2016) and same location (within 10-km buffer areas) have been compared here.

As mentioned previously, the cryospheric resources of south Asia are found to be unevenly distributed as shown in Table 20.1, indicating that the Himalaya is susceptible to glacial mass changed–induced subsurface earthquakes of varying frequency and intensity depending upon the areal and volumetric changes of ice masses either in terms of accumulation or of melting of ice sheets.

Glacial health for 2018 glaciers during 2001–11 was monitored by the Indian Space Research Organization (ISRO). The result shows that as many as 248 glaciers have retreated and 18 glaciers have shown advancement. Glaciers were chosen from Karakoram, Zanskar, Himachal, Uttarakhand,

Table 20.3 Glacial ice thickness change and earthquake magnitudes of specific locations.

Sr. No.	Location		Average ice cover change	Earthquake (Mw)	Depth
1	32.1 − 77.7	Himachal Pradesh	−0.283493661	3	10
2	34.3 − 75.7	Jammu & Kashmir	−0.077512546	4.5	65.2
3	34.0 − 75.5	Jammu & Kashmir	−0.135976548	3	37.4
4	34.2 − 75.5	Jammu & Kashmir	−0.131358806	2.9	0
5	27.7 − 88.1	Sikkim	−0.037561507	4.3	80.3
6	27.8 − 88.0	Sikkim	−0.051647797	3.1	84.7
7	27.8 − 88.1	Sikkim	0.019527431	3.4	0
8	27.6 − 88.1	Sikkim	−1.156820236	3.8	20
9	27.7 − 88.1	Sikkim	−0.162516502	4.3	80.3
10	27.9 − 88.1	Sikkim	−0.421615232	3.7	57.1
11	27.9 − 88.2	Sikkim	−0.347126015	3.2	25
12	28.8 − 83.0	Nepal	−0.26450494	4.2	10
13	29.8 − 82.5	Nepal	0.106767403	3	39.4
14	29.9 − 82.5	Nepal	0.041332486	3.6	10
15	29.9 − 82.6	Nepal	−0.223033389	3.6	0

Nepal, and Sikkim. Among these, only glaciers from Karakoram have shown advancement; the rest of the glaciers from Zanskar, Himachal, Uttarakhand, Nepal, and Sikkim have shown either no change or retreat (Fig. 20.6).

The mass balance study is a good measure for understanding the mass increase or decrease of respective glaciers for a certain period. The increase and decrease of the mass, that is, if the accumulation exceeds the ablation is known as positive mass balance and the reverse is negative mass balance. The glacial health through the glacier mass balance for the glaciers of Chandra, Bhaga, Miyar, Alaknanda, Baspa, Bhagirathi, Goriganga,

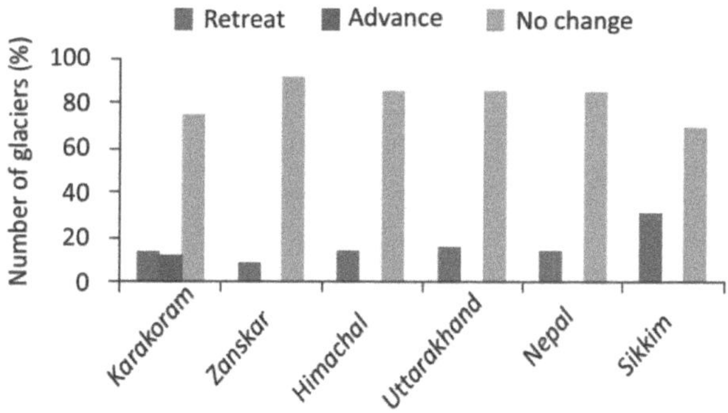

Figure 20.6 Measurement of glacial health, advancement, and retreat of The Himalayan glaciers. *Reproduced from ISRO (www.isro.gov.in).*

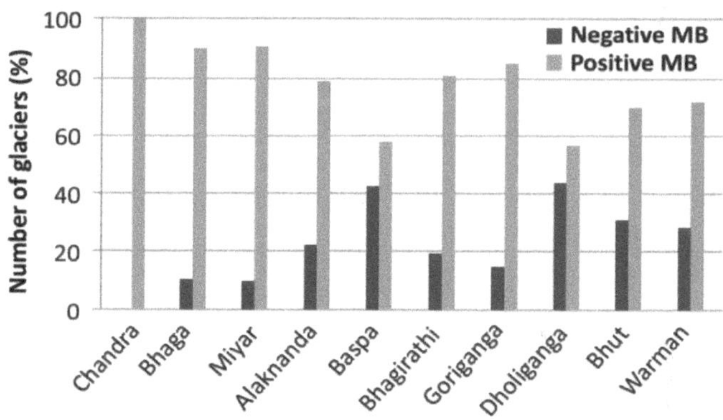

Figure 20.7 Glacial mass balance of glaciers from Himalayan origin river basins. *Reproduced from ISRO (www.isro.gov.in).*

Dhauliganga, Bhut, and Warman river basins were studied by ISRO (Fig. 20.7), and it is clearly observed that only except Chandra many of the glaciers in all other river basins are showing negative mass balance, which indicates more glacial melting than accumulation. Among these, the maximum negative mass balance is seen (Fig. 20.7) in Baspa and Dhauliganga river basin followed by Bhut, Warman, Alaknanda, and Bhagirathi. These rivers are the most important source of freshwater for our country. These glacial losses have the potential to lead to micro-to-moderate-level earthquakes in these areas.

The seismic activity found to occur due to the deglacial upliftment ranges from micro-to-moderate-range earthquakes. We showed the seismicity of magnitudes (Mb) ranging from 2.8 to 4.5 for the present study

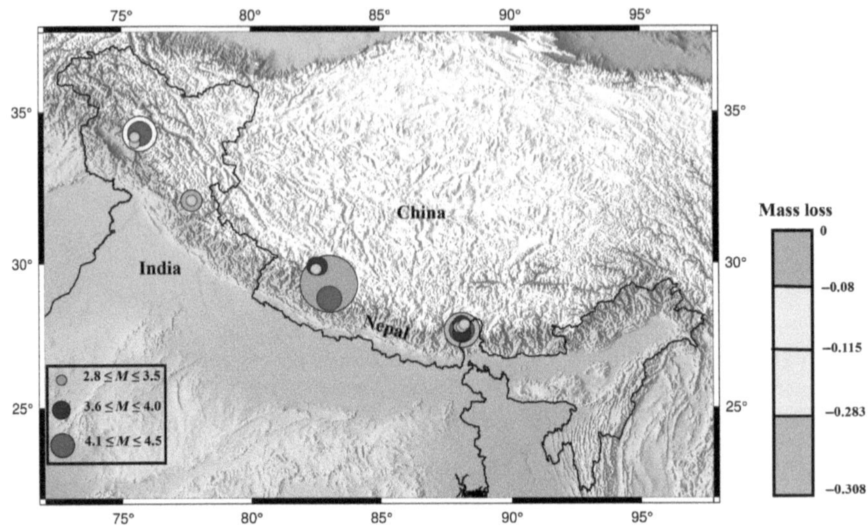

Figure 20.8 Distribution of significant earthquakes in different glacier zones of the Himalayas, demonstrating the glacial mass change–induced earthquakes (2.8 ≥ *M* ≤ 4.5).

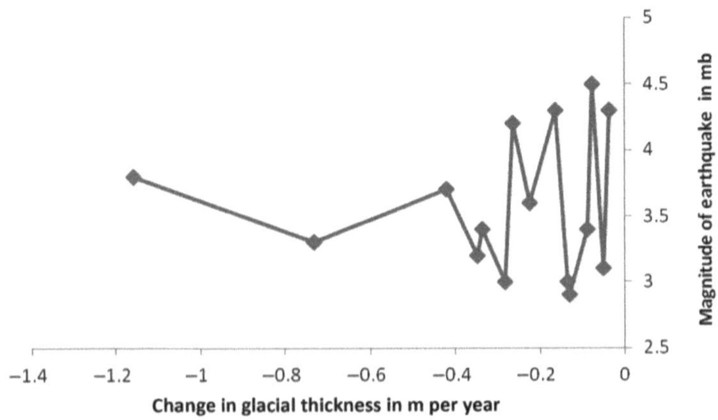

Figure 20.9 Changing magnitudes of the earthquake with changes in glacial ice thickness.

in association with glacial mass changes of some of the important glaciers located in different parts of the Himalaya (Fig. 20.8). The average glacial ice thickness change data show the changes in meter per year (Table 20.3; Fig. 20.9) and have a combination of both positive and negative values, which reveals that not only glacial mass loss but some of the places are gaining the glacial mass as well.

Both India (Sikkim) and Nepal are neighboring countries; therefore it is possible that both these places are experiencing similar climatic conditions, which is helping in glacial mass accumulation. These two places being in the middle Himalaya receives snowfall from both the South East Monsoon and Western Disturbances, which escalates the glacial mass gain. When the glacial ice is accumulated in one place and consolidates, it exerts immense stress on the rocks underneath, which ultimately can lead to small-scale earthquake when the stress goes beyond a certain level. On the other hand, when the glacial mass loss releases pressure from the bottom lying surface rock due to isostatic rebound, the underneath rock tries to regain its earlier position. This situation also leads to small-scale earthquakes. One small example can explain this process. If we take a spring and exert pressure on it, it will be compressed and compacted, but if we remove the pressure, it will try to return back to its earlier position. A similar process happens to the rocks beneath the glacier, but the elasticity of the rocks is less, therefore it generates fractures. When these fractures are on large scale, they can generate small-scale earthquakes. This is a common sight in the polar glaciers. An important fact is that the ISRO data of glaciers of Chandra, Bhaga, Miyar, Alaknanda, Baspa, Bhagirathi, Goriganga, Dhauliganga, Bhut, and Warman river basins show that during 2010 to 2012, the glaciers had either positive or negative mass balance (Table 20.4), which means in both ways, it can lead to seismic instability. Also, the alternative glacial accumulation and melting lead to weathering processes like block disintegration and exfoliation which could also lead to micro-scale earthquakes. Though this Himalayan region is seismically active being the plate boundary, seismic activity related to the plate movements or orogenesis is of higher magnitude. Figs. 20.8 and 20.9 demonstrate that the magnitude of the earthquake not only depends on the glacial ice thickness change but also on the characteristics related to the stress-bearing capacity of the underlying rock and the nature of the underneath seismogenic active faults.

Table 20.4 Glacial mass balance data.

Year	2012	2011	2010
No. of glaciers	1037	843	733
Negative	478	151	148
Positive	559	692	585

Source: www.isro.gov.in.

Table 20.4 conspicuously shows that there are significantly positive and negative ice mass changes in different years from 2010 to 2012 in the glaciated belts, which reflects two conditions of exerting thrust on the earth crust to promote loading due to thickening of ice and unloading due to melting of ice mass that instigate the failure of the seismogenic faults remained under critical stress-to-yield earthquakes.

20.3 Interpretation and discussions

Different types of glaciers (Fig. 20.4) are having a varying potential of ice accumulations and recession because of their diverse dynamics under the given cryospheric conditions. It is observed that a cirque glacier is associated with the accumulation of ice in small depressions on the side of a mountain, whilst the hanging glaciers flow down a mountainside but stop abruptly at a cliff (Fig. 20.2). After comparing the earthquake data with the glacial ice thickness change data for the same time frame and location (Table 20.3), it is observed that only 15 data points are in common. Magnitudes (Mb) of these 15 earthquakes vary from 2.9 (at 34.2°N and 75.5°E) to a maximum magnitude of 4.5 at 34.3°N and 75.7°E, and both the places are falling under Kashmir Himalaya. The positive ice thickness change is evident in Sikkim (27.8°N and 88.1°E) where an earthquake magnitude is found to occur of 3.4. Similarly some places of Nepal (29.8°N, 82.5°E and 29.9°N 82.6°E) also show a positive ice thickness change, and the earthquake magnitude is found varying from 3.0 to 3.6. The most important thing is that both positive and negative ice thickness changes lead to the earthquake of almost similar magnitude. The Earthquake magnitude of all 15 places varies between 2.9 and 4.5. In the present discussion, the earthquakes of magnitudes ($M > 5.0$) have not been considered since earthquakes associated with the change in glacial mass may correspond to low-to-moderate strengths due to lesser loads of ice masses on the brittle crust. Among the 15 places, only 3 places are showing the positive ice thickness and the rest of 12 places are associated with negative ice cover change, which proves that glacial ice cover loss has become common in Indian glaciers and also in neighboring countries, and this is leading to the small-scale earthquakes as evident from Table 20.3.

South Asian Himalayan countries (Bhutan, Nepal, India, and Pakistan) and even the adjacent areas of the Himalayas (e.g., Tibet) (Fig. 20.5) have ample set of evidence for occurrences of the hanging and cirque glacier in different parts of the Himalayas, which dictates the amount of changes in the lithospheric loads that can have sufficient strength to perturb the nature and extent of subsurface stress to bring brittle failures in the glaciated region of the Himalaya.

It is well documented that our planet suffered through unprecedented climatic transformation even before 20,000–50,000 years during geo-meteorological history of the Earth. Recently, climate change is being witnessed by frequent floods, drought, cyclone, and storm surge along with other climate change–induced disasters, occurring in the region, such as failure of monsoon or excessive rainfalls during the monsoon, glacial lake outburst floods (GLOF), and extreme variations in global and regional temperature (Mishra, 2014), besides excessive or scanty rains or precipitation, rather than snow falling even at higher altitudes that accelerates the melting of glaciers. Several studies shed light on the main concern that highlights the impacted climate change on seismogenesis. It is observed that the ice dissipated due to soaring temperatures; consequently the crust gets popped back up like a coiled spring released, and it causes brittle failure into seismogenic faults that trigger moderate-to-strong earthquakes in places where they are unheard of today (Mishra, 2014), which supports our interpretation of modulated seismicity referred to as GMCIMS due to melting of ice sheets in the glaciated belts of the Himalaya (Fig. 20.4).

Climate change will result in glacier melt–induced sea-level rise which will ultimately initiate catastrophes like landslides, snow avalanches, ice avalanches from glacier snouts, GLOF volcanic eruptions, earthquakes, and tsunamis (nsidc.org). Global warming–forced glacial melting has the capacity to bring seismically turbulent future scenarios. With the reduction of massive weight on the crust due to the melting of glacier, glacioisostatic rebound reactivates the faults and also lifts pressure on magma chambers, leading to the increased seismic activity (Masih, 2018), which is one of the intriguing issues to be studied in detail. The role of monsoon water has also been demonstrated by Iaffaldano et al. (2011) to show lateral variations in plate coupling, and their temporal changes become an efficient source to induce an uncommon form of plate motion, and it has been reported that monsoon speeds up Indian plate motion. Melting of the thicker ice sheets leads to the reduction in weight on the crust, allowing faults contained therein to slide more easily, which may promote increased earthquake activity in the Himalayan region during recent decades (Mishra, 2014). It is, however, felt that a detailed study on seismogenesis in the glaciated belts of the entire Himalaya needs to be carried out using archival earthquake data recently compiled by different institutes of South Asia based on data mining tools applied to neural network model (Dutta et al., 2013), which would furnish good correlation between glacial mass changes and earthquakes to ascertain GMCIMS as one of the additional efficient proxies of climate change along with temperature, precipitation, sea-level rise, global warming, and erratic extreme weathers, reflecting local impacts of regional climate change. Conversely, it has also been argued that great earthquakes ($M \geq 8.0$) are capable to tilt the axis of rotation of the earth, influencing the secular motion of the earth that significantly controls seasonal changes, which, in turn, contributes to climate change for a longer recorded period. Since the uneven rotation of the earth, the planet may have a profound bearing on the genesis of earthquakes (Mishra, 2014), which may be further corroborated with the seismic moment budget from all strong to great Himalayan earthquakes (Dutta et al., 2011).

A recent discussion by Olivieri et al. (Olivieri and G. Spada, 2015) for the seismicity of Greenland showed that ice mass unloading and earthquake suppression have an enigmatic relationship where a clear correlation between seismic activity and ice melting in Greenland is not found in the sense that the spatiotemporal distribution of earthquakes is mainly confined to regions where the thick layer of ice is absent and significant ice melting is presently occurring. Moreover, earthquake locations and corresponding depth distribution suggest two distinct governing mechanisms: postglacial rebound (PGR) that promotes moderate-size crustal earthquakes at Greenland's regional scale, whilst current ice melting promotes shallow low magnitude seismicity locally. It is established that intraplate earthquakes can also originate from processes that do not involve the horizontal movement of plates (Talwani, 2017; Mishra et al., 2008). PGR in response to the melting of the late-Pleistocene ice sheets (e.g., Turcotte and Schubert, 2002) may be a plausible factor for intraplate seismicity (Gutenberg and Richter, 1944). PGR also caused vertical and horizontal crustal deformation at high-latitude regions, inducing perturbation of stress within the lithosphere and the mantle (Spada et al., 2012). Ice unloading may be the cause of seismicity at the ice margins, and that this may be a possible cause of paleo-earthquakes, elsewhere in the world (e.g., Charle-voix; Wabash Valley) (Wu and Johnston, 2000). A similar study also showed for Sweden where rapid ice retreat was reported during the Pleistocene–Holocene and caused the earthquake (M 7.5) beneath the Lake Vattern earthquake (Jakobsson et al., 2014). It is interesting to note that four large earthquakes of 1811–12 in New Madrid (MO, United States) were found associated with PGR-induced earthquakes (Wu and Johnston, 2000) and concluded, "glacial unloading is unlikely to have triggered the large earthquakes (M 8.0) in New Madrid," which was further corroborated with other studies by Hough et al. (2000). Wu and Johnston (2000) showed that the load-induced stress-field variations reduce rapidly outside the ice margins, and their amplitude is too small to generate earthquakes of this size, which, in turn, suggests that glacial mass change–induced earthquakes beneath the Himalayas bear a similar mechanism of PGR (Steffen et al., 2014), where load-induced stress-field variations modulate the rate of seismicity in the glaciated belts of South Asia and its adjoining regions. In this context, it is worth mentioning that the role of rapid uplift in response to erosion rather than deglaciation may not be ruled out in triggering glacial mass change–induced earthquakes beneath the Himalayan arc, which is found to be in good agreement with the study carried out for the New Madrid earthquakes by Calais et al. (2010). Moreover, the role of PGR (Gutenberg and Richter, 1944) is now supported by modeling that provided a better interpretation of the postglacial origin of earthquakes (Steffen et al., 2014). It is evident that glacial melt–induced earthquakes are associated either with ice unload or rapid uplift due to erosion or with both.

One of the possible causes of earthquakes in Stable Continental Regions (SCRs) is associated with PGR, which has been well described by Schulte and Mooney (2005). As described before the role of PGR in the triggering of seismicity in deglaciated areas was proposed by various authors (Stein et al., 1979; Quinlan, 1984). Some of the authors like Muir-Wood (2000) propounded a relatively very complex mechanism of stress and strain interaction in deglaciated areas and it is argued that PGR has an important role in the stress redistribution across deglaciated areas (Steffen et al., 2012; Steffen et al., 2014) as well documented for Greenland and Antarctica that may be applicable to the glacial mass change–induced earthquakes in the Himalayan arc. This study provides the possible relationship between ice melting and the spatiotemporal distribution of seismicity for the Himalayan region associated with glaciers that are under melting associated with the impacts of climate change, which is very much corroborative with mass changes of the Himalayan glaciers (Tables 20.3 and 20.4 and Figs. 20.7–20.9) where a majority of demarcated glaciers in the different parts of the Himalaya are found associated with positive mass changes, while there are some glaciers that showed no change in ice mass, besides retreat of some of the glaciers due to negative ice mass changes, generating mass change–induced earthquakes beneath the Himalayas due to perturbation of stresses in the crust overlain by the thick pile of ice that allows the crustal faults to ooze the confined stress because of lessening of ice loads on the deglaciated belts. Our analyses of the present data set (Tables 20.3 and 20.4 and Figs. 20.6–20.8) clearly show the ice mass accumulation promotes the advancement of glaciers in the Himalayas that further exerts excessive loads on the crust and the in situ stress gets perturbed due to overloading, and the subcritical hidden seismogenic faults beneath the glacier belts become vulnerable to failure and resulted in micro-to-moderate earthquakes (Figs. 20.8 and 20.9). The relationship between deglaciation and the occurrence of earthquakes based on the recent models by Steffen et al. (2014) is well explainable for the Himalayan seismicity related to the glacial mass–induced earthquake as one of the proxies of climate change, which is also supported by studies, elsewhere in the world (Alley et al., 2010).

It is now reported that global warming due to regional climate change is responsible for inducing the polar ice cover melting, causing abrupt glacial mass loss or gain induced by the global or regional climate change that has the potential to trigger small-to-moderate earthquakes as observed in the Polar region (Khare et al., 2018). The Himalayan region is devoid of such studies related to glacial mass change–induced earthquakes. We studied the Himalayan glaciers for the first time and found that the Himalayan glaciers are also not spared from glacial mass losses or mass appreciation due to the advancement of ice accumulations in some parts of the Himalayas. We infer that the melting or mass loss of ice and mass sufficiency or accumulation of ice leads to the significant changes into the crustal stress perturbation that brings the failure to the subsurface seismogenic faults and resulted in earthquakes of varying strengths in the Himalayan region starting from Jammu and Kashmir to Arunachal Pradesh (Fig. 20.8). It is observed that the nature and extent of seismogenesis in the vicinity of the Himalayan glaciers gets dictated by the glacial mass change (ice melting or ice thickening) under the uncontrollable climate change scenario, which, in turn, suggests that GMCIMS associated with stress perturbation in the underneath seismogenic faults due to loading and unloading of ice masses in the glaciated belts of the Himalaya. The modulation in seismicity refers to the enhancement of the number of earthquakes, consisting of tectonic earthquakes and earthquakes triggered by glacial mass change in the Himalayan region. Nonetheless, GMCIMS may be considered as one of the proxies of climate change in the third pole of the Himalayas, besides its other adjudged indicators (temperature, precipitation, and sea-level rise) of the climate change in addition to global warming. In conclusion, we can comment that climate-driven abrupt glacial mass loss can generate instability in the entire Indian Himalayan region that yields a set of modulated seismicity as a potential proxy to indicate impacts of climate change on Himalayan glaciers and thereby genesis of modulated seismicity.

It is evident from the interpretation that there is a need to determine rebound stress in seismogenic fault stability with reference to making the assessment of mode of faulting to ascertain what types of faults getting associated with glacial mass change–induced earthquakes. There is still no common consensus on the issue of rebound stress to understand at what threshold value of stress rebound, tectonic earthquakes triggers. Comprehensive studies may shed important light on the debatable issue of differentiating glacial mass changes–induced earthquakes from tectonic earthquakes in the Himalayan region.

20.4 Conclusion

Global warming and regional climate change are responsible for inducing the polar ice cover melting because of abrupt glacial mass loss or gain induced by the global or regional climate change that led to small-to-moderate earthquakes, especially evident in the Polar region, whilst there are no such studies, so far made for the Himalayan region. We found that Himalayan glaciers are also not spared from glacial mass losses or mass appreciation due to the advancement of ice accumulations in some parts of the Himalayas. We infer that the melting or mass loss of ice and mass sufficiency or accumulation of ice leads to the significant changes into the crustal stress perturbation that brings failure to the subsurface seismogenic faults and resulted in earthquakes of varying strengths in the Himalayan region starting from Jammu and Kashmir to Arunachal Pradesh. The nature and extent of seismogenesis in the vicinity of the Himalayan glaciers gets dictated by the glacial mass change (ice melting or ice thickening) under the uncontrollable climate change scenario, which, in turn, suggests that GMCIMS associated with stress perturbation in the underneath seismogenic faults due to loading and unloading of ice masses in the glaciated belts of the Himalaya, and GMCIMS may be considered one of the proxies of climate change in the third pole of the Himalayas, besides its other adjudged indicators (temperature; precipitation, and sea-level rise) in addition to global warming. In conclusion, we can comment that climate-driven abrupt glacial mass loss can generate instability in the entire Indian Himalayan region that yields a set of modulated seismicity as a potential proxy to indicate impacts of climate change on Himalayan glaciers. In order to generate a comprehensive model of glacial mass change–induced earthquakes, we have to have a longer period earthquake archival catalog to understand the better correlation between glacial mass changes and the genesis of the corresponding earthquakes in the glaciated belts of the Himalaya that opens an avenue for future research. There is a need to determine rebound stress in seismogenic faults to ascertain their stability with reference to making the assessment of different modes of faulting to ensure what types of faults get associated with glacial mass change–induced earthquakes. There is still no common consensus on the issue of rebound stress to understand at what threshold value of stress rebound, tectonic earthquakes triggers. Further, a detailed study on it may shed important light on the issue of differentiating glacial mass changes–induced earthquakes from tectonic earthquakes, which can be studied in detail in the future. Findings of this study shall carry their bearing in understanding the Arctic glacier's dynamics and its linkage with ongoing global warming impacting the Arctic cryosphere.

Acknowledgments

Authors sincerely thank Dr. M. Rajeevan, Secretary, Ministry of Earth Sciences for his consistent motivation for conducting interdisciplinary study at the National Centre for Seismology (NCS), Ministry of Earth Sciences due to which such study has been carried out for the first time for the Himalayan glaciers using open data set from various sources, including Indian Space Research Organization (ISRO), Bengaluru, India. Pertinent references with information and figures received from Drs. Sameer Tiwari and Akshay Verma, Wadia Institute of Himalayan Geology (WIHG), Dehradun, India is thankfully acknowledged.

References

Alley, R.B., Andrews, J.T., Brigham-Grett, J., Clarke, G., Cuffey, K., Fitzpatrick, J., et al., 2010. History of the Greenland ice sheet: paleoclimatic insights. Quat. Sci. Rev. 29 (15), 1728–1756.

Amos, C.B., Audet, P., Hammond, W.C., Bürgmann, R., Johanson, I.A., Blewitt, G., 2014. Uplift and seismicity driven by groundwater depletion in central California. Nature 509 (7501), 483–486. Available from: https://doi.org/10.1038/nature13275.

Argus, D.F., Peltier, W.R., Drummond, R., Moore, A.W., 2014. The Antarctica component of postglacial rebound model ICE-6G_C (VM5a) based on GPS positioning, exposure age dating of ice thicknesses, and relative sea level histories. Geophys. J. Int. 198 (1), 537–563. Available from: https://doi.org/10.1093/gji/ggu140.

Bajracharya, S.R., Mool, P.K., Shrestha, B.R., 2006. The impact of global warming on the glaciers of the Himalaya. In: Proceedings of the International Symposium on Geodisasters, Infrastructure Management and Protection of World Heritage Sites, pp. 231–242.

Bajracharya, S.R., Maharjan, S.B., Shrestha, F., 2014. The status and decadal change of glaciers in Bhutan from the 1980s to 2010 based on satellite data. Ann. Glaciol. 55 (66), 159–166.

Bansod, S.D., Yin, Z.-Y., Lin, Z., Zhang, X., 2003. Thermal field over Tibetan Plateau and Indian summer monsoon rainfall. Int. J. Climatol. 23 (13), 1589–1605. Available from: https://doi.org/10.1002/joc.953.

Barnett, T.P., Adam, J.C., Lettenmaier, D.P., 2005. Potential impacts of a warming climate on water availability in snow-dominated regions. Nature. Available from: https://doi.org/10.1038/nature04141.

Beniston, M., 2003. Climatic change in mountain regions: a review of possible impacts. Clim. Change. Available from: https://doi.org/10.1023/A:1024458411589.

Bookhagen, B., Burbank, D.W., 2010. Toward a complete Himalayan hydrological budget: spatiotemporal distribution of snowmelt and rainfall and their impact on river discharge. J. Geophys. Res.: Earth Surf. Available from: https://doi.org/10.1029/2009JF001426.

Borsa, A.A., Agnew, D.C., Cayan, D.R., 2014. Ongoing drought-induced uplift in the western United States. Science 345 (6204), 1587–1590. Available from: https://doi.org/10.1126/science.1260279.

Calais, E., Freed, A., Van Arsdale, R., Stein, S., 2010. Triggering of New Madrid seismicity by late-Pleistocene erosion. Nature 466 (7306), 608–611.

Campbell, J.G., and Pradesh, H. (2005). Inventory of glaciers, glacial lakes and the identification of potential glacial lake outburst floods (GLOFs) affected by global warming in the mountains of India, Pakistan and China/Tibet Autonomous Region. International Centre for Integrated Mountain Development, GP O. Box, 3226.

Dutta, P.K., Mishra, O.P., Naskar, M.K., 2011. South Asia earthquake catalogue magnitude data regression analysis. Int. J. Statists. Analys. 1 (2), 161–170.

Dutta, P.K., Mishra, O.P., Naskar, M.K., 2013. Evaluation of seismogenesis behaviour in Himalayan belt using data mining tools for forecasting Cent. Eur. J. Geosci. 5 (2), 236–253.

Frei, A., Tedesco, M., Lee, S., Foster, J., Hall, D.K., Kelly, R., et al., 2012. A review of global satellite-derived snow products. Adv. Space Res. 50 (8), 1007–1029. Available from: https://doi.org/10.1016/j.asr.2011.12.021.

Geological Survey of India, 1999. In: Kaul, M.K. (Ed.), Inventory of the Himalayan Glaciers: A Contribution to the International Hydrological Programme. Special Publication, No. 34.

Ghazoui, Z., Bertrand, S., Vanneste, K., Yokoyama, Y., Nomade, J., Gajurel, A.P., et al., 2019. Potentially large post-1505 AD earthquakes in western Nepal revealed by a lake sediment record. Nat. Commun. Available from: https://doi.org/10.1038/s41467-019-10093-4.

Godard, V., Burbank, D.W., 2011. Mechanical analysis of controls on strain partitioning in the Himalayas of central Nepal. J. Geophys. Res.: Solid Earth. Available from: https://doi.org/10.1029/2011JB008272.

Groisman, P.Y., Karl, T.R., Knight, R.W., Stenchikov, G.L., 1994. Changes of snow cover, temperature, and radiative heat balance over the Northern Hemisphere. J. Clim. 7 (11), 1633–1656. Available from: https://doi.org/10.1175/1520-0442(1994)007%3C1633:COSCT%3E2.0.CO;2.

Gurung, T., 2005. An Overview of Glaciers, Glacier Retreat, and Subsequent Impacts in Nepal, India and China. WWWF Nepal Program.

Gutenberg, B., Richter, C.F., 1944. Frequency of earthquakes in California. Bull. Seismol. Soc. Am. 34 (4), 185–188.

Hampel, A., Hetzel, R., Maniatis, G., Karow, T., 2009. Three-dimensional numerical modeling of slip rate variations on normal and thrust fault arrays during ice cap growth and melting. J. Geophys. Res.: Solid. Earth. Available from: https://doi.org/10.1029/2008JB006113.

Hetzel, R., Hampel, A., 2005. Slip rate variations on normal faults during glacial–interglacial changes in surface loads. Nature 435 (7038), 81–84. Available from: https://doi.org/10.1038/nature03562.

Hough, S.E., Armbruster, J.G., Seeber, L., Hough, J.F., 2000. On the modified Mercalli intensities and magnitudes of the 1811-1812 New Madrid earthquakes. J. Geophys. Res. Solid Earth (1978-2012) 105 (B10), 23839–23864.

Iaffaldano, G., Husson, L., Bunge, H.P., 2011. Monsoon speeds up Indian plate motion. Earth Planet. Sci. Lett. 304, 503–510.

Immerzeel, W.W., Droogers, P., de Jong, S.M., Bierkens, M.F.P., 2009. Large-scale monitoring of snow cover and runoff simulation in Himalayan river basins using remote sensing. Remote Sens. Environ. 113 (1), 40–49. Available from: https://doi.org/10.1016/j.rse.2008.08.010.

Jakobsson, M., Ólafur, Ingólfsson, Spielhagen, Robert F., 2014. The dynamic Arctic. Quaternary Science Reviews 92, 1–8.

Johnson, J.B., Watson, L.M., Palma, J.L., Dunham, E.M., Anderson, J.F., 2018. Forecasting the eruption of an open-vent volcano using resonant infrasound tones. Geophys. Res. Lett. Available from: https://doi.org/10.1002/2017GL076506.

Khare, N., Mishra, O.P., Taguchi, S., 2018. Recent Advances in Climate Science of Polar Region. Polar Science (Elsevier), edited Spcl Issue, 18, 226P.

Liu, C.H., Shi, Y.F., Wang, Z.T., Xie, Z.C., 2000. Glacier resources and their distributive characteristics in China—a review on Chinese glacier inventory. J. Glaciol. Geocryol. 22 (2), 106–112.

Lutz, A.F., Immerzeel, W.W., Shrestha, A.B., Bierkens, M.F.P., 2014. Consistent increase in High Asia's runoff due to increasing glacier melt and precipitation. Nat. Clim. Change 4 (7), 587–592. Available from: https://doi.org/10.1038/nclimate2237.

Masih, A., 2018. An enhanced seismic activity observed due to climate change: preliminary results from Alaska. IOP Conf. Series: Earth Environ. Sci. 167, 012018. Available from: https://doi.org/10.1088/1755-1315/167/1/012018.

Mishra, O. P., Zhao, D., Wang, Z., 2008. The genesis of the 2001 Bhuj, India earthquake (Mw 7.6): A puzzle for peninsular India?, Spec. Iss. Ind. Min. 61(3&4), 149–170.

Mishra, O.P., 2014. Intricacies of the Himalayan seismotectonics and seismogenesis: need for integrated research. Curr. Sci. 106 (2), 176–187. Available from: http://www.jstor.org/stable/24099798.

Muir-Wood, R., 2000. Deglaciation seismotectonics: a principal influence on intraplate seismogenesis at high latitudes. Quat. Sci. Rev. 19 (14), 1399–1411.

Mukhopadhyay, B., Khan, A., 2015. A reevaluation of the snowmelt and glacial melt in river flows within Upper Indus Basin and its significance in a changing climate. J. Hydrol. 527, 119–132. Available from: https://doi.org/10.1016/j.jhydrol.2015.04.045.

Nagai, H., Fujita, K., Sakai, A., Nuimura, T., Tadono, T., 2016. Comparison of multiple glacier inventories with a new inventory derived from high-resolution ALOS imagery in the Bhutan Himalaya. Cryosphere 10, 65–85. Available from: https://doi.org/10.5194/tc-10-65-2016.

Olivieri, M., Spada, G., 2015. Ice melting and earthquake suppression in Greenland, Polar Science 9 (1), 94–106.

Qian, L., Laštovička, J., Roble, R.G., Solomon, S.C., 2011. Progress in observations and simulations of global change in the upper atmosphere. J. Geophys. Res.: Space Phys. Available from: https://doi.org/10.1029/2010JA016317.

Quinlan, G., 1984. Postglacial rebound and the foal mechanisms of eastern Canadian earthquakes. Can. J. Earth Sci. 21 (9), 1018–1023.

Raina, V.K., Srivastava, D., 2008. Glacier Atlas of India. Geological Society of India, Bangalore, 315 pp. ISBN13: 978-8-185-86780-9.

Rana, A.S., 2018. Summer School on Integrated Water Resources Management. PMD, Pakistan.

Schulte, S.M., Mooney, W.D., 2005. An updated global earthquake catalogue for stable continental regions: reassessing the correlation with ancient rifts. Geophys. J. Int. 161 (3), 707–721.

Sharma, P., Patel, L.K., Singh, A.T., Meloth, T., Ravindra, R., 2019. Glacier response to climate in Arctic and Himalaya during last seventeen years: a case study of Svalbard, Arctic and Chandra Basin, Himalaya. Climate Change and the White World. Springer, pp. 139–156.

Spada, G., Ruggieri, G., Sørensen, L.S., Nielsen, K., Melini, D., and Colleoni, F., (2012). Greenland uplift and regional sea level changes from ICESat observations and GIA modelling, Geophys J Int 189(3),1457.

Steffen, R., Wu, P., Steffen, H., Eaton, D.W., 2014. The effect of earth rheology and ice-sheet size on fault slip and magnitude of postglacial earthquakes. Earth Planet. Sci. Lett. 388, 71–80.

Stein, S., Sleep, N.H., Geller, R.J., Wang, S.-C., Kroeger, G.C., 1979. Earthquakes along the passive margin of Eastern Canada. Geophys. Res. Lett. 6 (7), 537–540.

Talwani, P., 2017. On the nature of intraplate earthquakes. J. Seismol. 21 (1), 47–68.

Terzago, S., von Hardenberg, J., Palazzi, E., Provenzale, A., 2014. Snowpack changes in the Hindu Kush–Karakoram–Himalaya from CMIP5 Global Climate Models. J. Hydrometeorol. 15 (6), 2293–2313. Available from: https://doi.org/10.1175/JHM-D-13-0196.1.

Turcotte, D.L., Schubert, G., 2002. Geodynamics. Cambridge University Press, Cambridge.

Vernant, P., Chery, J., 2006. Low fault friction in Iran implies localized deformation for the Arabia–Eurasia collision zone. Earth and Planetary Science Letters 246 (3–4), 197–206. Available from: https://doi.org/10.1016/j.epsl.2006.04.021.

Viste, E., Sorteberg, A., 2015. Snowfall in the Himalayas: an uncertain future from a little-known past. Cryosphere 9 (3), 1147–1167. Available from: https://doi.org/10.5194/tc-9-1147-2015.

Wu, P., Johnston, P., 2000. Can deglaciation trigger earthquakes in N. America? Geophys. Res. Lett. 27 (9), 1323–1326.

Wu, G., Zhang, Y., 1998. Tibetan Plateau forcing and the timing of the monsoon onset over South Asia and the South China Sea. Monthly Weather. Rev. Available from: https://doi.org/10.1175/1520-0493(1998)126%3C0913:TPFATT%3E2.0.CO;2.

Glossary

Accelerated thermokarsts It occurs when the thermal stability of the upper part of ice-rich permafrost is disrupted by an increase in active layer depth. *Thermokarst* can occur as either thaw subsidence or thermal erosion.

Accelerator mass spectrometry (AMS) It is a form of *mass spectrometry* that accelerates ions to extraordinarily high kinetic energies before *mass* analysis. AMS can outperform the competing technique of decay counting for all isotopes where the half-life is long enough.

Acetaldehyde (CH_3CH) *Acetaldehyde* is found in nature in many foods such as ripe fruits, cheese, and heated milk. *Acetaldehyde* is primarily used to produce other chemicals, including acetic acid and disinfectants, drugs, and perfumes.

Acetone ($(CH_3)_2CO$) Acetone, or propanone, is an organic compound with the formula $(CH_3)_2CO$. It is the simplest and smallest ketone. It is a colorless, highly volatile, and flammable liquid with a characteristic pungent odor.

Actinobacteria The *Actinobacteria* are a phylum of mostly Gram-positive bacteria. They can be terrestrial or aquatic. They are of great economic importance to humans because agriculture and forests depend on their contributions to soil systems.

Advanced Composition Explorer (ACE) It is a NASA's *explorers* program solar and space exploration mission to study matter, comprising energetic particles from the solar wind, the interplanetary medium, and other sources.

Advanced Pollution Instrument Inc. (API) Teledyne *API* (TAPI) designs and builds precision air quality and industrial gas monitoring *instrumentation* at our factory located in San Diego, CA, United States. We also provide a variety of precision Zero Air Generators and Calibration *instruments* for both standard and trace-level applications.

Advanced scatterometer (ASCAT) The *Advanced Scatterometer* (*ASCAT*) is a European Instrument developed by ESA and the only active instrument on the MetOp satellites. It is a C-Band radar in heritage to the Active Microwave Instruments (AMI) flown on the ESA ERS satellites.

Advanced Very-High-Resolution Radiometer (AVHRR) The *Advanced Very-High-Resolution Radiometer* (*AVHRR*) instrument is *a* space-borne sensor that measures the reflectance of the Earth in five spectral bands that are relatively wide by today's standards.

Aichi Biodiversity A set of 20 global targets under the Strategic Plan for *Biodiversity* 2011–20. They are grouped under five strategic goals: (1) Address the underlying causes of *biodiversity* loss by mainstreaming *biodiversity* across government and society. (2) Reduce the direct pressures on *biodiversity* and promote sustainable use. (3) Improve the status of biodiversity by safeguarding ecosystems, species and genetic diversity. (4) Enhance the benefits to all from biodiversity and ecosystem services. (5) Enhance implementation through participatory planning, knowledge management and capacity building.

Algae *Algae* are a diverse group of aquatic organisms that can conduct photosynthesis. Certain *algae* are familiar to most people; for instance, seaweeds (such as kelp or phytoplankton), pond scum, or the *algal* blooms in lakes.

Antarctic Treaty System The *treaty* entered into force in 1961 and currently has 54 parties. The *treaty* sets aside *Antarctica* as a scientific preserve, establishes freedom of scientific investigation, and bans military activity on the continent. The *treaty* was the first arms control agreement established during the Cold War.

Anthropogenic activities Anthropogenic effects, processes, objects, or materials are those that are derived from human or biological wastes that are produced as by-products of otherwise purposeful *human activities*.

Antifreeze proteins (AFPs) *Antifreeze proteins* (AFPs) or ice structuring *proteins* (ISPs) refer to a class of polypeptides produced by certain animals, plants, fungi, and bacteria that permit their survival in temperatures below the freezing point of water.

Arctic amplification Over the past 30 years, the *Arctic* has warmed at roughly twice the rate as the entire globe, a phenomenon known as *Arctic amplification*. Most scientists agree that this rapid warming is a signal of human-caused climate change. The floating sea ice cover of the *Arctic* Ocean is shrinking, especially during summer.

Arctic Climate Impact Assessment (ACIA) The *Arctic Climate Impact Assessment* (*ACIA*) is a *study* describing the ongoing *climate change* in the *Arctic* and its consequences: rising temperatures, loss of sea ice, unprecedented melting of the Greenland ice sheet, and many *impacts* on ecosystems, animals, and people.

Arctic Contaminants Action Program (ACAP) The *Arctic Contaminants Action Program* (*ACAP*) works to prevent and reduce pollution and environmental risks in the *Arctic*. *ACAP* encourages nations to strengthen policies and take *actions* to reduce *pollutants* and mitigate associated environmental, human health, and socioeconomic risks.

Arctic humidification There are several measures of the amount of water vapor in the air. The relative *humidity* is a measure often used by meteorologists and TV weather reporters. It is the ratio of water vapor in the air to the saturated water vapor content of the air. Overall, *humidity in the Arctic* atmosphere is low.

Arctic Monitoring and Assessment Programme (AMAP) The *Arctic Monitoring and Assessment Programme* is one of the six working groups of the *Arctic* Council. *AMAP* is mandated: To *monitor* and assess the status of the *Arctic* region with respect to pollution and climate change issues.

Arctic Ocean sea ice *Sea ice* is frozen seawater that floats on the *ocean* surface. It forms in both the *Arctic* and the Antarctic in each hemisphere's winter. It retreats in the summer but does not completely disappear. As the *ice* melts, it exposes *ocean* water to sunlight, spurring photosynthesis in phytoplankton.

Arctic oscillation (AO) The *Arctic oscillation* (*AO*) or *Northern Annular Mode/Northern Hemisphere Annular Mode* (*NAM*) is a weather phenomenon at the Arctic poles north of 20-degree latitude. It is an important mode of climate variability for the Northern Hemisphere. The Southern Hemisphere analog is called the Antarctic oscillation or Southern Annular Mode (SAM).

Baffin Bay An ice-clogged body of water between Greenland and the eastern islands of Canada's Arctic Archipelago, including Ellesmere, Devon, and *Baffin* Islands. It connects with the Arctic Ocean to the north and west and with the Atlantic Ocean to the south by way of Davis Strait.

Benthic foraminifera As indicated in the introductory page, foraminifera are separated into the planktonic and the benthic foraminifera based on their life strategy. Planktonic foraminifera is represented by many species with worldwide occurrence in broad latitudinal and temperature belts, floating in the surface or near-surface waters of the open ocean as part of the marine zooplankton. Benthic foraminifera is as successful as the planktonic foraminifera group and even more abundant in modern seas and can live attached or free, at all depths.

Biogenic silica (BSi) *Biogenic silica* (*bSi*), also referred to as opal, *biogenic* opal, or amorphous opaline *silica*, forms one of the most widespread *biogenic* minerals. *Silica* is an amorphous metal oxide formed by complex inorganic polymerization processes.

Biological soil crusts (BSCs) *Biological soil crusts* (*BSCs*), composed of eukaryotic algae, cyanobacteria, mosses, liverworts, fungi, and lichens, cover the uppermost mm of the *soil* surface in most arid and semiarid ecosystems throughout the globe and are one of the most conspicuous and important biotic components of these areas.

Black carbon (BC) Chemically, *black carbon* (*BC*) is a component of fine particulate matter ($PM \leq 2.5\ \mu m$ in aerodynamic diameter). The term *black carbon* is also used in soil sciences and geology, referring either to deposited atmospheric *black carbon* or to directly incorporated *black carbon* from vegetation fires.

Brown carbon (BrC) In chemistry, *brown carbon* (C_{brown}/*BrC*) is *brown* smoke released by the combustion of organic matter. It coexists with black *carbon* when released into the atmosphere. These two are the two most important light-absorbing substances in the atmosphere.

CAMS method The Collaborative Assessment and Management of Suicidality (CAMS) is a flexible therapeutic framework in which patient and provider work together to assess the patient's suicidal risk and use that information to plan and manage suicide-specific, "driver-oriented" treatment. It is a philosophy of clinical care that can be used for a wide range of suicidal patients across outpatient and inpatient treatment settings and in the context of various psychotherapies and treatment modalities.

Carbon dioxide (chemical formula CO_2) It is a colorless gas with a density about 53% higher than that of dry air. *Carbon dioxide* molecules consist of a *carbon* atom covalently double bonded to two oxygen atoms. It occurs naturally in the Earth's atmosphere as a trace gas.

Carbon monoxide (chemical formula CO) It is a colorless, odorless, and tasteless flammable gas that is slightly less dense than air. Approximately half of the *carbon monoxide* in Earth's atmosphere is from the burning of fossil fuels and biomass (such as forest and bushfires).

Carbonaceous aerosols *Carbonaceous aerosols* are the major, dominant component of the mass of suspended particles in polluted atmospheres. Their composition spans the range from "Organic *Carbon*" (often denoted "OC") to "Black" or "Elemental" *Carbon* (denoted "BC" or "EC").

Carbonyl compounds In organic chemistry, a *carbonyl group* is a functional *group* composed of a carbon atom double bonded to an oxygen atom: C=O. It is common to several classes of organic *compounds*, as part of many larger functional groups. A *compound* containing a *carbonyl group* is often referred to as a *carbonyl compound*.

Ceilometer A *ceilometer* is a device that uses a laser or other light source to determine the height of a cloud ceiling or cloud base. *Ceilometers* can also be used to measure the aerosol concentration within the atmosphere.

Census of Marine Life The *Census of Marine Life* was an international project that recorded the diversity, distribution, and abundance of *life* in the ocean. The *Census of Marine Life* was an international project spanning 10 years that recorded the diversity, distribution, and abundance of *life* in the ocean.

Central American Seaway (CAS) The Central American Seaway, also known as the Panamanic Inter-American and Proto-Caribbean Seaway, was a body of water that once separated North from South America. It formed in the Mesozoic (200–154 Ma) during the breakup of the supercontinent Pangaea and closed when the Isthmus of Panama was formed by volcanic activity in the late Pliocene (2.76–2.54 Ma).

Chroococcales The *Chroococcales* are an order of *Cyanobacteria* in some classifications that includes the harmful algal bloom *Microcystis aeruginosa*. Molecular data indicate that the members of the Chroococcales may not be a clade.

Circumpolar Biodiversity Monitoring Program (CBMP) The Circumpolar Biodiversity Monitoring Program (CBMP) is an international network of scientists, governments, indigenous organizations and conservation groups working to harmonize and integrate efforts to monitor the Arctic's living resources.

Conference of Parties 25 (COP 25) The 2019 United Nations Climate Change Conference, also known as COP25, is the 25th United Nations Climate Change Conference. It was held in Madrid, Spain, from December 2 to 13, 2019 under the presidency of the Chilean government. The conference incorporated the 25th Conference of the Parties to the United Nations Framework Convention on Climate Change (UNFCCC), the 15th meeting of the parties to the Kyoto Protocol (CMP15), and the second meeting of the parties to the Paris Agreement (CMA2).

Continental shelf A continental shelf is a portion of a continent that is submerged under an area of relatively shallow water known as a shelf sea. Much of these shelves were exposed by drops in sea level during glacial periods. The shelf surrounding an island is known as an insular shelf.

Deep Sea Drilling Project (DSDP) The Deep Sea Drilling Project (DSDP) was the first of three international scientific ocean drilling programs that have operated for more than 40 years. It was on June 24, 1966, that the Prime Contract between the National Science Foundation (NSF) and The Regents, University of California was signed.

Deglaciation Deglaciation describes the transition from full glacial conditions during ice ages to warm interglacials, characterized by global warming and sea-level rise due to change in continental ice volume.

Dehydrins Dehydrin (DHN) is a multifamily of proteins present in plants, which is produced in response to cold and drought stress. DHNs are hydrophilic, reliably thermostable, and disordered. They are stress proteins with a high number of charged amino acids that belong to the Group II Late Embryogenesis Abundant (LEA) family.

Desiccation Desiccation (from Latin de- "thoroughly" + siccare "to dry") is the state of extreme dryness or the process of extreme drying. A desiccant is a hygroscopic (attracts and holds water) substance that induces or sustains such a state in its local vicinity in a moderately sealed container.

Dissolved organic carbon Dissolved organic carbon (DOC) is the fraction of organic carbon operationally defined as that which can pass through a filter with a pore size typically between 0.22 and 0.7 μm. The fraction remaining on the filter is called particulate organic carbon (POC).

Environmental magnetism Environmental magnetism is the study of magnetism as it relates to the effects of climate, sediment transport, pollution, and other environmental influences on magnetic minerals. It makes use of techniques from rock magnetism and magnetic mineralogy.

Environmental protection Environmental protection is the practice of protecting the natural environment by individuals, organizations, and governments. Its objectives are to conserve natural resources and the existing natural environment and, where possible, to repair damage and reverse trends.

Eurasian Plate movement The Eurasian Plate is a tectonic plate that includes most of the continent of Eurasia (a landmass consisting of the traditional continents of Europe and Asia), with the notable exceptions of the Indian subcontinent, the Arabian subcontinent, and the area east of the Chersky Range in East Siberia. It also includes oceanic crust extending westward to the mid-Atlantic Ridge and northward to the Gakkel Ridge.

Excitation emission matrix (EEM) A measurement becoming more widely used in the field of fluorescence spectroscopy is the excitation emission matrix, or EEM. An EEM is a 3D scan, resulting in a contour plot of excitation wavelength versus emission wavelength versus fluorescence intensity.

Extreme weather events Extreme weather events such as tropical cyclones, floods, heavy rainfall, and droughts—well-known and expensive consequences of global warming—impacted many parts of the world.

Feedback Albedo and mechanisms Ice–albedo feedback is a positive feedback climate process where a change in the area of ice caps, glaciers, and sea ice alters the albedo and surface temperature of a planet. Warming tends to decrease ice cover and hence decrease the albedo, increasing the amount of solar energy absorbed and leading to more warming.

Fluorescence intensity (FI) The fluorescence intensity indicates how much light (photons) is emitted. It is the extent of emission and it depends on the concentration of the excited fluorophore. Fluorescence is created by the absorption of energy (light) by fluorescent molecules, called fluorophores.

Fluorescence spectrometer Fluorescence spectroscopy (also known as fluorimetry or spectrofluorometry) is a type of electromagnetic spectroscopy that analyzes fluorescence from a sample. It involves using a beam of light, usually ultraviolet light, that excites the electrons in molecules of certain compounds and causes them to emit light; typically, but not necessarily, visible light. A complementary technique is absorption spectroscopy. In the special case of single molecule fluorescence spectroscopy, intensity fluctuations from the emitted light are measured from either single fluorophores or pairs of fluorophores.

Formaldehyde (HCHO) Formaldehyde is an organic air pollutant containing one atom of carbon, one atom of oxygen, and two atoms of hydrogen. It has the chemical formula HCHO or CH2O. Formaldehyde is present in outdoor air but indoor air formaldehyde concentrations are usually much higher because of indoor formaldehyde sources.

Freeze–thaw stress Freeze–thaw events can affect plant hydraulics by inducing embolism. Ultrasonic activity started with the onset of freezing and increased with decreasing subzero temperatures, whereas no UE were recorded during thawing. The temperature at which 50% of UE were reached varied between −9.1°C and −31.0°C across species.

Frozen methane Frozen methane, also known as methane hydrates, is made up of methane gas molecules locked inside frozen water crystals. It looks like ice, forms at low temperatures and high pressures in the ocean, and is thought to contain 15%–40% of Earth's carbon.

Gas exploration Hydrocarbon exploration (or oil and gas exploration) is the search by petroleum geologists and geophysicists for deposits of hydrocarbons, particularly petroleum, and natural gas, in the Earth using petroleum geology.

Geoeconomics Broadly, geoeconomics (sometimes geoeconomics) is the study of the spatial, temporal, and political aspects of economies and resources. However, there is no widely accepted singular definition of geoeconomics as distinct from geopolitics.

Geomorphological mapping Geomorphological mapping has evolved from a purely field-based exercise aimed at accurate depiction of landforms on a map with limited interpretation to highly interpretive maps at different scales that are mainly based on digital remotely sensed data with limited ground-truthing in the field in some places.

Glacial fjord system A true fjord is formed when a glacier cuts a U-shaped valley by ice segregation and abrasion of the surrounding bedrock. According to the standard model, glaciers formed in preglacial valleys with a gently sloping valley floor.

Granulometry Granulometry is the measurement of the size distribution in a collection of grains. It is also called particle size distribution test. It is often characterized by the percentage of particles with certain diameter ranges (in micron meters).

Halogen chemistry Halogen, any of the six nonmetallic elements that constitute Group 17 (Group VIIa) of the periodic table. The halogen elements are fluorine (F), chlorine (Cl), bromine (Br), iodine (I), astatine (At), and tennessine (Ts).

Heartland theory In 1904, Sir Halford Mackinder published the Heartland theory. The theory proposed that whoever controls Eastern Europe controls the Heartland. It also supported the concept of world dominance. Explanation—A more revised version explains that whoever controls the heartland, controls the world island.

Himalayan region The Indian Himalayan Region (IHR) is the section of the Himalayas within India, spanning 11 Indian states and union territories, namely, UTs of Jammu and Kashmir and Ladakh, and States of Himachal Pradesh, Uttarakhand, Sikkim, Arunachal Pradesh, Nagaland, Manipur, Mizoram, Tripura, Meghalaya, Assam, and West Bengal.

Holocene Climatic Optimum The Holocene Climate Optimum (HCO) was a warm period during roughly the interval 9000–5000 years BP, with a thermal maximum around 8000 years BP.

Huttonian principle In Principles of Geology, 3 vol. (1830–33), Scottish geologist Sir Charles Lyell deciphered Earth's history by employing Huttonian principles and made available a host of new geologic evidence supporting the view that physical laws are permanent and that any form of supernaturalism could be rejected.

Hydroxyl radical (OH) The hydroxyl radical, •OH, is the neutral form of the hydroxide ion (OH^-). Hydroxyl radicals are highly reactive (easily becoming hydroxy groups) and consequently short-lived. They form an important part of radical chemistry.

Ice-rafted debris/detritus (IRD) Ice-rafted debris (IRD) is a terrigenous material transported within a matrix of ice and deposited in marine or lake sediments when the ice matrix melts (US National Climatic Data Center).

Ice Silk Road In 2017 Chinese and Russian leaders jointly proposed the "Ice Silk Road (ISR)," with an aim to promote cooperation and development in the Arctic. The ISR is an open initiative that abandons geopolitical thinking and advocates cooperation and a win-win perspective.

Ilulissat Declaration The Ilulissat Declaration was announced on May 28, 2008 by the five coastal states of the Arctic Ocean (United States, Russia, Canada, Norway, and The Kingdom of Denmark), meeting at the political level during the Arctic Ocean

Conference in Ilulissat, Greenland to discuss the Arctic ocean, climate change, and its protection.

Indian Ocean Dipole (IOD) The Indian Ocean Dipole (IOD) is defined by the difference in sea surface temperature between two areas (or poles, hence a dipole)—a western pole in the Arabian Sea (western Indian Ocean) and an eastern pole in the eastern Indian Ocean south of Indonesia. The IOD affects the climate of Australia and other countries that surround the Indian Ocean Basin and is a significant contributor to rainfall variability in this region.

Interstadials Interstadials are regarded as the relatively short-lived periods of thermal improvement during a glacial phase, when temperatures did not reach those of the present day and, in lowland mid-latitude regions, the climax vegetation was boreal woodland.

Isoprime Isoprime precisION is the highest performing stable isotope ratio mass spectrometer (IRMS) on the market, offering complete flexibility to move in any direction.

Joint Communiqué of Governments of the Arctic Countries on the Establishment of the Arctic Council Ministers and Senior Representatives of the Governments of Canada, Denmark, Finland, Iceland, Norway, the Russian Federation, Sweden, and the United States of America met in Ottawa, Canada, on September 19, 1996 and signed the Declaration on the Establishment of the Arctic Council. This inaugural meeting was attended by the leaders and senior representatives of three international Arctic indigenous organizations—the Inuit Circumpolar Conference; the Saami Council; and the Association of Indigenous Minorities of the North, Siberia, and the Far East of the Russian Federation, as Permanent Participants in the Council.

Kaolinite Kaolinite (/ˈkeɪəlɪnaɪt/) is a clay mineral, with the chemical composition $Al_2Si_2O_5(OH)_4$. It is an important industrial mineral. It is a layered silicate mineral, with one tetrahedral sheet of silica (SiO_4) linked through oxygen atoms to one octahedral sheet of alumina (AlO_6) octahedra.

Ketones Ketones (chemically known as ketone bodies) are by-products of the breakdown of fatty acids. The breakdown of fat for fuel and the creation of ketones is a normal process for everyone. In a person without diabetes, insulin, glucagon, and other hormones prevent ketone levels in the blood from getting too high.

Kongsfjorden–Krossfjorden Kongsfjorden–Krossfjorden and the adjacent West Spitsbergen Shelf meet at the common mouth of the two fjord arms. This paper presents our most up-to-date information about the physical environment of this fjord system and identifies important gaps in knowledge.

Land ice *Land ice* includes mountain glaciers and *ice* sheets, covering Greenland and Antarctica. These giant blocks of *ice* are melting and the water is flowing rapidly into the oceans. Schweiger: "Melting sea *ice* has no impact on sea-level rise because it's already floating in the ocean."

Last Glacial Maximum (LGM) The LGM was the coldest part of the last Ice Age, centered about 21 ka, which generally corresponds with the period of greatest ice advance. Sea level was depressed by about 125 m, exposing the Bering Land Bridge and the continental shelf.

Late embryogenesis abundant proteins (LEA proteins) Late embryogenesis abundant proteins (LEA proteins) are proteins in plants, and some bacteria and invertebrates that protect against protein aggregation due to desiccation or osmotic stresses associated with low temperature.

Laurentide Ice Sheet (LIS) The Laurentide Ice Sheet was a massive sheet of ice that covered millions of square miles, including most of Canada and a large portion of the

Northern United States, multiple times during the Quaternary glacial epochs, from 2.588 ± 0.005 million years ago to the present.

Leveling process Leveling is a process of determining the height of one level relative to another. It is used in surveying to establish the elevation of a point relative to a datum, or to establish a point at a given elevation relative to a datum.

Light absorption Light absorption is a process by which light is absorbed and converted into energy. An example of this process is photosynthesis in plants. Absorption depends on the electromagnetic frequency of the light and object's nature of atoms. The absorption of light is, therefore, directly proportional to the frequency.

Little Ice Age (LIA) Little Ice Age (LIA), climate interval that occurred from the early 14th century through the mid-19th century, when mountain glaciers expanded at several locations, including the European Alps, New Zealand, Alaska, and the southern Andes, and mean annual temperatures across the Northern Hemisphere declined by 0.6°C.

Luminescence dating Luminescence dating refers to a group of methods of determining how long ago mineral grains were last exposed to sunlight or sufficient heating. It is useful to geologists and archaeologists who want to know when such an event occurred. It uses various methods to stimulate and measure luminescence.

Magnitudes (Mb) An estimate of body-wave (mb) magnitude based on the size of the area over which the earthquake was felt, typically assigned to widely felt earthquakes that occurred before the invention of seismographs and to earthquakes occurring in the early decades of seismograph deployment for which magnitudes calculated from.

Mass balance A mass balance, also called a material balance, is an application of conservation of mass to the analysis of physical systems. By accounting for material entering and leaving a system, mass flows can be identified which might have been unknown, or difficult to measure without this technique.

Medieval Climate Anomaly (MCA) The Medieval Warm Period (MWP) also known as the Medieval Climate Optimum, or Medieval Climatic Anomaly was a time of warm climate in the North Atlantic region lasting from c.950 to c.1250.

Messinian salinity crisis The Messinian salinity crisis (MSC), also referred to as the Messinian event, and in its latest stage as the Lago Mare event, was a geological event during which the Mediterranean Sea went into a cycle of partial or nearly complete desiccation throughout the latter part of the Messinian age of the Miocene epoch, from 5.96 to 5.33 Ma (million years ago). It ended with the Zanclean flood, when the Atlantic reclaimed the basin.

Metal-tolerant bacteria Several species of *Bacillus*, especially from *cereus* and *subtilis* groups, are known for their metal ions resistance. Bacterial strains resistant and able to adsorb arsenic, lead, copper, cadmium, and zinc have already been described, usually isolated from heavy metal contaminated areas.

Microhabitats A microhabitat is a small area that differs somehow from the surrounding habitat. Its unique conditions may be home to unique species that may not be found in the larger region. Unfortunately, some habitats are threatened by pollution, extreme weather, or deforestation.

Milankovitch cycles A Milankovitch cycle is a cyclical movement related to the Earth's orbit around the Sun. There are three of them: eccentricity, axial tilt, and precession.

Natural resource stresses In regard to agriculture, Abiotic stress is produced by natural environment factors such as extreme temperatures, wind, drought, and salinity.

Neural network model The data first fed into the neural network from the source that is called the input. Its goal is to give the network data to make a decision or prediction about the information fed into it. The neural network model usually accepts real value sets of inputs and it should be fed into a neuron in the input layer.

North Atlantic Current North Atlantic Current, also called North Atlantic Drift, part of a clockwise-setting ocean-current system in the North Atlantic Ocean, extending from southeast of the Grand Bank, off Newfoundland, Canada, to the Norwegian Sea, off northwestern Europe.

North Atlantic Deep Water (NADW) North Atlantic Deep Water (NADW) is a deep water mass formed in the North Atlantic Ocean. Thermohaline circulation (properly described as meridional overturning circulation) of the world's oceans involves the flow of warm surface waters from the southern hemisphere into the North Atlantic. Water flowing northward becomes modified through evaporation and mixing with other water masses, leading to increased salinity. When this water reaches the North Atlantic it cools and sinks through convection, due to its decreased temperature and increased salinity resulting in increased density. NADW is the outflow of this thick deep layer, which can be detected by its high salinity, high oxygen content, nutrient minima, high $^{14}C/^{12}C$, and chlorofluorocarbons (CFCs).

Northern Hemisphere The Northern Hemisphere is the part of the planet that is north of the equator. All of North America and Europe are in the Northern Hemisphere. Most of Asia, two-thirds of Africa and 10% of South America are also in this hemisphere.

Nostocales The Nostocales are an order of cyanobacteria containing most of its species. It includes filamentous forms, both simple or branched, and both those occurring as single strands or multiple strands within a sheath.

Ocean circulation Ocean circulation derives its energy at the sea surface from two sources that define two circulation types: (1) wind-driven circulation forced by wind stress on the sea surface, inducing a momentum exchange and (2) thermohaline circulation driven by the variations in water density imposed at the sea surface.

Optically simulated luminescence (OSL) Optically stimulated luminescence is a late Quaternary dating technique used to date the last time quartz sediment was exposed to light. As sediment is transported by wind, water, or ice, it is exposed to sunlight and zeroed of any previous luminescence signal.

Orbital parameters Orbital elements are the parameters required to uniquely identify a specific orbit. In celestial mechanics these elements are considered in two-body systems using a Kepler orbit. There are many different ways to mathematically describe the same orbit, but certain schemes, each consisting of a set of six parameters, are commonly used in astronomy and orbital mechanics. A real orbit and its elements change over time due to gravitational perturbations by other objects and the effects of general relativity. A Kepler orbit is an idealized, mathematical approximation of the orbit at a particular time.

Oscillatoriales The taxonomic order Oscillatoriales includes many cyanobacteria species, all of which have coin-like or cylindrical cells joined end-to-end to form simple, unbranched filaments. Oscillatoriales are common in freshwater phytoplankton assemblages and some of the species can form dense blooms.

Paleoceanography Paleoceanography is the study of the history of the oceans in the geologic past with regard to circulation, chemistry, biology, geology, and patterns of sedimentation and biological productivity.

Paleoclimatic records Paleoclimatology (in British spelling, palaeoclimatology) is the study of climates for which direct measurements were not taken. As instrumental records only span a tiny part of Earth's history, the reconstruction of ancient climate is important to understand natural variation and the evolution of the current climate.

Paleoclimatologists Paleoclimatology is the study of previous climates that have existed during Earth's different geologic ages. Paleoclimatologists try to identify the causes of climate changes that have happened in the past in order to better understand our present and future climate.

Paleopalynology Palaeopalynology is the study of microfossils formed of organic matter, named palynomorphs; they are preserved in certain sedimentary rocks in all the geological periods.

Penetration depth Penetration depth is a measure of how deep light or any electromagnetic radiation can penetrate into a material. It is defined as the depth at which the intensity of the radiation inside the material falls to 1/e (about 37%) of its original value at (or more properly, just beneath) the surface.

Permafrost Permafrost is ground that continuously remains below 0°C for two or more years, located on land or under the ocean. Permafrost does not have to be the first layer that is on the ground. It can be from an inch to several miles deep under the Earth's surface.

Pipette method In the pipette method a sample is pipetted at different times and different depths of the suspension of the sample in a measuring cylinder. Time and depth are determined applying the Law of Stokes. The pipetted suspension is condensed and dried, and weighing determines the mass ratio of the pipetted fraction.

Radical oxidation Free radicals can cause large chain chemical reactions in your body because they react so easily with other molecules. These reactions are called oxidation. This causes the free radical to stabilize and become less reactive.

Radiocarbon dating Radiocarbon dating (also referred to as carbon dating or carbon-14 dating) is a method for determining the age of an object containing organic material by using the properties of radiocarbon, a radioactive isotope of carbon.

Rimland theory According to his rimland theory, the coastal areas or littorals of Eurasia are key to controlling the World Island, not the Heartland. As per Spykman, landlocked states usually faced security challenges from their immediate neighbors.

Respiration In physiology, respiration is the movement of oxygen from the outside environment to the cells within tissues, and the removal of carbon dioxide in the opposite direction.

Ross's gull Ross's gull (*Rhodostethia rosea*) is a small gull, the only species in its genus, although it has been suggested it should be moved to the genus *Hydrocoloeus*, which otherwise only includes the little gull. This bird is named after the British explorer James Clark Ross. Its breeding grounds were first discovered in 1905 by Sergei Aleksandrovich Buturlin near village of Pokhodsk in North-Eastern Yakutia, while visiting the area as a judge. The genus name *Rhodostethia* is from Ancient Greek rhodon, "rose", and stethos, "breast." The specific rosea is Latin for "rose-colored".

Saami Council The Saami Council is a voluntary Saami organization (a nongovernmental organization), with Saami member organizations in Finland, Russia, Norway, and Sweden. Since it was founded in 1956, the Saami Council has actively dealt with Saami policy tasks. For this reason the Saami Council is one of the indigenous peoples' organizations that have existed longest.

Saxifragaceae Saxifragaceae, the saxifrage family of flowering plants (order Rosales), comprising 36 genera and about 600 species of mostly perennial herbaceous plants.

The members are cosmopolitan in distribution but native primarily to northern cold and temperate regions.

Scanning electron microscopy (SEM) A scanning electron microscope (SEM) is a type of electron microscope that produces images of a sample by scanning the surface with a focused beam of electrons. The electrons interact with atoms in the sample, producing various signals that contain information about the surface topography and composition of the sample.

Scatterometer A scatterometer or diffusion meter is a scientific instrument to measure the return of a beam of light or radar waves scattered by diffusion in a medium such as air. Diffusion meters using visible light are found in airports or along roads to measure horizontal visibility.

Scheteligfjellet Scheteligfjellet is a mountain in Svalbard and has an elevation of 237 m.

Schirmacher Oasis The Schirmacher Oasis (or Schirmacher Lake Plateau) is a 25-km long and up to 3-km wide ice-free plateau with more than 100 freshwater lakes. It is situated in the Schirmacher Hills on the Princess Astrid Coast in Queen Maud Land in East Antarctica and is on average 100 m above sea level. With an area of 34 km^2, the Schirmacher Oasis ranks among the smallest Antarctic oases and is a typical polar desert.

Seafloor topography The ocean floor is almost entirely underlain by mafic oceanic crust (mostly basalt and gabbro, as described in more detail later), while the continental slopes are underlain by felsic continental crust (mostly granitic and sedimentary rocks).

Sea Power theory Sea power, a means by which a nation extends its military power onto the seas. Measured in terms of a nation's capacity to use the seas in defiance of rivals and competitors, it consists of such diverse elements as combat craft and weapons, auxiliary craft, commercial shipping, bases, and trained personnel. Aircraft used in the control of seaborne transportation function as an instrument of sea power even when they operate from a land base. Aircraft operating from carriers represent the extension of sea power even when they are attacking targets deep inland. Except for the great increase in bombardment of shore or inland targets from the sea, the functions of sea power were the same in World War II as they were in the 16th century, when warships specifically designed for fighting (as distinguished from armed merchantmen) first appeared.

Sea surface temperature (SST) Sea surface temperature (SST) (or ocean surface temperature) is the water temperature close to the ocean's surface. The exact meaning of surface varies according to the measurement method used, but it is between 1 mm (0.04 in.) and 20 m (70 ft.) below the sea surface.

Slant total electron content (STEC) Slant total electron content (STEC), the total number of free electrons on a ray path, is an important space weather observable. STEC is the main input for computerized ionospheric tomography (CIT). STEC can be estimated using the dual-frequency GPS receivers.

Solar EUV radiation Solar Extreme Ultraviolet (EUV) is solar radiation that covers the wavelengths 10–120 nm of the electromagnetic spectrum. It is highly energetic and it is absorbed in the upper atmosphere, which not only heats the upper atmosphere but also ionizes it, creating the ionosphere.

Tectonic movements Tectonic shift is the movement of the plates that make up Earth's crust. The Earth is made up of roughly a dozen major plates and several minor plates. The Earth is in a constant state of change. The Earth's crust, called the lithosphere, consists of 15–20 moving tectonic plates.

Terrestrial ecosystem A *terrestrial ecosystem* is a land-based community of organisms and the interactions of biotic and abiotic components in a given area. Examples of *terrestrial* *ecosystems* include the tundra, taigas, temperate deciduous forests, tropical rainforests, grasslands, and deserts.

Tertiary rocks The Tertiary is a system of rocks, above the Cretaceous and below the Quaternary, that defines the Tertiary Period of geologic time. EarthWords is an ongoing series in which we shed some light on the complicated, often difficult-to-pronounce language of science.

Trace elements A trace element is a chemical element whose concentration (or other measure of amount) is very low (a "trace amount"). They are classified into two groups: essential and nonessential. Essential trace elements are needed for many physiological and biochemical processes in both plants and animals.

Trace gases Trace gases are those gases in the atmosphere other than nitrogen (78.1%), oxygen (20.9%), and argon (0.934%) which, in combination, make up 99.934% of the gases in the atmosphere (not including water vapor).

UV-B radiation Invisible rays that are part of the energy that comes from the sun. UV-B radiation causes sunburn, darkening and thickening of the outer layer of the skin, and melanoma and other types of skin cancer.

UV–Vis spectrophotometer UV–Vis Spectroscopy (or Spectrophotometry) is a quantitative technique used to measure how much a chemical substance absorbs light. This is done by measuring the intensity of light that passes through a sample with respect to the intensity of light through a reference sample or blank.

US Geological Survey (USGS) The United States Geological Survey (USGS, formerly simply Geological Survey) is a scientific agency of the United States Government. The scientists of the USGS study the landscape of the United States, its natural resources, and the natural hazards that threaten it.

Vascular plants Vascular plants are plants that use specialized tissue for transporting food and water to different areas in the plant. Examples of vascular plants include trees, flowers, grasses, and vines. Vascular plants have a root system, a shoot system, and a vascular system.

Vegetation Vegetation is a general term for the plant life of a region; it refers to the ground cover provided by plants and is, by far, the most abundant biotic element of the biosphere. Vegetation serves several critical functions in the biosphere, at all possible spatial scales.

Vertical TEC (VTEC) The vertical TEC (VTEC) is determined by integration of the electron density on a perpendicular to the ground standing route, the slant TEC (STEC) is obtained by integrating over any straight path. To first order, the ionospheric radio effect is proportional to TEC and inversely proportional to the radio frequency f.

Volatile organic compounds (VOCs) Volatile organic compounds (VOCs) are emitted as gases from certain solids or liquids. VOCs include a variety of chemicals, some of which may have short- and long-term adverse health effects. Concentrations of many VOCs are consistently higher indoors (up to 10 times higher) than outdoors.

Warm Atlantic waters The Atlantic Ocean is the second-largest of the world's oceans, with an area of about this system of currents transport warm water into the North Atlantic, without which temperatures in the North Atlantic and Europe would plunge dramatically.

Warming hole The North Atlantic warming hole (NAWH) is referred to as a reduced warming, or even cooling, of the North Atlantic during an anthropogenic-driven global warming. A NAWH is predicted by climate models during the 21st century, and its pattern is already emerging in observations.

Western Spitsbergen Current The West Spitsbergen Current (WSC) is a warm, salty current that runs poleward just west of Spitsbergen (formerly called West Spitsbergen), in the Arctic Ocean. The WSC branches off the Norwegian Atlantic Current in the Norwegian Sea.

Winter war The Winter War also known as First Soviet-Finnish War was a war between the Soviet Union (USSR) and Finland. It began with a Soviet invasion of Finland on November 30, 1939, 3 months after the outbreak of World War II and ended 3.5 months later with the Moscow Peace Treaty on March 13, 1940.

X-ray computed tomography of sediments Continuous or dense shipboard measurements of physical properties are required for correlation with logging data and studies of high-resolution sedimentary environments. The X-ray computed tomography (CT) technique has the advantage of being nondestructive and providing high-resolution density data and images.

X-ray diffraction (XRD) X-ray powder diffraction (XRD) is a rapid analytical technique primarily used for phase identification of a crystalline material and can provide information on unit cell dimensions. The analyzed material is finely ground, homogenized, and average bulk composition is determined.

Younger Dryas (YD) The Younger Dryas (around 12,900 to 11,700 years BP) was a return to glacial conditions after the Late Glacial Interstadial, which temporarily reversed the gradual climatic warming after the Last Glacial Maximum (LGM) started receding around 20,000 BP. It is named after an indicator genus, the alpine-tundra wildflower *Dryas octopetala*, as its leaves are occasionally abundant in late glacial, often minerogenic-rich sediments, such as the lake sediments of Scandinavia.

Younger Dryas Stadial (YDS) The Younger Dryas (YD, c. 12.9–11.7 ka) represents the last return to cold, stadial conditions prior to early Holocene warming and is characterized by abrupt climate change. It marks the final phase of Termination I and is also referred to as Greenland Stadial 1 (GS-1).

Zombie Fires A "zombie fire" is a fire from a previous growing season that can smoulder under the ground which is made up of carbon-rich peat. The fires and record temperatures had the potential of turning the carbon sink into a carbon source and increasing global warming, the experts noted.

Zooplankton Zooplankton are the animal component of the planktonic community ("zoo" comes from the Greek word for animal). They are heterotrophic (other-feeding), meaning they cannot produce their own food and must consume instead other plants or animals as food. In particular, this means they eat phytoplankton.

Index

Note: Page numbers followed by "*f*" and "*t*" refer to figures and tables, respectively.

J

K

L

M

O

P

FSC
www.fsc.org
MIX
Papier | Fördert gute Waldnutzung
FSC® C083411

Kolibri 360 GmbH
Ferdinand-Jühlke-Straße 7
99095 Erfurt, Deutschland
produktsicherheit@kolibri360.de